INTEGRATION OF FUNDAMENTAL POLYMER SCIENCE AND TECHNOLOGY—2

The proceedings of the international meeting on polymer science and technology, Rolduc Polymer Meeting—2 held at Rolduc Abbey, Limburg, The Netherlands, 26–30 April 1987

INTEGRATION OF FUNDAMENTAL POLYMER SCIENCE AND TECHNOLOGY—2

Edited by

P. J. LEMSTRA

Eindhoven University of Technology, Eindhoven, The Netherlands

and

L. A. KLEINTJENS

DSM-Research, Geleen, The Netherlands

ELSEVIER APPLIED SCIENCE
LONDON and NEW YORK

ELSEVIER APPLIED SCIENCE PUBLISHERS LTD
Crown House, Linton Road, Barking, Essex IG11 8JU, England

Sole Distributor in the USA and Canada
ELSEVIER SCIENCE PUBLISHING CO., INC.
52 Vanderbilt Avenue, New York, NY 10017, USA

WITH 84 TABLES AND 322 ILLUSTRATIONS

© 1988 ELSEVIER APPLIED SCIENCE PUBLISHERS LTD
Softcover reprint of the hardcover 1st edition 1988

British Library Cataloguing in Publication Data

Integration of fundamental polymer science
and technology—2.
1. Polymer science
I. Lemstra, P. J.   II. Kleintjens, L. A.
547.7
ISBN-13: 978-94-010-7106-2   e-ISBN-13: 978-94-009-1361-5
DOI: 10.1007/978-94-009-1361-5

Library of Congress Cataloging in Publication Data applied for

FOREWORD

Polymer science has matured into a fully accepted branch of materials science. This means that it can be described as a 'chain of knowledge' (Manfred Gordon), the beads of the chain representing all the topics that have to be studied in depth if the relationship between the structure of the molecules synthesized and the end-use properties of the material they constitute is to be understood. The term *chain* indicates the connectivity of the beads, i.e. the multidisciplinary approach required to achieve the aim, *knowledge*, here defined as *quantitative* understanding of the relationship mentioned above in all its parts.

Quite a few conferences are being held at which the disciplinar beads themselves are discussed in detail, and new results within their framework are presented. In this respect, the IUPAC Microsymposia in Prague have made themselves indispensable, to mention one successful example. The bi-annual IUPAC Symposia on Macromolecules, on the other hand, supply interdisciplinary meeting places, which have the advantage and the disadvantage of a large attendance. Smaller-size conferences of a similar nature can often be found on a national level.

The organizers of the young, but already well-appreciated, Rolduc Meetings on the interplay between fundamental science and technology in the polymer field struck an interesting chord when they realized that focussing on the basic science behind technological problems would serve the purpose of concentration on insight along the chain of knowledge and avoid the surrender to too large a size for the meeting to really be a meeting. The first conference clearly emphasized the essential role basic research may play, building up understanding of individual steps in the immensely complex development stage, which still has to be carried out on a large and, consequently, expensive scale.

Basic research is both cheap and expensive. It is cheap in its running costs, no pilot plants being needed. Admittedly, it sometimes uses expensive equipment but industrial research institutes will have to have that anyway to support their development. Basic research is expensive for another reason which relates to investment in human effort and, thereby, indirectly in capital. In these days of highly advanced science it takes a talented Ph.D.

10 to 15 years to become the company's foothold in his or her particular field. In this situation, he or she has an accountability that may not be measurable with the usual criteria but is nevertheless comparable to that of the line hierarchy. In times of regression cost-reduction policy makers often go for easy short-term successes, ignoring the personal and financial aspects involved and thereby unwittingly jeopardizing their company's future.

The second Rolduc Meeting, the proceedings of which are collected in this volume, has realized its objective as well as the first did; the list of contents illustrates the point. The list also lets us surmise the usefulness of the meeting with respect to its supplying a place where scientists operating within the chain of knowledge can interact. Adding the ideal setting of the old Abbey with its beautiful park, one is not surprised at the success of these meetings and can only congratulate the organizers Ludo Kleintjens and Piet Lemstra, as well as the sponsor, DSM.

RONALD KONINGSVELD
Institute ΣΠ
Maastricht, The Netherlands

PREFACE

The aim of the Rolduc Polymer Meetings is to stimulate discussions between academic and industrial polymer scientists and engineers. The scientific committee of Rolduc Polymer Meeting—2 selected the following topics: Chemistry; Chain-Dynamics/Conformation; Thermodynamics/Blends; Networks/Gels; Crystallization/Structure/Morphology; Rheology/Processing; and Fibres/Composites. Each topic was introduced by invited experts after which contributed papers and posters were presented and discussed. About 250 participants took part in this RPM—2 meeting, which was held again in the Rolduc Abbey from 27 to 30 April 1987. All active participants were requested to submit a manuscript for this book.

We hope that the reader will enjoy this state-of-the-art report on a broad field of polymer technology ranging from chemistry via processing to properties. Unfortunately, it is quite impossible to have written on paper the ultimate goal of the Rolduc Polymer Meetings, being the *integration* of science and technology. These integrational aspects, hopefully, are already implanted in the numerous brain cells of the participants based on lively discussions during the meeting.

We thank all contributors to this volume.

P.J.L.
L.A.K.

CONTENTS

Part 1: Chemistry

Part 2: Chain-Dynamics/Conformation

Part 4: Networks/Gels

Part 5: Crystallization

Part 6: Structure/Morphology

Part 7: Rheology/Processing

Part 1
CHEMISTRY

REGIO- AND STEREOSPECIFICITY IN PROPYLENE POLYMERIZATION WITH CHIRAL CATALYTIC SYSTEMS

P. Pino
Swiss Federal Institute of Technology,
Institut für Polymere,
Universitätstrasse 6,
8092 Zürich, Switzerland

ABSTRACT

On the basis of the results obtained in the stereospecific polymerization of propylene with catalysts prepared from methylaluminoxanes and optically active, racemic and meso metallocene derivatives of Zr and Ti, the origin of the stereo- and regiospecificity in α-olefins polyinsertion is discussed.
The experiments prove that the chirality of the catalytic system and, under resonable assumptions, the conformation of the monomeric unit of the growing chain bound to the catalyst, control enantioface discrimination of the monomer and hence stereospecificity. Regiospecificity is connected with a double control of insertion reaction, the probability of a further chain growth being very small after an occasional 2-1 monomer insertion on an isospecific center. Further experiments are needed to obtain a better understanding of the role of the polymeric aluminoxanes in the "soluble" catalytic systems.

INTRODUCTION

The origin of stereospecificity and of the regiospecificity in α-olefins polymerization has been debated since the discovery of this reaction [1]. Natta and coworkers [2] postulated very early that as the catalytic system was able to distinguish between the two enantiofaces of α-olefins it must be chiral. As the stereospecific catalysts were prepared from non chiral components it was assumed that the chiral catalytic centers were racemic [2]. The origin of the chirality of the catalytic center can be in principle due to the asymmetric carbon atom of the last inserted monomeric unit (fig. 1a), to other chiral ligands bound to the metal atom of the catalytic center (fig. 1b) or finally, to the intrinsic molecular asymmetry of the catalytic center (fig.1c).

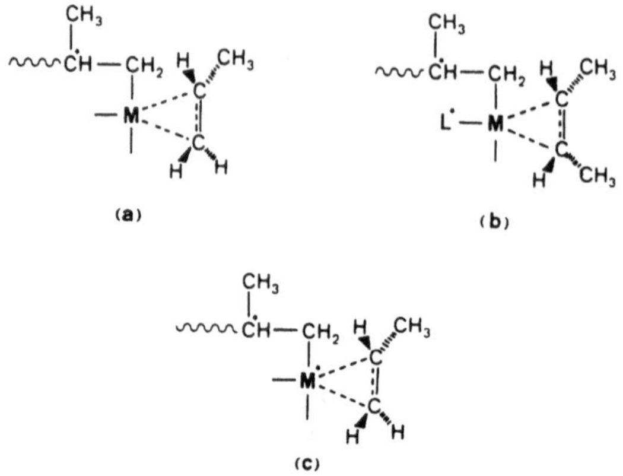

Figure 1. Possible origins of chirality in the catalytic centers for α-olefins stereospecific polymerization to isotactic polymers.

As in most catalytic systems no chiral ligands bound to the catalytic centers are present with the exception of the growing chain, the second possibility was discarded. Concerning the asymmetric carbon atoms of the growing chain, a strong influence was excluded in the case of the synthesis of isotactic polymers. In fact the chain growth occurs via a 1-2 insertion [3] and strong asymmetric induction by an asymmetric carbon atom in δ position with respect to the new chirogenic center to be formed, must be excluded considering the relatively high polymerization temperature (60°C-80°C) [4]. Asymmetric induction due to the asymmetric carbon atom of the growing chain is assumed to be the main stereoregulating factor in the synthesis of syndiotactic polypropylene in which the chain growth occurs via 2-1 insertion at low temperature (<0°C). The asymmetric center being in this case in α-position with respect to the metal atom of the catalytic center [5].

Regioselectivity is also very high in the polyinsertion reaction leading to isotactic polymer as pointed out already in the original patent [7]. In fact only in the case of vanadium catalysts more than 1% of structural inversions have been detected in the isotactic polymers [8]. This substantial regiospecificity has never been thoroughly discussed; it has been attributed both to electronic and to steric factors [9].

2) A SIMPLE STEREOCHEMICAL MODEL FOR THE TRANSITION STATES OF α-OLEFINS INSERTION INTO A METAL TO GROWING CHAIN BOND.

Based on the data obtained in the α-olefins polymerization with

$TiCl_4/MgCl_2/AlR_3$ catalytic systems a very simple stereochemical model has been proposed [10] for the transition states of the insertion step of α-olefins polymerization. This type of model, already successfully used in predicting the type of regio- and stereoselectivity in asymmetric hydroformylation [11], qualitatively explains the origin of regio- and stereoselectivity on the basis of steric effects only, under the assumption that the step determining structure and stereochemistry of the polymer chains is the insertion of the α-olefins into a metal-carbon bond. In this model of the transition state the α-olefin double bond is supposed to be coplanar and nearly parallel to the reactive metal to carbon bond; the substituent of the vinyl group is supposed to be accommodated in the least crowded portion of the space around the metal atom to which the growing chain is bound. In a planar representation of the above simplified transition state (fig.2) 4 quadrants are defined (1,2,-1 and -2) in which the space availability for the accommodation of the α-olefin vinyl group is different. Regioselectivity is determined by the relative size of the

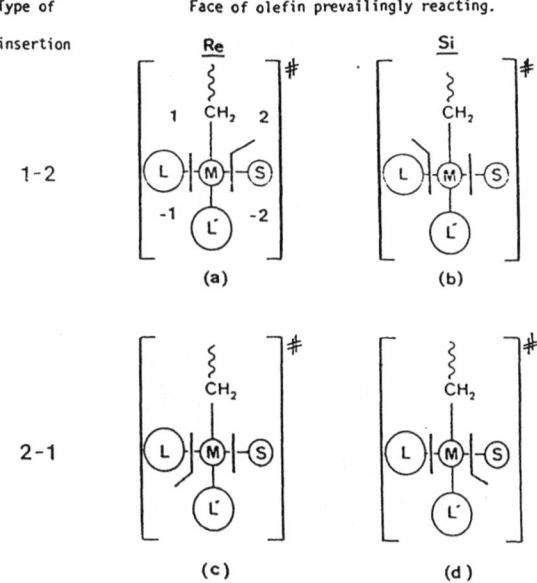

Figure 2. Planar representation of 4 possible transition states for propylene insertion into a metal -CH₂ bond in isospecific catalytic centers; the numbers 1,2,-1 and -2 in fig.2(a) indicate the four quadrants in which the methyl group of propylene can be situated.

terminal group of the growing chain and of the substituent, L', opposite
to it; stereospecificity of the isotactic type is determined by the
relative size of the two lateral substituents (L and S). This model, even
if very rough, has been extremely useful in the attempts to rationalize
the results obtained in the investigation of the polymerization of
different α-olefins, including racemic α-olefins, with heterogeneous
catalysts. It has also suggested new experiments contributing to the
understanding of the above catalytic systems [12] including the formation
of small amounts of syndiotactic polymers with catalytic systems producing
mostly isotactic polymers [13].

A more detailed representation of a transition state for α-olefin
insertion based on the crystalline structure of TiCl₃, and calculations of
non bonded interactions between growing chain and catalyst was proposed by
Corradini and coworkers [14]. In this model it is assumed that the least
crowded space region around the metal atom of the catalytic center is
occupied by the growing chain. The situation in term of our simplified
representation of the transition states is reported in fig.3. With respect

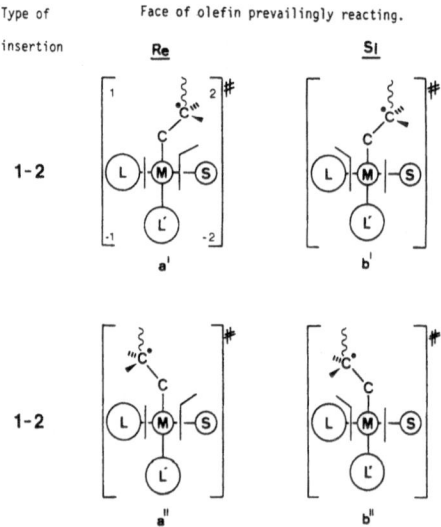

Figure 3. Planar representation of 4 possible transition states for
propylene insertion into a metal to CH₂ bond (see fig.1) taking
into account the conformation of the last monomeric unit of the
growing chain.

to the model reported in fig. 2 the relationship between enantioface of α-olefins prevailingly reacting and type of chirality of the catalytic center is opposite (e.g.lk instead of ul) in terms of the Prelog-Seebach nomenclature [15]. According to the transition state represented in fig. 2 the chirality of the catalytic centers should determine directly the enantioface discrimination of the incoming monomer. Taking into account the conformation of the growing chain (fig.3) the chirality of the catalytic center should influence the position in the space of the last monomeric unit of the growing chain. The enantioface discrimination of the incoming monomer should be determined both by the chirality of the catalytic center and by the conformation of the last monomeric unit of the growing chain.

From the above discussion it appears clearly that experiments using catalytic centers with known type of chirality might bring further interesting contributions to the understanding of the origin of regio- and stereospecificity in α-olefins polymerization.

3) POLYMERIZATION OF α-OLEFINS WITH CHIRAL CATALYTIC SYSTEMS.

A remarkable breakthrough in the field of α-olefins polymerization occurred at the end of 1984 when Ewen [16] using the Sinn-Kaminsky metallocenes/aluminoxanes catalytic systems and choosing a mixture of chiral and meso ethylene bis indenyl titanium dichloride, (first prepared by Brintzinger and coworkers [18]) as well as bis cyclopentadienyltitanium diphenyl as metallocene component, obtained isotactic polypropylene and substantially atactic polypropylene. Independently Kaminsky and coworkers, using racemic and optically active ethylene-bis-tetra-hydroindenyl zirconium dichloride ((EBTHI)ZrCl$_2$) as metallocene component, obtained exclusively highly isotactic polypropylene [19].

Pino and coworkers using meso-ethylene bis tetrahydroindenyl titanium dichloride as metallocene component obtained atactic polypropylenes with a substantially Bernoullian distribution of the isotactic and syndiotactic diads (20); furthermore using (-)(R) ethylene-bis-tetrahydroindenyl zirconium dimethyl and carrying out the polymerization in the presence of hydrogen they could isolate hydrogenated trimers and tetramers of propylene as well as a series of mixtures of optically active isotactic hydrooligomers and polymers with a number average molecular weight between 400 and 8000 [20].

The polymerization is believed to occur in solution; however the molecular

Table 1

Stereospecific polymerization of propylene with soluble or with $MgCl_2$ supported catalysts

Catalytic System	rac. (EBTHI)ZrCl$_2$/[-Al(CH$_3$)-O-]$_n$	TiCl$_4$/MgCl$_2$/AlR$_3$
Structure of active centers	Unknown. Zr^{IV} present; alkylation of Zr^{IV} probable. Dispersion of Zr$^+$ species in aluminoxane micelles possible. Substantially one type of catalytic centers present in solution	Unknown; Ti^{IV} species present; alkylation of Ti^{IV} probable. At least 5 families of catalytic centers present in the solid catalyst.
Typical catalyst productivity	3850 kg PP/Mol. Zr x h x mol C$_3$H$_6$ ℓ^{-1} at 10°C	2140 kg PP/Mol. Ti x h x mol C$_3$H$_6$ ℓ^{-1} at 60°C
Influence of monomer concentration on polymerization rate	rate proportional to monomer concentration:	rate proportional to monomer concentration: at least between 0,1 and 10 atm. C$_3$H$_6$
on molecular weight	very large	substantially none
Influence of temperature on $\bar{M}_n$	very large: $\bar{M}_n$ = 300.000 at -20; $\bar{M}_n$ =12.000 at 60°C	rather small between 0° and 60°C $\quad$ P$_{C_3H_6}$ >0,2 bar
Main chain termination process	β-hydrogen elimination	Transfer of hydrogen from the last monomeric unit of the growing chain to monomer
$\bar{M}_w/\bar{M}_n$	Large decrease of $\bar{M}_n$ in the presence of H$_2$ 1,5 - 2	Large decrease of $\bar{M}_n$ in the presence of H$_2$ > 5
Mode of insertion	substantially 1-2; occasional "2-1" insertions stop chain growth.	Mainly "1-2", evidences for "2-1" insertion in heptane soluble fractions and ether soluble fractions.
Regiospecificity	Very high; only traces of -CH(CH$_3$)-CH$_2$-CH$_2$--CH(CH$_3$)- groups shown by I.R. analysis	High for the heptane insoluble fraction; some irregularities are present in acetone soluble, diethylether soluble and heptane soluble fractions
Stereospecificity	Very high; in high polymers the content of isotactic pentads is higher than 95%; low polymers and even oligomers have a highly isotactic structure (less than 2% of steric inversions)	In the heptane insoluble fractions (40% of the polymer) the content of isotactic pentads is about 90%
Characterization of the highly crystalline high molecular weight ($\bar{M}_n$ > 10^4) Polypropylene	M.p. 151°C Stable crystalline form (after annealing): γ Insoluble in boiling heptane	M.p. 170°C Stable crystalline form (after annealing): α Insoluble in boiling heptane

weight of the methylaluminoxane catalyst component $(-Al(CH_3)-O-)_n$ seems to play a very important role in determining the catalyst activity indicating the possibility of formation e.g. of microaggregates (micelles) containing the catalytically active metallocene component. Furthermore when highly isotactic polypropylene is synthesized at low temperature, the polymer is insoluble in the reaction medium and chain growth can continue in heterogeneous phase.

As pointed out earlier by Kaminsky and Sinn [17] there is a number of remarkable differences between the polymerization of propylene with the usual supported heterogeneous catalytic systems (e.g. $TiCl_4/MgCl_2/AlR_3$) or with the metallocene/aluminoxanes catalytic systems, generally described as "soluble" catalytic systems as shown in table 1. The main differences concern the influence of monomer concentration and temperature on average molecular weight, the main polymer chain termination mechanism, the distribution of molecular weights, the catalyst productivity, the substrate selectivity [21], the regio- and stereoselectivity. Only the new indications concerning regio- and stereoselectivity arising from the above recent research will be discussed in this paper (see sections 4 and 5).

4) ORIGIN OF STEREOSPECIFICITY IN POLYPROPYLENES PREPARED WITH METALLOCENES/ALUMINOXANES CATALYTIC SYSTEMS.

According to Ewen [16] the asymmetric carbon atom of the last monomeric unit of the growing chains can control the enantioface selection in propylene polymerization to isotactic polymers if the polymerization is carried out at $-45°C$. In fact using a $Cp_2Ti(C_6H_5)_2/(Al(CH_3)-O-)_n/Al(CH_3)_3$ catalytic system a polypropylene has been obtained showing 85% of meso diads, the main stereochemical irregularities being present as isolated racemic diads (15%). At higher temperature this type of stereoregularity disappears and a substantially atactic polymer is formed. These data are very convincing and substantially confirm that at reaction temperature above $0°C$, asymmetric induction by growing chain plays substantially no role in enantioface selection in the synthesis of isotactic polymers. Very interesting are the previously discussed results obtained with meso, racemic and optically active titanocenes [18] and zirconocenes [22]. Under the reasonable assumptions that i) the catalytic species consists of a tight ion couple [23] formed according to scheme 1 and ii) that the

$$[Mt]X_2 + (-Al(CH_3)-O-)_n \rightleftharpoons [Mt]X^+ \ldots [X(-Al(CH_3)-O-)]^- \qquad (1)$$

geometry of the metallocene precursor is maintained in the catalyst, the
corresponding transition states for the olefin insertion taking into
account crystalline structure of the metallocene derivatives [22] can be
represented as shown in fig. 4 together with the corresponding simplified
models of the catalytic centers discussed in section 2.

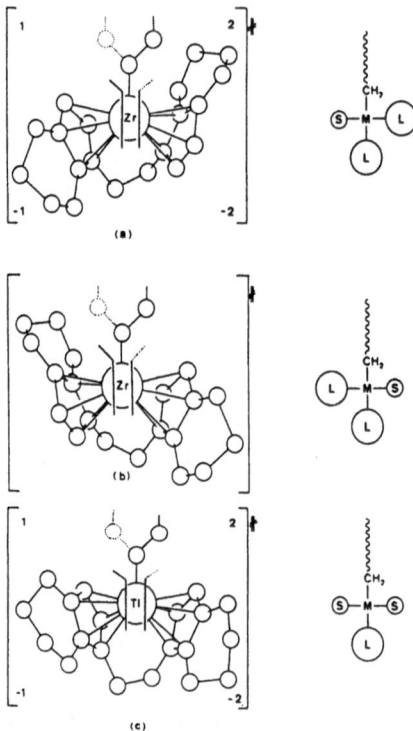

Figure 4. Representation of possible transition states for propylene
insertion with catalytic systems obtained from $(-)(R)(EBTHI)ZrX_2$
(a), $(+)(S)(EBTHI)ZrX_2$ (b) and meso $(EBTHI)TiX_2$ (c).

An inspection of fig. 4 shows that starting with the meso complex (fig.
4c) no enantioface selection can take place equal space availability
existing in quadrants 1 and 2. Starting with the (R) or the (S) complex
space available in quadrants 1 and 2 is largely different while very
little space is available in quadrants -1 and -2 due to the presence of
the ethylene bridge simulating a very large substituent opposite to the
growing chain. Considering the (R) antipode 2 situations can arise

depending on the occupation of the less crowded quadrant 1 by the α-olefin substituent (fig. 5a) or by the bulky growing chain (fig. 5b).

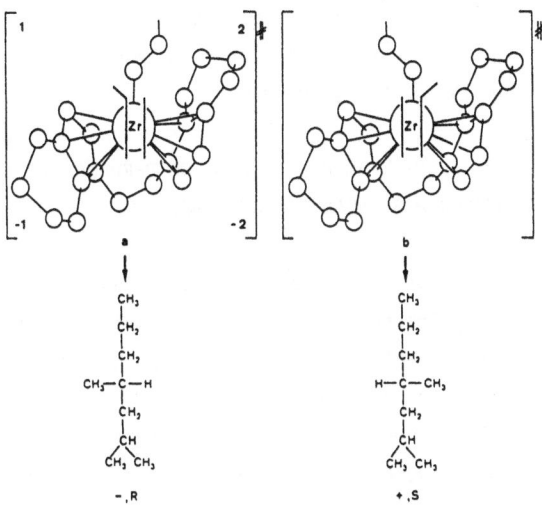

Figure 5. Transition states for propylene insertion with catalytic systems obtained from (-)(R)(EBTHI)Zr(CH$_3$)$_2$ leading to -(R) or (+)(S) hydrogenated trimer.

In the first case the Si enantioface and in the second case the Re enantioface of the olefin will prevailingly react. As starting with the (R)(EBTHI)Zr(CH$_3$)$_2$ the isolated propylene trimer obtained according to scheme 3 has the (S) configuration (~80% e.e) [20] the transition state represented in fig. 5b has a lower energy in agreement with the hypothesis by Corradini and coworkers [14].

The positive optical rotation of the oligomers with number average molecular weight up to 4000 and with known end groups [25], for which sign and value can be calculated according to Brewster [24], confirms that Re face of propylene prevailingly or exclusively react during the polymerization when the (R) metallocene is used as catalyst precursor. The stereospecificity is very high (>98% m diads) as shown by ^{1}H and ^{13}C NMR [25]. In conclusion, at least for the above catalytic system, the main stereoregulating factor in the propylene polymerization to isotactic polymer is the chirality of catalytic center as originally postulated [2] which compels the growing chain to occupy the most free portion of space around the catalytic center (Quadrant 1 in fig. 5) and induces the substituent of the vinylgroup to find a suitable spatial arrangement in the adjacent quadrant (Quadrant 2 in fig. 5) the other two quadrants being excluded because of the presence of the ethylene bridge between the two cyclopentadienyl groups. Therefore the high stereoselectivity seems to result from the cooperation of active center chirality and conformation of the last monomeric units of the growing chain.

5) ORIGIN OF REGIOSPECIFICITY IN POLYPROPYLENES PREPARED WITH METALLOCENES/ALUMINOXANES CATALYTIC SYSTEMS.

The hydrooligomerization of propylene with the (-)(R)(EBTHI)Zr$(CH_3)_2$/(Al (CH_3)-O-$)_n$ catalytic system has brought new very interesting information on the control of regioselectivity in propylene poly-insertion. In the presence of H_2 termination occurs practically only according to a hydrogenolysis of a Zr-carbon bond, and chains start always by insertion of monomer into a Zr-hydrogen bond (scheme 2). As terminal groups only n-

Scheme 2

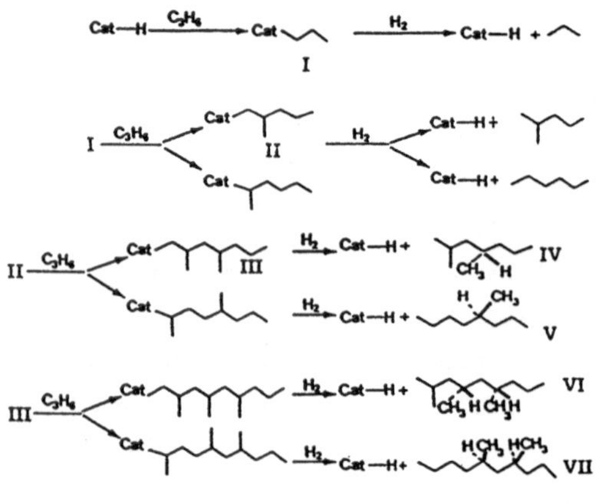

IV $[\alpha]_D^{20}$ = ~ + 10(4S) V $[\alpha]_D^{20}$ ~ +2(4S) .VI $[\alpha]_D^{60}$ = +18.4(neat)(4R:6S)

VII· ~ 2 (4S:6R)

propyl, isopropyl and n-butyl groups are present in the hydrooligomers and polymers, the concentration of n propyl groups corresponding to the sum of isopropyl and n butyl groups as shown by ^{13}C NMR. Furthermore, only traces of $-CH(CH_3)-CH_2-CH_2-CH(CH_3)-$ groups have been found by I.R. in the polymer having $\bar{M}_\eta$ = 17000.

These results show (scheme 2) that propylene polymerization occurs via 1-2 insertion and that if an occasional 2-1 insertion occurs, substantially no farther growth of the polymer chain takes place. In other words in isospecific centers after an occasional 2-1 insertion the activation energy for both 1-2 or 2-1 insertion is very high (scheme 3). This situation can be clearly accounted for considering the symplified

Scheme 3

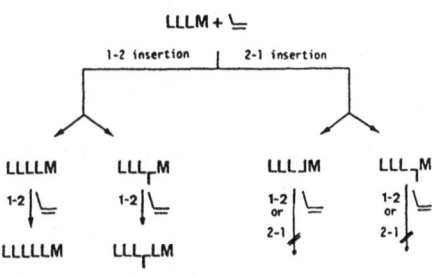

models of the transition state discussed in Section 2 (fig. 6). In fact no low energy transition state is possible in this case for steric reasons.

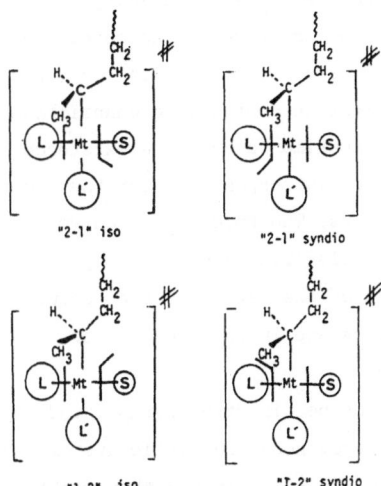

Figure 6. Transition states for the insertion of propylene into a metal to secondary carbon bond of an isospecific catalytic center.

As a consequence of the above facts we can expect that, particularly in the absence of hydrogen a remarkable number of catalytic centers bound to a secondary alkyl group are substantially inactive for the further insertion of monomer molecules. However these "dormant" catalytic centers may still insert other molecules, like e.g. CO which is used to "count" the number of catalytic centers in heterogeneous catalysts. Therefore the number of active centers determined by ^{14}CO method might be much higher than the real one if the heterogeneous catalysts behave like the metallocenes containing catalysts.

In conclusion the high regioselectivity observed with the catalytic system $(EBTHI)Zr(CH_3)_2/(-Al(CH_3)-O-)_n$ is due not only to the large difference between the energy of the transition states leading to the 1-2 or to the 2-1 insertion but also to the even higher energy of the transition states corresponding to 1-2 and to 2-1 insertions in isospecific catalytic centers in which a 2-1 insertion has occasionally taken place and in which the transition metal atom is bound to a secondary carbon atom.

CONCLUSIONS

The investigation of the stereospecific polymerization of propylene with catalytic systems prepared from optically active, racemic and meso metallocenes activated with aluminoxanes, has given the first direct experimental evidence concerning the factors responsible for enantioface discrimination and hence for stereospecificity in α-olefins polyinsertion. If the assumptions on the structure of the catalytic centers are correct, these experiments prove also the role of the conformation of the monomeric unit bound to the catalyst in controlling enantioface discrimination of the monomer. The same experiments show the existence of a double control of regioselectivity which is responsible for the exceptional structure regularity of the polymers. The role of the polymeric methylaluminoxanes is still not completely clarified and in our opinion a thorough investigation is needed of the nature of the "solutions" of aluminoxanes in toluene in the absence and in the presence of metallocenes, propylene and propylene oligomers.

The above results can not be directly extrapolated to the propylene polymerization with heterogeneous catalytic systems which behave differently as shown in Table 1. However they might stimulate new research on the structure of the numerous types of catalytic centers [26] formed on the solid surfaces.

REFERENCES

1. Natta, G., Pino, P., Corradini, P., Danusso, F., Mantica, E., Mazzanti, G., Moraglio, G., J. Am. Chem. Soc., 77, 1708 (1955).

2. Natta, G., Pino, P., Mazzanti, G., Gazz. Chim. Ital. 87, 528 (1957); Natta, G., J. Inorg. Nucl. Chem., 8, 589 (1958).

3. Natta, G., Pino, P., Mantica, E., Danusso, F., Mazzanti, G., Peraldo, M., Chim. Ind. (Milan) 38, 124 (1056).

4. Recently it has been found that isotactic polymers of propylene can be synthesized with non chiral soluble systems at -45°C. See ref. 16. However the same type of catalysts at room temperature yield atactic polymers. See ref. 17.

5. Zambelli, A., Sacchi, C., Makromol. Chem., 175, 2213 (1974); Zambelli, A., Locatelli, P., Rigamonti, E., Macromolecules, 12, 156 (1979).

6. Pino, P., Mülhaupt, R., Angew. Chem.Int. Ed. Engl., 19, 857 (1980).

7. Natta, G., Pino, P., Mazzanti, G., U.S. Patent 3112 300, (Nov. 26, 1963) Appl. date 8.6.1955.

8. Pino, P., Rotzinger, B., Makromol. Chem. Suppl., 7, 41 (1984).

9. Pino, P., Giannini, U., Porri, L., Polyinsertion, Encycl. Polym. Sci. Eng., New York a.o.: J. Wiley & Sons Inc., 2nd Ed. 1987 (in press).

10. Pino, P., Guastalla, G., Rotzinger, B., Mülhaupt, R., in Transition Metal Catalyzed Polymerizations, R.P. Quirk Ed., MMI Press Symp. Series 4, Harwood Acad. Publ. New York (1983) p. 435.

11. Consiglio, G., Pino, P., Top. Curr. Chem., 105, 77 (1982).

12. Pino, P., Rotzinger, B., von Achenbach, E., Makromol. Chem., Suppl., 13, 105 (1985).

13. Pino, P., Wei, J., unpublished results.

14. Corradini, P., Barone, V., Fusco, R., Guerra, G., Eur. Polym. J., 15, 1133 (1979), 16, 835 (1980).

15. Seebach, D., Prelog, V., Angew. Chem. Int. Ed. Engl., 21, 654 (1982).

16. Ewen, J.A., J. Am. Chem. Soc., 106, 6355 (1984).

17. Andresen, A., Cordes, H.G., Herwig, J., Kaminsky, W., Merk, A., Mottweiler, R., Pein, J., Sinn, H.J., Vollmer, H., Angew. Chem., 88, 689 (1976); Sinn, H., Kaminsky, W., Adv. Organomet. Chem., 18, 99 (1980).

18. Wild, F.R.W.P., Zsolnai, L., Huttner, G., Brintzinger, H.H., J. Organomet. Chem., 232, 233 (1982).

19. Kaminsky, W., Külper, K, Brintzinger, H.H., Wild, F.R.W.P., <u>Angew. Chem. Int. Ed. Engl.</u>, <u>24</u>, 507 (1985).

20. Pino, P., Cioni, P., Wei, J., Rotzinger, B., Arizzi, S., Lecture presented at the International Symposium Transition Metal Catalyzed Polymerization, June 16-20, 1986, Akron (USA).

21. Pino, P., Rotzinger, B., von Achenbach, E., in Catalytic polymerization of α-olefins, Keii, T. and Soga, K, Editors, <u>Elsevier Science Publishers</u>, B.V. Amsterdam (1986) p. 461.

22. Wild, F.R.W.P., Wasiucionek, M., Huttner, G., Brintzinger, H.H., <u>J. Organomet. Chem.</u>, <u>288</u>, 63 (1985).

23. Eisch, J.J., Piotrowski, A.M., Brownstein, S.K., Gabe, E.J., Lee, F.L., <u>J. Am. Chem. Soc.</u>, <u>107</u>, 7219 (1985).

24. Brewster, J.H., <u>J. Am. Chem. Soc.</u>, <u>81</u>, 5475 (1959).

25. Pino, P., Cioni, P., Wei, J., in press.

26. Pino, P., Fochi, G., Piccolo, O., Giannini U., <u>J. Am. Chem. Soc.</u>, <u>104</u>, 7381 (1982) See also ref. 10 and 21.

CHEMICAL ASPECTS IN THE SYNTHESIS OF

POLY(ε-CAPROLACTAM) FOR THE RIM PROCESS

Saverio Russo

Chemistry Department
University of Sassari, Italy

and

Giorgio Bontà, Amalia Imperato and Fabrizio Parodi[†]

C.N.R. Center on Macromolecules, Genoa, Italy

ABSTRACT

The activated anionic polymerization of ε-caprolactam has been found suitable for the reaction injection molding (RIM) technology. New types of functional activators, able to accelerate the reaction rate to a very great extent with an overall polymerization time of about 30 s., can also induce extensive crosslinking of the polyamide chains. The crosslinking reaction is strongly depending on the experimental conditions chosen for the synthesis.

Red phosphorus or its mixtures with synergistic compounds can provide flame retardancy to anionic poly(ε-caprolactam), under conditions suitable for the RIM technology.

Impact-resistant poly(ε-caprolactam) can be prepared by in situ formation of interpenetrating networks made of butadiene-acrylonitrile elastomers and polyamide. The two independent crosslinking reactions occur simultaneously in a very short time.

For all systems under study the RIM technology appears to be easily applicable.

[†]Present address: Vitrofil S.p.A., Vado Ligure, Savona, Italy

INTRODUCTION

Several new polymeric systems, mostly based on polyurethanes, poly-
ureas and aliphatic polyamides, have recently been found suitable for the
RIM (reaction injection moulding) technology.

These polymers display intrinsically good mechanical properties,
such as high strength and toughness, by virtue both of strong hydrogen-
bond and dipolar interactions and of extensive, segregated domains (either
crystalline or amorphous). Such polymers can become true engineering plastics for structural applications, by inducing some property improvements,
for instance by enhancing their elastic moduli, or heat distorsion temperatures, or thoughness at low-temperature, by decreasing creep compliances
or by providing them with additional features such as flame retardancy,
electrical conductivity, and so on.

The above performance improvements can be achieved by either molecular mass increase or crosslinking or chemical modification of the polymer itself, and/or by introducing in the polymer matrix suitable additives, such as fibres, whiskers, reinforcing structures, fillers, flame-
retardant agents, electro-conductive particles, rubbers, etc.

Nowadays, the aforementioned requirements are met at far larger
extent through conventional processing technology, by introducing the additives through melt mixing (mostly by screw extrusion), followed by moulding from the melt (injection moulding, sheets or laminates extrusion/
calendering, etc.). On the other hand, crosslinking can be carried out
on moulded parts as a post-treatment of heat- or photo- or high energy
radiation curing. The widely applied technology of melt-mixing and injection moulding, however, has many severe limitations. The melt viscosities
are unfavourably affected by solid additives, especially with long fibres
and at large volumetric contents. Therefore, high temperatures (leading
however to thermal degradation of either the polymer itself or various
additives) and/or very high shear rates (altering additives morphology

or even causing chain scission) are required. Similar problems are encountered when adopting very high molecular mass polymers, often untractable
being thermodegraded before softening. Post-crosslinking of moulded parts,
moreover, is a slow or incomplete process, which leads to dishomogeneous
shrinkages and deformations.

In principle, the RIM technology seems suitable to solve many of
the above problems, provided that some fundamental requirements are sati-
sfied, such as the absence of any adverse effect of the additive on the
polymerization kinetics. Moreover, the very low viscosities of liquid or
molten monomers allow to pre-disperse comparatively large amounts of solid
additives and even embed reinforcing structures (e.g. glass fibers) and
various inserts, leading to Reinforced RIM (R-RIM) (1).

Fast and uniformly distributed crosslinking can be obtained during
the RIM process, when suitable polyfunctional monomers or additives are
used. They can promote branching reactions, which may lead to IPN or semi-
IPN structures. Furthermore, the rather low reaction temperatures largely
eliminate thermal degradations, especially of polymeric additives, otherwise labile at the usual processing conditions.

RIM processes should be based on polymerization reactions free of
any by-products, completely avoiding vapours or gases evolution. Fast
initial polymerization rates are required so that the rapid viscosity raise prevents any particulate settling. High overall rates and monomer
conversions allow short cycle times (comparable to standard injection moulding). Furthermore, as mentioned above, organic or inorganic additives
have to be properly selected, in order to avoid any unfavourable influence
on polymerization mechanism and kinetics.

Fast reactions, free of by-products and suitable for RIM technology,
can be, for instance, the condensations between isocyanate and alcohol or
amine, and several ring-opening polymerizations, among which the activated
anionic polymerization of ε-caprolactam (CL) represents one of the most

promising synthetic routes (2-4).

Indeed, poly(ε-caprolactam) (PCL) is a well-known engineering plastics, characterized by high strength and good fatigue and wearing dura-
bility. Its main lacks can be depicted as rather poor impact strength at
low temperatures and large moisture absorption/desorption with intense
plastification/deplastification phenomena, strongly depending on environmental conditions.

Additional limitations are ease to undergo hydrolytic attack and
poor flame resistance because of dripping. Either the introduction of
finely dispersed rubbers or the copolymerization of CL with reagents containing macromolecular blocks (achievable at best by RIM process) largely
improve impact performances because of the relevant viscoelastic relaxation contribution at low temperatures/high frequencies. Block copolymers
of CL with various 'soft' prepolymers by anionic polymerization have been
thoroughly investigated (2,4) and eventually put on the market. RIM polymerization of CL (either pure or in presence of various additives), when
accomplished by carefully controlled crosslinking, can produce PCL with
improved stiffness (low creep) and strength, low plastification and better hydrolytic stability.

EXPERIMENTAL

Full details on the materials used, the polymerization procedure
and the samples characterization are given elsewhere (3,5-9). Table 1
summarizes sources of materials and typical experimental conditions for
the polymerization.

RESULTS AND DISCUSSION

i)   Slow and fast activation.

The anionic polymerization of CL (as well as of other lactams) is
a well known reaction, extensively studied by several research groups in
the world since 1941. It is initiated by strong nucleophilic reagents and

assisted by various activators (10), smoothly affording PCL through a complex series of main and side reactions, mechanisms and kinetics of which have been investigated in detail (11-15).

TABLE 1

Materials for the polymerization experiments, performed either in quasi-adiabatic reactor (a) or in the mold (b).

Materials	Source
- ε-caprolactam	- Enichem,Möntedipe, DSM
- initiator (Na caprolactamate)	- DSM, in-situ preparation
- activators (N-acetyl caprolactam, multifunctional carbamoyl-type activators)	- laboratory synthesis
- butadiene-acrylonitrile elastomers	- Enichem, Bayer, Polysar
- red phosphorus	- Saffa, Hoechst

a) quasi-adiabatic reactor, 60 ml., T_o = 155°C, mechanical stirrer, dry-nitrogen blanket.
b) RIM mold, 240 ml., T = 150, 160, 180, and 200°C.

The activated anionic polymerization of CL has recently been found suitable for RIM technology under specific experimental conditions and with new fast activators, leading to high-speed processes characterized by high polymer yield of ca. 97% in less than 60 s. (9, 16).

The most effective activators are acylating or carbamoylating reagents or, more conveniently, N-acyl and N-carbamoyl-lactams, very frequently adopting alkali metal caprolactamate as initiators.

The overall course of neat CL polymerization in quasi-adiabatic reactors has been investigated by our research group in presence of different amounts of sodium caprolactamate as initiator and of three activators, i.e. monofunctional N-acetyl-CL (I) and two difunctional N-carbamoyl-

CLs (hexamethylene-1,6-dicarbamoyl-CL (II) and 2,2,4-trimethyl-hexamethy-lene-1,6-dicarbamoyl-Cl (III)).

It has been reported in literature that the multi-functional N-acyl-ε-caprolactams induce crosslinking during polymerization (17) and that N-carbamoyl-ε-caprolactams (CL-blocked isocyanates) are much faster activators than N-acyl ones (18).

Our results, compairing the effectiveness of the above activators (I, II, and III), are given in Figure 1, where the overall polymerization times, t_p, are plotted as functions of the molar concentration of the activator A. Constant initiator concentration of 0.6 mol.% has been adopted. It is evident that (I) is a 'slow' activator, causing rather long t_p values, about ten times higher that the overall polymerization times related to (II) or (III). The very high activity displayed by the two N-carbamoyl-caprolactams ('fast' activation) makes them suitable for the RIM technology. The slight difference in reactivity between (II) and (III) are abviously due to the different backbone structure of the two 'blocked' disocyanates. A deeper insight on the role of structure on reactivity is in progress (9). According to their difunctionality, (II) and (III) induce extensive crosslinking (over 80%, depending on the experimental conditions), where (I) does not cause any gel formation.

Similar results on t_p have been obtained in the RIM polymerization experiments, whereas close-to-isothermal conditions, typical of RIM, strongly affect the extent of crosslinking (9).

ii)   Rubber-modified PCL.

In order to attain impact-resistant PCL, especially at low temperatures and dry conditions, we have dissolved various polar rubbers in CL monomer and followed the kinetic course of the polymerization reaction. Unlike other research groups, who have used telechelic rubbers, with functional groups able to act as polymeric activators, in order to synthe-

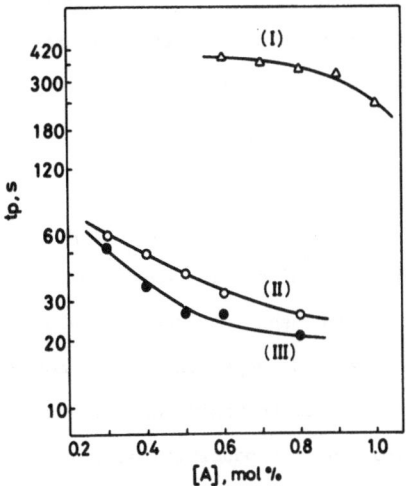

Figure 1. - Overall polymerization time as a function of activator concentration, for [I] = 0.6 mol%.(Quasi-adiabatic reactor).

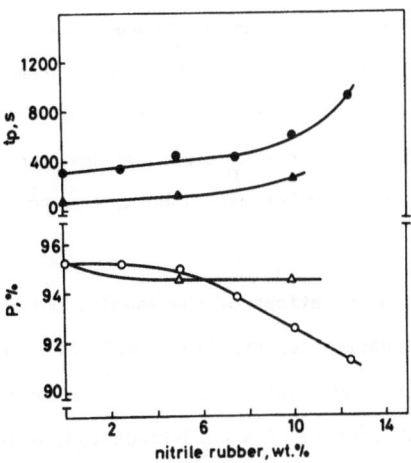

Figure 2. - Comparison between slow (O,●) and fast activation (Δ,▲). Effects on t_p and P.

size multiblock copolymers constituted of elastomeric and PCL sequences
(2,4,16), we have privileged high-molecular mass nitrile rubbers exempt
from any specific reactive group to be used as macro-activator. Various
amounts of butadiene-acrylonitrile elastomers (up to 12.5 wt.%) were
dissolved in CL and either slow (type I) or fast (II) activators used
for the polymerization runs.

In Figure 2 some relevant data on overall polymerization time, t_p,
and high polymer yield, P, are given. The AN content of the rubber is
39 wt.%. Similar results have been obtained with elastomers characterized
by different molecular mass and composition. For samples synthesized using
'slow' activators, only an intimate mixing of the various components of
the reacting mixture provides reproducible results both in terms of polymerization kinetics and polymer pair morphology. As compared to the values pertaining to neat CL polymerization, it can be seen that the nitrile
rubber causes some retardation and a sharp decrease of the high polymer
yield.

On the contrary, fast activators which induce a fivefold decrease of t_p for the neat system, show analogous effects in presence of large amounts of the rubber: t_p is ca. 200 s. for systems with 10 wt.% of
rubber. Also P is almost unaffected by the dissolved elastomer. High P
values, together with very short t_p times, ensure the suitability of the
systems, based on nitrile rubber dissolved in CL, to be utilized for the
RIM technology.

Morphological observations of the samples obtained either with
slow or fast activation essentially reveal the same phase organization, with minor differences which will be discussed further on. Figure
3 shows a typical SEM picture of a cold-fractured, gold-coated surface
of the as-polymerized material (rubber content: 10wt%, AN:39 wt%). The
morphology is constituted of a continuous PCL matrix containing large domains of variable size (in the range of 10-100 μm) which, in turn are made of many spherical sub-domains coated with a soft phase. A great number

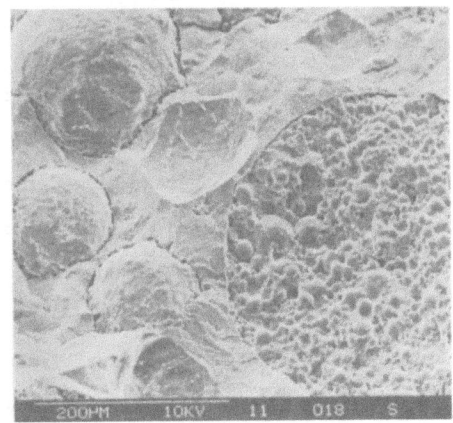

Figure 3. - SEM picture of a
cold-fractured surface of PCL/
nitrile rubber.

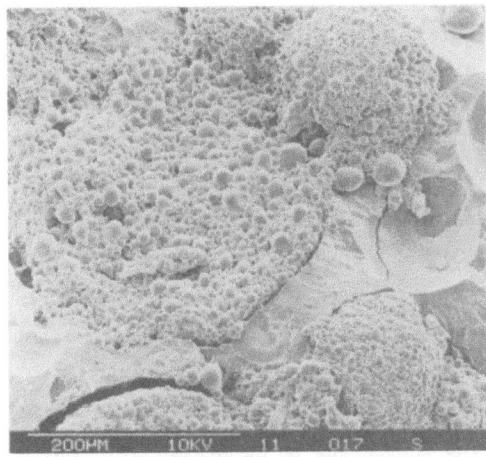

Figure 4. - SEM picture of the
above sample, after etching
with acetone.

Figure 5. - SEM picture of the
above sample, after extensive
treatment with hexafluorobutanol.

of interconnecting bridges between the macro-domains and the matrix are
also evident.

After surface etching with acetone, a solvent of the nitrile elastomer, the soft cover is removed, as shown in Figure 4 , and many small
spheres appear. The average diameter of the small spheres is of about
4 μm. By further etching with hexafluorobutanol, a good solvent of both
PCL and nitrile rubber, the spherical sub-domains collapse and a continous network phase appears (Figure 5). Molecular characterization of the
various extractable and residual fractions reveals that the sub-domains
are made of PCL spherulites, kept together by a crosslinked nitrile rubber, which constitutes the continuous network. The soft cover is made
of uncrosslinked elastomer. The crosslinking reaction is caused by the
basic reaction medium, as show by simulation experiments.

With fast activators, also PCL is crosslinked; thus, simultaneous
interpenetrating network (SIN) structures are formed. In such circumstances, the presence of cylindrical and spherical macrodomains has been
evidenced by SEM.

iii)   Flame-resistant PCL.

We have found that improved flame-retardant (FR) properties can
be provided to PCL by addition of red phosphorus to CL monomer (6). The
activated anionic polymerization process is not adversely affected by the
presence of the FR agent up to the concentration of 15 wt.%, both with
slow and fast activators. The overall phosphorus content can be reduced
by addition of magnesium oxide, which acts as a synergistic compound.
The FR additive remains homogeneously dispersed in the polyamide matrix.
From the characterization data of the resultant products in terms of
morphology, as well as of molecular, thermal and mechanical properties,
no significant modification has been induced by the FR additive. In
Table 2 some relevant data on the flame-retardant properties imparted
to PCL are given.

TABLE 2

Effect of red P and its mixture with MgO on flammability test for PCL
(UL-94).

FR agent,	wt.%	Synergistic agent, wt.%		UL-94 test data
none	0	none	0	FB[*]
red P	5	none	0	V-1
red P	8-15	none	0	V-0
red P	5	MgO	5	V-0

[*] FB = free burning

From the above data it can be inferred that the RIM technology, which requires both extremely fast polymerization and very high conversion, can
be usefully applied in order to synthesize FR poly(ϵ-caprolactam).

CONCLUSIONS

Fast activators, characterized by N-carbamoyl-lactam end groups,
provide overall polymerization times lower than 30 s., with monomer conversion of ca. 97% and high polymer yield of ca.96%. The extent of crosslinking depends on the experimental conditions (polymerization temperature,
initiator/activator ratio, active species concentration).

Flame resistant PCL can easily be obtained by in-situ polymerization of CL in presence of adequate amounts of red P or a 1:1 mixture of
red P and MgO.

Simultaneous interpenetrating networks, constituted of nitrile rubber and poly(ϵ-caprolactam), have been synthesized and characterized.
Depending on the experimental conditions, it is possible to obtain either
semi-IPN or intertwining polymer networks. Very peculiar morphologies and

properties are observed. AN content and molecular mass of the rubber can
be varied, depending on the required characteristics of the final product.
Slow or fast polymerization of the lactam can be adopted, in order to obtain a wide variety of morphologies. Other operational variables able to
strongly affect both polymerization kinetics and properties of the resultant materials are stirring stime, polymerization temperature, initiator and activator concentrations.

Immediately before and during the polymerization of ε-caprolactam
the nitrile rubbers undergo crosslinking reaction. As mentioned above
when fast activators are used, also the polyamide can crosslink. No
chemical links between the two polymers seem to be present. The extent of
the two crosslinking reactions can be varied at will by suitably playing
with the operational variables.

ACKNOWLEDGMENTS

Thanks are due to Drs.G.C.Alfonso, G.Costa, and A.Turturro for their
relevant contribution to several parts of the present work. Progetto
Finalizzato Chimica Fine e Secondaria -Bb and Bc-is gratefully acknowledged for financial support.

REFERENCES

1) J.E.Kresta,(Ed.) Reaction Injection Molding, Polymer Chemistry and
Engineering, ACS Symp. Ser.# 270, Washington, D.C., 1985.

2) R.M. Hedrick, J.D. Gabbert and M.H.Wahl, ibidem, pp. 135-162.

3) G.C.Alfonso, C.Chiappori, S.Razore and S.Russo, ibidem, pp.163-179.

4) J.L.M. van der Loos and A.A. van Geenen, ibidem, pp.181-191.

5) G.C.Alfonso, G.Costa, S.Russo and A.Turturro, Proc. 7th Italian Meeting
Macromolecular Sci., Galzignano, 1985, pp. 201-210.

6) G.C.Alfonso, G.Costa, M.Pasolini, S.Russo, A.Ballistreri, G.Montaudo
and C.Puglisi, J.Appl.Polym.Sci. 1986, 31, 1373.

7) G.C.Alfonso, G.Dondero, S.Russo, and A.Turturro, Morphology of Polymers, B.Sedlaček ed., W. De Gruyter, Berlin 1986, pp. 427-434.

8) G.C.Alfonso, G.Dondero, S.Russo, A.Turturro and E.Martuscelli, New Polym.Mat., in press.

9) E.Biagini, G.Bontà, A.Imperato, F. Parodi and S.Russo, to be published.

10) S.R.Sandler and W.Karo, Polymer Syntheses, Academic Press, New York 1974, vol. 1, ch.4, pp.89-115.

11) J.Šebenda, Comprehensive Chemical Kinetics, Edrs. C.H.Bamford and C.F.Tipper, vol. 15, ch.6, 'Lactams', Elsevier, Amsterdam 1976, pp.379.

12) H.K. Reimschuessel, J. Polym.Sci. Macromol.Revs. 1977, 12, 65.

13) J.Šebenda, Structural Order in Polymers, Edrs. F.Ciardelli and P.Giusti, pp.95-110.

14) H.Sekiguchi, Ring-Opening Polymerization, Edrs. K.J.Ivin and T. Saegusa, Elsevier Appl.Sci. Publ., Barking, Essex 1984, pp.809-918.

15) J.Šebenda, Makromol.Chem., Macromol.Symp. 1986, 6, 1.

16) J.D. Gabbert and R.M. Hedrick, Proc. 3rd Int. Conference on Reactive Processing of Polymers, Bischenberg 1984, pp. 137-148.

17) T.M. Frunze, V.V. Kurashev, V.I. Zaitsev, R.B. Schleifman and V.V. Korshak, Polym.Sci.USSR 1974, 16, 1452.

18) A.Ya. Malkin, V.G. Frolov, A.N. Ivanova, Z.S. Andrianova and L.A. Alekseichenko, Polym.Sci.USSR 1980, 22, 1097.

ANIONIC RING-OPENING POLYMERIZATION OF CARBONATES AND FORMATION OF BLOCKCOPOLYMERS

Helmut Keul and Hartwig Höcker
Lehrstuhl für Textilchemie und Makromolekulare Chemie
der RWTH Aachen
Veltmanplatz 8, D-5100 Aachen
West-Germany

ABSTRACT

Cyclic carbonates are accessible by reaction of diols with carbonic acid esters. With anionic initiators they can be polymerized under mild conditions to result in high molecular weight polymers. Under suitable conditions, a ring-chain equilibrium is established, i.e., beside acyclic polymers a homologous series of cyclic oligomers is obtained. The different cyclic carbonates discussed open a variety of secondary reactions, in particular cross-linking. Following the anionic mechanism, cyclic carbonates are suitable monomers for the preparation of blockcopolymers. Polystyrene-block-poly(2,2-dimethyltrimethylene carbonate) and multiblockcopolymers of 2,2-dimethyltrimethylene carbonate and ε-caprolactone were prepared.

INTRODUCTION

While the commercially highly important aromatic polycarbonates are prepared by polycondensation polymerizations, aliphatic polycarbonates may be obtained by ring-opening polymerization of the cyclic monomers. The cationic ring-opening polymerization follows the alkyl and the acyl scission mechanism and yields low molecular weight material (1). The anionic ring-opening polymerization with K_2CO_3 at 130°C also yields low molecular weight material in conjunction with decarboxylation (2).

In the present paper the anionic ring-opening polymerization of cyclic carbonates under mild conditions is described. Further, the preparation of blockcopolymers with styrene and ε-caprolactone is reported.

RESULTS AND DISCUSSION

Homopolymers

As cyclic carbonates, the following monomers were used:
2,2-Dimethyltrimethylene carbonate (1), 2-butyl-2-ethyltrimethylene

carbonate (2), 2-allyloxymethyl-2-ethyltrimethylene carbonate (3), 2-benzyl-oxymethyl-2-ethyltrimethylene carbonate (4), 2-oxo-2'-phenyl-5,5'-spirobi-(1,3-dioxane) (5), 2-oxo-1,3-dioxane-5-spirocyclohexane (6), dodeca-methylene carbonate (7), dihexamethylene dicarbonate (8), 2,2'-diethyl-2,2'-oxydimethylenedi(trimethylene carbonate) (9).

The different carbonates may be divided up into three classes with respect to their polymerization ability:

(i)   the six membered ring carbonates (1 to 6)

(ii) the monocarbonates with 15 ring atoms (7) and the dicarbonate with 18 ring atoms (8)

(iii) the bifunctional dicarbonate with an ether bridge (9).

For the first group of cyclic carbonates, the 2,2-dimethyltrimethylene carbonate is a representative member. With sec. butyllithium as an initiator at 10°C in toluene as a solvent and at an initial monomer con-centration of 10 weight-% a high molecular weight polymer with 90% yield is obtained after a reaction time of 5 minutes. Upon increasing the

temperature up to 20°C and the reaction time up to 48 h the ring-chain
equilibrium is established, i.e., beside high molecular weight polymer a
homologous series of cyclic oligomers is observed. The plot log x (x being
the degree of polymerization of the oligomers) vs. the elution volume
results in a straight line calibration curve. The double logarithmic plot
of the equilibrium concentration of the individual oligomers [M_x] vs. the
degree of the polymerization yields a straight line with a slope of -2,5
for the tetramer and the higher oligomers while the concentration of the
dimer and the trimer deviate from the straight line (3). Thus the prediction of the Jacobson-Stockmayer ring-chain equilibrium theory is
fulfilled (4).

From the yield of oligomers and polymers obtained at temperatures
between -10 and +25°C the ceiling temperature of the polymerization of 2,2-
dimethyltrimethylene carbonate can be estimated to be 30°C.

The polymer is highly crystalline and shows two peaks upon differential scanning calorimetry at 107°C and 123°C. From the X-ray wide angle
scattering it may be concluded that the small crystals disappear at 107°C
and very large crystals are formed which eventually melt at 123°C. The
pattern does not change in the respective temperature region; the homo-
geneity, however, of the Debye-rings disappears and individual diffraction
spots are observed.

The behaviour of 2 upon polymerization is very similar to that of 1.
The glass transition temperature of the polymer, however, is below room
temperature.

Monomer 3 again behaves very much like 1 and 2. The homopolymerization
of 3, however, yields a product which is insoluble because of cross-linking
via the pendent double bonds. The copolymerization of 3 (10 weight-%) with
1 (90 weight-%) yields a soluble copolymer.

Monomer 4 yields a high polymer; the establishment of the ring-chain
equilibrium - probably because of the bulky benzyl group - is kinetically
hindered. Fig. 1 shows the ^{1}H-NMR-spectrum of the monomer and the polymer.
It is seen that the diastereotopic protons d in the monomer become
equivalent in the polymer.

The spiro carbonates 5 and 6 may also be converted into their
polymers; in the case of 5 DMF must be used as solvent since the monomer is
insoluble in toluene and THF. It is expected that the acid cleavage of
acetale bonds (monomer 5) yields a hydrophilic polymer with geminal
methylol groups.

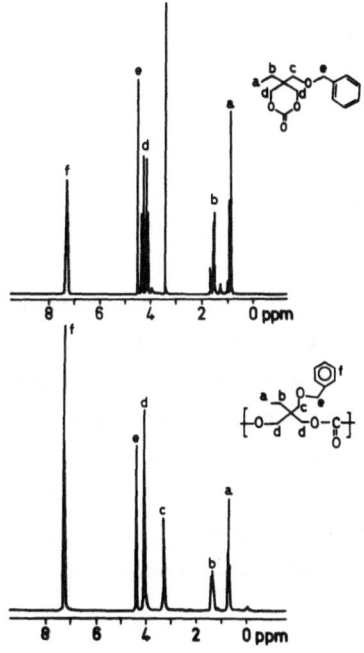

Figure 1. ^{1}H-NMR spectrum of $\underline{4}$ and poly(4).

Both monomers of the second type, $\underline{7}$ and $\underline{8}$, may be polymerized at elevated temperature (50-60°C) during 5 h. The polymer of $\underline{7}$ shows a melting point of 69°C while the polymer of $\underline{8}$ shows a melting point of ca. 50°C.

Monomer $\underline{9}$ is expected to yield a cross-linked polymer. Part of the polymer, however, is found to be soluble. Therefore it is assumed that upon intramolecular reactions (cyclopolymerization) the following structure is formed:

Blockcopolymers

The formation of blockcopolymers was studied with monomer $\underline{1}$. Instead of sec. butyllithium, polystyryllithium was used as an initiator. It is essential, however, to reduce the nucleophilicity of the living polystyryl-anion before reacting with the cyclic carbonate. This was achieved by first reacting polystyryllithium with ethylene oxid. The resulting alcoholate is effective in initiating the anionic ring-opening polymerization of $\underline{1}$.

The anionic ring-opening polymerization of ε-caprolactone is well
known as well as the formation of a ring-chain equilibrium (5). In the
kinetically controlled regime, however, a high molecular weight poly-ε-
caprolactone may be obtained. Thus, polystyryllithium may be used as
initiator for the anionic polymerization of ε-caprolactone (6).

On the other hand, blockcopolymers of 1 and ε-caprolactone may be
prepared starting from an anionically initiated polymer of 1 (A) which upon
addition of ε-caprolactone (B) forms the blockcopolymer. Since the anionic
ring-opening polymerization of 1 is much faster than that of ε-caprolactone
both monomers can be added to the initiator simultaneously to result in the
AB-diblockcopolymer. When a mixture of both monomers with given ratio is
added to the initiator in portions of given size blockcopolymers in a wide
variety may be obtained; according to block length and monomer distribution'
in the blockcopolymers, products with brittle to rubber elastic character
may be obtained. The ^{13}C-NMR-spectra of the homopolymers are given in
Fig. 2, those of an A-B blockcopolymer as well as of a multiblockcopolymer
are shown in Fig. 3.

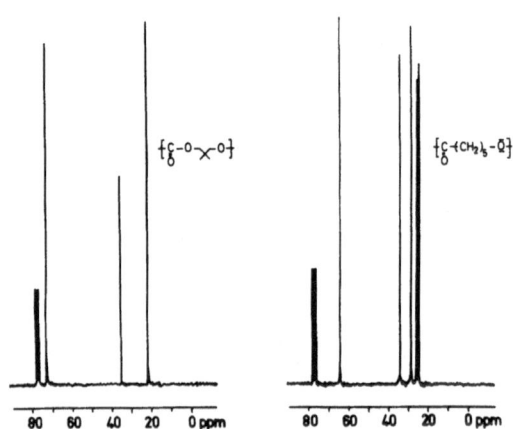

Figure 2.   ^{13}C-NMR spectrum of poly(1) and poly-ε-caprolactone.

Fig. 3a representing the NMR spectrum of an AB-diblockcopolymer with
the monomers added consecutively shows only the resonance lines of the
homopolymers. The blockcopolymer the spectrum of which is given in Fig. 3b
was obtained by adding both monomers simultaneously to the initiator
solution. The tapered structure, indicated by X, shows up in additional
resonance lines. For the multiblockcopolymer these lines have increased
significantly (Fig. 3c).

35

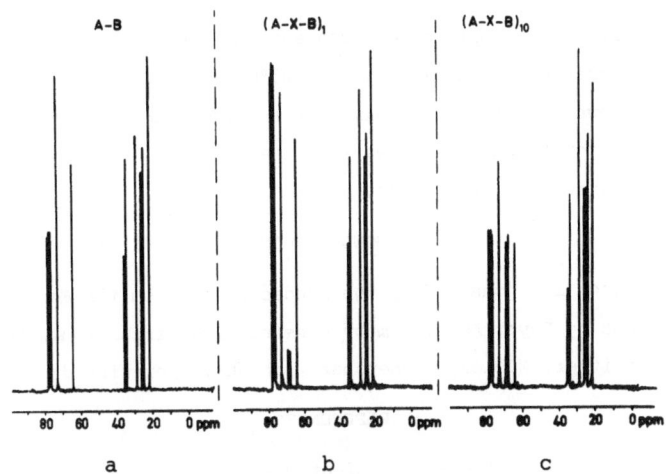

Figure 3. ^{13}C-NMR spectrum of blockcopolymers
a) poly($\underline{1}$)-$\underline{b}$-(ε-caprolactone); b) dto with tapered
sequence X; c) multiblockcopolymer (10 sequence units)
with tapered sequences X in between the sequence units.

The thermal properties of the diblockcopolymer with some tapered
structure are shown in Fig. 4 as a function of composition.

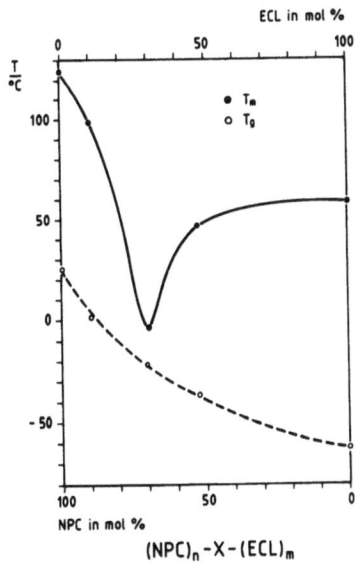

Figure 4.   Thermal properties of blockcopolymers poly(1)-
(ε-caprolactone) with tapered sequence X.

With increasing ε-caprolactone content the glass transition temperature decreases monotonically from 25°C to -65°C while the melting temperature shows a pronounced minimum of a composition of 70% NPC/30% ECL. Thus cyclic carbonates offer new ways to the formation of new products with a broad variety of properties.

ACKNOWLEDGEMENT

This project is sponsored by the Bundesministerium für Forschung und Technologie and by Bayer AG. The many constructive discussions with Drs. Ott, Morbitzer, Müller, Eichenauer are highly appreciated.

REFERENCES

1. Kricheldorf, H.R., private communication.

2. Bayer AG, private communication.

3. Keul, H., Bächer, R. and Höcker, H., Makromol. Chem. 1986, 187, 2579.

4. Jacobson, H. and Stockmayer, W.H., J. Chem. Phys., 1950, 18, 1600.

5. Yamashita, Y., Polym. Prepr. (Am. Chem. Soc., Div. Polym. Chem.) 1980, 21 (1), 51.

6. Keul, H. and Höcker, H., Makromol. Chem., 1986, 187, 2833.

PREPARATION OF BLOCK COPOLY(ARYLENE SULPHIDES) UTILISING PREFERENTIAL
POLYMER FORMATION IN POLYMERISATION OF COPPER(I) 4-BROMOTHIOPHENOXIDES

P.A. Lovell and R.H. Still

Department of Polymer Science and Technology, U.M.I.S.T., P.O. Box 88,
Manchester, M60 1QD, United Kingdom.

ABSTRACT

    Preferential polymer formation in polymerisation of copper(I) 4-bromo-
thiophenoxides has been used to advantage to prepare novel ABA block copolymers of poly(1,4-phenylene sulphide)(A) and poly(2-methyl-1,4-phenylene
sulphide)(B) from bis-(4-bromophenyl)sulphone by sequential reaction with
copper(I) 4-bromo-2-methylthiophenoxide and copper(I) 4-bromothiophenoxide.
The materials produced have been characterised by IR/NMR spectroscopy, elemental analysis, hot-stage microscopy and differential scanning calorimetry.
The preparation and properties of the block copolymers are compared to those
of the homopolymers and random copolymers comprising 1,4-phenylene sulphide
and 2-methyl-1,4-phenylene sulphide repeat units.

INTRODUCTION

    Polymerisation of metal(I) 4-halothiophenoxides yields high molar mass
polymer at conversions in the range 60-90% [1-4], in contrast to predictions
of classical step-growth polymerisation theory. *Preferential polymer formation* arises from the different activating influences of $-S^\ominus$ and $-S-$ on
nucleophilic substitution of the para-bromine atom in the substituted phenylene ring. Addition of monomer to monomer involves the unfavourable introduc-
duction of a second negative charge to one benzene ring (Scheme 1). Addition of monomer to either oligomer or polymer gives rise to a more stable σ-
complex one canonical form of which localises the negative charge on the
sulphide sulphur atom by d-orbital expansion of the sulphur octet (Scheme 2).
Thus, once dimer has formed it grows at the expense of further dimer formation and high molar mass polymer is obtained at low conversions.

    This paper reports the preparation of novel ABA block copolymers of
poly(1,4-phenylene sulphide)(PPS(A)) and poly(2-methyl-1,4-phenylene
sulphide)(PMPS(B)) via a reaction scheme (Scheme 3) which utilises preferential polymer formation to maximum advantage in limiting homopolymer and
random copolymer formation. The reaction scheme involves sequential addition
of equivalents of monomer to a growing activated-dibromide species deriving
from bis-(4-bromophenyl)sulphone. The first set of additions were of

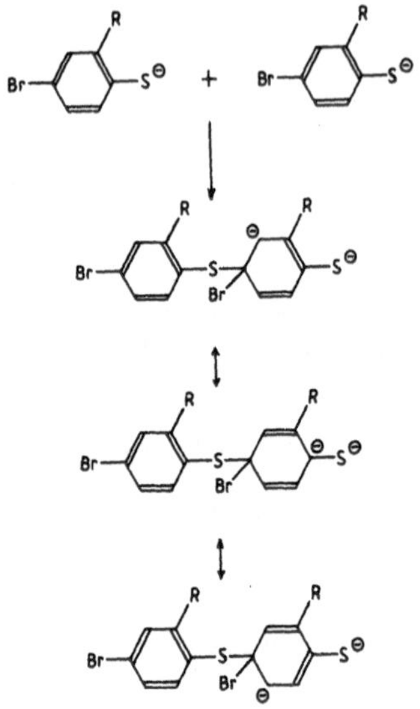

Scheme 1

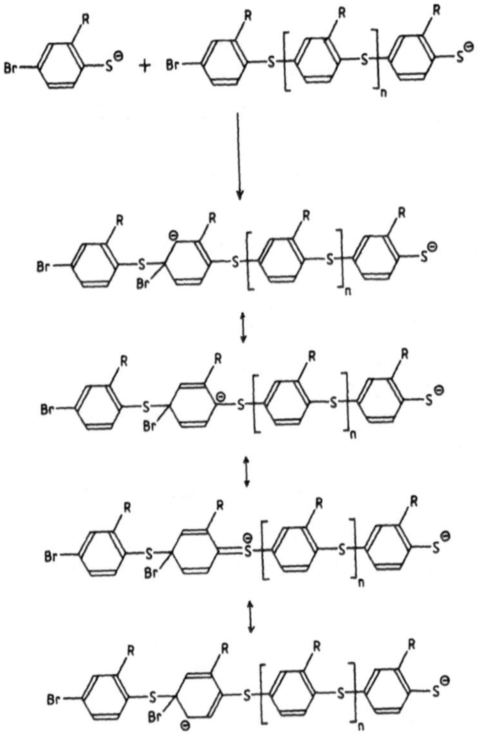

Scheme 2

copper(I) 4-bromo-2-methylthiophenoxide (CBMT) and the second were of
copper(I) 4-bromothiophenoxide (CBT).

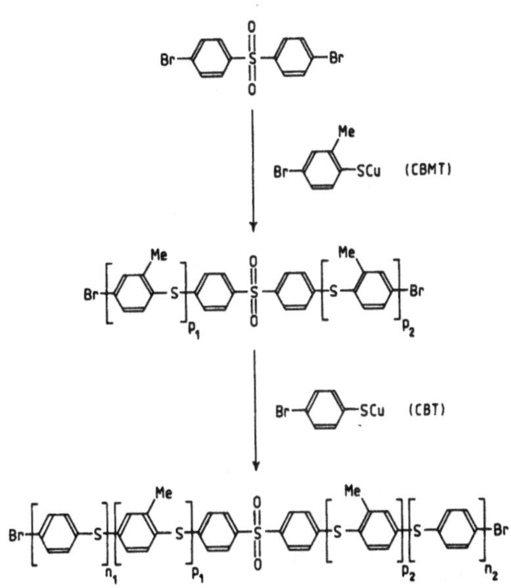

$$\overline{n} = (n_1 + n_2)/2 \quad \overline{p} = (p_1 + p_2)/2$$

Scheme 3

EXPERIMENTAL

PPS, PMPS and random and block copolymers comprising PPS and PMPS
repeat units were prepared employing solution polymerisation of CBT and/or
CBMT in boiling 10:1 (vol. ratio) quinoline:pyridine solvent mixture under
a dry nitrogen atmosphere [1-3,5]. The polymers were subjected to successive solvent extractions using solvents in order of increasing boiling point
to yield one or more of the following fractions: toluene soluble (TS), cold
toluene soluble (CTS), cold toluene insoluble (CTI), chlorobenzene soluble
(CBS), 2-chlorotoluene soluble (OCTS), 1,2-dichlorobenzene soluble (DCBS)
and 1-chloronaphthalene soluble (CNS).

Established procedures [1-3,5] were used to analyse polymer fractions
for composition (by elemental analysis, IR and 'H NMR spectroscopy),
melting ranges (hot-stage microscopy - HSM) and glass transition temperatures (Tg), cold-crystallisation temperatures (Tcc) and melting temperatures
(Tm) (by DSC).

RESULTS AND DISCUSSION

Homopolymer and copolymer preparation and fractionation are reported in
Table 1 together with data from compositional and HSM analysis.

TABLE 1

Homopolymer and copolymer preparation,fractionation and characterisation
data

Preparation	Fraction					
	Identity [a]	Yield [b] (%)	m [c]	$\bar{n}$ [d]	$\bar{p}$ [d]	Melting Range (°C)
PPS	CNS	89	1.00	738	–	271-279
	TS	11	1.00	738	–	135-230
PMPS	CTI	3	0.00	–	–	none [e]
	CTS	97	0.00	–	81	145-166
75%R [f]	DCBS	62	0.76	136	43	190-215
	CTI	18	0.76	60	19	165-177
	CTS	20	0.74	87	31	115-162
50%R [f]	CTI	6	0.58	29	21	138-147
	CTS	94	0.45	23	29	107-133
25%R [f]	CTS	100	0.27	18	49	115-137
75%B [g]	DCBS	34	0.80	13.0	3.5	239-256
	OCTS	21	0.79	12.2	3.5	225-240
	CBS	27	0.73	9.8	4.0	198-217
	CTI	10	0.64	7.5	4.8	197-212
	CTS	8	0.67	10.2	5.5	120-158
50%B [g]	OCTS	3	0.76	11.7	4.0	204-217
	CBS	28	0.61	9.3	6.6	165-197
	CTI	30	0.49	5.9	7.2	138-154
	CTS	39	0.38	4.5	8.9	102-112
25%B [g]	CTI	9	0.39	7.4	13.2	130-150
	CTS	91	0.24	2.1	9.7	105-120

a) abbreviations are defined in the experimental part
b) percentage of overall yield
c) m is the mole fraction of 1,4-disubstituted benzene units
d) $\bar{n}$ and $\bar{p}$ are the degrees of polymerisation with respect to PPS units and
PMPS units as estimated from bromine end-group analysis and m.
e) analysis to 345 °C
f) X%R = PPS/PMPS random copolymer preparation employing X mol% CBT with
100-X mol% CBMT
g) X%B = PPS(A)/PMPS(B) ABA block copolymer preparation employing, overall,
X mol% CBT with 100-X mol% CBMT.

The block copolymer preparation gave rise to much larger proportions of higher melting toluene-insoluble material than the equivalent random copolymer preparations. This observation has even more significance when consideration is given to the low molar masses of the block copolymer fractions and is consistent with the presence of extended PPS chain sequences in the copolymers obtained from the block copolymer preparations. Furthermore, the molar masses and IR spectra (showing sulphone, PPS and PMPS absorptions) of the block copolymer fractions are also in accord with those expected from their method of preparation. A distribution of composition was observed for each block copolymer preparation and arises from superposition of the second-stage PPS-block molar mass distribution (MMD) on the first-stage PMPS-block MMD.

The anomalous data for the TS PPS, CTS 75%R, and CTS 75%B fractions indicate that they contain significant proportions of cyclic polymer.

CNS PPS as prepared showed only a Tm (278 °C) but on shock-cooling from the melt gave a DSC trace in which Tg (85 °C), Tcc (134 °C) and Tm (279 °C) were observed. CTS PMPS as prepared showed a Tg (96 °C) and a weak melting endotherm (Tm = 162 °C) but on shock-cooling showed only a Tg (100 °C) thus revealing its essentially amorphous nature. The random copolymer fractions gave traces which were qualitatively similar to those for CTS PMPS, indicating them to be largely amorphous. This is consistent with disruption of the crystallisation of PPS repeat units by randomly-placed PMPS repeat units.

As prepared, each of the block copolymer fractions showed a Tm, with only CTS 25%B and CTS 50%B fractions showing a Tg. After shock-cooling, each block copolymer fraction showed a single Tg, indicating that PPS and PMPS are miscible in the melt phase. In addition, those fractions with m > 0.5 showed a Tcc and Tm. The Tcc increases as both m and $\bar{n}$ decrease in direct contrast with observations on PPS for which Tcc increases with $\bar{n}$. However, Tm increases with $\bar{n}$ as is normally observed for PPS. Thus the PMPS block has greatest effect on the process of crystallisation of the outer PPS blocks and does not unduly affect the properties of the crystalline phase once formed.

The cumulative evidence presented above indicates that, with the exception of CTS 75%B, the fractions from the block copolymer preparations approximate to their expected structures.

REFERENCES

1. A.B. Port, R.H. Still, J.Appl.Polym.Sci., 24, 1145 (1979)

2. A.B. Port, R.H. Still, Polymer Deg.Stabil., 1, 133 (1979)

3. A.B. Port, R.H. Still, Polymer Deg.Stabil., 1, 277 (1979)

4. R.W. Lenz, C.E. Handlovits, H.A. Smith, J.Polym.Sci., 58, 351 (1962)

5. Synthesis and characterisation of poly(arylene sulphides) parts 6 and 7, P.A. Lovell, R.H. Still, Makromol.Chem. (1987) in press.

RECENT ADVANCES IN CATIONIC POLYMERIZATION

Alessandro   Gandini
Laboratoire de Chimie Macromoléculaire et Papetière,
Ecole Française de Papeterie (INPG),
B.P. 65, 38402 Saint Martin d'Hères,
FRANCE

ABSTRACT

This short survey reviews the most significant contributions made to the advancement of cationic polymerization recorded in the last quinquennium. A critical appraisal is given of novel fundamental and applied concepts and results dealing successively with: (i) initiation mechanisms and initiator systems; (ii) propogation, with particular emphasis on the nature of the different active species; (iii) transfer reactions and their applications to the synthesis of functionalized oligomers and polymers; (iv) termination reactions; (v) recent claims to "living" polymerization systems, and (vi) copolymerizations leading to random, block and graft structures.

INTRODUCTION

At the beginning of this decade, the state of the art in cationic polymerization was thoroughly set forth in a series of monographs dealing with alkenyl (1,2) and heterocyclic (3) monomers. Since then, important contributions have come to enrich the knowledge of the fundamentals and the scope of this discipline. It seems therefore appropriate today to draw a critical appraisal of their relative merits and of the perspectives they open for further progress. Some of these reports appeared in the proceedings of the latest symposia on cationic polymerization (4,5), others are scattered in the recent literature. A condensed account of the situation up to 1984 is to be found in a chapter prepared for the new edition of the specialized encylopaedia (6).

Obviously, the present assessment cannot be exhaustive within the limited space available. It reflects instead a choise based on the author's priorities, as in most essays of this nature. The organization of this paper follows first the primary events of the chain reaction, then

deals with overall special situations and with the structure of the
products obtained therefrom.

INITIATION

<u>Mechanisms</u>

The reaction pathways leading to active species in cationic polymerization have been a long-standing source of debate and of some highly
elaborate experiments (1,3,6). In very broad terms and ignoring the
special instances of photo-,electro- and radiation-induced initiation,
two moieties can add onto the double bond (π-donor monomers) or onto the
heteroatom (n-donor monomers) to generate a chain carrier which will either
be a cation or a polarized ester molecule: (i) a proton arising from a
strong Brønsted acid (e.g. CF_3SO_3H) or from a Lewis acid/R<u>OH</u> complex
(e.g. $TiCl_4 \cdot H_2O$, but alcohols and carboxylic acids also function as proton
sources); (ii) an inorganic or organic cation arising from the self-
ionization of a "neutral" Lewis acid (e.g. $2\ AlCl_3 \rightleftharpoons AlCl_2^+ + AlCl_4^-$ or
$CH_3COClO_4 \rightleftharpoons CH_3CO^+ + ClO_4^-$), from a fully ionized saltlike species (e.g.
$Ph_3C^+\ SbF_6^-$ or $Et_3O^+\ BF_4^-$) or from a covalent molecule which only "disso-
ciates" upon electrophilic addition (e.g. $CH_3SO_3CF_3$). Clearly, the mechanisms of initiation can vary from single step direct additions to complex
series of events and the questions which often go unanswered have to do
with the moiety actually adding onto the monomer, the rates and energetics
of the processes and the efficiency of each step.

About thirty years ago, organic chemists discovered a unique and interesting property of 2,6-bulkily substituted pyridines, viz. their ability
to react with Brønsted (proton-generating) acids, but <u>not</u> with Lewis acids
because of steric hindrance (7). It took twenty-five years for polymer
chemists to recognize the usefulness of such compounds as diagnostic tools
to discriminate between reaction pathways (i) and (ii) (8). Since that
report, many groups have confirmed these features and in some instances
made a profitable use of them. The following two examples illustrate how
2,6-di-t-butylpyridine or its ·4-methyl homologue can readily provide mechanistic information about the mode(s) of initiation:
- BF_3 is unable to initiate the cationic polymerization of alkenes in the
  absence of a proton source (H_2O, CH_3OH, CH_3COOH,...), whereas $TiCl_4$ can
  in the same conditions (8). This suggests that the former Lewis acid
  does not self-ionize nor is it able to add as a molecule to the C=C bond,
  but that the latter probably initiates following self-ionization through

$TiCl_3^+$. These experiments were extended to other Lewis acids under varying conditions (solvent, temperature) and provided valuable information
as to the the relative contribution of direct initiation and cocatalysis
in each system as well as a means to determine the residual moisture in
systems where both types of initiation can occur (9).

- The complex $BF_3 \cdot Et_2O$ does not initiate the cationic polymerization of
alkenes in the absence of a proton source (9). This rules out a previously proposed mechanism based on initiation by ethyl cations arising
from the dry complex.

Other aspects dealing with the role of these hindered pyridines will
be discussed in the appropriate sections below but it is important to
underline that working with these compounds does not require the rigorous
experimental procedures otherwise used in fundamental studies on cationic
polymerization.

Interesting work has been undertaken recently to simulate initiation
in the cationic polymerization of both alkenes and heterocycles in order
to gain a better understanding of the kinetics and mechanisms of the real
systems. Dorfman and coworkers have extended their study on the reaction
of unpaired benzyl cations with alkenes (10) to open-chain and cyclic
ethers (11). The addition reaction to give the corresponding oxonium ions
was monitored and the bimolecular rate constants correlated with the basicity of the substrate. Mayr and coll. (12) have tackled the problem
through a more realistic approach in that the system used to simulate
initiation resembles more closely a polymerization situation, namely, the
addition reaction of a non-polymerizable alkene with a partly ionized
aryl halide/Lewis acid complex. The results obtained thus far have shed
some new light on the complicated issue of initiation by carbocations.

In a more fundamental vein, a group of gas-phase kineticists (13)
have carried out an interesting preliminary study of the reaction of BF_2^+
with alkenes using a selected-ion flow-tube apparatus. The rate constants
for the addition to ethylene, cis-2-butene, isobutene and styrene at room
temperature were very similar and indicate a rapid and non-discriminating
"initiation" in the gas phase. A few propagation steps and side reactions
of the resulting carbenium ions (unpaired) were also detected and characterized.

Cyclic siloxanes are readily polymerized by Brønsted acids, but only
recently it was shown that direct initiation by a Lewis acid without the

intervention of a proton is also possible (14): it remains to be ascertained whether the attack occurs on the Si or the O atoms.

Hall and co-workers (15) have examined in detail the initiation mechanism arising from the "spontaneous" reaction of N-vinylcarbazole with electrophilic alkenes. They clearly showed that old "charge-transfer" interpretations are invalid and that a zwitterionic intermediate is responsible for the polymerization.

New Initiators and Initiating Systems

A thorough investigation of the initiating capability of metal salts of strong Brønsted acids (16) has opened a new chapter in the catalogue of powerful initiators of cationic polymerization and indeed for most electrophilic reactions. This family includes perchlorates, triflates and salts with mixed anions (e.g. $(OTf)_2Al^+ SbF_6^-$), but also covalent coordinated molecules like $Et_3B(OTf)_3$. The mode of initiation of these compounds depends on the degree of drying that the system has undergone: selfionization gives direct initiation through the metal cation, but with moisture cocatalysis prevails, as shown by the use of sterically hindered pyridines and other specific mechanistic tests. The salts were found to be active in solution and in suspension both with alkenes and heterocycles.

It was shown in a recent publication that trialkylsilyl triflates can initiate the cationic polymerization of π- and n-donor monomers (17).

Aryldiazonium salts have been used as "thermal" initiators for over a decade. More recently (18) some of them have proved effective at room temperature and below with certain nucleophilic alkenes, and an on-going study (19) showed that initiation does not arise from the decomposition of the salts to give a Lewis acid related to the anion, but is instead caused by the direct addition of a cationic species issued from the diazonium moiety onto the monomer. Work is in progress to unravel this mechanism.

Homogeneous Ziegler-Natta catalysis has long been considered as being a cationic polymerization induced by Cp_2TiR^+ species which have as yet eluded isolation. Work by Jordan et al (20) has however shown that $Cp_2ZrCH_3^+$ BPh_4^- with or without THF stabilizing molecule is an initiator of ethylene polymerization in CH_2Cl_2. This important observation is most relevant to cationic polymerization and should provide ground for further studies.

Crivello's research into photo- and thermally initiated cationic polymerizations (21) has been extended to polymer-bound initiators (22) and to surface-adsorbed initiators capable of inducing interfacial reactions (23).

PROPAGATION

Rigorous investigations have led to the spectroscopic characterization
of the cationic species (chain carriers) derived from a number of styrene
homologues (24) and from several heterocyclic monomers (6). These contributions have widened considerably the basic catalogue of information concerning the active species in cationic polymerization. This is important
because the multiplicity of these species in most systems   represents
the crucial difficulty for a qualitative (mechanisms) and quantitative
(kinetics) understanding of propagation. Two aspects must be distinguished
in this context: the pseudocationic debate and the different types of ionic
carriers.

The fact that polarized ester molecules can propagate per se was first
recognized in systems involving styrene and other alkenes (1) and later
with n-donor monomers (3). Pseudocationic polymerization has since been
confirmed in numerous instances (6), some of which were presented as novel
types of propagation conducted by "invisible species" (6). Recent work on
styrene (25-27) and strong Brønsted acids has reproposed the problem of the
stability and activity of its esters and for the first time 1-phenylethyl
perchlorate was identified by ^{1}H-NMR spectroscopy (27). The basic problem
of the transient formation of the 1-phenyethyl carbenium ion and of its
lifetime as a function of its physical and chemical environment, in relation to the corresponding esters, should be reconsidered in cationic or pseudocationic polymerization in the light of Jencks' comprehensive studies of
the reaction pathways involving these types of "intermediate" (28). Interestingly, the two recent investigations on the possibility of attaining
"living" conditions in cationic polymerization (see below), have reproposed
pseudocationic mechanisms (29,30) identical to those originally put forward
over twenty years ago (1,6,31,32). Whereas propagation by an ester molecule derived from heterocyclic monomers has been an accepted mechanism
since it was first proposed in 1974 (3,6), it seems that the pseudocationic
mechanism with alkenes must be "rediscovered" at regular time intervals!

The relative abundance and activity towards propagation of ester molecules and different types of ionic species has been studied successfully
with a variety of n-donor monomers (3,6) and to a much lesser degree with
alkenes (6). Plesch and co-workers have made important contributions to
the latter topic and obtained k_p values for unpaired ions in nirobenzene
at room temperature (33) using a relatively simple technique which could

be exploited further. In the same spirit, but with a different experimental
approach, Sauvet et al. (34) have analysed in detail the complex situation
arising from the polymerization of p-methoxystyrene at 10°C in CH_2Cl_2 with
$SbCl_6^-$ as counterion. They proposed the existence of five different ionic
chain carriers with k_p values ranging from 8000 to 4×10^5 $1 \: mol^{-1} \: s^{-1}$.
In this study, another important point was raised within the context of the
mechanisms of propagation, namely the solvation (complexation?) of the
carbenium ions by a monomer molecule and/or a polymer moiety. This issue
would undoubtedly profit from a wider and closer investigation, e.g.
through spectroscopic studies with model compounds, owing to its relevance
to both fundamental and applied aspects of cationic polymerization and copolymerization.

The polymerization of cyclic ethers and acetals can be marred by the
formation of cyclic oligomers. An original solution to this problem was
applied by adding alcohols to the reaction medium with systems initiated
by protonic acids (35). The mechanism of propagation is radically changed
in that an activated (protonated) monomer molecule adds onto an OH-termi-
nated neutral polymer (or oligomer) chain (36,37). Thus, for example, the
polymerization of ethylene oxide in the presence of methanol (36) can be
adjusted to yield essentially linear products and very little 1,4-dioxane.
This interesting new type of inversed situation has been called "activated monomer propagation" and bears important consequences not only in minimizing cyclic products, but also in the synthesis of block copolymers as
discussed below.

An application of hindered pyridines in the context of propagation
(38) showed that it is possible to add such Lewis bases after protonic
initiation has taken place and thus not to alter the course of propagation
with trioxane, as this reaction involves a cation and a monomer molecule
but no protons. This type of experiment also shows that certain
chain carriers are unaffected by hindered pyridines and do not deprotonate
in its favour. Of course, this is not a generalizable statement and much
work remains to be done to elucidate the role of hindered pyridines in the
propagation stage of cationic polymerization as a function of the structure and lifetime of the active species. An investigation of this nature
is being pursued in the author's laboratory and for several systems involving vinyl ethers, substituted styrenes and some n-donor monomers there is
evidence suggesting little or no involvement of the hindered pyridine in
reactions with active species as such, i.e. cations or ester molecules (9).

TRANSFER REACTIONS

An important contribution to the understanding of colour formation and subsequent bathochromic shifts in the polymerization of vinyl ethers was provided in a study by Aoshima and Higashimura (39). They proposed that a polymer molecule is protonated at the side-oxygen atom and thereby loses a molecule of alcohol leaving an allylic-type carbenium ion on the chain in equilibrium with the corresponding neutral unsaturated structure. A succession of this sequence of events leads to polyunsaturated conjugated structures which are responsible for the colour evolution during polymerization. This transfer mechanism is similar to that encountered in the cationic polymerization of 2-vinylfuran (40), except for the loss of alcohol which is partly balanced by the formation of acetal end groups between an active species and an alcohol molecule.

Turning to useful exploitation of transfer reactions, the inifer approach (2,6) comes to mind. A recent assessment of this very useful mechanism for the preparation of telechelics (41) underlines the advantages and drawbacks of Kennedy's original systems and offers interesting modifications to improve the main features and suppress unwanted reactions. The use of hindered pyridines proved again profitable because it scavenged the polymerization media by removing HCl, i.e. a source of protonic initiation. Another important mechanistic contribution arising from this study (41) was the fact that chlorine-terminated poly(isobutene) and model compounds were shown to initiate the polymerization of isobutene in the presence of BCl_3 and give chain extension with the former.

Another way of synthesizing telechelics by transfer reactions calls upon the addition of furan compounds like 2-methylfuran to an alkene cationic polymerization (42). A judicious choice of experimental parameters leads to predominant transfer by electrophilic substitution at the C5 position of the furan ring and therefore polymer (or oligomer) chains with a functionality very close to unity. The terminal furan moiety can be transformed into more reactive functions through rather simple chemistry such as the Diels-Alder reaction.

Hindered pyridines have been used to confirm the existence of "spontaneous"(unimolecular) transfer reactions in the polymerization of certain alkenes by Brønsted acids (6,9). An active species (ester or ions) can give an acid molecule in a reaction which is the reverse of initiation. This transfer mechanism would obviously be quenched by the hindered pyridine, whereas the corresponding bimolecular transfer involving the same active species and a monomer molecule should not be affected by its presence. In the polymerization of α-methylstyrene by CH_3SO_3H in CH_2Cl_2 at room temperature, transfer reactions predominate. It was shown (9) that samples withdrawn at different intervals after initiation and treated with a hindered pyridine lost their activity very rapidly. This proves that acid expulsion is a frequent event in this system, i.e. spontaneous transfer is an important chain breaking reaction. It is important to underline here that the intervention of the hindered pyridine is a <u>termination</u> reaction (to prove the occurrence of a transfer reaction): such Lewis bases cannot in fact be transfer agents because whenever they intervene they eliminate a proton from the system and therefore diminish its polymerization activity.

"LIVING" SYSTEMS

π - and n,π -Donor Monomers

Only those polymerizations which are characterized (or claimed to be characterized) by the absence or a negligible intervention of transfer and termination reactions and by $R_i \gg R_p$ (or by initiation completed separately from propagation) will be analyzed here, since these criteria are the indispensible prerequisites for the living behaviour of a system, as clearly shown by Szwarc's hystorical contribution to anionic polymerization. These stringent conditions imply that the DP of the product grows linearly with the amount of monomer consumed (in a given reaction, but also upon second monomer addition, etc.), the proportionality constant between these two parameters being the (constant) concentration of chain carriers (considered here as monofunctional). The large output of papers dealing with "quasi-living" polymerization issued as a joint effort from Akron and Budapest (43,44) does not satisfy these requirements and indeed could be interpreted otherwise as pointed out by Sigwalt(45). Attention will be focussed instead on more recent work from the same authors and from Higashimura's laboratory which deals now with "living cationic polymerization" of alkenes and vinyl ethers.

In 1983 Sawamoto and Higashimura proposed the use of I_2/HI in nonpolar solvents as a suitable system to induce the living polymerization of vinyl ethers. Since then numerous papers and reviews have appeared on this topic (29, 46) and polar solvents have been used successfully by increasing the relative HI concentration. These systems are undoubtedly the closest situation in cationic polymerization suggesting a living cha-

racter. As already pointed out the mechanism is a pseudocationc propagation
in which the monoiodide initially formed by electrophilic addition of HI
onto the C=C bond is activated by solvation of an iodine molecule. Thus, a
chain end oscillates continuosly between an inactive and an active mode, viz.
reversible termination occurs (a situation which is not incompatible with a
living polymerization). The natural consequences of this discovery are the
preparation of functionalized polymers and of block copolymers as reviewed
below. DP remain relatively low and this could mask the existence of some
transfer, as already discussed (45). The extension to other monomers is
not straightforward, although the use of I_2 with p-methoxystyrene gave an
interesting behaviour. A recent report from Kyoto (47) shows encouraging
results with a different system: p-methylstyrene/AcClO$_4$/CH$_2$Cl$_2$/-78°C particularly in the presence of n-Bu$_4$N$^+$ClO$_4^-$. Again propagation seems to be mostly pseudocationic after the suppression of the ionic species, or else ion
pairs are the dominant chain carrier.

Faust and Kennedy (30) have recently claimed "for the first time the
truly living polymerization of isobutylene". The initiating system involved
a mixture of tertiary acetates and BCl$_3$. The results and arguments presented to substantiate this claim seem rather preliminary and more evidence is
necessary, particularly in view of the abnormally high values of M_w/M_n obtained in the only experiment which purports a living character.

The same precautionary comments seem relevant for a paper by Zsuga and
Kelen (48) on the living polymerization of isobutene and styrene in the
presence of phosgene.

<u>n-Donor Monomers</u>

Several heterocyclic monomers are known to give living cationic polymerizations under specific conditions, particularly with respect to the
stability of the anions (3,6). Tetrahydrofuran, 1,3-dioxolane, t-butylazi-
ridine and some oxazolines are good representatives of this behaviour for
different families of heterocycles. Most of the recent studies in this
domain have concentrated on the preparation of telechelics and block copolymers rather than on more fundamental aspects related to the characterization of new living systems. This work is therefore examined in the next
section.

In conclusion, a real progress has been achieved in the search for
living systems, particularly with alkene-type monomers for which nothing
encouraging could be said at the end of the last decade.

TELECHELICS AND COPOLYMERS

Apart from the functionalization through predominant transfer reviewed above and Kennedy's continuing contributions in the synthesis of macromers and telechelics, the living systems elaborated by the Kyoto school provide an excellent ground for polymer and oligomer structures with specific reactive sites (46,49), including macromers such as methacrylate functions attached to a polyvinylether chain which underwent radical polymerization (49).

With heterocyclic monomers, a good deal of research has been devoted to the preparation of functional (mono-, di-, poly-) oligomers and polymers (3). Recent contributions include the work of Yasuda et al. (50) on symmetrical telechelics of tetrahydrofuran prepared in the presence of diols and triols and of oxiranes; the studies of McGrath and coll. (51) on functionally terminated siloxane oligomers; the thorough investigations at Ghent on aziridine-derived macromolecules (52) and the research of Saegusa and co-workers on cyclic imino ethers and cyclic trivalent phosphorus monomers (53), including the preparation of macromers (54).

Random copolymers by cationic polymerization are by no means a novelty, but recent original studies have appeared. Thus, Sawamoto et al. (55) have synthesized a random copolymer of isobutyl vinyl ether with p–methoxystyrene which was almost monodisperse and Cesca has summarised the excellent work conducted in his laboratory on the copolymerization of isobutene with conjugated dienes (56).

Alternated copolymers can be obtained by zwitterionic polymerization of a cationic-sensitive monomer with an anionic-sensitive monomer. Odian and Gunatillake (57) recently analysed this topic and conducted a specific study on the 2–methyl–oxazoline/acrylic acid pair.

Block copolymerization by cationic initiation with alkenes is still a difficult feat (6), but the living conditions for vinyl ethers (see above) have yielded several new materials bearing sequential structures (49). Block copolymers arising from a heterocyclic and an alkene monomer have been prepared (3,6) and a recent interesting addition to this type of macromolecules involved the preparation of living dicationic polydioxepane and blocking each end with 1,2–dimethoxyethylene (58). As for block copolymers obtained from two heterocyclic monomers, the numerous examples reported up to a few years ago (3,6) have continued growing. A novel concept in this field comes from the application of the activated monomer priciple (see above) to acetals in the presence of macroglycols (59).

Cationic grafting has received much impetus in Sigwalt's laboratory where the clearest and most fundamental studies have been carried out (60). This continuing investigation has also provided new interesting materials. An original approach to grafting describes the use of carbon black as substrate (61) on which active groups were previously attached. Finally, the first cationic grafting of cellulose was carried out by esterifying the polysaccharide with a strong protonic acid and by adding oxazolines which were grafted by pseudocationic polymerization promoted by the ester groups (62).

REFERENCES

1. Gandini, A. and Cheradame, H., Adv. Polym. Sci., 1980, 34/35.
2. Kennedy, J. P. and Maréchal, E., Carbocationic Polymerization, Wiley-Interscience, New York, 1982.
3. Penczek, S., Kubisa, P. and Matyjaszewski, K., Adv. Polym. Sci., 1980, 37 and 1985, 68/69.
4. Goethals, E. J., Ed., Cationic Polymerization and Related Processes, Academic Press, London, 1984.
5. Heublein, G., Ed., Makromol. Chem., Macromol. Symposia, 1986, 3.
6. Gandini, A. and Cheradame, H., Encyclopedia of Polymer Science and Engineering, John Wiley and Sons, New York, 1985, Vol. 2, 729.
7. Brown, H. C., and Kanner, B., J. Am.Chem. Soc., 1953, 75, 3865 and 1966, 88, 986.
8. Moulis, J. M., Collomb, J., Gandini, A., Cheradame, H., Polym. Bull., 1980, 3, 197.
9. Arlaud, P., Roudet, J., Martinez, A. and Gandini, A., unpublished work to be presented at the 8th International Symposium on Cationic Polymerization, Munich, August 1987.
10. Wang, Y. and Dorfman, L. M., Macromolecules, 1980, 13, 63.
11. Reed, D. T. and Dorfman, L. M., Macromolecules, 1984, 17, 32.
12. Mayr, H., Schneider, R. and Pock, R., Ref. 5, p.19.
13. Forte, L., Lien, M. H., Hopkinson, A. C. and Bohme, D. K., Makromol. Chem. Rapid Commun., 1987, 8, 87.
14. Boileau, S., Cheradame, H., Gandini, A., Jordan, E. and Lestel, L., Preprints of Communications presented at the 5th International Symposium on Ring-Opening Polymerization, Blois, France, June 1986, p. 79.
15. Gotoh, T., Padias, A. B. and Hall, H. K. Jr., J. Am. Chem. Soc., 1986, 108, 4920.
16. Collomb, J., Arlaud, P., Gandini, A. and Cheradame, H., Ref. 4, p.49.
17. Gong, M. S. and Hall, H. K. Jr., Macromolecules, 1986, 19, 3011.
18. Hall, H. K. Jr., and Howey, M. A., Polym. Bull., 1984, 12, 427.
19. Roudet, J. and Gandini, A., unpublished results to be presented at the 8th International Symposium on Cationic Polymerization, Munich, August 1987.
20. Jordan, R. F., Bajgur, C. S., Willett, R and Scott, B., J. Am. Chem. Soc., 1986, 108, 7410.
21. Crivello, J. V., Adv. Polym. Sci., 1984, 62, 1.
22. Crivello, J. V. and Lee, J. L., Polym. Bull., 1986, 16, 243.
23. Hult, A., MacDonald, S. A. and Willson, C. G., Macromolecules, 1985, 18, 1804.
24. Matyjaszewski, K. and Sigwalt, P., Macromolecules, in press.

25. Szwarc, M., Macromolecules, 1984, 17, 1993.
26. Matyjaszewski, K. and Sigwalt, P., Makromol. Chem., 1986, 187, 2299.
27. Matyjaszewski, K., Polym. Prepr., 1986, 27 (2), 112.
28. Ta-Shma, R. and Jencks, W. P., J. Am. Chem. Soc., 1986, 108, 8040, and references therein.
29. Higashimura, T., Miyamoto, M. and Sawamoto, M., Macromolecules, 1985, 18, 611.
30. Faust, R. and Kennedy, J. P., Polym. Bull., 1986, 15, 317.
31. Gandini, A. and Plesch, P. H., J. Chem. Soc., 1965, 4826.
32. Giusti, P. and Andruzzi, F., Chim. Ind., (Milan), 1966, 48, 435.
33. Plesch, P. H., ref. 4, p.1.
34. Sauvet, G., Moreau, M. and Sigwalt, P., ref. 5., p. 33.
35. Penczek, S., Kubisa, P. and Symanski, R., ref. 5., p. 203.
36. Brzezinska, K;, Symanski, R., Kubisa, P. and Penczek, S., Makromol. Chem. Rapid Commun., 1986, 7, 1.
37. Wojtania, M., Kubisa, P. and Penczek, S., Preprints of Communications presented at the 5th International Symposium on Ring-Opening Polymerization, Blois, France, June 1986, p. 51.
38. Stasinski, J. and Kmowska, G., ref. 37, p.197.
39. Aoshima, S. and Higashimura, T., Polym. J., 1984, 16, 249.
40. Alvarez, R., Gandini, A. and Martinez, R., Makromol. Chem., 1982, 183, 2399.
41. Nuyken, O., Pask, S. D., Vischer, A. and Walter, M., Makromol. Chem., 1985, 186, 173 and ref. 5., p. 129.
42. Razzouk, H., Bouridah, K., Gandini, A and Cheradame, H., ref. 4., p.355.
43. Kennedy, J. P. and Kelen, T., J. Macromol. Sci.-Chem., 1982-3, A18(9).
44. Gyor, M., Kennedy, J. P., Kelen, T. and Tüdös, F., J. Macromol. Sci.-Chem., 1984, A21(10), pp. 1295, 1311 and 1323.
45. Sigwalt, P., Polym. J., 1985, 17, 57.
46. Sawamoto, M., Enoki, T. and Higashimura, T., Macromolecules, 1987, 20, 1, and references therein.
47. Tanizaki, A., Sawamoto, M. and Higashimura, T., J. Polym. Sci., Polym. Chem. Ed., 1986, 24, 87.
48. Zsuga, M. and Kelen, T., Polym. Bull. 1986, 16, 285.
49. Sawamoto, M. and Higashimura, T., ref. 5., pp. 83 and 99.
50. Yasuda, K., Yokoyama, Y., Matsumoto, S. and Harada, K., ref.4,p.379.
51. Sormani, P. M., Minton, R. J., McGrath, J. E., Ring Opening Polymerization Kinetics, Mechanisms and Synthesis, McGrath, J.E., Ed., ACS Symposium Series No. 206, 1985, 147.
52. Goethals, E.J., Van de Velde, M.and Munir, A., ref.4, p.387 and ref.37, p.35
53. Kobayashi, S. and Saegusa, T., ref.5., p.179.
54. Kobayashi, S.,Kaku,M.,Sawada,S. and Saegusa,T.,Polym.Bull.1985,13,447.
55. Sawamoto,M., Ohtoyo,T., Higashimura,T., Guhrs,K.H.and Heublein G., Polym.J., 1985,17, 929.
56. Cesca, S., ref.4, p. 105.
57. Odian, G. and Gunatillake, P.A.,Macromolecules,1984,17,1297 and 2236.
58. Reibel,L.C., Durand,C.P. and Franta,E.,Can.J.Chem.,1985,63, 264.
59. Reibel,L., Zouine,H. and Franta,E.,ref.5., p.221.
60. Pary,B., Tardi,M., Polton,A. and Sigwalt,P.,Eur.Polym.J.,1985,21,393.
61. Tsubokawa,N.,Jian,Y.,Yamada,A. and Sone,Y.,Polym.Bull.,1986,16,249.
62. Cheradame,H., Upe,T.A. and Gandini, A., Makromol.Chem., Macromol. Symp., 1986, 6, 261.

FAST ION CONDUCTION IN COMB SHAPED POLYMERS

J M G Cowie
Department of Chemistry, University of Stirling, Stirling, Scotland.

ABSTRACT

Studies of fast ion conduction in solid amorphous polymer-salt mixtures have increased rapidly in the last decade since it was demonstrated that their application in dry battery manufacture was feasible.

The most successful type of polymer matrix found so far is poly(ethylene oxide) PEO which can dissolve a wide range of salts to form solid homogeneous solutions. A major drawback in these systems is that PEO tends to crystallise readily and as ionic conduction takes place in the amorphous regions of the polymer a reduction in the crystalline content of the host polymer is desirable, if reasonable conduction levels are to be attained at ambient temperatures.

One way of achieving this is to synthesise comb-shaped polymers with short side chains of poly(ethylene glycol) or poly(propylene glycol) (PPG) which do not crystallise but are long enough to coordinate with the metal cations of the added salt.

Structures with this architecture based on poly(methacrylic acid), poly(itaconic acid) and polyphosphazene backbones with either PEO or PPG side chains have been prepared by several groups of workers. These form amorphous solutions with a range of alkali metal salts and conductivity levels of up to 10^{-3} Scm^{-1} have been recorded in several systems. The factors influencing ionic conductivity in these comb branch systems such as salt selection, glass transition temperatures, and temperature of measurement, will be discussed with particular reference to the polymers based on poly(itaconic acid), and ethylene oxide macromers.

INTRODUCTION

One of the first major studies on the effect of mixing salts with a polymer was reported by Moacanin and Cuddihy [1]. They observed that LiClO$_4$ could be dissolved in poly(propylene oxide), PPO, to give homogeneous, amorphous, mixtures and that the glass transition temperature (Tg) rose as more salt was added to the mixture. It was later shown that several other polymers were capable of dissolving salts with a concomitant enhancement of Tg [2].

Real interest in these polymer/salt solutions only developed after it was demonstrated by Wright et alia [3,4] that mixtures of poly(ethylene oxide), PEO, and various inorganic salts displayed conductivity levels which were sufficiently large to merit consideration for use as solid polymeric electrolytes in high energy density batteries. Subsequent investigations have shown that in addition to the polyalkyloxides there are several systems suitable for consideration, containing donor atoms, such as O, N and S, which are capable of complexing with the cations of inorganic salts. Among those studied so far are poly(ethylene succinate) [5,6], poly(ß-propiolactone), [7], poly(ethylene adipate) [8], poly(ethylene imine) [9-11], poly(alkylene sulphides) [12] and poly(vinyl methyl ether) [13], but PEO still seems to be the best structure for solvating cations. This appears to be related to the favourable spacing between the ether oxygens and the tendency for the PEO chain to adopt a helical structure. It has been shown that PEO has a $(ttg)_7$ conformation in two turns of a helix whose repeat distance is 19.25Å. Papke et alia [14] have suggested that on complexing with metal cations PEO has a $(ttg^+ ttg^-)$ conformation which forms a tighter helix. This also provides an oxygen lined cavity which can easily accommodate certain cations of a given size, eg Na^+ can fit comfortably into such a cavity and coordinates with four oxygens in this geometry. This favourable conformation is most easily achieved with PEO. Construction of space filling models demonstrates that even PPO can only form a much more open helical structure because of steric hindrance and that this is probably a less efficient cation binding conformation. This difference in structure can be seen in Figures 1(a) and (b) for short segments of PEO and PPO attached as an ester unit to a

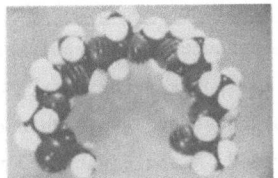

Figure 1(a): Molecular model of polyethylene oxide trimer.

Figure 1(b): Molecular model of polypropylene glycol trimer.

monomer unit.

Possible Structural Modifications.

On the basis of published work it is now possible to draw some general conclusions concerning effective polymer/salt systems for use as ion conducting media. When seeking a suitable system for study the initial criteria to consider are:

(i) the polymer should be capable of dissolving a range of inorganic salts effectively, so as to produce a high density of charge carriers;

(ii) an amorphous polymer/salt solution must be formed which also has a low Tg, to maximise chain flexibility and facilitate ion transport.

In this respect PEO suffers from a major drawback as it is a semi crystalline polymer at ambient temperatures and this restricts the conductivity until the temperature rises above the melting point of the polymer. This can be overcome in a number of ways. One method, successfully employed by several research groups [15-22] is to shorten the length of the PEO segment and to link these together using urethane [15-17] or siloxane units [19-21] to form block copolymer structures or crosslinked networks, or to interrupt the regularity of the PEO sequence using PPO blocks or methylene oxide units [17]. A second method involves the synthesis of comb-shaped polymers comprising short side chains of PEO attached to poly(methacrylic acid) [23,24], poly-(phosphazene) [25], poly(itaconic acid) [26] or poly[oxy(methylsilene)] [27] chains. If the side chains are kept sufficiently short (8-9 ethylene oxide units) then crystallization can be avoided and amorphous systems result. These structures are shown in Table 1, and the behaviour of 2,4,5 and 6 will now be discussed in greater detail.

Glass Transition Temperatures

The conductivity in polymer/salt systems tends to improve when Tg is low and the polymer chains are more flexible. Addition of salts to the polymers increases Tg and this dependence is illustrated in Figure 2 for structures 2, 4 and 6 with n=3 when $NaClO_4$ is the added salt. The polyphosphazene and polymacromer samples are very similar and so should exhibit better conductivities than the polyitaconate based system. This general trend is confirmed for conductivities measured at a $[Na^+]/[EO$ unit] ratio of 0.125 although the polymacromer appears to have a slightly higher level of conductivity than the polyphosphazene in spite of having a marginally larger Tg at this salt concentration, as can be seen in Figure 3.

TABLE 1
Comb-shaped polymers used for conductivity studies

Structure	Tg/K	Structure	Tg/K
1. $\sim(CH_2-C)\sim$ with CH_3 and $COO(CH_2CH_2O)_nCH_3$	215 (n=8)	4. $\sim(CH_2-C)\sim$ with $CH_2COO(CH_2CH_2O)_nCH_3$ and $COO(CH_2CH_2O)_nCH_3$	218 (n=3)
2. $\sim(N=P)\sim$ with $O(CH_2CH_2O)_nCH_3$ and $O-(CH_2CH_2O)_nCH_3$	207 (n=3)	5. $\sim(CH_2-C)\sim$ with $CH_2COO(CH_2CHO)_{17}H$, CH_3, $COO(CH_2CHO)_{17}H$, CH_3	220
3. $\sim(Si-O)\sim$ with CH_3 and $O-(CH_2CH_2O)_nCH_3$	195 (n=8)	6. $\sim(CH_2-CH)\sim$ with $O-(CH_2CH_2O)_nCH_3$	206.5 (n=3)

The elevation of Tg on addition of salt tends to have the same general trend in all the samples and this suggests that a structure which has a low Tg in the undoped state will have a correspondingly lower Tg

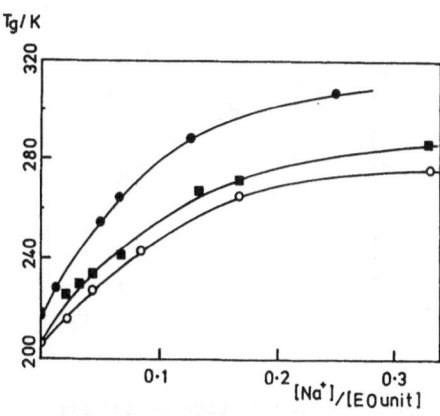

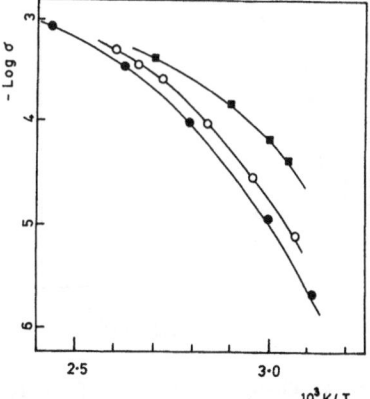

Figure 2: Variation of Tg with added NaClO$_4$ for n=3, structure 2 (-O-); structure 4 (-●-); and structure 6 (■).

Figure 3: Temperature dependence of log (conductivity) for structure 2 (-O-); structure 4 (-●-); and structure 6 (■), all with n=3 and [Na$^+$]/[EO unit] = 0.125.

when salt is added over the whole concentration range, than one with a larger initial Tg. Consequently the former will exhibit higher conductivities at corresponding temperatures. Unfortunately this generalization does not apply to all polymer/salt mixtures.

(i) Nature of added Salt

It might be expected that if the increase in Tg caused by the addition of salt to the polymer could be minimized, higher conductivities could be achieved. The effect of changing the type of salt was examined first using structure 5, the polyitaconate with polypropylene glycol side chains. Four salts were used all of which formed homogeneous solutions with the polymer but the Tg was enhanced to differing degrees as can be seen in Figure 4. Values of Tg were 96K, 76K, 57.5K and 9K for $NaClO_4$, $LiClO_4$, $ZnCl_2$, and LiCl respectively at a [cation]/[PG unit] ratio of 0.3. Similar trends have been observed by Wetton et alia [28] who also showed that LiCl had little effect on the Tg of poly(propylene oxide). It would be natural to assume that the use of

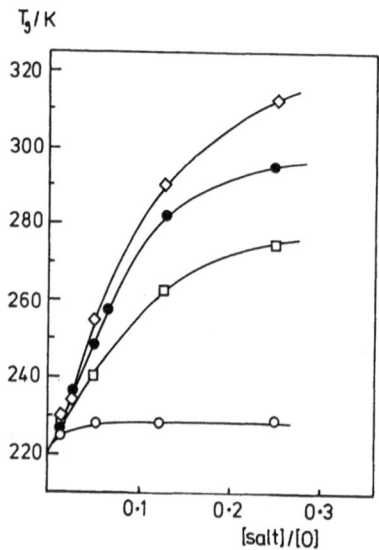

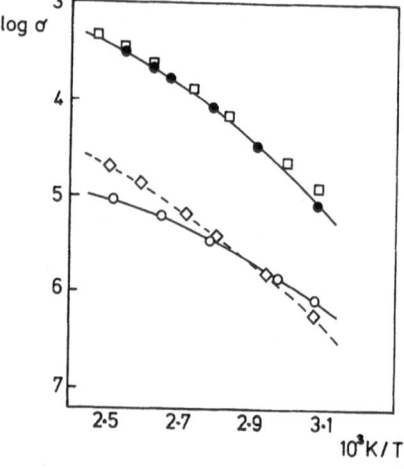

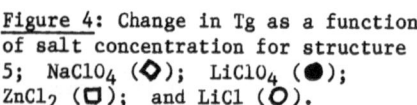

Figure 4: Change in Tg as a function of salt concentration for structure 5; $NaClO_4$ ($\Diamond$); $LiClO_4$ ($\bullet$); $ZnCl_2$ ($\square$); and LiCl (O).

Figure 5: Temperature dependence of log (conductivity) for LiCl (O) $ZnCl_2$ ($\Diamond$); $LiClO_4$ ($\square$); and $NaClO_4$ ($\bullet$) dissolved in structure 5 at [salt]/[0] ratios of 0.05.

mixtures containing LiCl would give the best levels of conductivity. This was not found to be the case and the conductivity data plotted in Figure 5 show that superior conductivity levels are obtained when the Tg enhancement is largest.

Thus it appears that there are two conflicting consequences of adding salts to a polymer. When there is effective solvation of the cation by the polymer, the Tg rises because complexing leads either to chain stiffening or crosslinking, or both. When the Tg is not greatly enhanced by the salt, this indicates that there is very little complexing of the cation itself and the salt tends to exist mainly in the form of ion pairs. In the latter case there will be relatively few free ions to act as charge carriers and so the conductivity will be low thereby negating any advantage gained from the low Tgs.

(ii) Effect of side chain length. Dissolution of salts in an alkylene oxide matrix will be easier if the salts have a low lattice energy and the anion is a weak ligand, but the length of the ethylene oxide unit is also important. The variation in Tg on addition of $LiClO_4$ to structure 4 when n is varied from 1 to 7 is shown in Figure 6. There is little effect when n=1 but the Tg is already high and best results are obtained when n is between 3 and 7 units long. Crystallization tends to become a factor when the side chain length is extended much beyond this so no advantage is gained by increasing the chain length further. Conductivities are plotted in Figure 7, illustrating that these are best when n=7, for mixtures with $LiClO_4$ at a $[Li^+]/[EO$ unit] of 0.250. Interestingly, the conductivity levels are relatively better in the n=3 sample than for n=7 if a reduced $(1/T-Tg)$ temperature plot is examined.

Conductivity

(i) Temperature dependence. It is clear from the curvature observed in Figures 3, 5 and 7, that the temperature dependence of conductivity is non Arrhenius. The data can be linearized by using the semiempirical Vogel-Tamman-Fulcher (VTF) equation

$$\sigma = AT^{-1/2} \exp[-B/(T-T_0)] \tag{1}$$

where A and B are constants, and T_0 is the temperature at which the configurational entropy of the system approaches zero [29]. A number of groups have treated their data according to equation (1) by assuming T_0 = Tg either of the undoped polymer or of the actual polymer/salt mixture

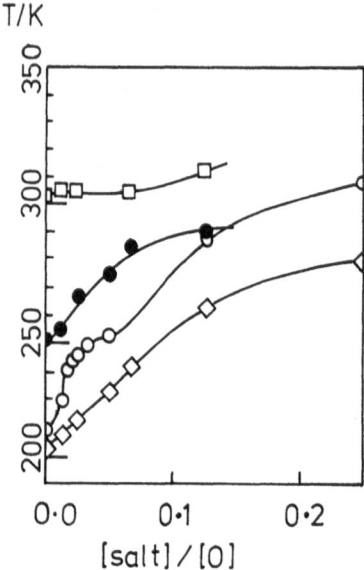

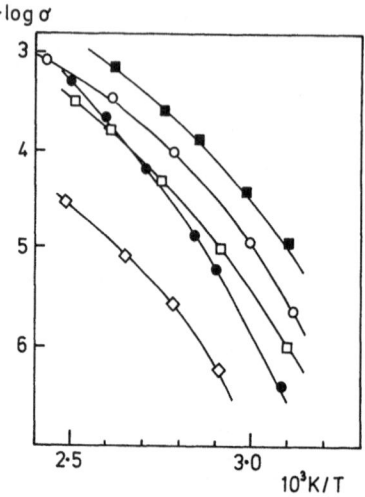

Figure 6: Changes in Tg when LiClO₄ is added to structure 4 for n=1 (□); n=2 (●); n=3 (○) and n=7 (◇).

Figure 7: Temperature dependence of log (conductivity) for n=1 (◇); n=2 (□); n=3 (○) and n=7 (■), Structure 5 (●).

and they have then assumed that B can be transposed into an apparent activation energy. Alternatively a non linear least squares analysis of the data can be made which gives values of A, B and T_o commensurate with the "best fit".   It is also possible to interpret B more formally using the Adam–Gibbs approach [30] through equation 2.

$$B = \frac{T_o S_c^* \Delta\mu}{k_B \Delta C_p T} \qquad (2)$$

where $\Delta\mu$ is the potential energy barrier opposing a segmental rearrangement in the polymer, S_c^* is the minimum configurational entropy for such a cooperative segmental rearrangement which can be assumed to be $= k_B \ln 2$, ΔC_p is heat capacity change on moving from the glass to the liquid and k_B is Boltzmann's constant. Values of T_o are not easily obtained by direct experimentation but are theoretically predicted to lie about 50K lower than Tg.   These parameters, calculated from equations 1 and 2 for the comb-shaped polymer/salt systems studied here are gathered in Table 2.   While the values of (Tg–T_o) are not

TABLE 2

Activation energies and T_0 values for comb-shaped polymer/salt systems
calculated from equations 1 and 2

Structure	$\frac{[Salt]}{[0]}$	T_0/K	$(Tg-T_0/K)$	$\Delta\mu/kJmol^{-1}$
4/LiClO₄	0.0125	212.5	17.5	–
	0.0500	180.9	72.1	85.7
	0.125	236.3	51.7	–
	0.250	239.7	64.8	–
4/NaClO₄	0.0125	195.7	33.3	35.8
	0.0500	201.7	53.8	52.9
	0.125	233.3	54.7	61.0
	0.250	241.0	65.5	80.7
5/LiClO₄	0.0125	187.3	41.2	47.8
	0.050	201.5	47.0	63.4
	0.125	234.5	48.0	79.9
	0.250	250.3	44.7	119.0
5/NaClO₄	0.050	213.4	40.6	61.0
	0.125	223.2	67.3	123.3
6/LiClO₄	0.0125	188.1	28.5	–
	0.050	197.9	34.6	–
	0.125	214.5	46.0	–
	0.250	232.3	48.1	–
			Average 47.7	

always consistent in each set there is an encouraging preponderance in
the region of the expected 50K and an average over all the data gives
$(Tg-T_0) \approx 47.7K$.

These data suggest that the analysis using the Adam–Gibbs theory is
acceptable and that the activation energies derived might be more
meaningful than those estimated by other methods.   The values of $\Delta\mu$ tend
to increase with rise in salt content, and of course Tg, and vary from 36
to 123 kJ mol^{-1}.   Estimates for the energy required to break a sodium
ion-ether oxygen link in the gas phase are about 60 kJ mol^{-1} and if
four of these were broken in cation transport from one site to the next
one would require $\sim$ 250 kJ mol^{-1} for the process.   Energies required
for ion transport in the condensed state are likely to be much less and
if the ion exchange mechanism was an associative one, rather than a
dissociative one, then the energies estimated here would not be
unreasonable.   In the associative method of ion transport there is a
sharing of donor-cation interaction between the ligand groups as the
cation is moved from one set of coordinating salts to the next by
segmental motion of the polymer chain, thereby assisting movement by
forming an intermediate species.   This would obviate the need to break
all contacts at once with a high energy input.   The rise in the

activation energies as the salt increases may reflect the fact that as Tg will be higher, chain flexibility is lower and the movement of the cation attached to the chain requires more energy. Thus the activation energy may be a composite of the ability to move the polymer chain segment and the ion interchange energy as it moves from one site to the next.

More detailed investigations of the mechanism are necessary if we are to be able to improve the design of such systems to produce better ion conductors.

REFERENCES

1. Moacanin, J. and Cuddihy, E.F., J.Polym.Sci. Part C, 1966, 14, 313.

2. Hannon, M.J. and Wissbrun, K.F. J.Polym.Sci., Polym.Phys.Ed., 1975, 13, 113.

3. Fenton, D.E., Parker, J.M. and Wright, P.V., Polymer, 1973, 14, 589.

4. Wright, P.V., Br.Polym.J., 1975, 7, 319.

5. Watanabe, M., Rikukawa, M., Sanui, K., Ogata, H., Kato, H., Kobayashi, T., and Ohtaki, Z., Macromolecules, 1984, 17, 2902.

6. Dupon, R., Papke, B.L., Ratner, M.A. and Shriver, D.F., J.Electrochem. Soc., 1984, 131, 586.

7. Watanabe, M., Togo, M., Sanui, K., Ogata, N., Kobayashi, T. and Ohtaki, Z., Macromolecules, 1984, 17, 2908.

8. Armstrong, R.D., and Clarke, M.D., Electrochim. Acta, 1984, 29, 1443.

9. Harris, C.S., Shriver, D.F. and Ratner, M.A., Macromolecules, 1986, 19, 987.

10. Chang, C.K., Davis, G.T., Harding, C.A., and Takahashi, J., Solid State Ionics, 1986, 18/19, 300.

11. Takahashi, T., Davis, G.T., Chiang, C.K. and Harding, C.A., Solid State Ionics, 1986, 18/19, 321.

12. Clancy, S., Shriver, D.F. and Ochrymowycz, L.A., Macromolecules, 1986, 19, 606.

13. Cowie, J.M.G. and Martin, A.C.S., Polym.Bull., 1987, 17, 113.

14. Papke, B.L., Ratner, M.A. and Shriver, D.F., J.Phys.Chem.Solids, 1981, 42, 493.

15. Killis, A., Le Nest, J.F., Gandini, A., and Cheradame, H., J.Polym.Sci.Polym.Phys.Ed., 1981, 19, 1073.

16. Killis, A., Le Nest, J.F., Gandini, A., and Cheradame, H., Makromolek.Chem., 1982, 183, 1037.

17. Gandini, A., Le Nest, J.F., Leveque, M. and Cheradame, H., "Integration of Fundamental Polymer Science and Technology", Eds. Kleintjens and Lemstra, Elsevier Appl.Sci.Pub., 1986, p.250.

18. Watanabe, M., Nagano, S., Sanui, K. and Ogata, N., Solid State Ionics, 1986, 18/19, 338.

19. Nagaoka, K., Naruse, H., Shinohara, I., and Watanabe, N., J.Polym.Sci.Polym.Lett.Ed., 1984, 22, 659.

20. Bouridah, A., Dalard, F., Deroo, D., Cheradame, H., and Le Nest, J.F., Solid State Ionics, 1985, 15, 233.

21. Adamic, K.J., Greenbaum, S.G., Wintersgill, M.C. and Fontanella, J.J., J.Appl.Phys., 1986, 60, 1342.

22. Giles, J.R.M. and Greenhall, M.P., Polymer Commun., 1986, 27, 360.

23. Bannister, D.J., Davies, G.R., Ward, I.M. and McIntyre, J.E., Polymer, 1984, 25, 1600.

24. Xia, D.W., Soltz, D. and Smid, J., Solid State Ionics, 1984, 14, 221.

25. Blonsky, P.M., Shriver, D.F., Austin, P. and Allcock, H.R., Solid State Ionics, 1986, 18/19, 258.

26. Cowie, J.M.G. and Martin, A.C.S., Polymer Commun., 1985, 36, 298.

27. Fish, D., Khan, I.M. and Smid, J., Makromolek.Chem.Rapid Commun., 1986, 7, 115.

28. James, D.B., Wetton, R.E. and Brown, D.S., Amer.Chem.Soc., Polym. Preprints, 1978, 19(2), 347.

29. Gibbs, J.H. and Di Marzio, J.Chem.Phys., 1958, 28, 373.

30. Adam, G. and Gibbs, J.H., J.Chem.Phys., 1965, 43, 139.

64

STABILIZATION OF POLY(ETHYLENE TEREPHTHALATE) (PET)
AGAINST HYDROLYSIS BY CARBOXYLIC END GROUP CAPPING

C. Tintel
Enka Research Institute,
P.O. Box 60, 6800 AB ARNHEM,
The Netherlands

ABSTRACT

Stabilization of poly(ethylene terephthalate) (PET) against
hydrolysis can be performed by lowering the carboxylic end group
concentration. Therefore a polycarbodiimide and a bisepoxy compound have
been used. For a proper stabilization an excess of polycarbodiimide should
be added, so that newly formed carboxylic end groups are also removed. The
bisepoxide used shows a relatively low reactivity towards carboxylic end
groups. Sodium benzoate and triphenyl phosphine catalyze the reaction,
albeit that the first, being relatively basic, strongly induces
cross-linking reactions, in contrast to the last catalyst.

INTRODUCTION

Poly(ethylene terephthalate) (PET) is widely used for the production
of high-strength fibres, monofilaments, films, bottles and
fibre-reinforced plastics. For some of these applications improved thermal
and hydrolytic stability of PET is required.

The nature of the used catalysts plays an important role in
determining the thermal stability, whereas the concentration of carboxylic
end groups in the polymer largely determines the hydrolytic stability [1].
In the hydrolytic degradation reaction new carboxylic end groups are
formed, so that this reaction is autocatalytic with respect to carboxylic
end groups. Thus it will be clear that the addition of a carboxylic end

group capping agent must improve the hydrolytic resistance of the PET. A host of reactive compounds can be applied for this purpose [2], albeit that often undesirable side reactions prevent practical use.

Carbodiimides are highly reactive towards carboxylic end groups of PET [3,4]. Reaction results in the formation of an intermediate adduct, which, after rearrangement, leads to the amidation of the carboxylic end group and the formation of an isocyanate [5]. Liberation of isocyanates from the melt must be considered as a serious draw-back of the use of low-molecular weight carbodiimides. On the other hand, when polycarbodiimides are used this problem can be avoided.

Epoxy compounds can also be used for the protection of carboxylic end groups [2,4]. Upon reaction the epoxide ring is opened and a hydroxyl group is formed. This hydroxyl group, however, can give rise to undesirable side reactions, e.g. transesterification, which can possibly lead to cross-linked systems.

In this article we shall describe the results of our experiments with a polycarbodiimide and a bisepoxy compound as carboxylic end group capping agents for PET.

MATERIALS AND METHODS

The PET used possesses an I.V. of 1.01 dl.g^{-1} and a carboxylic end group concentration of 18.5 mmole.kg^{-1} (Arnite A06 300, supplier Akzo Plastics, The Netherlands). The polycarbodiimide is supplied by Rhein-Chemie (FRG). 1,4-Butanedioldi(2,3-epoxypropyl)ether was supplied by Pharmacia (Danmark). Catalysts have been supplied by Jansen Chimica (Belgium) and tris(2,4-di-tert-butylphenyl)phosphite (Irgafos 168) by Ciba Geigy (Switzerland).

The stabilizers have been extruded into the PET using a single-screw extruder, equipped with a Barr screw. The chips have been dried at 140°C under vacuum. The PCD-stabilized materials have been extruded into monofilaments, which have subsequently been stretched (stretch ratio 5). These monofilaments (ca. 250 tex) have been tested for their hydrolytic resistance by exposition to saturated steam of 120°C. Melt-flow index (MFI) measurements were performed at 290°C and with 2.16 kgf load.

RESULTS AND DISCUSSION

A. Polycarbodiimide (PCD)

Various amounts of PCD have been added to the PET by way of melt-extrusion. Figure 1 represents the influence of exposition to saturated steam (120°C) on the different materials.

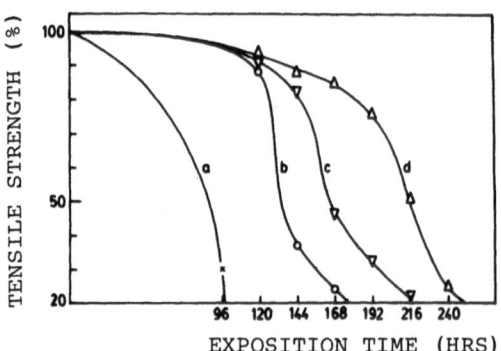

Figure 1. Tensile strength vs. time of exposition to saturated steam for PCD-stabilized PET.
a. pure PET
b. PET + 1.25 % PCD ($\approx$52 mmole.kg^{-1} NCN)
c. PET + 1.85 % PCD ($\approx$76 mmole.kg^{-1} NCN)
d. PET + 2.50 % PCD ($\approx$103 mmole.kg^{-1} NCN)

As expected, increasing amounts of PCD lead to an improved hydrolytic resistance. It should be noted that for effective stabilization a much higher carbodiimide (NCN) concentration than the initial carboxylic end group concentration is required. Thus unreacted PCD will be able to react with carboxylic end groups which are formed during the hydrolytic degradation reaction. Moreover, PCD can also react with water itself under the used circumstances. Which of these reactions prevails is yet unknown, however.

B. Bisepoxide (BEP)

1,4-Butanedioldi(2,3-epoxypropyl)ether (BEP) has been added to PET during melt-extrusion (25 mmole.kg^{-1}). The influence of the addition of sodium benzoate (NaOBz; 3.2 mmole.kg^{-1}) and triphenyl phosphine (Ph$_3$P; 3.2 mmole.kg^{-1}) as catalysts and Irgafos 168 (3.3 mmole.kg^{-1}) as stabilizer against transesterification, has been investigated and the results are shown in Figure 2.

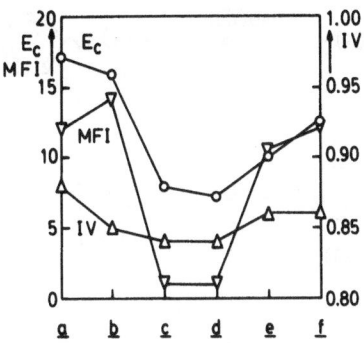

Figure 2. Carboxylic end group concentration (E_c, mmole.kg^{-1}), intrinsic viscosity (IV, dl.g^{-1}) and melt-flow index (MFI, g.10min^{-1}) of PET with 0.5 % of BEP ($\approx$ 25 mmole.kg^{-1}). Influence of catalysts.
   a. pure PET
   b. PET + 0.5 % BEP
   c. PET + 0.5 % BEP + NaOBz
   d: PET + 0.5 % BEP + NaOBz + Irgafos 168
   e: PET + 0.5 % BEP + Ph$_3$P
   f: PET + 0.5 % BEP + Ph$_3$P + Irgafos 168

The bisepoxy compound proves to be relatively unreactive towards carboxylic end groups in the absence of a catalyst (b). However, the addition of small amounts of NaOBz (c), as well as Ph$_3$P (e), leads to an increased reactivity. This results in a decreased carboxylic end group concentration. However, the addition of NaOBz to the system leads to a strongly decreased MFI, indicating the occurrence of a cross-linking reaction. On the contrary, Ph$_3$P does not induce such a drastically decreased MFI. Apparently in this case the cross-linking reaction does not play an important role. Another difference between the two catalysts is that NaOBz gives rise to a viscosity drop, in contrast to Ph$_3$P. To our opinion the differences in behaviour between the two catalysts can be ascribed mainly to the difference in basicity.

The addition of NaOBz to the system leads to the formation of the polar, basic intermediate 1, which will be apt to react further, thus leading to cross-linked systems. Ph$_3$P does not give rise to the formation of a highly reactive, ionized intermediate (see 2). Thus it can be understood that in this case the cross-linking reaction will be much less pronounced, rendering this system more suitable for extrusion purposes.

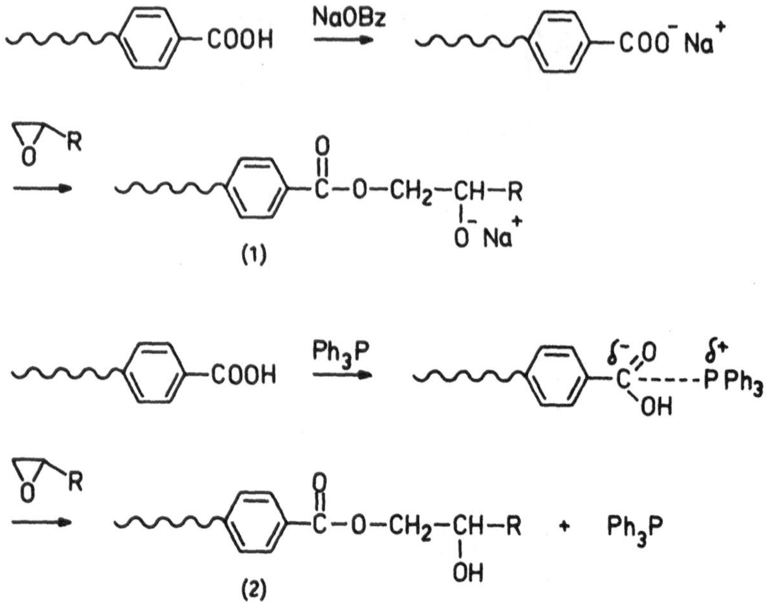

68

(1)

(2)

Scheme 1. Reaction mechanisms for the NaOBz and Ph_3P catalyzed reactions.

REFERENCES

1. Zimmerman, H. and Kim, N.T., Investigations on thermal and hydrolytic degradation of poly(ethylene terephthalate). Polym. Eng. Sci., 1980, 20, 680-3.
2. Inata, H. and Matsumura, S., Chain extenders for polyesters. I. Addition-type chain extenders reactive with carboxyl end groups of polyesters. J. Appl. Polym. Sci., 1985, 30, 3325-37.
3. Neumann, W., Holtschmidt, H., Odenthal, J.P. and Fischer, P. (Bayer AG), Stabilization of polyesters with polycarbodiimide. US Patent no. 3,193,522 (1965).
4. Bhatt, G.M., Monofilaments of low carboxyl content for use in fabricating a paper machine dryer fabric. Int. Pat. Appl. WO 83/01253 (1983).
5. Van Guldener, D.B. and Sikkema, D.J., Isocyanates from reaction of carbodiimides with carboxylic acids at 280°C. Chem. Ind. (London), 1980, (15), 628.

GRAFT COPOLYMERIZATION OF STYRENE-
MALEIC ANHYDRIDE AND CIS-POLYBUTADIENE

Lingyun Li, Wenxuan Zhu, Jingen Zhang and Xiangming Xu
Polymeric Materials Research Institute,
Shanghai Jiao Tong University,
Shanghai, P.R. China

ABSTRACT

Copolymerization of styrene-maleic anhydride in the presence of cis=
polybutadiene and toluene has been studied. The existence of various donor=
acceptor complexes in the propagation reactions has been confirmed. Attentions were paid to the effect of reaction parameters (agitation speed,
polymerization temperature, initiator concentration, rubber content, MA
content) on phase inversion, and influence of addition of MA to polymerization system on structure and properties of graft copolymers.

INTRODUCTION

Rubber modified styrene-maleic anhydride copolymer, which is formed
by the dispersion of elastomer in a glassy matrix of polymer, is one class
of heterophase polymers with commercial significance after HIPS and ABS.
Rubber has the function to raise up impact strength,whereas maleic anhydride can increase the heat resistance of polymer. Since the nature of copolymerization of maleic anhydride and styrene tends to be alternate,
which is unacceptable for the requirements of structural materials , it is
necessary to control the sequence structure of copolymer into random form.

This paper is devoted to study such a kind of polymerization system:
during the polymerization of styrene with cis-polybutadiene dissolved in
it, a solution of maleic anhydride in toluene is dropped gradually. Polymerization is initiated by a free radical initiator. This is a complex
multi-component system. Attentions are paid to the presence of various
donor-acceptor complexes in system, factors to affect phase inversion, and
the influence of addition of maleic anhydride to the polymerization system
on the structure and properties of graft copolymers.

EXPERIMENTAL

Reactor used was a 1.5 l glass reactor, with stirrer of anchor type.

Rubber was dissolved in predistilled styrene under nitrogen atmosphere. After the temperature was raised to a chosen degree, initiator BPO
was added. During the course of polymerization, toluene solution of MA was
added through a metering pump. Before the end of reaction, a stabilizer
was added. Samples were withdrawn frequently to determine the conversion
degree. The product was vacuum dried at 180°C.

In the polymerization process, both quantity of toluene added and
feeding rate of metering pump were maintained to be constant.

DONOR–ACCEPTOR COMPLEXES IN STYRENE–MALEIC
ANHYDRIDE–CIS–POLYBUTADIENE–TOLUENE POLYMERIZATION SYSTEM

The mechanism of incorporation of maleic anhydride to the polymerization system of styrene has been studied extensively for many years(1).
For a multi-component system comprising styrene–maleic anhydride with cis=
polybutadiene and toluene, we first studied the presence of donor–acceptor
complexes by the use of H' NMR.

According to the H' NMR spectra of St, MA and St+MA,when both St and
MA are present in the system, the chemical shift of MA is toward high
field. This verifies the formation of donor–acceptor complex of St and MA.

Similarly, we have verified that between toluene and MA,cis–polybutadiene and MA, donor–acceptor complexes are also formed.

We have used UV spectra to determine the complex equilibrium constant
K of St–MA complex system. The result is 0.23 (25 C), which is in fair
agreement with those from Seymour (2).

Based upon above observations, it is quite evident, St has a strong
tendency to donate electron. This is why St can undertake alternate copolymerization with MA readily. However, toluene and cis–polybutadiene
are also electron donors to certain extent, so they can compete with St
to form complexes with MA so.as to reduce the concentration of St–MA complex, and to lessen the ability of electron accepting of MA, which is
favorable for the formation of random copolymer.

COPOLYMERIZATION OF STYRENE–MALEIC ANHYDRIDE
IN THE PRESENCE OF CIS–POLYBUTADIENE AND TOLUENE

TABLE 1

Effect of agitation speed on phase inversion

Agitation speed,rpm	80	110	140	170	200
Phase inversion point, % conversion	24.2	21.5	21.3	21.1	20.9
Rubber phase volume fraction %	18.5	18.0	17.5	17.0	17.0
Average particle size,μ	4.5	3.5	3.1	——	——

TABLE 2

Effect of polymerization temperature on phase inversion

T,°C	75	80	85	90
Phase inversion point,% conversion	20.6	21.2	26.2	——
Rubber phase volume fraction, %	17.8	15.8	15.3	14.8

TABLE 3

Effect of initiator concentration on phase inversion

BPO, 10^{-2}M	2.62	3,94	5.25
Phase inversion point, % conversion	23.2	22.0	21.2

TABLE 4

Effect of rubber content on phase inversion

PB, % wt	7	10	12
Phase inversion point, % conversion	21.2	23.2	25.1

TABLE 5

Effect of MA content on phase inversion

MA, % wt	7.5	10.0	12.5
Phase inversion point, % conversion	24.1	23.3	20.5

Effect of reaction parameters on phase inversion

During the polymerization, the viscosity of the system increases gradually . After the attainment of a maximum value, viscosity drops down up to the minimum value, which indicates the end of inversion.

The agitation speed has a tremendous influence on phase inversion, and then on the morphology of product.

As shown in Table 1, increase of agitation speed results in earlier phase inversion, which is in agreement with the result observed in HIPS

TABLE 6
Effect of MA content on polymer structure

MA,%wt	0	5.0	7.5	10.0	12.5
$\overline{M}w \times 10^{-4}$	8.6	9.3	12.2	——	16.6
$\overline{M}w/\overline{M}n$	2.6	2.6	2.7	——	3.4
Swelling index	3.6	5.7	6.2	6.9	6.9
Apparent grafting degree,%	242	232	205	203	159

TABLE 7
Effect of MA content on polymer properties

MA, % wt	0	5.0	7.5	10.0	12.5
Notched izod impact strength, kg cm/cm^2	5.6	7.7	20.6	27.3	23,7
Unnotched izod impact strength, kg cm/cm^2	10.9	18.9	64.8	107.6	77.9
Melt index 200°C, 5kg, g/10 min	4.4	3.6	3.3	1.8	1.0
Vicat temperature,°C	80	88	98	106	110

polymerization(3). Under the action of shearing agitation, spherical rubber particles with resin occlusion were obtained. With the increase of agitation speed, the particles of dispersed phase are reduced in size; meanwhile, the rubber phase volume fraction is reduced.

Table 2 illustrates the effect of elevation of temperature on delaying phase inversion. This is due to the reluctance of rubber phase to break by shearing at elevated temperature. Table 2 shows also the effect of elevation of temperature on decreasing the rubber phase volume fraction. From the morphological observation on rubber particles, more wrinkles are found on rubber film. Although the particle sizes are relatively large, the resin occlusion contents are relatively small.

Table 3 illustrates increase of initiator concentration results in earlier phase inversion. This result is somewhat different from those of HIPS (3). Probably it is due to difference in polymerization system.

Increase of rubber content leads to delay of phase inversion (Table 4) which is in agreement with those of HIPS (3).

Increasing the MA content causes the earlier appearance of phase inversion (Table 5).

It is understood that the combination of these factors further comlicate the appearance of phase inversion.

Effect of addition of MA to polymerization system on structure and properties of graft copolymers

Table 6 illustrates molecular weight and its distribution index increase with the increase of MA content.

The increase of MA content leads to the increase of swelling index (Table 6) ,corresponding to the decrease of crosslink density of rubber phase.

The apparent grafting degree of rubber decreases with increase of MA content (Table 6). This is probably due to the number of graft points on the rubber is lessened.

Table 7 illustrates the influence of MA content on the properties of rubber modified SMA copolymer. When MA content is increased, the heat resistance of material increases, while melt index decreases. The impact strength of material exhibits a maximum at 10% MA content, which probably corresponds to the relatively suitable match of distribution of rubber particle size, crosslink density, molecular weight, its distribution and sequence structure of copolymer.

CONCLUSIONS

With the addition of cis-polybutadiene and toluene to the polymerization system of St and MA, St-MA donor-acceptor complex as well as toluene-MA and cis-polybutadiene-MA complexes is formed. The competition of these complex reactions is favorable to the formation of SMA random copolymer.

Due to the addition of MA,the phase inversion appearing in the prepolymerization stage may be somewhat defferent with those resulting from HIPS polymerization.

Increasing MA content causes advance of phase inversion, increase of swelling index and decrease of apparent grafting degree. Increasing MA content also increases the molecular weight and its distribution index of resin phase.

Increase of MA content leads to the increase of heat resistance of material, and decrease of melt index. There is a maximum value of impact strength at 10% MA content, this probably is due to relatively good match of various structural parameters of material.

REFERENCES

1. Deb, P.C., J. Polym. Sci.: Polym. Letters Ed., 1985, 23, 233-9.

2. Seymour,R.B. and Garner D.P.,J.Coating Tech.,Jan. 1976, 48(612),41-5.

3. Riess G. and Gaillard P.,Preparation of rubber-modified polystyrene. In Polymer Reaction Enginnering,ed. K.H. Reichert and W. Geiseler,Hanser Publishers, New York, 1983, pp.221-36.

A STUDY OF INFRARED LASER-INITIATED
POLYMERIZATION OF METHYL METHACRYLATE

Chuanlong Yue, Xiangming Xu
Polymeric Materials Research Institute
Shanghai Jiao Tong University

and

Hanyun Lu, Chuanhua Yue
Shanghai Institute of Laser Technology
Shanghai, P.R.China

ABSTRACT

This work aims to explore the application of infrared laser in the polymerization chemistry. By using a continuous-wave CO_2 laser with wave length of 10.6 μ and output power of 100 w/cm^2, it has been found that MMA can be polymerized. The IR spectrum of the polymerized product is shown to be nearly the same as those of PMMA from conventional polymerization. Further, it has been shown that this reaction belongs to free anionic polymerization. The distribution of molecular weight appears with two peaks. The tacticity of the polymer is similar to those of free-radical polymerization.

INTRODUCTION

Most of experiments in the laser-initiated polymerization dealt with using UV lasers(1). This article deals with using infrared laser to initiate the polymerization of methyl methacrylate. The infrared CO_2 laser is best suited in terms of energy, power, availability, and ease of use. It is a clean and low-maintenance instrument operating in the 9.2–10.9 μ range.

EXPERIMENTAL

The laser used was a continuous-wave single mode operated CO_2 laser with wave length of 10.6 μ and output power of 100 w/cm^2.

Exp.1.   20ml purified MMA was placed in a flask being submerged into an ice-water bath. The CO_2 laser beam was directed into the sample for sixteen times with each of 15 sec irradiation and 15 sec suspension. The product was then precipitated by adding methanol,filtered, washed and dried.

Exp.2. Using a mixture of 10 ml purified MMA and 10.8 ml purified styrene, experimented as Exp.1.

Exp.3. Using 20 ml purified styrene, experimented as Exp.1.

Exp.4. 20 ml purified MMA with 0.1% BPO was placed in a flask. After heating at 75°C for 30 min, the product was isolated as Exp.1.

Exp.5 Using a mixture of 10 ml purified MMA and 10.8 ml purified styrene with 0.1% BPO, experimented as Exp.4.

The profucts obtained above were studied on IR spectra, and those of Exp.1 was further studied on H'NMR-spectrum and the molecular weight distribution.

RESULTS AND DISCUSSION

Polymerization of MMA can be initiated by infrared laser

It is well known that photochemical effect of infrared light is based on the excitation of molecular vibrational levels. But the intensity of conventional sources of monochromatic infrared radiation is too weak to initiate polymerization. The situation changed fundamentally with the advent of lasers, which have strong intensity and high selectivity. If the wavelength of an infrared laser is in keeping with the vibrational frequency of some bond in a molecule,strong resonance will happen and the bond will be dissociated(2).

The CO_2 laser we used was highly monochromatic. Its wavelength is 10.6μ, equivalent to 943.4 cm^{-1}. The out-of-plane C-H bending vibration of $=CH_2$ in MMA is 945 cm^{-1}. This means that under radiation of 10.6 μ laser, the C-H bond in $=CH_2$ group will be dissociated, thereby polymerization may be initiated.

In Exp.1, some white powder product has indeed been obtained, its IR spectrum is nearly the same as those of free-radical polymerized PMMA obtained from Exp.4, as shown in Figure 1(a) and (c), respectively. So it has been confirmed that MMA can be polymerized by radiation of 10.6 μ laser.

But in Exp.3, while styrene was irradiated by 10.6 μ laser, no polymerized product could be found. This is due to the absence of bond vibration near 943.4 cm^{-1} in styrene.

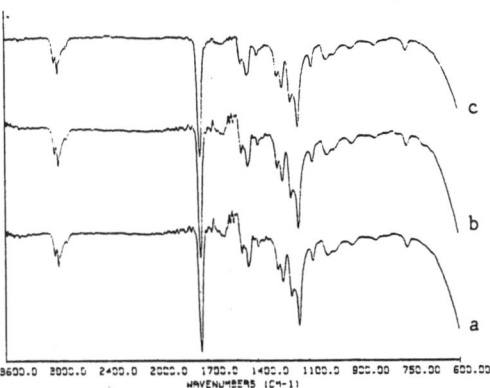

Figure 1. IR spectra of products obtained from Exps.1,2 and 4.
(a)From Exp.1; (b) From Exp.2; (c) From Exp.4.

Type of the polymerization

Generally, MMA may be polymerized through free-radical or anionic polymerization. In order to identify the type of polymerization in Exp.1, we designed Exp.2, where white powder product was also produced. Its IR spectrum is shown in Figure 1(b), which is the same as Figure 1 (a).

As mentioned above, styrene cannot form any free-radicals or ions during the 10.6 μ laser irradiation. Hence the polymerization in Exp.2 must be initiated by either the free-radicals or the anions from MMA.

If the polymerization is initiated by the free-radicals of disso-ciated MMA,the styrene will be activated by these free-radicals and the final product will be composed of MMA with styrene in proportion nearly as 1:1(3). For comparison, we designed Exp.5, in which the MMA and styrene were polymerized through conventional free-radical process. Its IR spectrum is shown in Figure 2, where a distinct peak appears near 700 cm^{-1}, which represents the out-of-plane C-H bending vibration of benzene ring. How-ever,in Figure 1(b) such a peak does not appear. This means that MMA is not polymerized through free-radical mechanism.

Now, we consider the possibility of anionic polymerization. Becau-se the MMA anion is not sufficiently nucleophilic to initiate the poly-merization of styrene, the end product will be almost pure PMMA(3). Accor-ding to the identity of Figure 1(b) and (a), it can easily be verified. Hence the polymerization of MMA initiated by 10.6 μ laser belongs to the

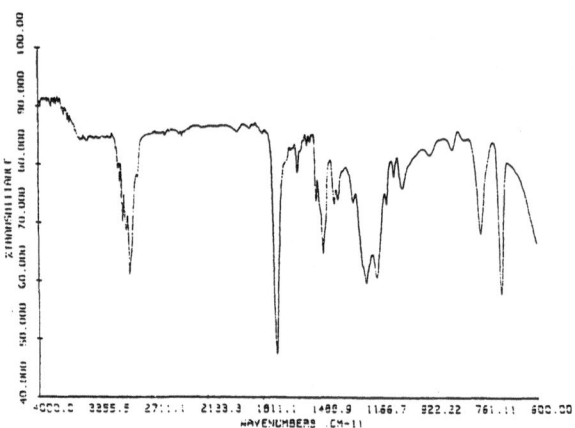

Figure 2. IR spectrum of product obtained from Exp.5 anionic polymerization.

Mechanism of the polymerization

The polymer obtained from Exp.1 was analysed by HPLC and the result is shown in Figure 3. The distribution of molecular weight appears with two peaks, like those initiated by two initiators.

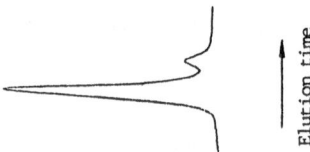

Figure 3. Distribution of molecular weight of product obtained from Exp.1 (Varian 5060, GMH 6)

From the discussion above, MMA is dissociated during irradiation to form two ion species H^+ and $^-CH=C(CH_3)COOCH_3$. The anion is an active site to attack the MMA monomer and bring about anionic polymerization, whereas the proton can react with the MMA monomer to form carbonium ion

$$CH_2=\overset{CH_3}{\underset{+}{C}}--\overset{OH}{\underset{}{C}}-O-CH_3$$

Both proton and carbonium ion may terminate the growing polymer chains. Because the carbonium ion is more stable than H^+, it will play a dominant role in the termination competition. But H^+ is more active and can terminate the growing chain more quickly. Therefore there exist two different mole-

cular weight ranges. The small peak of smaller molecular weight range
corresponds to the polymer chains terminated with proton, whereas the
large peak of larger molecular weight range corresponds to those terminated
with carbonium ion.

Tacticity of the polymer

Generally, ion pairs with more or less degree of association exist
always in the growing polymer chain in anionic polymerization initiated
by initiators(3). But in laser-initiated polymerization mentioned above,
the anion is formed by dissociation of the monomer itself, no ion pairs
exist there. Hence the infrared laser-initiated polymerization of MMA
belongs to free anionic polymerization. Consequently, the tacticity of
the polymer will be similar to those of free-radical polymerization.
Figure 4 shows the H'NMR spectrum of polymer obtained from Exp.1. The
peaks at δ 1.255,1.025 and 0.853 correspond separately to the proton absorption of α-methyl in isotactic(I), heterotactic (H) and syndiotactic (S)
placements. From the integration curve, the ratio I:H:S = 1.8:23.2:75, and
the value $4IS/H^2$ = 1.04. This value is in the range of free-radical polymerization of MMA(3).

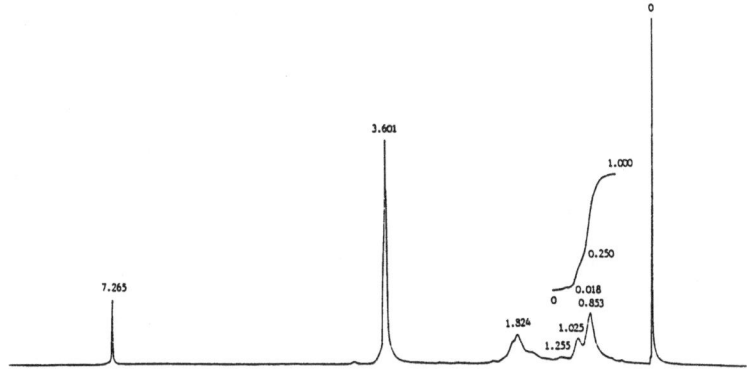

Figure 4. H'NMR spectrum of product obtained from Exp.1
(solvent: $CDCl_3$)

CONCLUSIONS

MMA can be polymerized by continuous-wave CO_2 laser with wavelength
of 10.6 μ and output power of 100w/cm^2. This reaction belongs to free anionic
polymerization. The distribution of molecular weight appears with two peaks.
The tacticity of the polymer is similar to those of free-radical polymerization.

REFERENCES

1. Sadhir, R.K., Smith, J.D.B. and Castle, P.M.,J.Polym. Sci.: Polym Chem. Ed, 1985, 23,411-427.

2. Ambartzumian, R.V. and Letokhov, V.S., Multiple photon infrared laser photochemistry. In Chemical and Biochemical Applications of lasers,Vol.3, ed. C.B. Moore, Academic Pr., New York, 1977, pp. 167-314.

3. Jenkins, A.D. and Ledwith, A, Reactivity, Mechanism and structure in Polymer Chemistry, John Wiley & Sons, London, 1974

CATIONIC TRANSANNULAR OLIGOMERIZATION
OF 1,5-CYCLOOCTADIENE

Cuiming Yuan, Deyue Yan, Xiangming Xu
polymeric Materials Research Institute,
Shanghai Jiao Tong University,

and

Fosong Wang
Changchun Institute of Applied Chemistry,
Academia Sinica,
P.R. China

ABSTRACT

Cationic transannular oligomerization of 1,5-cyclooctadiene initiated by BF_3OEt_2 and $AlCl_3$ is presented in this paper. The number-average molecular weight of the resultant polymer is about 1,500, and the softening temperature range is 140-160°C. By means of NMR, IR and PC-MS analyses, the chain structure of poly(1,5-cyclooctadiene) proposed earlier by Marvel et al. is corroborated.

INTRODUCTION

In 1963, Marvel et al. (1) reported that 1,5-cyclooctadiene (COD), in the presence of triisobutyl-aluminum and titanium tetrachloride yielded an essentially saturated polymer. The authors suggested a mechanism of transannular polymerization. Recently, the authors of this paper carried out the transannular polymerization of 1,5-cyclooctadiene with $AlCl_3$ as a catalyst (2). However, the structure of the oligomer has not been definitely confirmed by experimental data. This work concerns the cationic transannular polymerization of 1,5-cyclooctadiene initiated by BF_3OEt_2 and $AlCl_3$. The effects of reaction conditions such as the concentrations of the catalyst and the monomer, etc. on the polymer yield are investigated. Spectrographic (PC-MS) studies ascertain that the resulting polymer is poly-2,6-bicyclo-(3,3,0)octane.

EXPERIMENTAL

Materials

Cis,Cis-1,5-cyclooctadiene was supplied by Aldrich Chemical Company. Prior to polymerization, 1,5-cyclooctadiene was purified by repeated shaking with concentrated sodium hydroxide solution for three times at room temperature, then was washed with dilute sulfuric acid and with water. Predried solvent with calcium hydride was distilled under reduced pressure. Dichloromethane was refluxed over calcium hydride for 24 hr. and distilled under an atmosphere of nitrogen. The water content in COD was determined by means of Karl-Fischer method which should be within 15 ppm, and for one in dichloromethane should be less than 6 ppm.

Polymerization

When using such liquid catalyst as BF_3OEt_2, the polymerization was carried out in dry 20 ml ampoules to which the monomer, the solvent, and the catalyst were added by hypodermic syringes. Then the ampoules were sealed and left alone at the desired temperature. The reaction was terminated by addition of alcohol and hydrochloric acid. The products were dissolved in chloroform, then precipitated, filtered and washed three times with alcohol, and dried to constant weight in vacuo at 40°C. The polymer yield was determined by weighing.

For initiation by $AlCl_3$, the polymerizations were carried out in a dry serum bottle, in which the bulb containing $AlCl_3$ was previously placed. The monomer and the solvent were added through the rubber cap to the serum bottle with hypodermic syringes, and then the bulb can be shattered by means of stirring the bottle, at the same time the reaction began. The termination of reaction and the purification of polymers were the same as those mentioned above.

RESULTS AND DISCUSSION

Effect of the molar ratio of catalyst to monomer on polymer yield

1,5-cyclooctadiene was polymerized with BF_3OEt_2 and $AlCl_3$ in CH_2Cl_2 under fixed reaction conditions. The results are shown in Figures 1 and 2, which indicate that the polymer yield increases with the elevation of the molar ratio of catalyst to monomer, and finally reaches a limited polymer yield.

Effect of solvent

It is well kown that the rate of cationic polymerization will increase greatly as one increases the solvating power of the reaction medium(3,4).

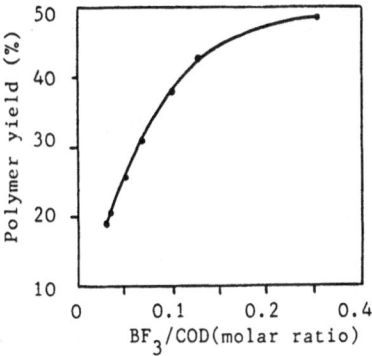

Figure 1. Relationship between the polymer yield and BF$_3$.
(COD) = 4.1 mol/L; Solvent:CH$_2$Cl$_2$; 168 hr.;Room temperature

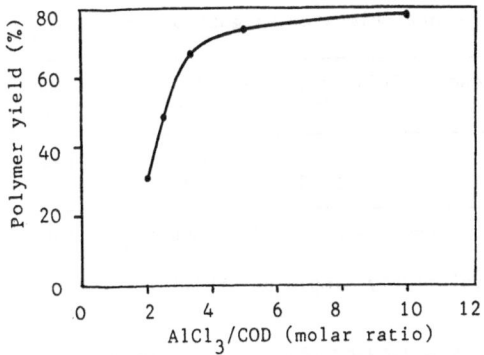

Figure 2. Relationship between the polymer yield and AlCl$_3$.
(COD) = 3 mol/L; others cf. Figure 1.

Figure 3 shows the relationship betweeen the polymer yield and molar ratio of CH$_2$Cl$_2$/COD. These plots exhibit a maximum polymer yield at certain molar ratio of dichloromethane to monomer. The polymer yield increases with increasing CH$_2$Cl$_2$/COD, then reaches a maximum, and finally either decreases or levels off.

Characterization of polymers

The data of number-average molecular weghts of polymers obtained are listed in Table 1, which manifests the products are oligomers.

In the course of our studies on transannular polymerization, it is of interest to examine the mechanism of COD polymerization. This monomer

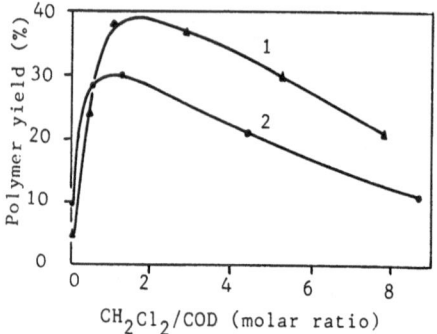

Figure 3. Relationship between the polymer yield and ratio of CH$_2$Cl$_2$/COD.
1. AlCl$_3$/COD (molar ratio) = 0.033;
2. BF$_3$/COD (molar ratio)=0.1, BF$_3$/H$_2$O (molar ratio)=10; others cf. Figure 1.

TABLE 1

Number-average molecular weights of polymers

Catalysts	Monomer conc. (mol/L)	Polymer yield (%)	$\overline{M}n$ (VPO)	Softening temp. range (°C)
AlCl$_3$	2.8	50.7	1540	138–160
BF$_3$OEt$_2$	5.4	28.5	891	140–160

Solvent: CH$_2$Cl$_2$; 72 hr.; Temp.:24°C; Catalyst/COD (molar ratio) = 0.03

offers at least three relatively simple possibilities for homopolymerization:

1. COD ⟶ $-(CH_2-CH=CH-CH_2)_n$ ring-opening polymer

2. COD ⟶   cross-linked polymer

3. COD ⟶ (I) or (II)   transannular polymer

   (I)   (II)

The NMR and IR spectra of the polymer show that there are few residual carbon–carbon double bonds in the polymer chains, and this fact indicates that COD does bot undergo ring-opening polymerization with AlCl$_3$ and BF$_3$OEt$_2$ as catalysts. Since the polymer obtained is a white powder, soluble in bezene, THF, chloroform, and dichloromethane, it can be concluded that the polymer is linear, otherwise the independent polymerization of each single double bond of COD would lead to cross-linkage. All of these results verified that poly(1,5-cyclooctadiene) must be the transannular polymer as shown in (I) or (II).

In order to determine the structure of poly(1,5-cyclooctadiene) synthesized by cationic catalysts, we examined the PC–MS spectra of the polymer which indicate that the predominant compound of the pyrolytic products of poly(1,5-cyclooctadiene) is bicyclo-(3,3,0)-octane. Therefore, PC–MS analysis of poly(1,5-cyclooctadiene) has verified that the polymer has structure (I).

REFERENCES

1. Reichel, B. and Marvel, C.S., J. Polymer Sci.; Part A, 1963, 1, 2935.

2. Deyue Yan, Qigui Guo, Ping Xia, Xueqing Wu, Seng Li, Kexue Tongbao, (in Chinese) 1984, 29, 1560.

3. Ledwith, A. and Sherrington, D.C., Reactivity and Mechanism in Cationic Polymerization, Chap. 9, In Reactivity, Mechanism and Structure in Polymer Chemistry, ed. A.D. Jenkins and A. Ledwith, Wiley-Interscience, New York, 1974.

4. Kennedy, J.P. & Feinberg, S.C., J. Polymer Sci.: Polym. Chem. Ed., 1978, 16, 2191.

ALTERNATING COPOLYMERS
VIA COMPLEXED MONOMERS

C. D. Eisenbach
Lehrstuhl Makromolekulare Chemie
Universität Bayreuth
P.O. Box 101251
D-8580 Bayreuth, FRG

ABSTRACT

Alternating copolymers of acrylics and olefins can be synthesized at
-70°C in dichloromethane by using acrylic monomers complexed with alkylalu-
minum halides; monomer conversion as well as molecular weights are control-
lable parameters by varying the composition of the reacting system and the
reaction conditions. The copolymers exhibit unusual material properties and
seem to be interesting materials in view of the development of new areas of
technological application.

INTRODUCTION

The material properties of polymers are basically controlled by the
primary structure of the polymer chain and the resulting superstructure. The
variation of the primary structure is essential for obtaining polymeric materials with special and/or new properties; in this context, alternating copolymers, in which different monomer units are always linked in the same
way and alternate in a regular fashion along the chain, are of particular
interest: First, they can be considered as new homopolymers with a repeat
unit of two monomers; second, the properties of alternating copolymers are
not predictable as can be those of statistical copolymers, i.e., alternating
copolymers open a gate to new polymer materials with unusual properties and
potential new areas of application.

However, there exists only a very limited number of pairs of monomers
with suitable copolymerization parameters required for the formation of
strictly alternating copolymers. A completely new approach to alternating
copolymers is to use comonomer pairs in which one of the comonomers is complexed by a third, non-polymerizable agent [1]. The monomers are classified

[2] as electron donor (olefins) and electron acceptor monomers (acrylics), and the already low electron density of the double bond of the latter co-monomer is further reduced by complex formation with Lewis acids. The role of the complexing agent on the course and the elementary steps of the poly-reaction are by far not clear, and are subject of controversial views so far [3, 4].

COPOLYMERIZATION

The polyreaction of the ternary system olefin/acrylic/alkylaluminum halide does not start spontaneously [5] after combining the three components in, e.g., dichloromethane solution as reported in literature [1] ; it is only upon addition of oxygen or organic peroxides that the copolymerization is initiated in a controlled way, as was extensively investigated with ethylene or propylene and acrylonitrile complexed with ethylaluminum halides:

$$CH_2=CH_{\underset{CH_3}{|}} \;+\; CH_2=CH_{\underset{\substack{CN \\ \downarrow \\ al \\ /|\backslash}}{|}} \xrightarrow{\text{O}_2 \text{ or Peroxide}} \left[CH_2\,CH_{\underset{CH_3}{|}}-CH_2\,CH_{\underset{\substack{CN \\ \downarrow \\ al \\ /|\backslash}}{|}} \right]_n$$

Highly pure, aluminum free and strictly alternating copolymers are iso-lated after decomplexation of the initially formed complexed copolymer and decomposition of the aluminum organyl, followed by reprecipitation of the copolymer. Depending on the reaction conditions with regard to the molar ratio and absolute concentrations of the various components, the type of Lewis acid and reaction time, both the conversion and the molar mass of the resulting copolymer can be adjusted, (Fig. 1 a,b). The role of aluminum organyl is dual in that it acts as a coinitiator and as complexing agent for the acrylic monomer as well as the resulting copolymer: There are strong evidences for the formation of a radical initiator species from the redox system aluminum organyl/oxygen (or peroxide) and a radical chain pro-pagation reaction (cf. [4]); the carbon-carbon double bond of the acrylic monomer, i.e., the α-C-atom of the terminal acrylic unit of the propagat-ing chain have to be activated to a certain extent by the interaction nitrile group - Lewis acid in order to result in an alternating copoly-merization. High Lewis acid strength of the ethyl aluminum chloride is advantageous with regard to the yield and the molar mass of the copolymers, i.e., ethyl aluminum dichloride is complexing agent of choice (cf. [5]).

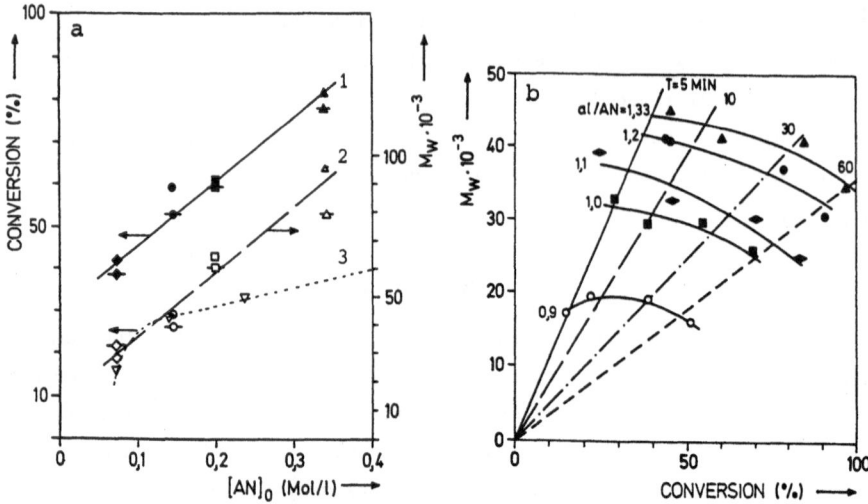

Fig. 1: Conversion and molar mass M_w in the alternating copolymerization of acrylonitrile (AN) with propylene (Pr, curve 1,2) and ethylene (Et, curve 3) in the presence of ethyl aluminum dichloride (EADC); Pr(Et)/AN/EADC 3:1:1; O_2 initiator; CH_2Cl_2 solvent; -70 (-80)°C; 10 (150) min reaction time; variation of [AN]$_o$ (Fig. 1a), molar ratio EADC(al) / [AN] and reaction time t (Fig. 1b)

COPOLYMER STRUCTURE AND PROPERTIES

Among the various comonomer pairs investigated so far, alternating copolymers are always obtained only for the systme acrylonitrile/propylene (or i-butylene) irrespective of the nature of the aluminum organyl employed and the initial molar ratio of the comonomers; more or less pronounced deviations from an equimolar composition of the copolymers are observed with other comonomer pairs. The strictly alternating copolymer structure has been proved unequivocally for acrylonitrile/propylene- and in most cases also for acrylonitrile/ethylene-copolymers by ^{13}C-NMR spectroscopy.

The alternating ethylene/acrylonitrile-copolymer differs from all the other systems by a relatively high degree of crystallinity of about 60 %, whereas copolymers with other olefins are completely amorphous; this also holds true for the acrylamide- or acrylic acid-copolymer obtained from the acrylonitrile-copolymers by polymer analogous reaction. These differences in morphology are also reflected in the mechanical properties of the co-polymers and their solubility characteristics, which in some cases unex-pectedly and drastically differ from thoses of the corresponding homopoly-

mers. These unexpected and unusual properties of alternating copolymers may
be illustrated exemplarily with the mechanical properties of the acryloni-
trile-propylene and -ethylene alternating copolymers (Fig. 2): The temperature dependency of the storage modulus G' of the semicrystalline ethylene
copolymer is very similar to isotactic polypropylene; the modulus of the
completely amorphous propylene copolymer drops extremely sharp when exceeding the glass transition temperature, followed by a highly viscous melt.

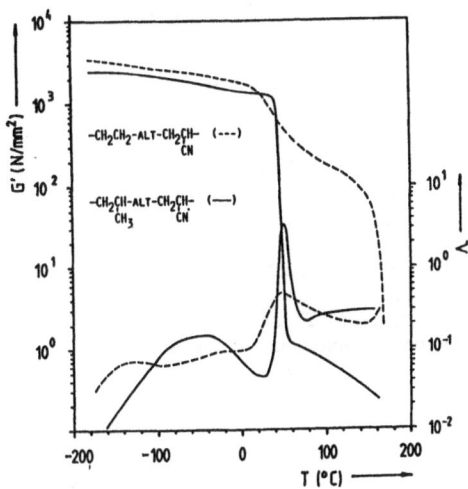

(Fig. 2) Dynamic mechanical properties of poly(ethylene-alt-acrylonitrile)
and poly(propylene-alt-acrylonitrile).

ACKNOWLEDGEMENT

The author is grateful to his coworkers U. Bülow, U. Daum, B. Fürderer
and Dr. F. G. Schmidt who did most of the experimental work. Financial support of this work by the Bundesminister für Forschung und Technologie (Grant
No. 03C201A2) is gratefully acknowledged.

REFERENCES

1) M. Hirooka, H. Yabuuchi, S. Moritan, S. Kawasumi, K. Nakaguchi, J. Polym.
   Sci., Polym. Lett. B 5, 47 (1967)
2) M. Hirooka, H. Yabuuchi, S. Kawasumi, K. Nakaguchi, J. Polym. Sci., Polym.
   Chem. Ed. 11, 1281 (1973)
3) H. Hirai, J. Polym. Sci. Macromol. Rev. 11, 47 (1976)
4) C. H. Bamford, X. Han, R. J. Malley, Plaste Kautschuk 29, 137 (1982)
5) C. D. Eisenbach, U. Bülow, U. Daum, H. Fischer, B. Ulmschneider,
   Angew. Makromol. Chemie 145/146, 125 (1986)

NEW FLUORINATED SILICONES AND POLYSTYRENES

R. Dorigo, A.M. Garnault, D. Teyssié and S. Boileau [1]
Collège de France, 11 Place Marcelin Berthelot
75231 Paris Cédex 05, FRANCE

ABSTRACT

New fluorinated allyl ethers and thioethers were prepared in excellent yields from fluorinated alcohols and thiols using phase transfer catalysis. These precursors were used for the synthesis of fluorinated silicones either by chemical modification of polymethyl hydrosiloxanes or by ring opening polymerization of fluorinated cyclosiloxanes. Moreover polystyrenes with fluorinated side-chains were prepared by free-radical polymerization of fluorinated monomers. Those new polymers were characterized by IR and NMR and their thermal behaviour was examined.

INTRODUCTION

One of the main interests of fluorinated polysiloxanes arises from their applications in high performance coatings and sealants with high resistance to heat and chemicals [2]. Several methods have been used to prepare fluorinated silicones including polycondensation of fluorinated dichlorosilanes [3]. Anionic polymerization of fluorinated cyclosiloxanes has also been reported [4,5] but hydrosilylation of fluorinated olefins with polymethylhydrosiloxanes is among the most frequently used procedures [6,7].

As far as the nature of the olefinic compound is concerned, the preparation of allyl perfluoroethers $(C_nF_{2n+1}CH_2-O-CH_2-CH=CH_2)$ by refluxing R_FCH_2OH with allyl bromide in dry acetone in the presence of K_2CO_3 has been reported but the yields are low [8]. Therefore we applied phase transfer catalysis to the preparation of allyl and the corresponding perfluoroallylethers starting from a series of alcohols [9]. The same procedure was used to prepare fluorinated compounds from chloromethylstyrene, which were subsequently homopolymerized and copolymerized with styrene [10].

RESULTS AND DISCUSSION

Fluorinated Silicones

We have prepared a series of fluorinated allyl ethers and thioethers starting from the following alcohols and thiol : $C_6F_{13}C_2H_4OH$, $C_8F_{17}C_2H_4OH$, $H(CF_2)_nCH_2OH$ (with n=1,2,3) and $C_6F_{13}C_2H_4SH$. The alcohol or thiol was reacted with allyl chloride (used as the solvent) in the presence of concentrated aqueous sodium hydroxide with tetrabutylammonium hydrogen sulfate (TBAH) as the catalyst, leading to nearly quantitative yields.

The fluorinated thiol was also grafted onto a polymethylvinylsiloxane ($\overline{M}_n$=36.000) prepared by anionic polymerization of tetramethyltetravinyl-cyclotetrasiloxane with Li^+ + [211] as the counterion [11]. The modification was carried out in toluene with AIBN as a catalyst. The Tg of the resulting polymer was -40°C and a melting point was observed at -10°C.

The hydrosilylation of the olefinic ethers was performed with polymethylhydrosiloxane ($\overline{DP}_n$=35) and with poly(methylhydro)(dimethyl)-siloxane statistical copolymers containing respectively 50% and 18% SiH units, according to the following scheme :

$$(CH_3)_3Si-O-\left[(Si(CH_3)_2-O)_{100-x}(Si(CH_3)(H)O)_x\right]_{35}-Si(CH_3)_3$$

with x = percentage of SiH units

$$+ \left| \begin{array}{l} CH_2=CH-CH_2-O-R_F \\ \\ 60°C, \text{ toluene, } N_2, H_2PtCl_6 \end{array} \right.$$

$$(CH_3)_3Si-O-\left[(Si(CH_3)_2O)_{100-x}(Si(CH_3)(C_3H_6-O-R_F)O)_x\right]_{35}-Si(CH_3)_3$$

The progress of the reaction was monitored by the disappearance of the 2160 cm^{-1} infrared band corresponding to the SiH bond. After purification the polymers were characterized by ^1H and ^{13}C NMR and the molecular weights measured by GPC were in good agreement with the expected values.

The thermal behaviour of the modified polysiloxanes was investigated and the Tg values are reported in Table I. As expected, the glass transition temperatures raised on substitution of SiH with a bulkier group (Tg of polymethyltrifluoropropylsiloxane is about -74°C). There is also a melting transition in the case of the alkylether modified polymers as in the case of the sulfur containing fluorosilicones.

TABLE I

Glass transition and melting temperatures (in °C) of modified siloxane polymers and copolymers with the general formula :

$$(CH_3)_3Si-O\left[\underset{\underset{R}{|}}{\overset{\overset{CH_3}{|}}{(Si-O)}}_x \; \underset{\underset{CH_3}{|}}{\overset{\overset{CH_3}{|}}{(Si-O)}}_{100-x}\right]_n Si(CH_3)_3$$

R	x(%)	T_g(°C)	T_m(°C)
-H	x	-119	–
$-C_3H_6-O-C_8H_{17}$	100	-78	-52
	50	-97	-74
	18	-110	–
$-C_3H_6-O-C_2H_4-C_6F_{13}$	100	-66	–
	50	-88	–
	18	-108	–
$-C_2H_4-C_6H_4-CH_2Cl$	50	-70	–
$-C_2H_4-C_6H_4-O-C_2H_4-C_6F_{13}$	50	-78	–

Finally, the preparation of fluorinated silicones by ring opening polymerization of fluorinated cyclotetrasiloxane was attempted. This compound was obtained by hydrosilylation of a perfluoroallylether with D_4H. The polymerization was carried out in bulk, under vacuum, with KOH + [222] as an initiator at room temperature. The resulting homopolymer was insoluble in all common solvents. A copolymerization with a high proportion of D_4 (87% in the starting reaction mixture) was therefore attempted. The resulting copolymer contained c.a. 10% of fluorinated units as checked by ^{1}H NMR and was clearly a mixture of cyclics and linear polymers as shown by G.P.C.

Fluorinated polystyrenes

Two styrene compounds modified with perfluoro alcohol and thiol were synthetized by reaction of $C_6F_{13}C_2H_4OH$ or $C_6F_{13}C_2H_4SH$ with chloromethylstyrene (also used as a solvent). The reaction was carried out in the presence of TBAH as a phase transfer catalyst and sodium hydroxide as the aqueous phase. The homopolymerization of those two compounds with AIBN as an initiator only lead to insoluble polymers. Their copolymerization with styrene was thus investigated and lead to soluble copolymers when the proportion of fluorinated derivative was less than 50%. The reactivity ratios of the two monomers (M_1) respective to styrene (M_2) were determined according to the Kelen-Tüdos method and the following values were obtained :

$$CH_2=CH-C_6H_4-CH_2-O-(CH_2)_2-C_6F_{13} \quad (r_1=0.7; \ r_2=1.0);$$
$$CH_2=CH-C_6H_4-CH_2-S-(CH_2)_2-C_6F_{13} \quad (r_1=0.9; \ r_2=0.9).$$

The grafting of a fluorinated alkyl chain onto the aromatic ring did not affect the reactivity of styrene. But the proportion of fluorinated compound in the copolymer affected its thermal behaviour : the introduction of 30% fluorinated thioether units lowered the Tg from 100°C (polystyrene) to 40°C.

As a conclusion, several methods were used to prepare new poly-siloxanes containing fluorinated side groups linked to the backbone through an ether or thioether group. Fluorinated derivatives of polystyrenes were also synthetized and the physico-chemical properties of those polymers as well as their applications are currently being investigated.

REFERENCES

[1] Laboratoire de Chimie Macromoléculaire associé au CNRS : UA 24.

[2] Boutevin,B. and Pietrasanta,Y., Progress in Organic Coatings, 1985, 13, 297-331.

[3] Kim,Y.K., Rubber Chem. Technol., 1971, 1350-1362.

[4] Lee,C.L., Fr. Pat. 6943553, 1969.

[5] Yuzhelevskii,Yu.A., Kagan,Ye.G. and Fedoskeyeva,N.N., Vysoko. Soed., 1970, A 12, 1585-1587.

[6] Atherton J.H., Brit. Pat. 1418465, 1972.

[7] Chujo Y. and McGrath,J.E., Am. Chem. Soc. Pol. Prep., 1983, 24 n°2, 47-49.

[8] Steward,O.W. and Pierce,O.R., J. Org. Chem., 1961, 26, 2943-2947.

[9] Youssef,B., Boutevin,B., Garnault,A.M. and Boileau,S., J. Fluor. Chem., 1987, 35, 399-410.

[10]Garnault A.M., Thèse de Docteur-Ingénieur, Paris, 1986.

[11]Hubert,S., Hémery,P. and Boileau,S., Makromol. Chem., Macromol. Symp., 1986,6,247-252.

IONOMER MODEL SYSTEMS - SYNTHESIS-CHARACTERIZATION-RHEOLOGY

Reimund Stadler, Martin Möller, Jürgen Omeis,
Josef Burgert, Liane de Lucca Freitas

Hermann Staudinger Haus - Institut für Makromolekulare Chemie -
Stefan Meier Str.31 - D-7800 Freiburg

SYNOPSIS

The synthesis and characterization of two types of ionomeric model systems is discussed. (A) Polystyrenes and polybutadienes with few 2-vinyl-pyridine units at one chain end form micellar type associates after quaternization. Viscoelastic data in the flow region are in agreement with the picture of a three-dimensional ramificated structure of ionic clusters. (B) In polybutadienes with statistically distributed 4-(3'-nitro-4'-hydroxy)-phenyl-1,2,4-triazolidine-3,5-dione groups, thermoreversible networks are formed by either hydrogen bonds or by ion pair interactions. The different behaviour of dimeric complexes or multiple ionic clusters is evident in the stress strain behavior.

INTRODUCTION

The properties of polymers with few ionic functional groups (ionomers) can be varied by the type of ion pair (anion, cation), the degree of substitution and the chain topology, i.e. whether the ionic groups are distributed statistically along the chain or located at the chain ends[1,2]. The ionic groups of statistically modified polymers form clusters of varying size while for end-functionalized polymers, a well defined cluster structure similar to that of inversed micells have been observed[3,4]. In this paper, some new results on both kinds of ionomers are presented.

A. "MONOFUNCTIONAL" IONOMERS

Polystyrenes and polybutadienes with short 2-vinyl-pyridine blocks (scheme A) (PS-b-2VP and PB-b-2VP) were obtained by stepwise anionic polymerization. The number average 2-VP block length was varied from n = 1 to 10 monomer units. These polymers were ionized by reacting the pyridine units with HCl or CH_3I. In dilute solution, these polymers form inverse micells with a highly polar ionic core and a nonpolar shell. In Figure 1, the weight average molecular weights determined by light scattering are plotted as a function of the number of 2-VP units for the non-quaternized and quaternized PS-b-2VP samples. The molecular weights of the non-quaternized materials were about 20,000. Upon quaternization, a sharp increase in the weight average molecular weight is observed. If the number of 2-VP units is small (n<4), the number of chains forming a micelle (N) is about 10-12, which is similar to the value obtained for polystyrenes with a single terminal sulfonate .group[3],[4]. For the polymers with longer 2-VP end-blocks, the degree of association is strongly enhanced (N~ 50-60). Clearly, this indicates a different structure in the two types of systems.

Similar and consistent information on the structure was obtained from dynamic mechanical experiments in the linear viscoelastic region. For the non-quaternized materials with short end blocks (N<9), a 'normal' behavior is observed in the transition from the rubbery plateau to the flow. This indicates that the 2VP blocks are too short to form a separate phase as is known for block copolymers. After quaternation with HCL, a different behavior is observed. In Figure 2, the storage modulus mastercurves are plotted as a function of frequency for polybutadienes of approximately constant molecular weight (10,000 - 15,000), with one, five and nine 2VP-units (PB-2VP1*HCl, PB-2VP5*HCl, PB-2VP9*HCL) respectively. The primary molecular weight is about 5 times the critical entanglement molecular weight M_e in these polybutadienes. For comparison, the data of two linear polybutadienes (PB1=26,000 and PB2=211,000) are also reported in Figure 2. For the temperatures used in the dynamic experiments of the ion-containing polymers, the rubbery plateau zone of the polybutadiene was not reached. Only the decay from the rubbery plateau is observed at high frequencies. With increasing length of the ionic block, this decay is only slightly shifted to lower frequencies. However, a second plateau is observed. The level of this second plateau increases with the length of the 2VP-block and is extended to lower frequencies. In contrast to this behavior, the rubbery plateau of ionomers with functional groups distributed statistically along the chains is broadened to higher temperatures (lower frequencies)[5],[6]. This effect is similar to an increase in the molecular weight (PB1 --> PB2). The difference between the mechanical behavior of endfunctionalized and statistically substituted ionomers must be explained by a different association structure of the different ionomers.

The bulk behavior of the ionomers with terminal 2VP*HCl blocks differs
blocks differs also from the behavior observed in dilute solution, where star like inverse micelles have been reported for the sample with short ionic blocks [3]. It is known that the terminal relaxation times for star molecules with arm length > 2*M_e increases at a constant level of the plateau zone. Thus, even for the PB-2VP1*HCl sample, the dynamic mechanical data of the bulk sample are not in agreement with a star-like structure. It is more likely that in the bulk where the molecules interpenetrate each other, an infinitely extended ionic structure is formed either by extended ionic clusters or by closely packed ionic clusters. This supermolecular structure could account for the observed second plateau in the storage modulus.

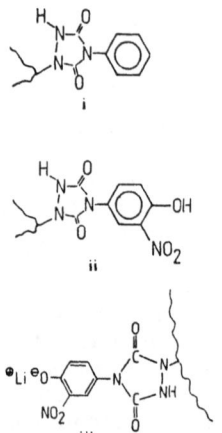

A: Synthesis of polysty-
rene/polybutadiene with few
2-vinylpyridine end groups.

B: Structures of urazole functional groups in polybutadienes.

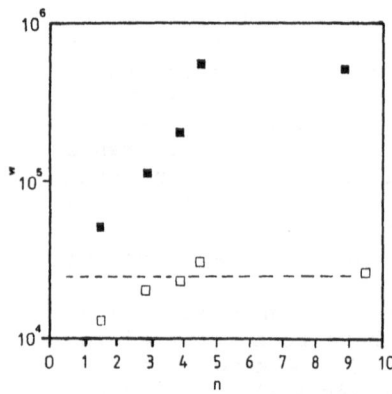

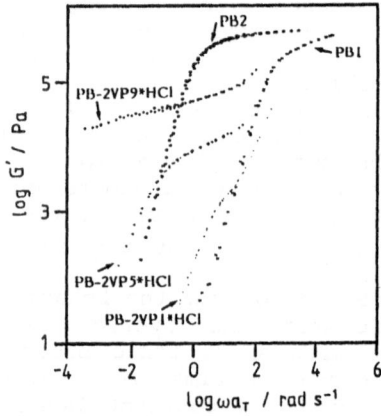

Fig.1: Molecular weights M_w
(light scattering) for non-
quaternized   and quaternized   PS-2VP samples for
various 2VP lengths.

Fig.2: Storage modulus master-
curves for two linear polybutadienes (PB1 and PB2) and linear
polybutadienes with short 2VP
end blocks of various lengths
(1,5,9 2-VP units)

B. o-NITRO-PHENOLE-POLYBUTADIENE IONOMERS

Elastomeric ionomers with statistically distributed ionic groups are of interest as single phase thermoplastic elastomers. The most important systems are the sulfonated EPDM's[1,2]. In contrast, sulfonation of polybutadiene results in a partially crosslinked material. Because polybutadienes are available as narrowly distributed polymers by anionic polymerization, polybutadiene ionomers may be of considerable help in studing the rheological properties of ionomers with statistically distributed ionomeric groups. In the following discussion preliminary results concerning a new type of ionic group are presented.

By the reaction of 4-phenyl-1,2,4-triazoline-3,5-diones with polydienes, phenylurazole groups are introduced into unpolar polymers[7]. These urazole groups form hydrogen bonded complexes[8,9]. In addition, by introducing a 3'-nitro-4'-hydroxy-substituent into the phenyl group[10] and subsequently reacting with butyllithium, polybutadiene ionomers become available (scheme B). The corresponding low molecular weight model compounds 1,2 form salts with defined amounts of crystalline water. One acidic hydrogen is detected per phenylurazole , while two acidic hydrogens are present in the substituted derivative (Fig.3). The pK_a values obtained from the titration curves (~5.1) are similar to those of the carboxyl groups.

Figure 4 shows the stress-strain curves for a technical grade polybutadiene (CB-10) with 2% of either i,ii or iii. In the first two cases the thermoreversible network is formed by the hydrogen bond interaction between two urazole groups[9]. The difference between the polymers modified with i and ii is only small. The same result was obtained from dynamic-mechanical measurements in the linear viscoelastic region[11]. This similar behavior can be explained by the formation of an intramolecular hydrogen bond of the o-nitro-hydroxy group. These samples show considerable permanent set. In contrast, the ionized sample shows a considerably higher modulus and a much reduced permanent set. Figure 5 shows the corresponding stress-strain curves for samples with different degrees of modification, that were transformed to the ionomers by reacting with n-butyllithium. The stress-strain properties of these samples are different from the thermoreversible networks based on hydrogen bonds only. Besides a higher Young modulus, the ionic networks behave like covalently crosslinked networks at room temperature. Flow is observed only after heating to high temperatures.

It can be concluded that by structural modification of 1,2,4-triazoline-3,5-diones, polydiene ionomers can be obtained by a well defined chemical reaction. The properties of the ionomers strongly differ from the corresponding hydrogen bonded networks.

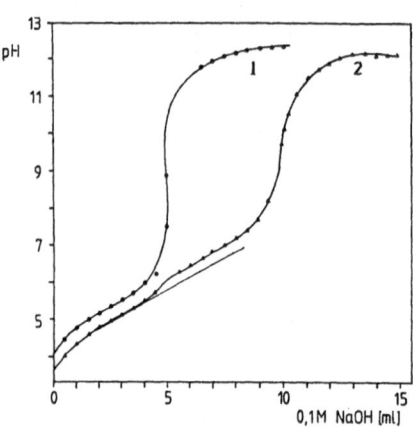

Fig.3: Titration curves of
model compounds **1** and **2**

Structures of low molecular
weight models compounds **1,2**

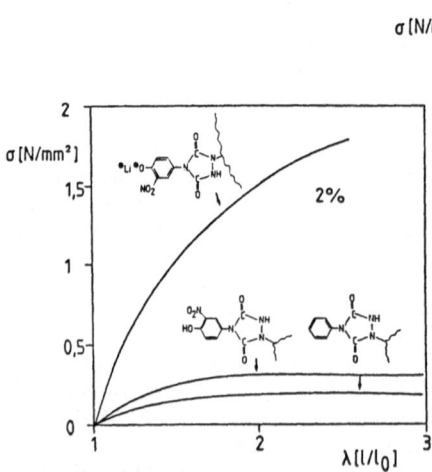

Fig.4: Stress strain curves
for polybutadiene modified
with 2% of **i,ii,iii**

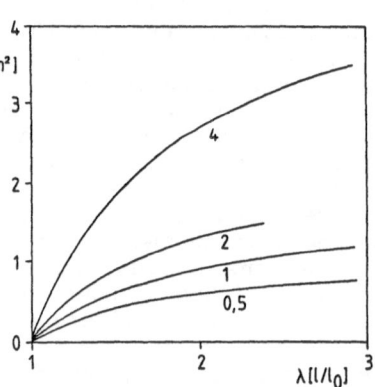

Fig.5: Stress strain curves for
polybutadiene modified with various amounts of **iii** after
metallation with Li$^+$

ACKNOWLEDGMENT

This work has been supported by the DFG through SFB60 "Funktion durch Organisation in Makromolekularen Systemen" and by DAAD (L.F.) and Landesgraduiertenförderung Baden-Württemberg (J.B.).

REFERENCES

1) Eisenberg A.,"Ions in Polymers", Adv.Chem.Ser.**187**, Am.Chem.Soc., Washington D.C. 1980

2) Eisenberg A., Bailey F.E. eds. "Developments in Ion Containing Polymers", Appl.Sci.Publ. London 1983

3) Möller M., Omeis J., Mühleisen E. in "Reversible Gelation in Polymers" P.Russo ed., Adv.Chem.Ser., in press

4) Williams C.E., Russel T.P., Jerome R., Horion J., in ref.2

5) Agarwal P.K., Makowski H.S., Lundberg R.D., Macromolecules **13** (1980) 1679

6) Agarwal P.K., Lundberg R.D., Macromolecules **17** (1984) 1918, ibid. p.1757

7) Leong K.W., Butler G.B., J.Macromol.Sci.**A-14** (1980) 287

8) Stadler R., Burgert J., Makromol.Chem. **187** (1986) 1681

9) Stadler R., de Lucca Freitas L., Coll.&Polym.Sci. **264** (1986) 778

10) Burgert J., Stadler R., Chem.Ber. in press

11) de Lucca Freitas L, Burgert J., Stadler R., Polym.Bull. in press

STUDY ON THE KINETICS OF EMULSION POLYMERIZATION BY REGRESSIVE SIMULATION

Mian Chang
Dalian Institute of Light Industry,
No.1, Baoding Street,
Dalian, China

ABSTRACT

This paper, based on a series of emulsion polymerization experiments and on the concept of the inherent law of discrete variables correlated function, introduces physical concepts into the Euclidean Space and sketches a mathematical model to substitute the Bernstein polynomial equation.

According to this, so that the approximation process can be greatly simplified and the general kinetics equation on emulsion polymerization can be derived, as follows:

$$-\frac{dC}{dt} = KC^{\alpha(D)}$$

where $\quad k = \theta\, C_o$

$$\alpha(D) = \frac{1}{A} \left\{ \frac{\sum\limits_{n=1}^{k^*} \frac{1}{n}\left[\,(-1)^{n-1}\,(D-1)^n - (aD)^n\,\right]}{1 - \frac{1}{A}\sum\limits_{n=1}^{k^*}\frac{1}{n}D^n} \right\}$$

$(A=\ln C_o)$

(where D, C, t are the conversion, monomer concentration and reaction time,respectively). A conversion--time history can be described exactly with this model and some micro-parameters of polymerization kinetics can be obtained.All these favor the control of reactions and shed light to further study of the reaction mechanism.

INTRODUCTION

As there are many components entering the emulsion polymerization process directly or indirectly,it is difficult to ascertain the reaction mechanism. Smith and Ewart[1] had put forward a hypothesis to describe the process quantitatively, but it was valid only within a narrow application range.Later,though Stockmayer[3],O'Toole[4] and Bataille[2] had provided some supplements, there were still discrepancies between the theory and experiments. For example, there is a deviation amounting 15-20% for the emulsion polymerization of styrene[2].It is obvious that such large a deviation should not be due to the experimental errors but should be attributed to the flaw of the theory.Surveying the results of emulsion polymerization and using the vector analysis and approximation theory, we have derived a kinetic formula for emulsion polymerization as follows:

$$-\frac{dC}{dt} = KC^{\alpha(D)}$$

$$(k=\theta C_o, \quad \alpha (D)=\log \frac{D(1-\alpha D)}{(1-D)C_o})$$

Tests have proved that errors are less than 1.6%.

As is well known, when using conventional kinetic methods to treat emulsion polymerization reaction, one may encounter many difficulties. Therefore, we sketch out a model in Euclidean space according to the point-spread rule of experimental data, and to reduce the inner product -space of the sum of the squared residuals to a minimum. Set column vectors $\vec{X}$, $\vec{Y}$, $\vec{6}$ and $\hat{E}$ (Assuming that $\dot{X}$ has no significant relationship with E) respectively as follows:

$$\vec{X} = \begin{bmatrix} x_1 \\ x_2 \\ \vdots \\ x_r \end{bmatrix} \qquad \vec{Y} = \begin{bmatrix} y_1 \\ y_2 \\ \vdots \\ y_n \end{bmatrix} \qquad \vec{E} = \begin{bmatrix} 1 \\ 1 \\ \vdots \\ 1 \end{bmatrix} \qquad \vec{6} = \begin{bmatrix} 6_1 \\ 6_2 \\ \vdots \\ 6_n \end{bmatrix}$$

where $X_i = \exp(-\theta t_i)$ and $Y_i =$ observation

They are the subset Span=${X,E}$ of the n-dimensional Euiclidean space. We choose parameters a, b and θ (where a, b are determined parameters and θ is a pseudo-rate constant) and make the module

$$\vec{6} = \vec{Y} -a\vec{E} -b\vec{X}$$

Then, we obtain easily

$$\vec{6} =(\vec{Y}-a\vec{E}-b\vec{X})' \ (\vec{Y}-a\vec{E}-b\vec{X})$$

If

$$\begin{cases} \frac{\partial}{\partial a}(\|\vec{6}\|)=0 \\ \frac{\partial}{\partial b}(\|\vec{6}\|)=0 \end{cases}$$

then, we solve equations above to obtain the values of a and b.

EXPERIMENTAL

1. The emulsion polymerization of styrene
(1) Materials; Styrene, the monomer, was washed with a solution of 10% NaOH to remove the inhibitor (tert-butyl catechol). The excess of NaOH was removed by washing with distilled water, and then distilled at reduced pressure at 40°C. The initiator, $K_2S_2O_8$, is analytical pure reagent; $CH_3(CH_2)_{11} SO_3Na$-emulsifier; H O-reaction medium.

TABLE 1
Experimental Sets

Wt% / Components	expNo. 1	2	3
Styrene	40	30	30
$K_2S_2O_8$	0.24	0.24	1.00
$CH_3(CH_2)_{11}$ --SO_5Na	0.67	0.67	1.00
H_2O	100	100	0.67

(2) Procedure. In a 1000ml glass reactor, styrene is added after other

ingredients (except the monomer) have been mixed thoroughly. Then N is bubbled through to removed air and disperse the monomer. The polymerization reaction is carried out at 65℃. Samples are with-drawn at intervals and the emulsion is broken down by alum. After separating the products, the conversion of reaction is calculated.
(3) Results: The experimental D vs. t curve (Fig. 1) is shown.

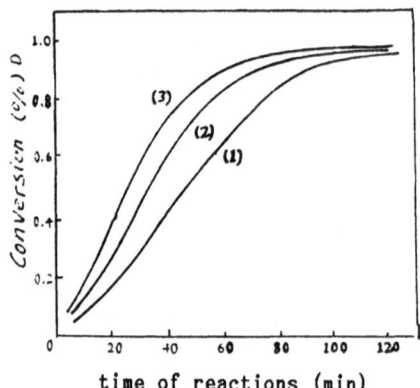
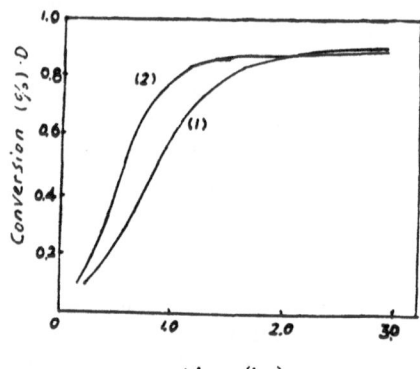

time of reactions (min)　　　　　　　time (hr)
Fig.1 The curve of D-t for styrene　　Fig.2 The curve of D-t for D₄
2. The emulsion polymerization of octamethylcyclotetrasiloxane(D)

TABLE 2.
Experimental Sets

	D₄	MOH	Dodecyl-benzyl-halideamide	H_2O
1	30	0.5(NaOH)	0.8(dodecyl-benzyl-chloromide)	100
2	30	0.5 (KOH)	1.0(dodecyl-benzyl-bromineamide	100

(1) Producedure: The reactions are carried out at 85℃. Reaction conversions are obtained after breaking down the emulsion with ether.
(2) Results: The experimental D-t curve (Fig.2) is shown.
3. The emulsion copolymerization for butadiene-styrene

TABLE 3
Experimental Sets

Components ＼ Wt% ＼ Order	No. 2	SBR₁₀₀₀
Butadiene	71	75
Styrene	29	25
H O	180	180
$CH_3(CH_2)_{11} SO_3Na$	4.5	4.5
$K_2S_2O_8$	0.3	0.3

(1) Procedure: Refer to reference[7].
(2) Results: The experimental D-t cuve (Fig.3) is shown.

103

4. Soapless emulsion polymerization of methyl methacrylate in the presence of calcium sulfite.
(1) Procedure: Refer to the reference[9].
(2) Results: The experimental D-t curve (Fig.4) is shown, as follow:

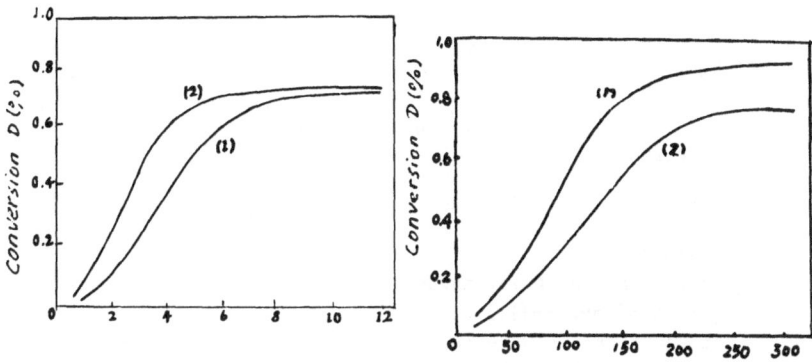

time (hr)

Fig.3 The curve of D-t for the   Fig.4 The curve of D-t for the
butadiene-styrene   methylmethacrylate

DERIVATION OF FORMULA

Based on Weierstrass Theorem, we sketch a functional model having the significance of Tchebcheff-theory, e.i. $y=1/(a+be^{-\theta x})$, to substitute the Bernstein polynomial in the Euclidean space through the analysis of the "S"-shaped curve.

Linearizing the vectors, we have
$$\overrightarrow{Y_i} = 1/y_i = 1/D_i$$
$$\overrightarrow{X} = e^{-\theta x_i} = e^{-\theta t_i}$$

Relating the factors of kinetics, we have

$$\begin{bmatrix} t_1 & & & \\ & t_2 & 0 & \\ & 0 & t_3 & \\ & & & \ddots \\ & & & t_n \end{bmatrix} \begin{bmatrix} \theta_1 \\ \theta_2 \\ \vdots \\ \theta_n \end{bmatrix} = (-1) \begin{bmatrix} \ln X_1 \\ \ln X_2 \\ \vdots \\ \ln X_n \end{bmatrix}$$

Determination of average value of θ, as follows:

$$\theta = \frac{1}{k-1}\Sigma \theta_{ji} = \frac{1}{x \ \Sigma (j-i)} \ln \left\{ \prod_{0}^{k^*} \frac{y_i \ (1-y_i \ k^* / \Sigma^{k^*} y_{(f-k^*)})}{y_i \ (1-y_j \ k^* / \Sigma_{5} y_{(f-k^*)})} \right\} \tag{1}$$

where k^*---experimental point number.

    f ---same above (generally speaking $f > k^*$)

The detailed step on the average value of θ is in Ref.[10] The values of a and b can be obtained by regression.

From $D=1/(a+be^{-\theta t})$, we have

$$-\frac{dD}{dt} = \frac{\theta \ be^{-\theta t}}{(a+be^{-\theta t})^2} \tag{2}$$

$$- \frac{dD}{dt} = \theta\, D(1-aD) \qquad (3)$$

For $C=(1-D)C_0$, and if $[(1-D)C_0] \equiv C^{\alpha(D)} = D(1-aD)$ then, we obtain

$$- \frac{dC}{dt} = KC^{\alpha} \qquad (4)$$

where $k = \theta\, C_0$, $\alpha(D) = \log \dfrac{D(1-aD)}{(1-D)C_0}$

$$\alpha(D) = \frac{1}{A} \left\{ \frac{\sum\limits_{n=1}^{k^{\ast}} \frac{1}{n} [(-1)^{n-1}(D-1)^{n} - (aD)^{n}]}{1 - \frac{1}{A}\sum\limits_{n=1}^{k^{\ast}} \frac{1}{n} D^{n}} \right\}$$

$(A = \ln C_0)$

where   k --experimental point number

n --natural number$(1, 2, \ldots, k^{\ast})$

k--complex rate constant, $\alpha(D)$ --pseudo-order of reaction, a--reciprocal of maximum conversion. It can be seen that the exact values of these parameters can be determined.

EXAMPLES

Using the data of D-t, from the Eq.(1), we choose f=13(the last term) k=0,1,2,3, i-j are 6-7,6-8,7-8,7-9 respectively.

Substitute the known data into Eq.(1), we have

$$\theta = 0.06627$$

Hence we obtain

$$a = 1.048$$
$$b = 27.03$$

Substitute the values of a, b, θ into Eq.(1), and we have

$$D_1 = \frac{1}{1.048 + 27.03 e^{-0.06627\,t}}$$

TABLE 4
Examination

exp No.	1	2	3	4	5	6	7
D exp	0.06636	0.1202	0.2075	0.3331	0.4854	0.6360	0.7582
D calcu	0.06675	0.1215	0.2105	0.3383	0.4922	0.6421	0.7637
$\triangle$D/D exp(%)	0.5907	1.081	1.446	1.558	1.401	0.9591	0.7254

Examination

exp No.	8	9	10	11	12	13
D exp	0.8420	0.8933	0.9383	0.9500	0.9553	0.9559
D calcu	0.8455	0.8949	0.9377	0.9497	0.9536	0.9542
$\triangle$D/D exp(%)	0.4157	0.1791	0.064	0.032	0.178	0.183

It can be seen that there is a good coincidence between the modle and the experimental results. Similarly, with other experimental data we can obtain:

$$D_2 = \frac{1}{1.043+25.33e^{-0.6923\tau}} \tag{6}$$

$$D_3 = \frac{1}{1.042+23.77e^{-0.7551t}} \tag{7}$$

$$D_1' = \frac{1}{1.148+17.37e^{-3.+90\tau}} \tag{8}$$

$$D_2' = \frac{1}{1.154+19.68e^{-5.200t}} \tag{9}$$

$$D_1'' = \frac{1}{1.336+32.05e^{-0.8500t}} \tag{10}$$

$$D_2'' = \frac{1}{1.334+27.68e^{-1.200t}} \tag{11}$$

$$D_1''' = \frac{1}{1.098+12.46e^{-0.0350\tau}} \tag{12}$$

$$D_2''' = \frac{1}{1.266+28.51e^{-0.02650t}} \tag{13}$$

Examination proved that the above results were satisfactory, too. For the sake of brevity we have omitted the process of data treatment to Eq. (6)-(13). Knowing the values of a, b and θ, we can obtain the kinetic formula in the form of Eq.(3) and (4),and the kinetic parameters such as complex rate constant and pseudo-order of reactions can be determined.

Acknowledgement: The author is indebted to Professor Tang Aoqing of the Jilin university who has shown warm interest in this work.

ACKNOWLEDGEMENT

I wish to thank Prof. Robert W.Lenz University of Massachusetts for helpful discussion on this work.

REFERENCES

1. Smith,W.V. and Ewart,R.H., J.Chem.Phys., 1948, 16, 592.
2. Bataille,P., Van,B.T. and Pham,Q.B., J.Polym. Sci.Polym. Chem.Ed, 1982, 20, 795.
3. Stockmayer,W.H., J.Polym.Sci., 1957, 24, 314.
4. O'Yoole,J.T., J.Appl.Polym.Sci., 1965, 9, 1291.
5. Timan,A.F.: Theory of Approximation of Function of a Real Variable, New York, Macmillan 1963.
6. Mian Chang, KEXUE TONGBAO(Eng. Ed.), 1984, 29, 10, 1417.
7. Feng Xinde "macromolecular chemistry" Science Press(Beijing, China), 1981, 45.
8. Zhang Xinhua etc. J.Polymer Communication,Science Press(2), 1982,154
9. Arai,M., Arai,K. and Saito,S., J.Polym. Sci.Polym.Chem. Ed., 20, 1982, 1021.
10. Mian Chang, KEXUE TONGBAO(Eng. Ed.), 1985, 30,7, 894-899.

POLYSILOXANES WITH PENDANT CARBAZOLE GROUPS.
PREPARATION AND PHOTOCONDUCTIVITY

P. Strohriegl, H. Domes and D. Haarer

Lehrstuhl für Experimentalphysik IV and Bayreuther Institut

für Makromolekülforschung (BIMF), Universität Bayreuth,

Postfach 101251, D-8580 Bayreuth, FRG

ABSTRACT

The synthesis and the photoconductive properties of four novel polysiloxanes are reported. In the polymers 3a-d carbazole groups are attached
to the polysiloxane backbone with flexible spacers of various length. The
polymers were prepared from the carbazole derivatives 1a-d and poly(hydrogenmethylsiloxane) 2 in a polymer analogous reaction; they were charate-
rized by DSC and various spectroscopic methods.
Time-of-flight measurements show that polysiloxane 3a has an effective
charge carrier mobility of 9.8×10^{-7} cm^2/Vs which is roughly the same as
has been found for the well known polymeric photoconductor poly(N-vinylcarbazole) (PVK). The absolute values of the transient photocurrent are
somewhat smaller for polymer 3a. This is a consequence of the different
charge carrier quantum yield in the two materials. Polymer 3a and PVK were
also investigated by the xerographic discharge method, which yields similar
values of the dark decay and the photo decay of both polymers.

INTRODUCTION

Polymers with pendant carbazole groups like poly(N-vinylcarbazole)
(PVK) are of interest for basic investigations as well as for technical
applications. Since 1957 when H. Hoegl described the photoconductive
properties of PVK for the first time [1][2], there has been increasing
interest in the preparation of polymers with similar properties for applications such as photocopiers, laser printers and electrophotographic
printing plates [3]. Recently polymers with pendant carbazole groups have
also been used in the preparation of conductive polymers [4][5], and in

fundamental studies of electron donor–acceptor complexes and polymer miscibility [6].

Although a lot of photoconductive polymers have been synthesized and investigated during the past 30 years [7][8], there is yet no clear picture as to the correlation between the chemical structure of a polymer and its photoconductive properties. Following our interest in the structure – property relationship of organic photoconductors, we decided to prepare a series of novel polysiloxanes [9] and to investigate their photoconductive properties. In these polymers the carbazole groups are separated from the polysiloxane backbone by alkylene spacers of various length.

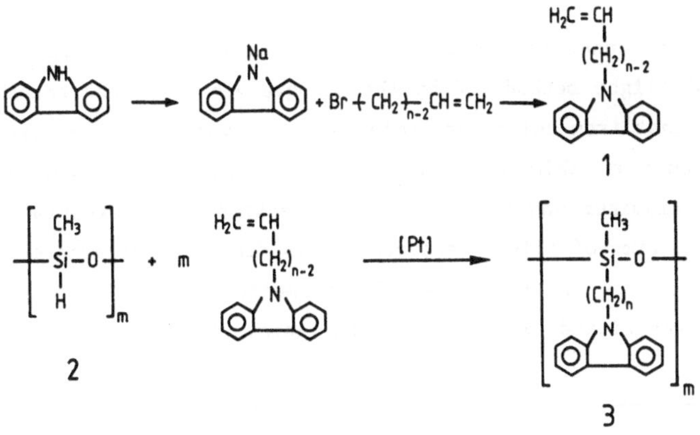

Scheme 1

n	3	5	6	11
	1a	1b	1c	1d
	3a	3b	3c	3d

RESULTS AND DISCUSSION

The preparation of the novel polysiloxanes is outlined in scheme 1. Compounds 1a–d were prepared from 9-carbazolyl sodium and the corresponding ω–alkenyl bromides [10]. In a second step 1a–d were reacted with poly (hydrogenmethylsiloxane) 2 to yield the polysiloxanes 3a–d. The polymer analogous hydrosilylation reaction yields polymers with a high degree of substitution. More than 95 % substitution was calculated by evaluating the nitrogen content of the polymers. A detailed analysis of the ^{1}H-NMR spectra also leads to a degree of substitution of more than 95 % for all the polysiloxanes. IR- spectroscopy shows that there are no residual Si-H groups in the products. Further evidence is provided by the fact that the polymers are not crosslinked after treatment with water. The thermal properties of the novel polysiloxanes were investigated by DSC. Polymer 3a is partially crystalline and has a melting point of 66 °C and a glass transition temperature of 51 °C. The glass transition temperatures decrease with increasing spacer length from 7 °C (polymer 3b) to -5 °C (3c) and -45 °C (3d). All the polysiloxanes are soluble in THF, toluene and chloroform.

The photoconductive properties of the polymers were investigated by the time-of-flight method and by the xerographic discharge technique. Thin films were cast from chloroform solutions onto poly(ethylene terephthalate) foils coated with thin aluminum layer. A thin, semitransparent layer of evaporated aluminum was used as second electrode. The experimental set-up used for the time-of-flight measurements is described elsewhere [11]. Fig. 1 shows the results of the time-of-flight measurements. The bend in log I-log t plot was fitted with two straight lines. Their intersection point

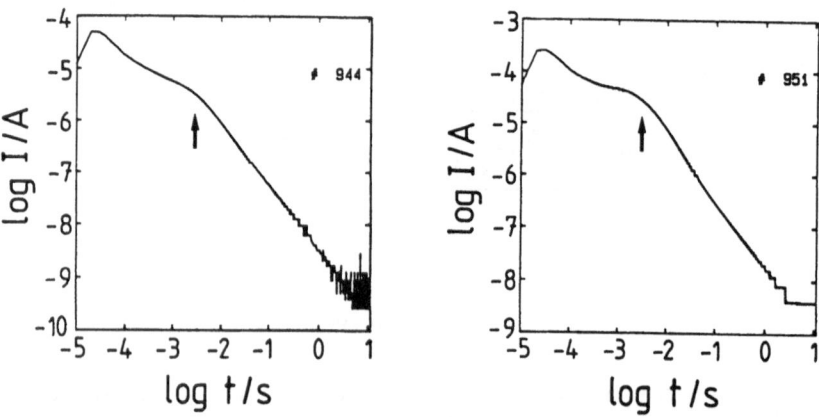

Fig. 1: Typical time-of-flight transients for polysiloxane 3a (left) and PVK (right) plotted on a log-log scale. The arrows indicate the transit times when 300 V are applied across the 10 micron thick samples.

marks the transit time t_T from which the effective carrier mobility μ_{eff} was calculated. We found a carrier mobility of 9.8×10^{-7} cm^2/Vs for polysiloxane 3a whereas PVK has a mobility of 9.0×10^{-7} cm^2/Vs. The absolute value of the transient photocurrent is somewhat smaller for the polysiloxane 3a than for PVK. This is a consequence of the different charge carrier quantum yield in the two materials.

The results of the xerographic discharge measurements are shown in Fig. 3. The films are charged in a corona during the first 20 seconds. In the time interval from 20 to 40 seconds the films are kept in the dark. A dark decay of 22 % for polymer 3a and 31 % for PVK was found. After 40 seconds the films are illuminated with a Xenon arc lamp for 1 second and a photodecay of 81 % for polymer 3a and 92 % for PVK was found. The shape of the curves remains unchanged after 100 cycles of charging and discharging for both polymers. We therefore conclude that there is no tendency to accumulate charges in these materials.

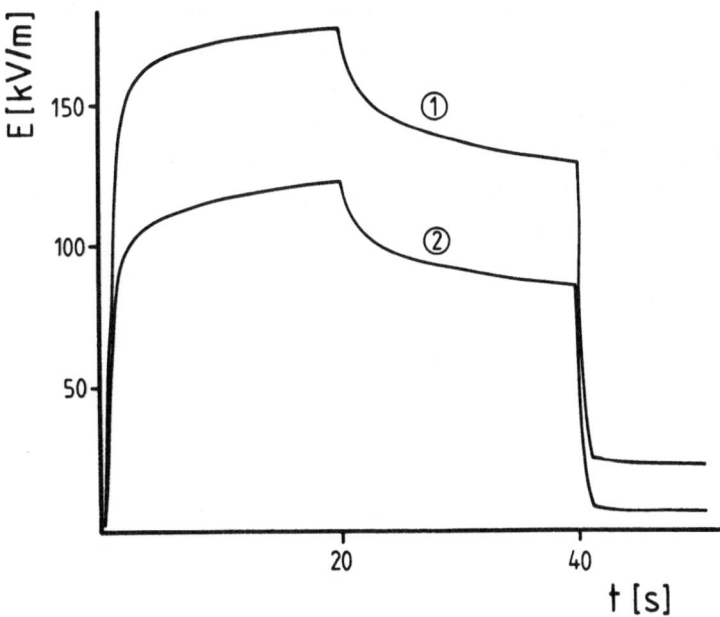

Fig. 2: Xerografic discharge measurement of polysiloxane 3a (1) and PVK (2). Both samples were sensitized for visible light with a dye mixture.

ACKNOWLEDGEMENT

We would like to thank Dr. Distler, Dr. Jäckel and Dr. Leyrer (BASF AG, Ludwigshafen) for their help in preparing the xerographic measurements and thin films and the Wacker Chemie, Burghausen for delivering poly(hydrogenmethylsiloxane) and the catalyst. The technical assistance of W. Joy is gratefully acknowledged.

This work was supported by the Deutsche Forschungsgemeinschaft as part of SFB 213.

REFERENCES

1. Ger. 1068115 (1957), Kalle AG, invs.: H. Hoegl, O. Süs, W. Neugebauer, Chem.Abstr. 55, 20742a (1961)

2. H. Hoegl, J.Phys.Chem. 69, 755 (1965)

3. D.M. Burland, L.B. Schein, Physics Today, May 1986, 46

4. H. Kanega, Y. Shirota, H. Mikawa, J.Chem.Soc., Chem.Commun. 1984, 158

5. B.R. Hsieh, M.H. Litt, Macromolecules 19, 516 (1986)

6. J.M. Rodrigues-Parada, V. Percec, Macromolecules 19, 55 (1986)

9. P. Strohriegl, Makromol.Chem., Rapid Commun. 7, 771 (1986)

10. J. Heller, W.A. Hewett, Makromol.Chem. 73, 48 (1964)

11. E. Müller-Horsche, D. Haarer, H. Scher, Phys.Rev. B 35, 1273 (1987)

^{13}C-NMR STUDY OF THE MICROSTRUCTURE OF ETHYLENE-VINYL ALCOHOL COPOLYMERS

H. Ketels
Laboratory of Polymer Technology
Eindhoven University of Technology
P.O. Box 513, 5600 MB   Eindhoven
The Netherlands

and

G. v.d. Velden
DSM-Research BV
P.O. Box 18, 6160 MD   Geleen
The Netherlands

ABSTRACT

High-resolution NMR methods are extremely useful for microstructural analysis of ethylene copolymers. In the present study ethylene-vinyl alcohol (E-VOH) copolymers with an increasing content of ethylene were synthesized and the microstructure was investigated using ^{13}C NMR. The methylene sequence distributions in these copolymers were investigated. The occurence of anomalous linkages in E-VOH copolymers i.e. 1,2- and 1,4-diol structures were analyzed. The presence of 1,4-diol structures could readily be detected in all E-VOH copolymers. The 1,2-diol structure could be detected in E-VOH copolymers with a low content of ethylene.

INTRODUCTION

Both ^{1}H and ^{13}C NMR have been applied to study the microstructure of ethylene-vinyl acetate (E-VA) [1,2] and ethylene-vinyl alcohol (E-VOH) copolymers [3,4]. The sequence distribution has been determined, mainly by ^{13}C NMR [2,4]. Tacticity or configurational sequence placements have

only been analysed in detail using ^{1}H NMR of E-VOH copolymers [3]. In
this study the merits of ^{13}C NMR applied to a series of E-VOH copolymers
have been exploited, emphasising compositional sequences (methylene
sequence distribution), and the occurence of anomalous linkages.
Anomalous linkages, i.e. head-to-head, tail-to-tail and tail-to-head can
occur besides the normal head-to-tail linkage. In previous studies, it
was possible to detect specific resonances of 1,2-diol and 1,4-diol in
poly(vinyl alcohol-co-crotonic acid) and poly(vinyl alcohol) (PVOH) using
ultra-high field NMR, respectively 125 [5] and 100 MHz [6] ^{13}C NMR.
1,4-Diol structures have been detected in E-VOH copolymers (not in E-VA
copolymers) using ^{13}C NMR [4,7]. In spite of several attempts, resonances
due to 1,2-diol structures have not been found [4,7]. A tentative
explanation could be that E-VOH copolymers do not contain a significant
amount of 1,2-diol structures (relative to PVOH). Moreover, the 1,2-diol
content should decrease with increasing ethylene content. Probably E-VOH
copolymers, analyzed in previous studies ($\leq$ 30 mole% ethylene), contained
at the detection limit of a 22.5 MHz ^{13}C NMR spectrofotometer, a too
small percentage of 1,2-diol structures. In this study 50 MHz ^{13}C NMR has
been used to study the anomalous linkages in E-VOH copolymers.

MATERIALS AND METHODS

Four samples of E-VOH copolymers were prepared by hydrolysis of the
corresponding E-VA copolymers. The E-VA copolymers were synthesized in a
homogeneous system, with tertiair-butyl-alcohol (TBA) as solvent and
α,α'-azobis (isobutyronitril) as initiator. The E-VA copolymers were
first hydrolyzed in methanol using sodiumhydroxide and subsequently in
methanol/water. PVAL (Mowiol 66-100) was obtained from a commercial
source (Hoechst), as was the case for the E-VOH copolymer E (molar
ethylene content = 32%) obtained from Kuraray (EPT). 50 MHz ^{13}C NMR
spectra were obtained on a Varian XL-200 spectrometer.
Conditions: lock-solvent mixture phenol-D_2O (80-20, weight ratio),
temperature = 40°C.

RESULTS AND DISCUSSION

1. Composition

The copolymer composition of E-VOH copolymers has been determined via [1]H
NMR [3] and crosschecked via ^{13}C NMR [4] using standard methods.

2. Sequence Analysis

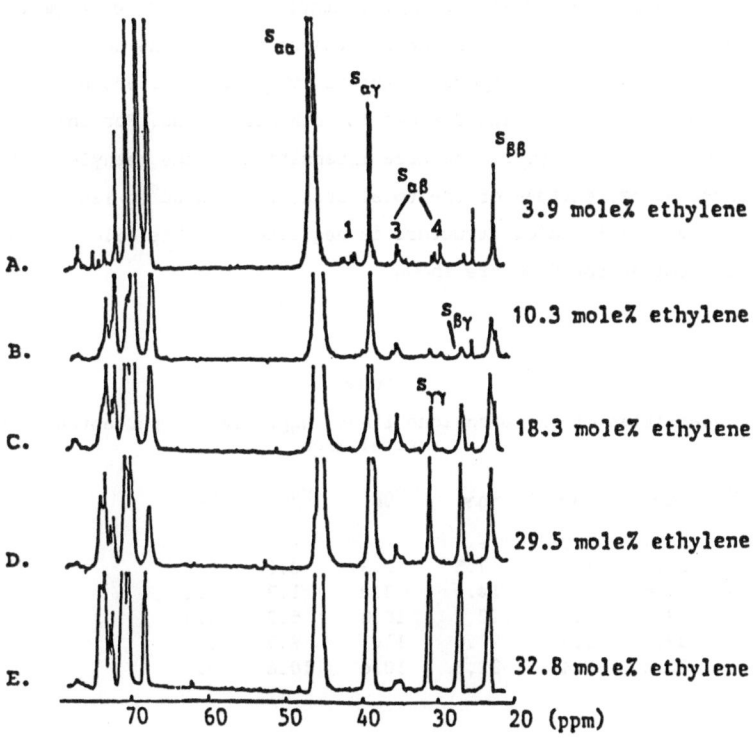

Figure 1. ^{13}C NMR spectra of the E-VOH copolymers

Figure 1 shows the ^{13}C NMR spectra of a series of E-VOH copolymers.
The spectra can be separated in two spectral regions. The low field
region of the spectra shows five to six methine carbon resonances which
can be assigned to the six VOH-centered triad sequences: E-VOH-E,
mE-VOH-VOH, mmVOH-VOH-VOH, rE-VOH-VOH, mr- and rrVOH-VOH-VOH [4]. E
and VOH are ethylene and vinylalcohol units while m and r denote the
meso and racemic configurations. In the high field region of the spectra
at least six well resolved resonance patterns for the methylene carbons
are observed. The assignment of the methylene resonances in the E-VOH
copolymers has been performed in an analogous manner as described in
[4] but we have adopted the same nomenclature for E-VOH copolymers as
suggested by Carman and Wilkes [8] for ethylene-propylene copolymers.
Each of the methylene, i.e. secondary (S) carbons in the sequences can
be identified by a pair of Greek letters. The Greek letters denote the
nearest methine neigbouring C atoms in both directions. For the
different E-VOH copolymers the area intensities of the methylene lines,
expressed as percentages of the total area, are listed in Table 1. The
presence of the 1,2-diol structure is neglected in this calculation.
Also the values for PVOH are shown.

TABLE 1

Area intensities of the methylene lines, expressed as percentages of the
total area

	$S_{\alpha\alpha}$	$S_{\alpha\beta}$	$S_{\alpha\gamma}$	$S_{\beta\beta}$	$S_{\beta\gamma}$	$S_{\gamma\gamma}$
PVOH	98.7	1.3	–	–	–	--
A	85.6	1.9	7.5	3.3	0.6	1.1
B	71.8	3.6	14.9	7.2	1.7	1.1
C	53.1	4.7	22.2	10.8	5.2	4.0
D	37.0	3.7	29.6	11.1	9.7	8.8
E	35.6	2.4	31.3	10.6	10.6	9.6

With the results as shown in Table 1 it is possible to calculate the
number average methylene sequence length distribution i.e. n_1, n_2,
n_3 and n_{4+}. In Figure 2 the methylene sequence distribution has been
plotted for all copolymers as a function of the methylene sequence
length. Obvious conclusions are: approximately constant level of n_2,
and the increase of long methylene sequences with increasing ethylene
content.

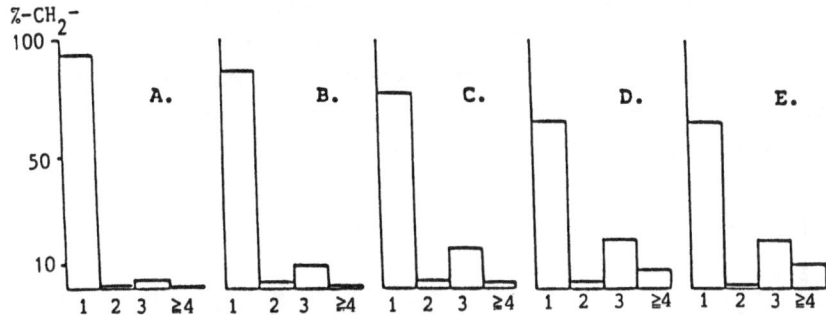

Figure 2. -CH$_2$- sequence distribution of the E-VOH copolymers

3. Defectstructures

As was pointed out in the introduction, anomalous linkages can occur in PVOH and E-VOH copolymers, the so-called 1,2- and 1,4-diol structures. The calculated ^{13}C chemical shifts of the methylene carbons in the diol structures are listed in Table 2.

TABLE 2

Calculated ^{13}C chemical shifts of the methylene carbons in the diol structures

	fragment	type	δ (ppm)	
I	a b c C-C-C-C-C-C-C-C-C-C OH OH OH OH OH	1,2+1,4	42.2 30.5 34.2	a b c
II	a a b C-C-C-C-C-C-C-C-C OH OH OH OH OH	1,2	42.2 47.0	a b
III	b a a b C-C-C-C-C-C-C-C-C-C OH OH OH OH	1,4	34.2 47.0	a b

The 1,2- and 1,4-diol structures in PVOH can easily be detected with ^{13}C NMR, as is evident in Figure 3, where a 50 MHz ^{13}C NMR spectrum of PVOH (Mowiol) is represented. The resonances 1, 3 and 4, which can be attributed to a fragment of coupled 1,2- and 1,4-diol structures (a, b, c of fragment I in Table 2), are observed.

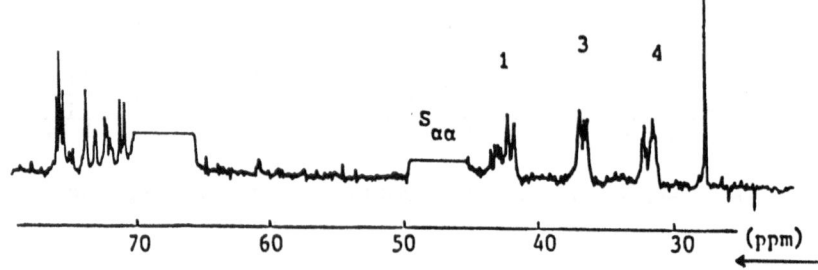

Figure 3. ^{13}C NMR spectrum of PVOH (Mowiol)

In the ^{13}C NMR spectra of all E-VOH copolymers, evidence for the presence of the 1,4-diol structure can easily be derived from the presence of a resonance at 34.2 ppm. The contents of the 1,4-diol structure can be calculated using the formula for copolymers containing only 1,4 diol:

$$\text{mole\% 1,4-diol} = \frac{I_{S\alpha\beta}/2}{I_t}$$

Where $I_{S\alpha\beta}$ is the area intensity of the $S_{\alpha\beta}$ line at 34.5 ppm and I_t is the total area intensity of the methine and methylene lines. The factor 2 has to be left out for copolymers containing 1,2- and 1,4-diol structures. The calculated contents of 1,4- and 1,2-diol structures are listed in Table 3.

TABLE 3
Diol content (mole%) of PVOH and E-VOH copolymers

	PVOH	A	B	C	D	E
1,4-diol	1.2	1.0	1.9	1.1	0.9	0.8
1,2-diol	1.2	1.0	±0.5	–	–	–

As can be concluded from Table 3, the content of the 1,4-diol structure in PVOH and E-VOH copolymers are in the same order. Coupled 1,2- and 1,4-units (fragment I, Table 2) have been observed in the ^{13}C NMR spectrum of polymers A and B. Isolated 1,4-units are observed for the polymers C, D and E. No evidence for the presence of isolated 1,2-diol structures has been found for all polymers (absence of a resonance at 42.2 ppm). However in the polymers A and B where coupled 1,2 and 1,4-units are present, the presence of an additional isolated 1,2-unit can not be excluded (overlapping ^{13}C NMR resonances). The presence of isolated 1,4-diol structures has to be attributed to the insertion of ethylene fragments (at least for the polymers C, D and E). For the polymers where adjointed 1,2- and 1,4-diol structures are observed (A,B) both possibilities have to be considered (experiments with ^{13}C enriched ethylene would in principle unravel this phenomenon). Thus it is possible to detect the 1,2-diol structure in E-VOH copolymers with an ethylene content to approximately 10 mole%.

ACKNOWLEDGEMENT

The authors express their gratitude to Mr. J. Beulen for recording the ^{13}C NMR spectra.

REFERENCES

1. Wu, T.K., J. Polym. Sci., Part A-2, 1970, 8, 167.
2. Wu, T.K., Ovenall, D.W., Reddy, G.S., J. Polym. Sci., Polym. Phys. Ed., 1974, 12, 901.
3. Wu, T.K., J. Polym. Sci., Polym. Phys. Ed., 1976, 14, 343.
4. Moritani, T., Iwasaki, H., Macromolecules, 1978, 11, 1251.
5. Amiya, S., Uetsuki, M., Macromolecules, 1982, 15, 166.
6. Ovenall, D.W., Macromolecules, 1984, 17, 1458.
7. Amiya, S., Iwasaki, H., Fujiwara, Y., Nippon Kagaku Kaishi (J. Chem. Soc. Japan), 1977, 11, 1698.
8. Carman, C.J., Wilkes, C.E., Rubb. Chem. Tech., 1971, 44, 781.

Part 2

CHAIN-DYNAMICS/CONFORMATION

MONTE CARLO STUDY OF CYCLIC POLYMERS

G. ten Brinke
Department of Polymer Chemistry,
University of Groningen,
Nijenborgh 16, 9747 AG Groningen,
The Netherlands

ABSTRACT

The results of Monte Carlo calculations of closed random walks are discussed. A comparison is made between the random walk model and the freely jointed model with a Lennard-Jones potential. Numerical evidence for a lowering of the intramolecular θ-temperature of cyclic polymers compared to linear polymers is given.

INTRODUCTION

Topological constraints can strongly affect static and dynamic properties of macromolecules. Well known is the additional contribution of trapped entanglements to the elasticity of rubbers. Even if the constraints are not really permanent they may act as such. An example is given by tight-knots which can be introduced by crystallization at sufficient undercooling or in elongational flow fields, provided loose knots are already present [1]. Memory effects in polymer melts and defects in polymer fibres can be attributed to it. Here, we will concentrate on the most simple case of self-knots. Furthermore, in order to make the discussion as rigorous as possible, only cyclic polymers will be considered. In that case knots are mathematically well defined entities.

RANDOM WALK MODEL

In a recent paper [2] we presented the results of a Monte Carlo simulation of closed random walks on a body centered cubic lattice. The probability for the occurrence of a knot and the influence of knots on the coil dimension as well as on the particle scattering function were determined. The presence of a knot was established using the Alexander polynomial and the connection with real polymers was based on the equivalent chain concept. Applied to ring polystyrene the data indicated that for a molecular weight in the range of 10^5 approximately 30% of all rings will be knotted provided the linking of the ends takes place under θ-conditions for the linear precursor.

Experimentally, scattering experiments can be used to verify, at least semi-quantitatively, these predictions. To show this we observe first that the reaction product will contain both linear and cyclic molecules. These can be separated by fractional precipitation or by size exclusion chromatography. The first method is based on the experimental fact that the θ-temperature for ring polymers is several degrees lower than for the corresponding linear polymers [3]. With the help of size exclusion chromatography it is possible to select the most compact coils, which in this case corresponds not only to ring molecules but even to mainly knot containing ring molecules. Scattering experiments performed on a dilute solution of these molecules should be compared with similar experiments on a dilute solution of mostly unknotted ring molecules.

A sample of this kind will be obtained if ring closure takes place under good solvent conditions. Our calculations [2] show that in a Kratky plot the particle scattering function for knot containing molecules will have a more pronounced maximum due to a more compact spherical-like structure.

The closed walks obtained by the Monte Carlo simulation as a subset of all possible random walks satisfy the random walk statistics. Apparently this is also the case for the much smaller subset of all closed unknotted walks [4]. On the other hand, we know from experiments [3] that the inter-molecular θ-temperature, defined as the temperature at which the second virial coefficient $A_2 = 0$, is several degrees lower for ring molecules than for linear molecules. A similar result [5] is found for the intramolecular θ-temperature, defined as the temperature for which the mean square radius of gyration is proportional to the number of monomers. This indicates that the random walk model is not the most appropriate model for the kind of problems discussed here. A in some respect related observation was made by Olaj and Pelinka [6]. They showed that, contrary to the original Flory approach [7], the second virial coefficient of linear molecules becomes zero at the θ-temperature due to a compensation between negative and positive contributions to the binary cluster integral defining the excluded volume. However, just as in the case of a real gas, these contributions correspond to different intermolecular distances. As a consequence, even at the θ-temperature, strong overlap of polymer molecules in dilute solutions is excluded. Experimentally, this result was confirmed recently by Chang and Morawetz [8]. It is due to the finite thickness of the molecules. Numerically this can be taken into account by using a freely jointed model with a van der Waals type interaction instead of a random walk model.

FREELY JOINTED MODEL

The model consists of N beads connected in the case of ring molecules by N bonds of length 1. The angles between neighbouring bonds are unrestricted. The interaction between the beads is described by a Lennard-Jones potential

$$U_{ij} = \varepsilon[(1/r_{ij})^{12} - 2(1/r_{ij})^6]$$

where r_{ij} is the distance between bead i and j and ε the interaction parameter. Starting from an arbitrary chosen initial conformation new conformations are created by selecting a bead k at random followed by a rotation over an arbitrary angle φ around the axis connecting beads k-1 and k+1, as indicated in Figure 1.

The energy difference between the two conformations determines whether the new conformation is accepted (Metropolis scheme). Baumgärtner [9] showed that for linear molecules the θ-temperature of this model corresponds to $k\theta/\varepsilon = 3.70 \pm 0.01$.

Figure 2a presents preliminary results for the mean square radius of gyration R_g^2 of ring molecules for different values of kT/ε. It shows that the θ-temperature is somewhat lower than for linear molecules. This is demonstrated more clearly by a log-log plot of R_g^2 versus the number of monomers, N for $kT/\varepsilon = 3.7$ given in Figure 2b. The slope is 1.169 ± 0.01 in good agreement with the theoretical value of 1.176 ± 0.002 for linear self-avoiding walks [10], but slightly smaller than most values found in the literature for closed self-avoiding walks [11-16].

This result suggests that the excluded volume effect is still important if ring closure takes place under θ-conditions for the linear precursor. As a consequence, less knotted molecules may be present than predicted by the random walk calculations. However, the data should be handled with great care, since all ring molecules investigated so far were unknotted. This occurred because the initial conformations were unknotted, and the conformation generating procedure preserves, at least in practice, the topological state [17]. Further calculations using a different but related generating procedure, with an enhanced probability of passing from one topological state to another, are in progress [17].

REFERENCES

1. de Gennes, P.G., Macromolecules, 1984, 17, 703.

2. ten Brinke, G. and Hadziioannou, G., Macromolecules, 1987, 20, 480.

3. Roovers, J. and Toporowski, P.M., Macromolecules, 1983, 16, 843.

4. des Cloizeaux, J. and Metha, M.L., J. Phys. (Paris), 1979, 40, 665.

5. Hadziioannou, G., Cotts, P.M., ten Brinke, G., Han, C.C., Lutz, P., Strazielle, C., Rempp, P. and Kovacs, A.J., Macromolecuels, 1987, 20, 493.

6. Olaj, O.F. and Pelinka, K.H., Makromol. Chem., 1976, 177, 3413.

7. Flory, P.J. and Krigbaum, W.R., J. Chem. Phys., 1950, 18, 1086.

8. Chang, L.P. and Morawetz, H., Macromolecules, 1987, 20, 428.

9. Baumgärtner, A., J. Chem. Phys., 1980, 72, 871.

10. le Guillon, J.C. and Zinn-Justin, J., Phys. Rev. Lett., 1977, 39, 95.

11. Baumgärtner, A., J. Chem. Phys., 1982, 76, 4275.

12. Kumbar, M. and Windwer, S., J. Chem. Phys., 1968, 49, 4057.

13. Naghizedah, J. and Sotobayashi, H., J. Chem. Phys., 1974, 60, 3104.

14. Bruns, W. and Naghizedah, J., J. Chem. Phys., 1976, 65, 747.

15. Chen, Y., J. Chem. Phys., 1981, 74, 2034.

16. Bishop, M. and Michels, J.P.J., J. Chem. Phys., 1985, 82, 1059.

17. ten Brinke, G. and Michels, J.P.J., in progress.

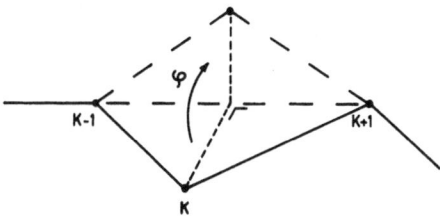

Figure 1. Creation of a new conformation by rotating k over an
arbitrary angle φ around the axis connecting beads k-1
and k+1.

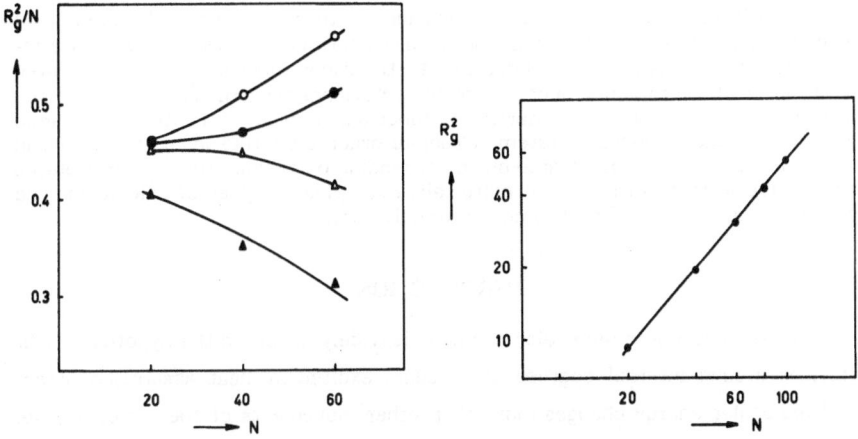

Figure 2a. Chain length dependence Figure 2b. Log-log plot of chain
and temperature dependence
of mean square radius of
gyration R_g^2: $kT/\varepsilon=2.0$ (▲),
$kT/\varepsilon=3.0$ (△), $kT/\varepsilon=3.7$ (●)
and $kT/\varepsilon=4.0$ (o).
length dependence of
mean square radius of
gyration R_g^2 for $kT/\varepsilon=3.7$.

CONFORMATIONAL ANALYSIS AND ORDER-DISORDER TRANSITIONS IN FLEXIBLE MOLECULAR CHAINS

L. Barino and R. Scordamaglia
Molecular Modeling Area,
Istituto Guido Donegani S.p.A.
28100 Novara
ITALY

ABSTRACT

The effect linked to the transformation of thermal energy absorbed by a compound into internal rotational energy are analyzed through a new statistic-conformational model. The application of statistical mechanics to all allowed conformational microstates belonging to the rotational hypersurface of a molecule leads to the calculation of partition function, entropy and thermodynamic potentials of the molecular system. Order-disorder conformational transitions in flexible chains can then be pointed out as depending on temperature and correlated to phase transitions in the bulk of materials. Examples are given of calculation and prediction about Tg and Tm of some known polymers.

INTRODUCTION

For flexible linear molecular chains it is widely accepted the hypothesis that rotations around internal degrees of freedom excited by heat absorption affect total molecular energy changes more than other movements of the molecule, i.e. roto-translations and vibrations.

Applications of statistical mechanics to all allowed conformational microstates belonging to the rotational hypersurface leads to the calculation of partition function for the system and hence of various molecular thermodynamic properties and steric features.

Conformational transitions versus temperature can be pointed out together with molecular shapes and correlations can be searched between transitions in a single chain and phase transitions in the bulk of corrisponding molecular or polymeric materials.

THEORETICAL METHOD CSD AND RESULTS

The studied molecular system is simply formed by a molecule or a fragment of linear chain, considered as ensambles of linked atoms able to assume a statistical set of configurations (the molecular conformations), of which we know geometries and total (non-bonded) energy. To such a system the well known Boltzmann's distribution laws can be applied giving the values of thermodynamic functions at constant volume and in equilibrium conditions.

CSD (Conformations Statistical Distribution) method (1-2) is formed by three programs subsequent one to the other:

ERHYCA <u>Entire Rotational HYpersurface Conformational Analysis</u> calculates the non-bonded energy (Lennard Jones 6-12 function plus electrostatic term) of all the conformations generated through step by step rotations of molecular portions around free internal rotational bonds.

BOLSTAT <u>BOLtzmann STATistics</u> calculates the statistical distribution function of conformations on discrete energy levels E_i: $Z = \sum_i n_i \cdot \exp(-1/KT \cdot E_i)$, where n_i is the number of accepted conformations on i-level, T temperature (^0K), K Boltzmann factor.

It allows moreover to choose the energy threshold value under which to scan the hypersurface from ERHYCA.

CLUDIS <u>CLUster DIStribution</u> represents in quite an original and compact way the conformational hypersurface, by defining ensambles of conformations (clusters) separated by rotational energy barriers higher than energy threshold and corresponding to conformational minima. Clusters populations and shapes are determined at various temperatures.

The following quantities are calculated for each cluster j:

- partition function $Z_j = \sum_i n_{ji} \cdot \exp(-1/KT \cdot E_i)$
- cluster probability in the hypersurface $\quad P_j = Z_j/Z$
- Gibbs entropy $S_j = -K \sum_i p_{ji} \cdot \ln p_{ji}$, where p_{ji} is the probability of conformation i in cluster j

- internal energy $U_j = \bar{E}_j = \sum_i p_{ji} \cdot E_i = \dfrac{\sum_i n_{ji} \cdot \exp(-1/KT \cdot E_i) \cdot E_i}{\sum_i n_{ji} \cdot \exp(-1/KT \cdot E_i)}$

- heat capacity at constant volume $C_{V_j} = \left(\dfrac{\partial \bar{E}_j}{\partial T}\right)_V$ and $\left(\dfrac{\partial S_j}{\partial T}\right)_V = \dfrac{C_{V_j}}{T}$

- Helmholtz free energy $F_j = U_j - TS_j = -KT \cdot \ln Z_j$

As examples in figures 1 and 2, ΔS curves vs. temperature are reported for polyethylene and polystyrene molecular fragments. Experimental transitions are indicated in the graphs.

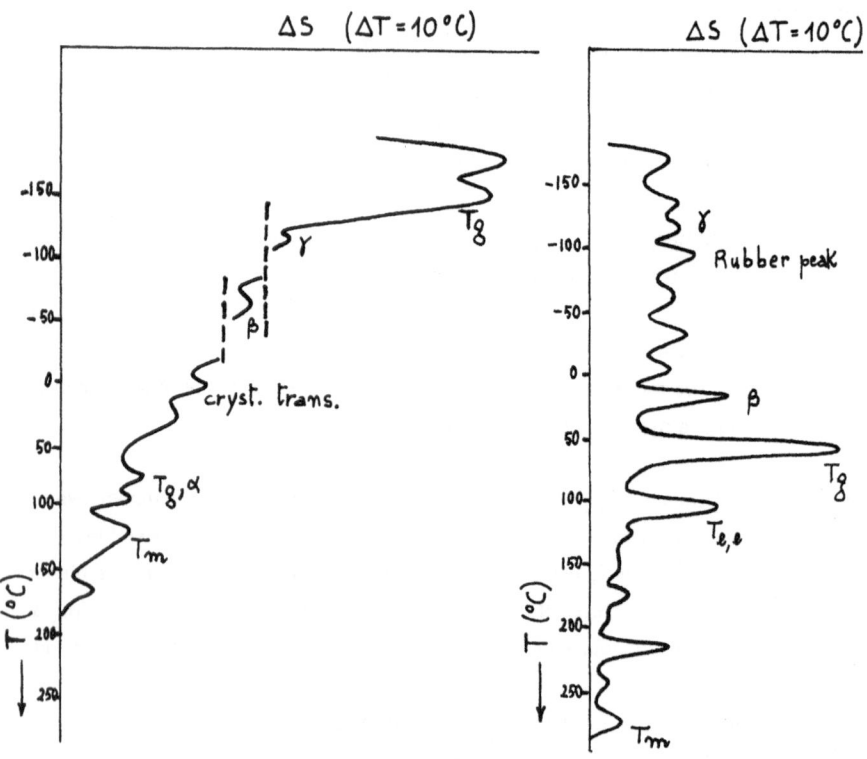

Fig. 1 - ΔS curve for polyethylene fragment $CH_3 + CH_2 \rightarrow_7 CH_3$ with 8 rotational angles.

Fig. 2 - ΔS curve for p-styrene $CH_3 + CH_2 - C H \rightarrow_7 H$ with 14 rotational angles

Both macroscopic phase transitions and structural changes happening inside a phase show a clear correspondence with conformational transitions.

Calculated spectra appear moreover well suited to make comparison between different substituted compounds belonging to a congeneric series. Tg and Tm peaks move with respect to the ones of target molecule, as Table 1 shows for a number of compounds in poly-p-xylylene series.

TABLE 1

Dependence of transition temperature on substituent groups in poly-p-xylylene

R	R'	Tg (°C) exp / calc	Tm (°C) exp / calc
H	CN	90 / 86	270 / 270
H	CH_3	50 - 60 / 57	200 - 210 / 240
H	C_2H_5	25 / 25	160 - 170 / 160
H	H	60 - 80 / 60	$\sim$ 420 / 380
H	Cl	80 / 75	290 / 280
Cl	Cl	110 / 112	$\sim$ 380 / 350

Note that calculated temperatures are expected to be lower than experimental ones, since theory lets the molecular system get its equilibrium state at each temperature, which is not experimentally true. Moreover the intermolecular interactions between chains have the effect of opposing intramolecular disorder and of arising transition temperatures.

MOLECULAR DESCRIPTORS

Thermodynamic theoretical spectra cannot however distinguish between the two different groups of transitions, i.e. transitions between two different phases and inside a phase. Other parameters must be searched that can affect packing between molecules and whose changes with temperature show meaningful variations in the phase transitions temperature ranges for bulk.

130

Physical quantities such as molecular volume, torsions on free dihedral angles in the most probable clusters, point charge distributions and dipole moment have been tested as "descriptors" characterizing a molecule. Their variation with temperature may provide a distinguishing feature for the conformational transitions pointed out by thermodynamic curves. In figure 3 ΔS, $T \cdot \Delta S$ and volume variation curves vs. temperature are reported for the polystyrene isotatic fragment of four monomers. Calculated values resulted from weighted averages over all conformational possibilities of the fragment at each temperature. One can see that meaningful indications can be derived: in rubber peak and glass transition ranges important volume changes happen; moreover calculations have evidenced for the same temperatures an abrupt increase in rotational movement of atomic groups at the ends of fragment.

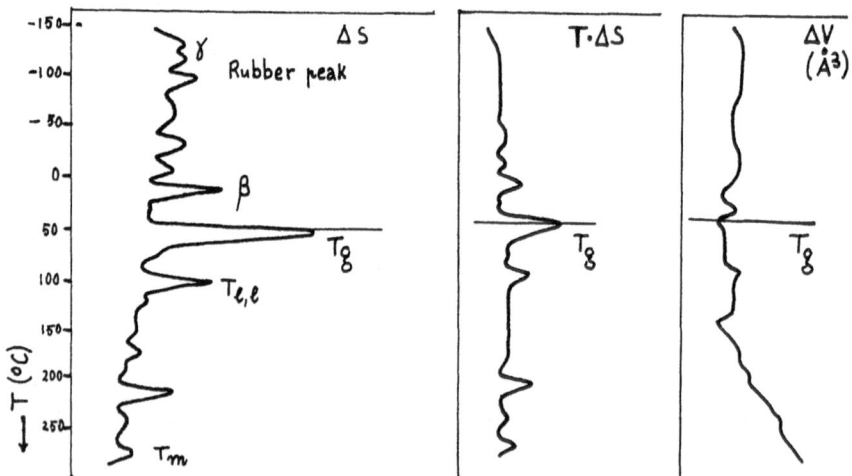

Fig. 3 - ΔS, $T \cdot \Delta S$, ΔV curves for polystyrene, with $\Delta T = 10°C$

1. Castellani, G. and Scordamaglia, R., A fast computer program for conformational analysis, Comp.& Chem., 1984, 8, 127.

2. Scordamaglia, R. and Barino, L., New Geometrical and Electronic Descriptors of Molecules for Structure - Activity Relationships, QSAR and strategies in the design of bioactive compounds (Procedings of the Fifth European Symposium on Quantitative Structure - Activity Relationships, Bad Segeberg - FDR, 1984), J.K. Seydel ed., VCH, Weineim, 1985, pp. 299-304.

THE ITERATIVE CONVOLUTION TECHNIQUE AND ITS APPLICATION TO THE EXCLUDED VOLUME PROBLEM IN THE CONFIGURATIONAL PROPERTIES OF POLYMERS

Clive A Croxton
Department of Mathematics
University of Newcastle, NSW 2308,
AUSTRALIA

The iterative convolution (IC) technique is reviewed in the context of the lattice-based and renormalization group approaches. A wide variety of applications of the IC technique are presented including the geometrical properties of isolated chains, dense entanglements, solvent effects, terminally attached sequences at a boundary, the development of loops trains and tails, ring and star polymers. Wherever possible comparisons are made with Monte Carlo simulations. In all cases the emphasis is upon the incorporation of excluded volume effects within the system.

INTRODUCTION

The iterative convolution (IC) technique has been extensively presented elsewhere [1,2], and it is inappropriate to review that discussion here. It is, however, worthwhile discussing the particular objectives of the technique in the context of the principal alternative approaches - the lattice-based and renormalization group treatments - to determine the extent to which the IC technique complements these approaches and to identify its particular strengths and shortcomings.

The primary objective of the IC technique is the description of the configurational properties of finite, continuum self-interacting sequences such that the principal dimensional features such as mean square separations, radii of gyration, segment distribution functions and scattering functions could be readily determined. Versatility was also a major objective: the description of *heterogeneous* systems was an important requirement, as was the description of non-linear geometries such as ring, star and comb structures. The only input quantity was to be the specification of the interaction matrix $[\Phi_{ij}]$ which defines the complete set of central, non-saturating pair potentials developed between any segment pair i,j within the system. In many of these respects the IC technique differs substantially in application from the lattice-based and renormalization group approaches. Apart from the necessary preoccupation with the $N \to \infty$ limit in the latter approaches, the description of heterogeneous systems poses some difficulties for these treatments, as does the incorporation of solvent effects. Ag ain, the resolution of fine structure within such systems is not easily achieved in the infinite molecular weight limit, particularly in cases of restricted geometry such as at boundaries. On the other hand, the IC technique is primarily restricted to systems of low to intermediate molecular weight ($N<60$), and does not readily extend to the description of, for example, the exponent representations of mean square length and radius of gyration in the limit $N \to \infty$, as do the alternative approaches. Again, whilst self-interaction is explicitly incorporated through the specification of $[\Phi_{ij}]$, the IC technique is nevertheless an approximation having many generic similarities to the modern structural theories of dense liquid systems.

The iterative convolution approach seeks to determine the complete set of normalised spatial probability distribution functions Z(ij|N) between segments i,j within the N-mer.

This is achieved on the basis of the coupled set of integral equations [1]

$$Z(ij|N) = H(ij)\prod_{k}{}'\int Z(ik|N)Z(kj|N)dk \qquad (1)$$

where H(ij)=exp (-Φ(ij)/kT and $\prod'$ represents the formation of the geometric mean of the product of convolution integrals via the field point k≠i,j. The H(ij) function explicitly introduces self-interaction within the system, whilst sequential connectivity and preservation of contour length of the sequence is ensured by setting all Z(i,i±1|N)= δ(r-σ), where σ is the segment diameter. Additional features of the sequence such as fixed bond angles, stiffness and torsion constants and elastic sequential connectivity are readily incorporated by including further restrictions on the neighbouring Z-functions.

Physically, eq(1) is an ansatz expressing the spatial correlation between segments i and j in terms of a direct component H(ij), and the geometric mean $\prod'$ of all indirect routes of propogation between i and j via the remaining segments in the sequence. Eq(1) may be substituted into itself, thereby casting it into iterative form [1] and solved by fast Fourier transform techniques which, incidentally, yield the scattering function permitting direct comparison with experimentally determined quantities. As mentioned above, the IC technique is an approximation, and whilst the interaction function is explicitly included, the reduction from an N-fold integral to an N-fold product of integrals inevitably results in a loss of fidelity in the expression of self-interaction within the system. It is primarily this feature of the approximation which restricts its application to systems of small to intermediate molecular weight. Nevertheless, as we shall see below, the technique is capable of successfully describing many of the configurational properties of such systems, and in a number of cases, identifying features not previously reported, but which have been subsequently confirmed on the basis of Monte Carlo simulation.

RESULTS

Whilst any form of central, non-saturating pairwise interaction may be specified in [Φ(ij)], all results presented here relate to perfectly flexible hard sphere systems. The adquate description of such primitive systems is an essential precursor to the description of more realistic assemblies, and as such follows the development of successful theories of dense fluid systems. Moreover, the computer simulation of a hard sphere system is substantially easier than for its more realistic counterpart, and at this stage the use of more complex interactions is unwarranted.

We shall now present a variety of examples of application of the IC technique. In cases which have been previously reported we shall identify only the basic features of the analysis, presenting in more detail those systems which have not been previously described.

Isolated linear chains [1]   The interaction matrix [Φ(ij)] is specified for a system of hard spheres of unit diameter, and the iterative form of eq (1) solved for the complete set of spatial distributions Z(ij|N) from which all internal and end-to-end mean square separations may be determined. Thus, in Fig 1 we show the mean square end-to-end length $<R_N^2>$ and mean square radius of gyration $<S_N^2>$ of the 'pearl necklace' model as a function of number of segments, N. Also shown are the corresponding Monte Carlo results. The agreement is seen to be good over the range of chain lengths investigated, the $<S_N^2>$ result being particularly gratifying since it involves the complete set of internal distributions and provides a stringent test of the technique. Satisfactory description of this most basic of systems is considered an essential prerequisite before analysis of more exotic systems is warranted.

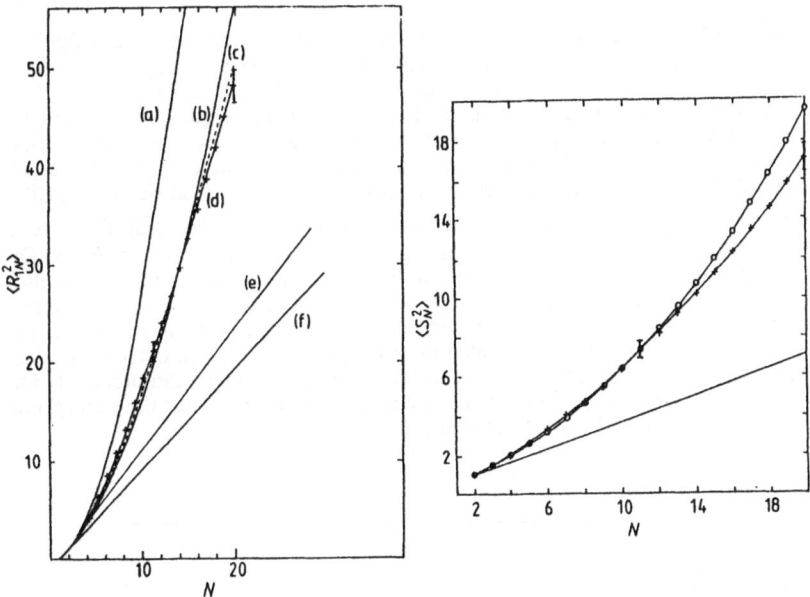

Figure 1. The mean square end-to-end distance $<R_{1N}^2>$ as a function of chain length on the basis of various approximations and compared with the Monte Carlo estimate. (a) Percus-Yevick (Curro), (b) geometric convolution, (c) arithmetic convolution, (d) Monte Carlo, (e) non-iterative convolution, (f) random walk.

The radius of gyration $<S_N^2>$ as a function of chain length on the basis of various approximations. (Crosses, Monte Carlo; open circles, geometric convolution; full curve, random walk.)

The above treatment has been extended to the description of dense entanglements of chains [3], and the equation of state determined as a function of packing fraction. In this analysis such a chain is treated as being in a solvent of its own segments, and the hard sphere interaction between segments within a given chain replaced by a potential of mean force $\Psi(ij)$. This latter function is familiar from the theory of liquids, and describes the effective interaction between two particles immersed in a solvent at given packing fraction: of course, $\Psi(ij) \rightarrow \Phi(ij)$ as the solvent packing fraction $\eta \rightarrow 0$. Agreement between the IC prediction [3], and recent Monte Carlo simulations [4] is good at intermediate to high packing fractions, and this potential of mean force approach will form the basis of other solvent descriptions below.

As we have emphasised above, any sequence of central pairwise interactions may be specified in the IC approach, and we are therefore able to model a terminally attached linear hard sphere sequence at a rigid boundary by allowing the diameter of the first segment $\sigma_0 \rightarrow \infty$ [2,5]. In fact it is found that the asymptotic properties of the sequence are rapidly achieved as increases, and for the short to intermediate length sequences (N<20) investigated here we find σ_0=64 is adequate. The segment density

profile $\rho(z|N)$ normal to the boundary is shown in Fig 2, and is seen to exhibit a pronounced discontinuity at one segment diameter from the boundary. The overall structure is confirmed by Monte Carlo simulation, including the discontinuity which may be directly attributed to the distribution of the second segment about the one terminally attached to the boundary. Furthermore, the existence of this discontinuity has been confirmed analytically [6]. Introduction of a hard sphere solvent [7] via the potential of mean force concept outlined above substantially modifies the density

profile (Fig 3). $\rho(z|N)$ appears to develop oscillations whose period equals the solvent diameter and whose amplitude develops with increasing solvent packing fraction. Of particular interest is the contact value of the density profile at the boundary, since this affects directly the development of loop and train configurations. We observe that the contact value develops dramatically with increasing solvent packing fraction, to an extent which appears inversely related to solvent diameter [7]. In other words, for small diameter solvents at high packing fractions there is a much greater propensity for the formation of loops and trains than in the absence of solvent, when the hard sphere sequence is predisposed to the formation of tails.

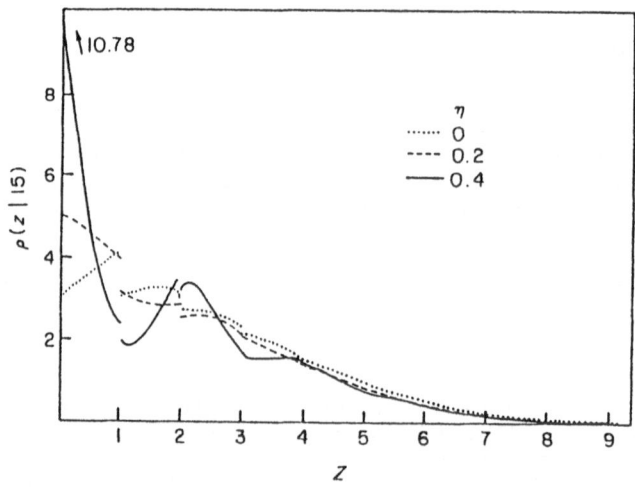

Figure 2. The mean square end-to-end length $<R_{1N}^2>$ and radius of gyration $<S_N^2>$ as a function of chain length, solvent packing fraction and diameter.

135

The development of loops, trains and tails as a function of solvent composition has also been investigated on the basis of the IC approximation [8], and in the zero solvent case corresponding Monte Carlo simulations have been conducted. Additional analysis beyond that of the IC approximation is required, and under the circumstances the agreement is considered good. In Fig 3 for example, we show the IC and MC estimates of the percentage of the chain in the form of loops trains and tails as a function of chan length. A more complex behaviour develops with the introduction of solvent, however, the dependence upon solvent composition closely relates to the development of the contact value of $\rho(z|N)$ described above [9].

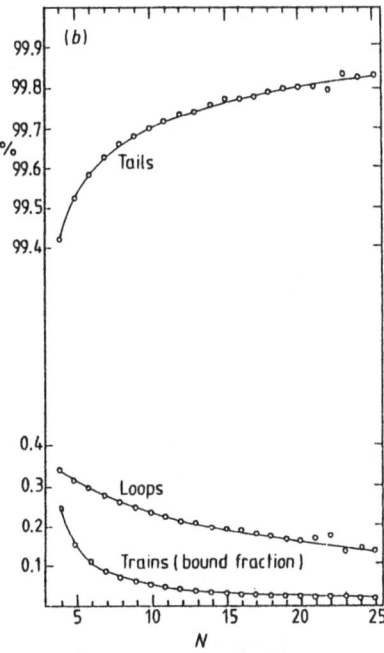

Figure 3. Component fractions as a function of chain length, expressing percentage of chain in loop train and tail form.

The versatility of the IC technique readily permits the description of self-avoiding systems of non-linear geometry such as rings [10] and stars [11]. In the latter case, for example, a uniform star of f branches having n monomers per branch with a central vertex monomer (mol.wt.$N = nf+1$) is generated by severing the sequential connectivity interaction at every nth monomer in the linear sequence, and requiring monomers $1,n+1,2n+1,3n+1,...$ to be in sequential contact with the central vertex particle. All other interactions are hard sphere and the entire geometry is specified by the interaction matrix $[\Phi(ij)]$.

Lattice-based and renormalization group analyses of uniform self-avoiding hard sphere star systems have been reported recently [12,13] following an initial proposal by Daoud and Cotton [14] in which various scaling proposals were made relating the structure and dimensional properties of stars to the degree of branching f and the number of monomers per branch n. Daoud and Cotton propose three interior structural regimes - an inner core region of constant density in which the branches are fully extended, an intermediate regime in which inter- and intra-branch interference results in theta-like behaviour, and an outer region in which the branches effectively behave like isolated self-avoiding sequences.

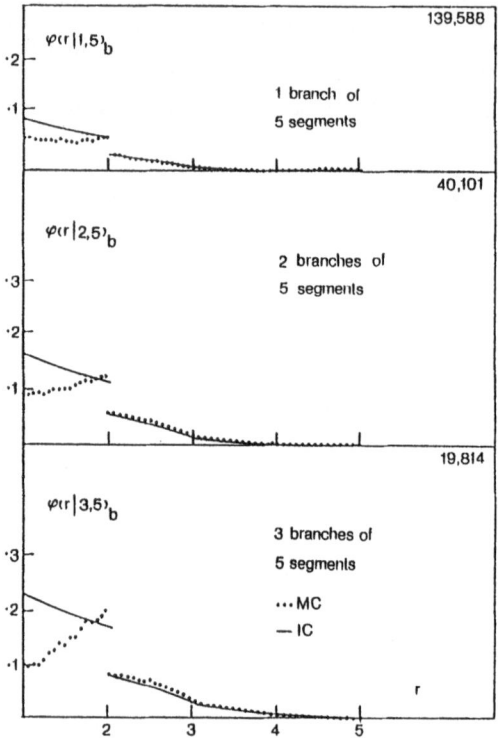

Figure 4.

Of course, the lattice-based and renormalization group analyses relate to the limit $N \rightarrow \infty$, and so the two interior core regimes remain inaccessible to these treatments. In Fig 4 we present the radial concentration profiles for small to intermediate sized uniform stars, and immediately observe the progressive development of a discontinuity with increasing degree of branching f. Such a structure was not previously predicted, but is nevertheless confirmed by Monte Carlo simulation. Again, the central core density appears from from constant as proposed by Daoud and Cotton, both on the basis of the MC and IC analyses. We observe, however, that the IC approximation allows a degree of hard sphere interpenetration at high packing densities and accordingly overestimates the monomer concentration within the central core. At larger radial distances, however, there appear further structural features, and the MC and IC treatments are in close agreement.

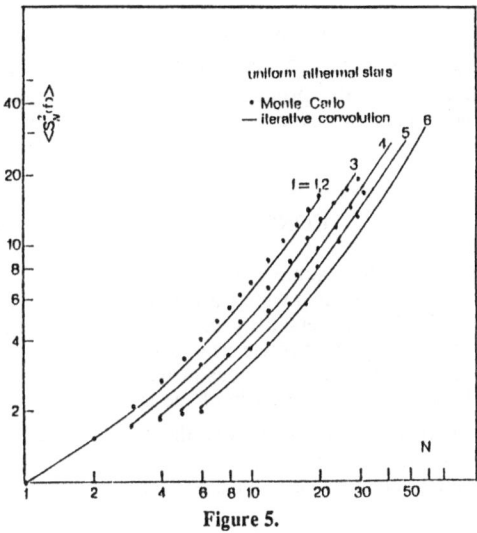

Figure 5.

A plot of $\ln<S_N^2(f)>$ vs ln N (Fig 5) shows that the radius of gyration develops as $\sim N^{1.2}$ and appears independent of the degree of branching, f. This behaviour was also observed in the lattice-based analyses. The agreement between the IC and Monte Carlo techniques is also seen to be excellent. The ratio $g(f)=<S_N^2(f)>/<S_N^2(1)>$ relating the radii of gyration of a star to a linear polymer of the same molecular weight has been extensively investigated, and we present our results in Table I.

TABLE I

The ratio g(f) as a function of degree of branching.

	f=	3	4	5	6
this work (IC)		.721	.545	.429	.341
this work (MC)		.692	.536	.447	
lattice-based [12]		.76±0.01	.60±0.01	.51±0.01	.43±0.01
renormalization group [13]		.798	.667	.580	.519
random walk [15]		.778	.625	.520	.444

It is apparent that the IC and MC estimates are substantially lower than the others, the discrepancy increasing with branching number, f. It should also be said that these are limiting estimates as N→∞, extrapolated from relatively small systems in the case of IC and MC analyses. Nevertheless, the latter two estimates are in good agreement, and already yielded g values substantially below the others at the highest N

One final prediction of the Daoud-Cotton model is the branching dependence of $g(f)$ which, they propose, is $\sim f^{-2}$ for small stars, $\sim f^{-1/2}$ for intermediate stars, and $\sim f^{-4/5}$ for large stars. The latter proposition appears confirmed on the basis of lattice-based treatments [12], however, on the basis of IC and MC analyses we find $g(f) \sim f^{-1}$ more appropriate.

CONCLUSIONS

We have presented here some examples of the application of the iterative convolution technique as applied to a wide variety of problems. We believe we have demonstrated the accuracy and versatility of the technique in the description of continuum systems of small to intermediate size. The technique extends to other more complex systems, perhaps the most ambitious of which is the prediction of the 3-D structure of globular proteins on the basis of the specification of the amino-acid interaction matrix $[\Phi(ij)]$. It is envisaged that the method is capable of describing many other systems such as polynematics, the gel-glass transition, electrical conductivity in polymers and block co-polymer systems.

REFERENCES

1.   Croxton, CA, J Phys A: Math Gen, (1984) *17* 2129.
2.   Croxton, CA, in *Fluid Interfacial Phenomena* Ed. C A Croxton,
        John Wiley & Sons, New York and Chichester, 1986 Ch 7.
3.   Croxton, CA, J Phys A: Math Gen, (1979) *12* 2497.
4.   Dickman, R 1986 (private communication).
5.   Croxton, CA, J Phys A: Math Gen, (1986) *19* 987.
6.   Croxton, CA, Phys Lett A, (1985) *IIIA* 453.
7.   Croxton, CA, Polymer Comm, 1987 (in press).
8.   Croxton, CA, J Phys A: Math Gen, (1986) *19* 2353.
9.   Croxton, CA, Polymer Comm, 1987 (in press).
10.   Croxton, CA, J Phys A: Math Gen, (1985) *18* 995.
11.   Croxton, CA, Macromolecules, 1987 (in press).
12.   Whittington, S, Lipson, JEG, Wilkinson, MK and Gaunt, DS,
        Macromolecules (1986), *19*, 1241.
13.   Miyake, A and Freed, KF, Macromolecules (1983) *16* 1228.
14.   Daoud, M and Cotton, JP, J Phys (Paris) (1982) *43* 531.
15.   Stockmeyer, WH and Zimm, BH, J Chem Phys (1949) *17* 301.

MECHANICAL AND THERMOPHYSICAL PROPERTIES OF POLYMERS AND CHEMICAL STRUCTURE: FROM COMPUTER SIMULATIONS TO EXPERIMENTAL DATA

WITOLD BROSTOW

Institut für Physikalische Chemie, Johannes Gutenberg Universität, D-6500 Mainz, Federal Republic of Germany; and Department
of Materials Engineering, Drexel University, Philadelphia, PA
19104, U.S.A.

ABSTRACT
Chemical structure is related to properties on the basis of
results of computer modelling of mechanical behavior. An equation for the shift factor $a_{T,c}$ dependent on temperature T and
concentration c is provided, connection to the chain relaxation
capability noted, and potential application range of the approach discussed. Constituents of free volume are defined, and
effects of individual constituents on mechanical properties

1. INTRODUCTION

With so many classes of polymeric materials in rapidly increasing use, their service performance becomes the bottom line.
The key question is: will a material serve for a reasonable period of time, or will it fail prematurely? Often a second question is: can we make a material with better properties? In spite of the fact that these questions are asked so often, answer-
ing them still involves much more art than science. Thus, when
one tries to make a new material with improved mechanical properties by blending existing polymers, the standard procedure
is experimental, by trial-and-error. Luck has a much larger
role here than guidance provided by existing theories.

Continuum concepts, in particular linear-elastic fracture
mechanics (LEFM), eminently successful for metals, are also
pertinent to polymers; for a lucid review see Pascoe [1]. However, a simple transfer of LEFM concepts to polymers has in
general a little chance ot success. As noted by Wolf [2] in a
somewhat different context, even for a fixed fraction of each
kind of monomeric units in a given sample, each different way
of assembly of these constituents leads to molecules of different length and stereoregularity, different branching and cross
linking characteristics, etc. At the same time, even detailed
knowledge of molecular structure is insufficient for the determination of macroscopic properties. Interaction dynamics is important, particularly as chains respond collectively to external stresses from the service environment.

Given this complexity of structure, our best chance of understanding polymeric systems should come from computer simulations. Important characteristics such as the length and shape
of a representative part of the chain, lengths of branches and
their locations along the main chain, or angles between various

chain fragments can be simulated [3]. Even representation in terms of individual segments is possible. Structural parameters can then be varied one at a time, and effects of each upon properties studied. We shall discuss such studies in Section 2.

While computer modeling should provide a long term solution,, a partial elimination of the trial-and-error procedures, but achievable on a shorter time scale, is also of interest. Among the multitude of polymer chain parameters, we have opted for free volume v^f - in conjunction with chain conformations - as the key characteristics. Thus, the number of parameters to be varied is reduced drastically from the start. At the same time, parameters not accounted for explicitly are reflected in free volume; the average branch length is an example par excellence.

Our choice of free volume finds strong support in the approach to viscoelasticity developed by Ferry and analyzed in detail in his book [4]. As for thermodynamic properties, introduction of an explicit v^f term in the partition function by Flory [5-7] produced a drastic improvement in the accuracy of theoretical predictions [5-10]. Moreover, a long series of connections between chemical structure and macroscopic properties is analyzed by van Krevelen [11]. We conclude that our free volume approach offers the opportunity for successful dealing with rheological and mechanical properties on one hand, thermophysical on the other, and also with connections between these two broad classes of properties. This approach is discussed in Sections 3 & 4.

2. COMPUTER MODELING

There is a variety of procedures for simulating polymer chain dynamics; for an analysis see Fixman [12]. Relatively much less effort was invested in modeling of mechanical properties; this in contrast to crystals and metals, for which work on fracture was started earlier [13].

Pertinent for us is a Monte Carlo study of a chain with excluded volume interactions in the presence of an external stretching force by Webman, Lebowitz and Kalos [14]. They found the linear response for low and moderate forces and nonlinear for strong forces. A chain of atoms under stress was studied by Weiner and Pear [15]. They used a double-welled potential developed first by Ryckaert and Bellemans for n-alkanes [16]; see also Swindoll and Haile [17]; the potential provides good representation of rotational energy barriers.

While work on single chain is quite illuminating, we have stressed above the importance of collective response. Theodorou and Suter [18,19] simulated deformation of polymeric glasses in the elastic region. They find that the deformation is accompanied by concerted displacement of chain segments approximately 10 bonds long.

A third type of modeling, the most pertinent for us, deals with large deformations, involving bond breaking and failure. Such work was initiated by Cook and Mercer (C-M) [20], with somewhat related and nearly simultaneous studies by Brown and Clarke [21], and by Termonia e.a. [22]. The latter use a Monte

Carlo procedure, however, while the C-M algorithm represents an adaptation of the method of molecular dynamics [23]. The C-M method, used also in subsequent studies [24,25], can be characterized by the relation

$$F_{ij}(t) = -(\partial U^c(t)/\partial R_{ij}) + F^{ext}(t) \qquad (1)$$

where F_{ij} is the force acting in the i-th chain on the j-th interacting element (segment), t = time, U^c = configurational energy of the system, R_{ij} the respective positional coordinate and F^{ext} represents the external forces.

For instance, the C-M method can be used to study stress relaxation [25]. Exemplary results will be presented at the Meeting. The reliability of a polymeric material is determined by an interplay of nondestructive and destructive processes [26]. The key nondestructive response is the dissipation of externally furnished mechanical energy by chain relaxation; chain relaxation capability (CRC) has been defined [27]. The stress relaxation modeling shows that a crossover exists between the region of smaller F^{ext} in which CRC is sufficient and the region where the breaking of a small number of bonds leads to crack propagation.

The same C-M method can be used to study stress-strain behavior. By using the Ryckaert-Bellemans potential [16], it turns out that modeling reproduces well key features known from the experiment [25].

Other experimental procedures can be also modelled by the C-M procedure; thus, creep at constant stress is another possibility. Moreover, as in the renormalization group method [28], a point can represent a segment (a repetitive unit), a sequence of segments, or on a much larger scale even, say, a crystalline polymer block. Uniaxial as well as multiaxial stresses can be applied. Clearly, much more can and should be done.

3. CHAIN RELAXATION CAPABILITY AND SHIFT FACTORS

Chain relaxation can take more than one form, partly in function of the type of the mechanical force applied. We shall mean by relaxation both transmission of externally furnished energy along the chain (intensified segment vibrations, for instance) as well as conformational changes executed by the chains under the influence of the force. We define [27] CRC as the amount of external energy dissipated by relaxation in a unit of time per unit weight of a polymer. Clearly, CRC is related to v^f. A convenient measure of CRC is the temperature shift factor a_T, based on the principle of time-temperature equivalence [4,11,29]. The shift factor can be related to free volume via [26]

$$\ln a_T = A' + \frac{B}{\tilde{v}-1} \qquad (2)$$

Eq. (2) is based on the Doolittle viscosity-free volume relation [30] and B is the Doolittle constant; the reduced volume $\tilde{v} = v^*/v$, where v is a characteristic (hard-core, incompressible) volume, and both v^* and v pertain to, say, 1 g of the material; $v = v^* + v^f$.

The use of eq. (2) enabled the derivation of a successful formula relating the impact transition temperature for polymers to the stress concentration factor [26]. Moreover, a_T for shear creep compliance of a two-phase adhesive can be described above, at and below the glass transition region by a single equation [27]. Further, eq. (2) can be used à rebours, and then it furnishes information on temperature dependence of polymer density from the knowledge of mechanical properties.

Eq. (2) was obtained for a one-component material. Important for dealing with polymer blends, multicomponent melts, solutions, as well as materials absorbing liquid condensates from the environment, is the concentration dependence of a_T. Knauss and Kenner [31] (see also [29]) have found that absorbed moisture affects creep rates similarly as a temperature change. A shift factor dependent on both temperature and concentration c can be derived (32), namely

$$\ln a_{T,c} = A + \ln \frac{v}{v_{ref}} + \ln \frac{T_{ref}}{T} + \frac{B}{\tilde{v}-1} \qquad (3)$$

The use of eq. (3) involves a reliable $\tilde{v}$ (T) equation. A good candidate is the Hartmann equation of state

$$\tilde{P}\tilde{v}^5 = \tilde{T}^{\frac{3}{2}} - \ln \tilde{v} \qquad (4)$$

where $\tilde{P} = P/P_{ref}$, $\tilde{T} = T/T_{ref}$, and the index ref pertains to a reference state. Further, a formula relating $\tilde{v}$ to concentration is needed; from the formalism of walks on graphs [33], for a binary i + j system [32]

$$\tilde{v} = \varphi_i w_i \tilde{v}_{ii} + 2\varphi_i w_j \tilde{v}_{ij} + \varphi_j w_j \tilde{v}_{jj} \qquad (5)$$

where φ_i and w_i are segment and weight fractions of the i-th component; $\tilde{v}_{v_x v_y}$, are terms related to pairwise interactions. Eq. (5) belongs to mean-field approximations; Stockmayer [34] analyzes relation of such approximations to integrals over distribution functions.

Still further possibilities exist. The fact that shift factors are measures of CRC can be also used outside of the usual range of dynamic mechanical properties of polymers tested as functions of time. Thus, work is in progress on explanation and prediction of slow crack growth in plastic pipes as studied by Fleißner [35], and also on the times-to-burst of plastic pipes under an internal gas pressure [36].

4. CONSTITUENTS OF FREE VOLUME

While many researches achieve good results by employing free volume concepts, other - following apparently similar lines - encounter difficulties. There are at least two reasons for this. First, a number of mutually incompatible definitions of free volume exist [37]. Second, it should be realized that the volume (e.g. specific) of a material represents in fact a sum of several constituents. Two free volume constituents were recognized by Frenkel (38) and analyzed by Kanig (39), and three by Roszkowski [10]. At least, we have

$$V = v_w + v_n + v_i + v_d \qquad (6)$$

Here v_w is the proper volume of the atoms, calculable from their Van der Waals radii; v_n is the non-accessible volume between atoms, calculable also from the radii and from the interatomic distances for a given structure (location of branches, etc.); v_i represents segmental vibrations; neutron scattering studies indicate that the frequency of such vibrations does not change with the temperature, but the amplitude does [40]. One infers that v_j consists of free volume elements attached to atoms and polymer segments. By contrast, v_d is not attached (independent holes). One can decompose v_d further, namely $v_d = v_r + v_t$ (7), where v_r represents space available to rotational movements of segments and v_t to translations. The last term appears only at and above the glass transition. However, the translational space is at least in part available also for rotational transitions.

The existence of volume components enumerated above has important consequences for mechanical properties. Typically, one expects that a v^f increase (more branching, for instance) improves the properties; of course, CRC increases at the same time. However, high v^f has also detrimental effects. Consider CRC - which is partly individual and partly collective, as well as collective chain cohesion (CCC). At high densities we have high CCC but poor CRC, with typical brittle fracture. At intermediate densities we have still satisfactory CCC and already good CRC, a happy medium. At low densities the cohesion is lost and even very good CRC is insufficient to prevent failure. A somewhat similar analysis applies to thermal stability of polymers.

ACKNOWLEDGEMENTS

Discussions with a number of colleagues are appreciated, in particular with: Dr. Robert Cook, Lawrence National Laboratory, Livermore, CA; Dr. Manfred Fleißner, Hoechst AG, Frankfurt/Main; Dr. Franz Fujara and Prof. Bernhard A. Wolf, Johannes Gutenberg University; and Prof. Nicholas W. Tschoegl, California Institute of Technology, Pasadena, CA. A Research Award from the Deutsche Forschungsgemeinschaft, Bonn, is acknowledged.

REFERENCES

1. Pascoe, K.J., in Failure of Plastics, ed. W. Brostow & R.D. Corneliussen, Hanser Publishers, Munich-Vienna-New York, 1986, Chapter 7.

2. Wolf, B.A., Pure & Appl. Chem., 1985, 57, 323.

3. Brostow, W., Gruda, J., Sochanski, J.S. and Turner, J.E., J. Chem. Phys., 1979, 10, 3268.

4. Ferry, J.D., Viscoelastic Properties of Polymers. 3rd Ed., Wiley, New York, 1980.

5. Flory, P.J., Orwoll, R.A. and Vrij, J., J. Am. Chem. Soc., 1964, 86, 3507, 3515.

6. Flory, P.J., J. Am. Chem. Soc., 1965, 87, 1833.

7. Flory, P.J., _Disc. Faraday Soc._, 1970, 49, 7.

8. Patterson, D. and Delmas, G., _Disc. Faraday Soc._, 1970, 49, 98.

9. Patterson, D., _Pure & Appl. Chem._, 1972, 31, 133.

10. Roszkowski, Z., _Mater. Chem. Phys._, 1981, 6, 455.

11. Van Krevelen, D.W., _Properties of Polymers_, 2nd Ed., Elsevier Amsterdam - Oxford - New York, 1976; 3rd Ed. in preparation by Van Krevelen, D.W. and Brostow, W.

12. Fixman, M., _J. Chem. Phys._, 1978, 69, 1527, 1538.

13. Ashurts, W.T. and Hoover, W.G., _Phys. Rev._ B 1976, 14, 1465; Monti, A.M. and Savino, E.J., _Interam. Conf. Mater. Technol._, 1978, 5, 13; Mullins, M., _Acta Metall._, 1984, 32, 381.

14. Webman, I., Lebowitz, J.L. and Kalos, M.H., _Phys. Rev. A_, 1981, 23, 316.

15. Weiner, J.H. and Pear, M.R., _Macromolecules_, 1977, 10, 317.

16. Ryckaert, J.-P. and Bellemans, A., _Chem.Phys. Letters_, 1975, 30, 123; idem, _Faraday Disc._, 1978, 66, 95.

17. Swindoll, R.D. and Haile, J.M., _J. Comput. Phys._, 1984, 53, 289.

18. Theodorou, D.N. and Suter, U.W., _Macromolecules_, 1985, 18, 1206; idem, _ibid._, 1986, 19, 139.

19. Theodorou, D.N. and Suter, U.W., _Macromolecules_, 1986, 19, 379.

20. Cook, R. and Mercer, M.B., _Mater. Chem. Phys._, 1985, 12,571.

21. Brown, D. and Clarke, J.H.R., _J.Chem.Phys._, 1986, 84, 2858.

22. Termonia, Y., Meakin, P. and Smith, P., _Macromolecules_, 1985 18, 2246; idem, _ibid._, 1986, 19, 154.

23. Alder, B.J. and Wainwright, T.E., _J. Chem. Phys._, 1957, 27, 1208.

24. Brostow, W. and Turner, D.P., _J. Rheology_, 1986, 30, 767.

25. Cook, R., preprint from the Lawrence Livermore National Laboratory.

26. Brostow, W., in _Failure of Plastics_, ed. W. Brostow & R.D. Corneliussen, Hanser Publishers, Munich-Vienna-New York, 1986, Chapter 10.

27. Brostow, W., _Mater. Chem. Phys._, 1985, 13, 47.

28. Pakula, T., _Polymer Bull._, 1980, 3, 415.

29. Kenner, V.H., in _Failure of Plastics_, ed. W. Brostow & R.D. Corneliussen, Hanser Publishers, Munich-Vienna-New York, 1986, Chapter 2.

30. Doolittle, A.K., _J. Appl. Phys._, 1951, 22, 1741.

31. Knauss, W.G. and Kenner, V.H., _J. Appl. Phys._, 1980, 51, 5131.

32. Brostow, W., in preparation.

33. Brostow, W., Phys. Chem. Liquids, 1972, 3, 91.

34. Stockmayer, W.H., Macromol. Chem., 1960, 35, 54.

35. Fleißner, M., Kunststoffe, 1987, 77, 45.

36. Palermo, E.F. and DeBieu, E.I., Proc. Plastic Fuel Gas Pipe Symp., 1985, 9, 215.

37. Brostow, W., Polymer, 1980, 21, 1410.

38. Frenkel, J.I., Kinetische Theorie der Flüssigkeiten, Berlin 1957, p. 194.

39. Kanig, G., Kolloid-Z., 1963, 190, 1; idem, J. Polym. Sci.C, 1967, 16, 1957; idem, Kolloid-Z., 1969, 233, 829.

40. Fujara, F., communication from Johannes Gutenberg University, Mainz, 1987.

INFLUENCE OF CHAIN SEMIFLEXIBILITY ON THE PROPERTIES OF POLYMER LIQUID CRYSTALS

T. Odijk
Department of Physical and Macromolecular Chemistry,
Gorlaeus Laboratories, Leiden University,
P.O.box 9502, 2300 RA Leiden,
The Netherlands

ABSTRACT

In practice slight bending fluctuations in the configurations of semiflexible chains exert a considerable influence on the properties of their nematic solutions. This effect is explained in regard to the bend and twist moduli and the surface tension. A qualitative analysis is presented of the pitch of cholesteric polymers. The inclusion of semiflexibility is imperative if we want to explain the experimental power laws discovered for poly- γ-benzyl-L-glutamate liquid crystals.

INTRODUCTION

In the field of polymer liquid crystals one can discuss a large number of salient problems deserving the attention of both pure and applied scientists. Of particular importance is the precise effect of chain semiflexibility on for instance the isotropic-liquid-crystalline phase transition, the elastic moduli and rheological phenomena. It has long been asserted that relatively short chains, about a persistence length long say, ought to behave like rigid rods but this contention has come under increasing attack. Here I argue that the behaviour of confined semiflexible macromolecules rarely if ever approximates that of nonflexible rods. Theoretical ideas about the concentration dependence of the elastic moduli and the surface tension of polymer nematics are surveyed along with a brief confrontation with experiments. I also calculate the pitch of cholesteric solutions.

Several theorists have independently pointed out that very slight fluctuations in the configurations of stiff chains could render the rod model useless in describing the properties of confined semiflexible polymers [1-4]. Specifically, incorporation of these fluctuations into the classical theory of polymer nematics [5] goes a long way towards explaining the molecular weight dependence of the onset of liquid crystal formation [6]. I first summarize these arguments before dealing with the influence of external fields.

NEMATIC RODS

Consider a nematic solution of rods of length L and diameter D. Their average orientation is denoted by the director $\vec{n}$ with respect to which we can define the solid angle Ω of a test rod (Ω = polar angle θ, azimuthal angle ϕ). For a nematic we introduce a normalized, single-rod distribution function $f(\theta) = f(\pi-\theta)$. In order to understand the Onsager theory [5] qualitatively, it suffices to suppose that f is a simple Gaussian depending on the parameter α ($\alpha \gg 1$)

$$f(\theta) \approx \left(\frac{\alpha}{4\pi}\right) \exp - \frac{1}{2} \alpha\theta^2 \tag{1}$$

This implies

$$\langle\theta^2\rangle = \frac{2}{\alpha} \tag{2}$$

It is useful to focus on this second moment when applying scaling considerations. Thus, the orientational ordering per rod is immediately written as

$$S_{or}(\alpha) = - \int f(\theta) \ln f(\theta) \, d\Omega \approx \ln \langle\theta^2\rangle \approx - \ln \alpha \tag{3}$$

because $\langle\theta^2\rangle$ is proportional to the average orientational freedom of a rod.

The orientational entropy must be balanced against the excluded-volume effect scaling as $L^2 D \langle|\sin \gamma|\rangle \rho_L$ where ρ_L is the number density (number of macromolecules per unit volume) and γ is the angle between two probe rods. In view of the fact that $\langle|\sin \gamma|\rangle \approx \langle\theta^2\rangle^{1/2} \approx \alpha^{-1/2}$ by eq(2), we have to minimize the free energy per rod

$$\frac{\Delta F}{k_B T} \approx - S_{or}(\alpha) + L^2 D \, \rho_L \, \alpha^{-1/2} + \ln \rho_L \tag{4}$$

with respect to α. This yields

$$\alpha \approx \left(L^2 D \, \rho_L\right)^2 \tag{5}$$

Although convincing, this theory [5] does not agree with experiment. In retrospect, this is not surprising since the complete neglect of chain fluctuations is an untenable supposition.

NEMATIC WORMLIKE CHAINS

A realistic model for semiflexible polymers is the wormlike chain characterized by a persistence length P. A fast method of understanding how chain fluctuations affect eqs(3-5) is by way of scaling arguments [6,7].

A standard result from the theory of wormlike chains - in fact, an application of the central limit theorem in disguise - states that the mean-square angle between two points s and t along a chain contour and separated by a distance $|s-t|$ is given by [8]

$$\langle \psi^2(s-t) \rangle \simeq \frac{|s-t|}{P} \qquad\qquad (6)$$

This is valid when the distance $|s-t|$ is short enough. However, in a nematic solution the restriction eq(2) must come into play when s and t are separated by more than a certain distance i.e. when $|s-t| \gtrsim \lambda$ where

$$\langle \psi^2(\lambda) \rangle \simeq \frac{\lambda}{P} \simeq \langle \theta^2 \rangle \simeq \alpha^{-1} \qquad or \qquad \lambda \equiv P/\alpha \qquad (7)$$

Thus, the spatial correlations are ideally wormlike for $|s-t| \lesssim \lambda$ but, otherwise, must be in accordance with the nematic field imposed by eq(2). Because this field deflects a test chain towards the director about once every λ on average, λ is called a deflection length. A long chain of length L can be regarded as a sequence of L/λ links of length λ .

We assign an orientational free energy k_BT to each unit or link [6]

$$\Delta F_{or}/k_B T \simeq \frac{L}{\lambda} = \frac{L\alpha}{P} \qquad (L \geq \lambda) \qquad (8)$$

The total free energy per macromolecule is still given by eq(4). On minimizing ΔF with respect to α, we get

$$\alpha \simeq (LPD\,\rho_L)^{2/3} \qquad (L \geq \lambda) \qquad (9)$$

Thus, chain flexibility forces the concentration dependence of the ordering parameter α to decrease from a second to a two-thirds power. In practice, P is about 50 nm for many stiff polymers and α is of order 10 so that a typical value for λ is 5 nm. Accordingly, eqs 3 and 5 can seldom be used legitimately because semiflexibility dominates.

In the table I compare a quantitative version of the above ideas [1,2,6] with the number densities $\rho_i \equiv \rho_{L,i}$ and $\rho_n \equiv \rho_{L,n}$ at the isotropic-nematic transition for schizophyllan in H_2O as given by Enomoto et al. [9]. We see that the theory has predictive power - there are no adjustable parameters - although $\rho_{L,n}$ is overestimated. This is probably due to the approximate description of the excluded-volume effect as well as the complete neglect of van der Waals forces. For a detailed comparison with other results, see [6].

TABLE

Contour-length dependence of the ratios X_i and X_n of experimental densities [9] to theoretical ones [6] pertaining to the I-N transition for schizophyllan in H_2O

L/P	0.30	0.52	0.80	1.08	2.24	4.0	10.0
$X_i = \dfrac{\rho_{i,exp}}{\rho_{i,th}}$	0.91	0.91	0.90	0.89	0.89	0.88	0.87
$X_n = \dfrac{\rho_{n,exp}}{\rho_{n,th}}$	1.06	1.18	1.25	1.27	1.30	1.31	1.30

Persistence length P = 200 nm; Diameter D = 1.67 nm [9]

SCALING RECIPE

In a statistical sense we can view a nematic solution of N long semiflexible chains as composed of NL/λ effective quasi-rodlike units. Therefore, results derived for rods are extended to stiff polymers by the substitution

$$L \to \lambda = P/\alpha$$
$$\rho_L \to \rho_\lambda = \alpha_{,}\rho_P \qquad (10)$$

Here, ρ_ℓ is the concentration of segments chosen to have length ℓ, i.e. $\ell\rho_\ell = L\rho_L$. Note that the prescription eq(10) enables us to obtain eq(9) from eq(5). We next apply eq(10) to more complicated problems.

ELASTIC MODULI

When we deform a nematic liquid crystal, the director $\vec{n}$ becomes a function $\vec{n}(\vec{r})$ of position $\vec{r}$, so that the orientational distribution f is altered to $f+\delta f$, implicitly dependent on $\vec{r}$. For weak strains the orientational entropy is invariant to a first approximation [10] although the excluded-volume effect does change. By eq(4) we can express the increment in free energy density $N\delta\Delta F/V$ as follows [11]

$$\frac{N\delta(\Delta F)}{k_B TV} \approx L^2 D \ \rho_L^2 \int \int \ d\Omega_1 d\Omega_2 \ |\sin\gamma| \ \delta f_1 \ \delta f_2 \qquad (11)$$

Now, in the bend mode two rods interact when separated by a distance less than L so that the typical angle of deformation $\varepsilon \approx L|\vec{n} \times \vec{\nabla} \times \vec{n}|$ and $\delta f \approx \alpha\theta\varepsilon f$. Hence, eqs(2) and (5) yield for the bend moduli [12]

$$K_3 = \frac{2N\delta(\Delta F)}{V|\vec{n} \times \vec{\nabla} \times \vec{n}|^2} \approx k_B T \ L^4 D \ \rho_L^2 \ \alpha^{\frac{1}{2}} \approx (\frac{k_B T}{D}) \ (L^2 D \rho_L)^3 \qquad (12)$$

Analogously, we can derive the twist modulus [12]

$$K_2 \approx k_B T \ L^4 D \ \rho_L^2 \ \alpha^{-\frac{1}{2}} \approx (\frac{k_B T}{D}) \ L^2 D \ \rho_L \qquad (13)$$

For semiflexible chains somewhat longer than the deflection length eqs(12) and (13) together with the substitution eq(10) lead to [12]

$$K_3 \approx (\frac{k_B T}{D}) \ P^2 D \rho_P \qquad (14)$$

$$K_2 \approx (\frac{k_B T}{D}) \ (P^2 D \rho_P)^{1/3} \qquad (15)$$

Again, chain semiflexibility has a drastic influence on the concentration dependence. These scaling relations are in good agreement with experimentally determined moduli for poly-γ -benzyl-L-glutamate (PBLG) in various solvents. For example, fig.5 of ref.[13] would give $K_2 \sim \rho_P^{0.36}$ i.e. very close to eq(15) whereas a linear dependence of K_3 on ρ_P has been reported only recently[14]. For an alternative analysis of these moduli yielding similar results, see [15]. I have purposely not discussed the splay deformation because it is subtler and outside the scope of this review, see [15-17].

SURFACE TENSION

The orientational freedom of a rod decreases as it approaches an interface. Provided the interface is inert the rod experiences a repulsive force of entropic origin. Hence, if van der Waals and other long range forces are weak, a depletion layer should adjoin the air-solution interface. Since the orientational freedom of a rod within a nematic is given by eq(2), the thickness of the depletion layer scales like

$$d \approx L\langle\theta^2\rangle^{\frac{1}{2}} \approx L\alpha^{-\frac{1}{2}} \tag{16}$$

The director is postulated to be parallel to the interface. Now, the increase in surface tension equals the free energy of the depletion layer per unit area [18]

$$\Delta\tau \approx \left(L^2 D\rho_L \alpha^{-\frac{1}{2}}\right)\left(d\rho_L\right)k_B T \approx \frac{k_B T}{LD} \tag{17}$$

As before, we extend this result by using eq(10). For semiflexible chains, we get [19]

$$\Delta\tau \approx \frac{k_B T}{\lambda D} \approx k_B T \left(\frac{P^{1/3}\rho_P^{2/3}}{D^{1/3}}\right) \quad (L \gg \lambda) \tag{18}$$

I am not aware of any experiment with which eq(18) can be compared.

PITCH OF POLYMER CHOLESTERICS

As argued by Straley[20], chiral interactions between rodlike macromolecules would lead to a change in the free energy δF linear in δf, the variation of the distribution function with respect to the hypothetical nematic (i.e. with the chiral forces switched off). He proposed the following expression

$$\frac{N\delta F}{k_B TV} \approx L^2 D\rho_L^2 \iint d\Omega_1 d\Omega_2 \quad \cos\gamma \sin\gamma \, c(\cos\gamma) \, f_1 \delta f_2 \tag{19}$$

where c is some unspecified function and absolute quantities are implied so that we need not worry about the sign of pseudoscalar quantities. If we model the macromolecules as rods enveloped by helical threads of thickness Δ, the angle ε characterizing the chiral deflection of two interacting rods is given by [20]

$$\varepsilon \approx \Delta|\vec{n}.\vec{\nabla} \times \vec{n}| \tag{20}$$

Furthermore, the approximation c = constant is reasonable.

It is now straightforward to calculate the pitch in the Gaussian approximation. We have $\delta f \approx \alpha\varepsilon\delta f$ so that the chiral coefficient K_t can be approximated by

$$\left| K_t \right| \equiv \left| \frac{N\delta F}{k_B T V \left| \vec{n} \cdot \vec{\nabla} x \vec{n} \right|} \right| \approx k_B T \, \rho_L^2 \, L^2 D\Delta \tag{21}$$

The pitch $p = \left| \frac{K_2}{K_t} \right|$ is obtained via eq(13)

$$p \approx (D\Delta \rho_L)^{-1} \tag{22}$$

For semiflexible polymers eqs(9,10) and (22) yield

$$p \approx \Delta^{-1} P^{-4/3} D^{-5/3} \rho_p^{-5/3} \tag{23}$$

Even though this scaling analysis has been set up for repulsive interactions, it is not unreasonable to suppose that eqs(22) and (23) are not too bad for macromolecules interacting via very weak chiral forces of an attractive nature. Indeed, the concentration dependence in eq(23) is very near the $\rho^{-1.8}$ variation discovered for PBLG by DuPré and Duke [13]. (The slight molecular weight dependence of both K_2 and K_t could be caused by end effects [2,6]).

On the whole, incorporation of chain semiflexibility enables us to resolve a large number of puzzles that have plagued theorists for years. How well these new ideas allow for an explanation of nematodynamics remains to be seen.

REFERENCES

1. Khokhlov, A.R. and Semenov, A.N., Liquid-crystalline ordering in a solution of long persistent chains. Physica, 1981, 108A , pp. 546-56.
2. Khokhlov, A.R. and Semenov, A.N., Liquid-crystalline ordering in a solution of partially flexible macromolecules. Physica, 1982, 112A , pp. 605-14.
3. Odijk, T., On the statistics and dynamics of confined or entangled stiff polymers. Macromolecules, 1983, 16 , pp. 1340-4.
4. Doi, M., Effect of chain flexibility on the dynamics of rodlike polymers in the entangled state. J.Polym.Sci.Polym.Symp., 1985, 73 , pp. 93-103.
5. Onsager, L., The effects of shape on the interaction of colloidal particles. Ann. N.Y. Acad.Sci, 1949, 51 , pp. 627-59.
6. Odijk, T., Theory of lyotropic polymer liquid crystals; Macromolecules, 1986, 26 , pp. 2313-29.
7. Odijk, T., Scaling theory of the isotropic-liquid crystalline phase transition in a solution of wormlike polymers, Polym.Comm., 1985, 26 , pp. 197-8.
8. Yamakawa, H., Modern Theory of Polymer Solutions, Harper & Row, New

York, 1971.

9. Enomoto, H., Einaga, Y. and Teramoto, A., Viscosity of concentrated solutions of rodlike polymers, Macromolecules, 1985, $\underline{18}$, pp. 2695-702.

10. Priest, R.G., Theory of the Frank elastic constants of nematic liquid crystals, Phys.Rev.A, 1973, $\underline{7}$, pp. 720-9.

11. Straley, J.P., Frank elastic constants of the hard-rod liquid crystal, Phys.Rev.A, 1973, $\underline{8}$, pp. 2181-3.

12. Odijk, T., Elastic constants of nematic solutions of rodlike and semiflexible polymers, Liq.Cryst., 1986, $\underline{1}$, pp. 553-9.

13. DuPré, D.B. and Duke, R.W., Temperature, concentration and molecular weight dependence of the twist elastic constant of cholesteric poly-γ -benzyl-L-glutamate, J.Chem.Phys., 1975, $\underline{63}$, pp. 143-8.

14. Taratuta, V., Lonberg, F and Meyer, R.B., Concentration study of the viscoelastic properties of a polymer nematic liquid crystal. Abstract PY-06 of the 11th International Liquid Crystal Conference, Berkeley, 1986.

15. Grosberg, A. Yu and Zhestkov, A.V., Dependence of coefficients of elasticity of a nematic polymer liqud crystal on the rigidity of the macromolecules, Vysokomol. Soed., 1986, $\underline{28}$, pp. 86-91.

16. de Gennes P.G., Polymeric liquid crystals: Frank elasticity and light scattering, Mol.Cryst.Liq.Cryst.Lett., 1977, $\underline{34}$, pp. 177-82.

17. Meyer, R.B., Macroscopic phenomena in nematic polymers. In Polymer Liquid Crystals, eds. A. Ciferri, W.R. Krigbaum and R.B. Meyer, Academic, New York, 1982, pp. 133-63.

18. Doi, M. and Kuzuu, N., Structure of the interface between the nematic and istropic phases consisting of rodlike molecules, J.Appl.Polym.Sci. Appl.Polym.Symp., 1985, $\underline{41}$, pp.65-8.

19. Odijk, T., The depletion layer of a lyotropic polymer liquid crystal at the air-solution interface, Macromolecules, 1987, in press.

20. Straley, J.P., Theory of piezoelectricity in nematic liquid crystals, and of the cholesteric ordering, Phys.Rev.A, 1976, $\underline{14}$, pp. 1835-41.

CONFORMATIONALLY DISORDERED MESOMORPHIC PHASES OF CYCLODODECANE AND CYCLOTETRADECANE

H.Drotloff, G.Kögler, D.Oelfin and M.Möller[*]
Institut für Makromolekulare Chemie Universität Freiburg,
Herman-Staudinger-Haus, Stefan-Meier-Str. 31,
D 7800 Freiburg, Federal Republic of Germany

ABSTRACT

Solid state structure of cyclododecane and cyclotetradecane and their dynamics have been investigated by temperature dependent CP/MAS-^{13}C- and solid state ^{2}H-NMR spectroscopy. NMR resonances are averaged by high molecular mobility at temperatures above an order-disorder transition, which was also detected by differential scanning calorimetry. The onset of molecular motion is however already evident in the ordered crystal. Different symmetry may explain the remarkably high difference in the changes in entropies at the transition from a fully ordered crystal to an orientationally disordered state.

RESULTS AND DISCUSSIONS

Cyclododecane and cyclotetradecane are examples of flexible ring molecules which undergo solid-solid transitions from the fully ordered crystal to a phase of intermediate order. In contrast to plastic crystals and liquid crystals, the positional and orientational orders are preserved to a great extent. A sharp transition in the DSC-trace demonstrates, that a cooperative molecular motion occurs at the transition from the ordered arrangement.

In spite of the similar molecular structure, the transition entropies of cyclododecane and cyclotetradecane are remarkably different. The solid-solid transition of cyclododecane is 135 K below its melting transition at 333.8 K. The transition entropy of 2.8 J/K·mol is small compared to the melting entropy of 44 J/K·mol [1]. In contrast, the disordering transition of cyclotetradecane occurs only 7 K below its melting point, and involves a change in entropy that is nearly twice as large as the increase in entropy upon melting [1].

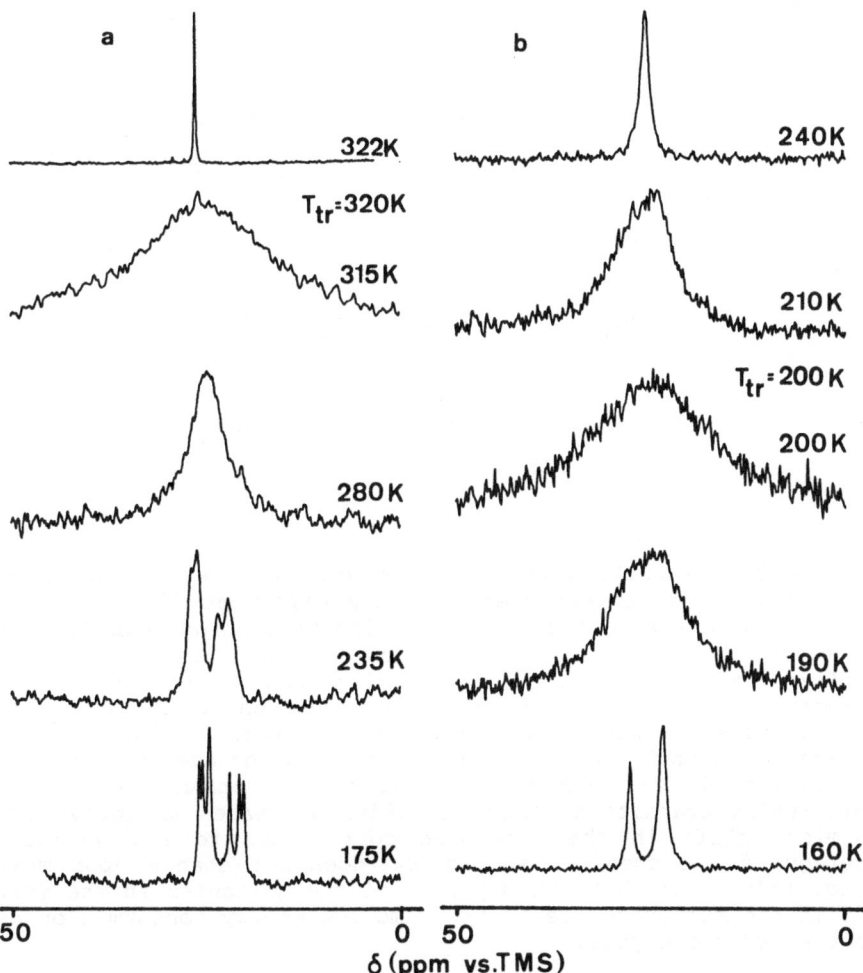

Figure.1: ^{13}C-NMR-spectra   of   cyclotetradecane   (a)   and cyclododecane (b)

Fig. 1 shows the ^{13}C-NMR spectra of $C_{12}H_{24}$ and $C_{14}H_{28}$ as a function of temperature. At low temperatures, well resolved resonance lines can be assigned to C-atoms in conformationally different ring positions (Table 1). As demonstrated by X-ray data   and   force   field   calculations,   the   lowest   energy conformations of cyclododecane (ggagga)$_2$ and cyclotetradecane (ggaggaa)$_2$ are similar [2,3].

TABLE 1
Chemical shifts (ppm) of the C-atoms in conformationally different ring positions .

	$C_{12}H_{24}$	$C_{14}H_{28}$
ga.gg	23.58	23.57
		23.03
ga.ag	-	24.87
ag.ga	27.91	27.45
gg.aa	-	28.90
		28.76

Below the transition to the mesophase, line width increases with increasing temperatures with the different lines finally collapsing to a single broad peak. Line broadening results from a transition from slow to fast exchange of atoms between the different topological sites of the ring. Coalescence of the NMR resonances apparently occurated with the onset of molecular interconversion. Above the coalescence temperature the spectra of both compounds show an additional line broadening, due to modulations in C-H decoupling by molecular motion [5]. At the solid-solid transition, the line width decreases abruptly. The chemical shift in the mesophase corresponds to the weighted average of the chemical shifts and does not change upon melting, indicating that the majority of the molecules in the melt and in the mesophase are in the same low energy conformation as in the ordered crystal.

Fig. 2 shows the ^{2}H-NMR-spectra of partially deuterated cyclododecane and cyclotetradecane. At low temperatures, a broad Pake-spectrum with $\Delta\nu_\perp$ = 125 kHz, typical of completely rigid molecules, is observed for both compounds. In agreement with the ^{13}C-NMR-results, molecular motion sufficiently fast to average the spectra is already observed below the solid-solid transition, leading to a line narrowing.

Additionally, within the high temperature phase of cyclododecane the total line width $\Delta\nu_\parallel$ decreases from about 100 kHz at 205 K to about 32 kHz at 325 K, indicating a change of molecular motion within the mesophase. This may be explained by the

onset of a 2-site jump like rotation around one or both symmetry axes in the ring plane at the transition to the mesophase.

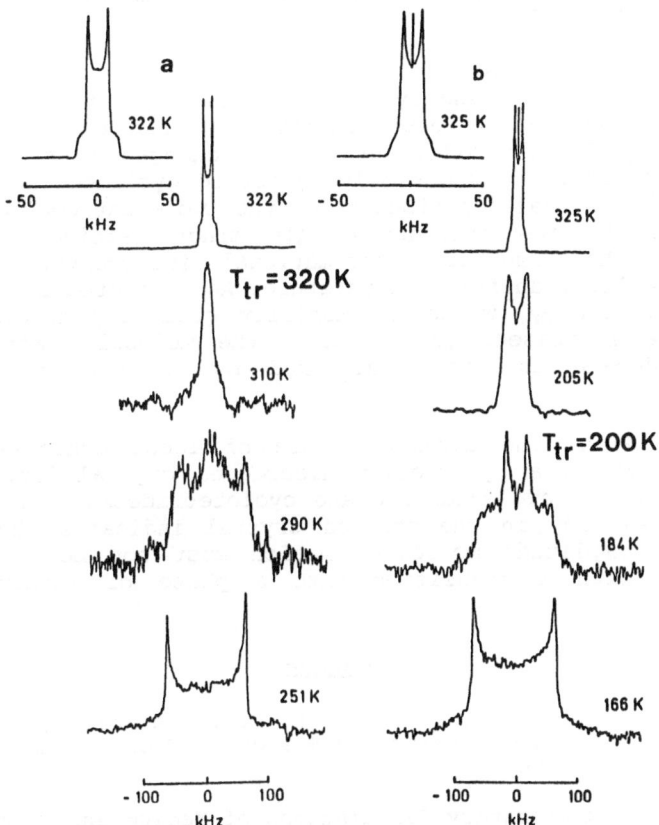

Figure.2: ^{2}H-NMR-spectra of cycotetradecane (a) and cyclododecane (b).

The ^{2}H-NMR spectra of $C_{14}H_{26}D_2$ are not essentially different.
However, compared with cyclododecane, the change in line width
occurs in a narrower temperature range. At higher temperatures,
the ^{2}H-NMR spectra of cyclododecane and cyclotetradecane have
the same line width $\Delta\nu_\parallel = 32$ kHz and the same axialsymmetric
shape, indicating that molecular motion is the same in both
compounds.

As demonstrated by X-ray diffraction, the molecules in the
mesophase of both compounds exist in a regular packing [3,4].
Ring inversion occuring within this regular packing by pseudo-
rotation or successive conformational changes may explain the
averaging of the NMR-spectra, which is observed already below
the transition temperature. While inversion of cyclododecane

converts one enantiomeric form into the other, e.g. $(g^+g^+ag^+g^+)_2$ into $(g^-g^-ag^-g^-)_2$, a given conformation of cyclo-tetradecane and its mirror image resulting from ring inversion are identical [2,3].

Due to different symmetry in the lowest energy conformation of cyclododecane (D_4) and cyclotetradecane (C_2), the same reorientational motion can result in different degrees of disorder and thus explain the different entropy changes at the solid-solid transition. For the highly symmetric cyclododecane molecules, a statistical distribution of the two enantiomeric forms within the lattice introduces only minor lattice defects. Therefore, the ease of conformational interconversion may explain the low transition temperature. As indicated by ^{13}C and ^2H-NMR-spectroscopy, molecular mobility gradually develops as temperature increases. In contrast, the molecular motion of cyclotetradecane is highly hindered by an ordered lattice packing.

In conclusion, the dynamic nature of local conformational changes leads to a dynamically disordered crystal lattice in the mesophase of cyclododecane and cyclotetradecane. The onset of motion already in the ordered crystal indicates, however, that large amplitude molecular motion must not be directely correlated with a transition into a phase of intermediate order.

REFERENCES

1. Drotloff, H., Emeis, D., Waldron, R.F. and Möller, M., Chain Folding and Mesomorphic States of Cycloalkanes, Polymer in press.

2. Dale, J., Exploratory Calculations of Medium and Large Rings, Part 1. Conformational Minima of Cycloalkanes, Acta Chem. Scand. 1973, 27, pp 1115-1129.

3. Dunitz, J.D. and Shearer, H.M.M., Die Strukturen der mittleren Ringverbindungen, III. Die Struktur des Cyc-lododecans, Helv. Chim. Acta., 1960, 43(1), pp 18-35.

4. Drotloff, H., Rotter, H., Emeis, E. and Möller, M., The Mesomorphic State of Cyclotetradecane, submitted to J. Am. Chem. Soc..

5. Rothwell, W.P. and Waugh, J.S., Transverse Relaxation of Dipolar Coupled Spin Systems under rf Irradiation: Detecting Motion in Solids, J. Chem. Phys. 1981, 74(5), pp 2721-2732.

DYNAMICS AT INTERFACES

W. Gronski and G. Stöppelmann

Institut für Makromolekulare Chemie der Universität Freiburg,

Hermann-Staudinger-Haus, D-7800 Freiburg

ABSTRACT

Poly(isoprene-b-styrene) blockcopolymers with deuterated styrene or isoprene segments at the block boundary were investigated by ^{2}H NMR in a broad temperature range. In each case two thermal transitions are observed, one for segments at the interface, the other for the core of the microdomains. The spectra and the nonexponential spin lattice relaxation of both kinds of systems are discussed with respect to the structure of the interface.

INTRODUCTION

The nature of the interface in polymer multiphase systems has been studied extensively in the case of blockcopolymers from a structural point of view by SAXS(1,2) by DSC(3) and dynamic mechanical measurements(4). The results were interpreted in general in terms of a smooth homogeneous transiton between the microphases. Recently a model of a locally sharp boundary of a statistical structure was proposed to explain the difference between theoretical and experimental values of the width of the transition(5). Because additional information can be gained by experiments which are sensitive to segmental motions at the interface we started a ^{2}H NMR study of poly(styrene-b-isoprene) blockcopolymers having alterna-

tively deuterated styrene or isoprene segments of well defined length at the block boundary.

MATERIALS AND METHODS

Poly(styrene-b-isoprene) blockcopolymers were synthesized by
sequential anionic polymerization in benzene. They are characterized by 50 wt% styrene, 80% cis-1,4 content and M_n =
10000 g/mol. In the case of the material published here the
major part of the blocks, 40 monomer units, was labeled with
1,4-d_4 isoprene or 1,2-d_3 styrene at the block
junctions(SI$_D$I and SS$_D$I). Deuterated polystyrene and polyisoprene of M_n = 20000 g/mol was also prepared as reference
systems. ^{2}H NMR spectra were taken with a Buker CXP 300 NMR
spectrometer at 46 MHz using the solid echo pulse technique. Spin lattice relaxation times were measured with a
saturation recovery sequence consisting of five saturation
pulses followed by the solid echo sequence.

RESULTS

Fig. 1 shows that the motional behaviour of the isoprene
blocks in the deuterated blockcopolymer(SI$_D$I) and polyiso-
prene(PI$_D$) as seen by ^{2}H NMR is distinctly different. The
dynamical glass transition on the ^{2}H NMR scale occurs at a
mean motional frequency of 4×10^5 Hz corresponding to the
inverse width of the spectrum in the rigid limit. The transition is associated with a minimum of the intensity and
occurs at 240 K for the homopolymer and at 257 K for the
blockcopolymer and are shifted by ca 40° to higher temperature relative to the DSC glass transitions. While a homogeneous line narrowing in a narrow temperature interval is
observed for PI$_D$ the transition in SI$_D$I occurs in two
stages. The appearence of a motionally narrowed component at
257 K indicates the main transition of the polyisoprene
domains. In a subsequent broad temperature interval up to
270 K a broad second component of nearly constant width of

ca. 108 kHz persists simultaneously with the narrow component. With increasing temperature the intensity of the latter increases at the expense of the first. Thus there is
direct spectroscopic evidence of the existence of motionally
restricted isoprene sequences at the block boundary.

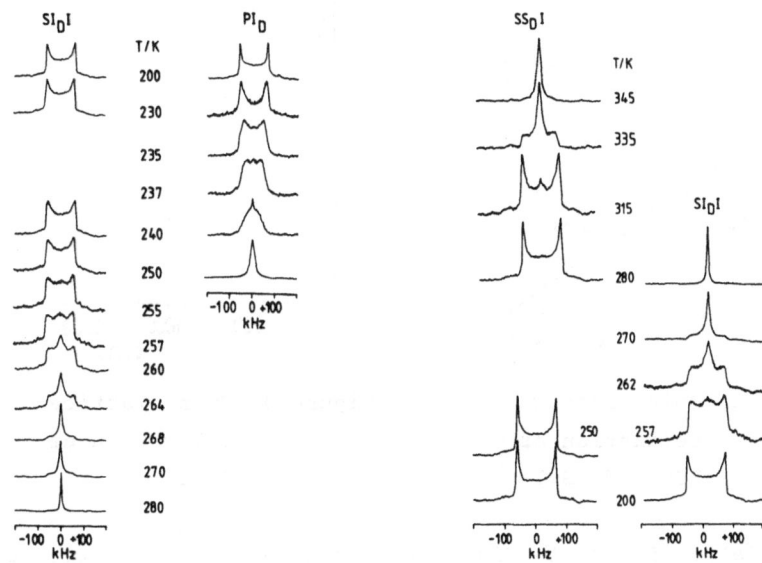

Figure 1. ^{2}H spectra of   Figure 2. ^{2}H spectra of
    SI$_D$I and PI$_D$ SI$_D$I and SS$_D$I

A similar behaviour is observed for the blockcopolymer with
deuterated styrene segments at the block boundary(SS$_D$I)
which is compared to SI$_D$I in Fig. 2. The main transition of
the inner core of the hard domains occurs at ca 335 K at the
intensity minimum. The onset of formation of a motionally
narrowed component at 315 K corresponds to the onset of
molecular motion of styrene segments at the domain boundary.
It has its correspondance to the softening of the isoprene
boundary sequences occurring at approximately the same frequency but at a lower temperature of 270 K. The apparently
low degree of motional coupling between unlike sequences at
the domain boundary is evident even more at constant temperature. At 280 K, for instance, all isoprene sequences,
including those at the domain boundary are fully mobile and
a narrow line is observed whereas the styrene component is

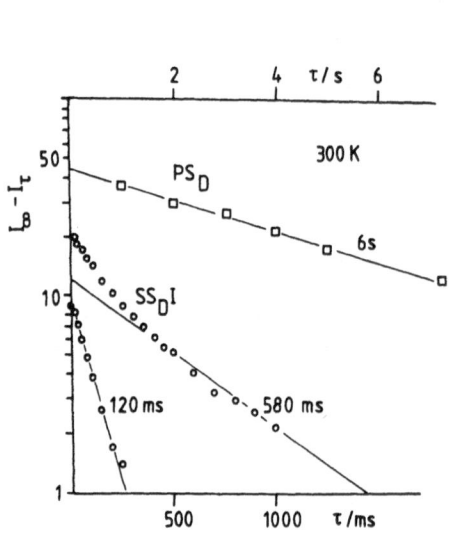

Figure 3. Spin lattice
relaxation of
PS$_D$ and SS$_D$I

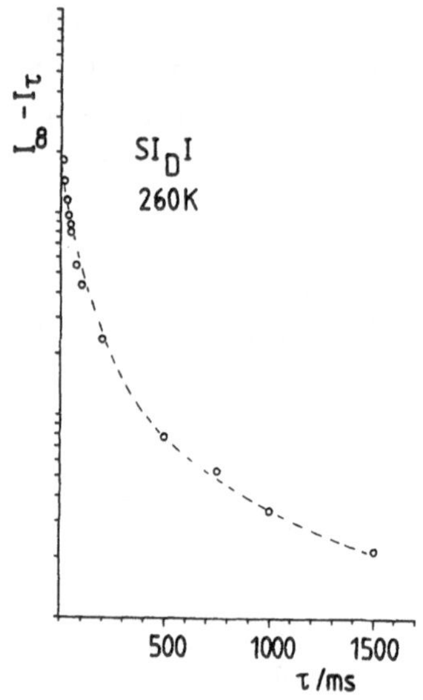

Figure 4. Spin lattice
relaxation of
SI$_D$I

essentially rigid. In Fig. 3 the spin lattice relaxation of
polystyrene is compared to the relaxation of the styrene
deuterated blockcopolymer at 300 K. The latter exhibits a
clear two component behaviour whereas the relaxation of the
homopolymer can approximately be described by a single T_1.
This two component behaviour is present over a temperature
range from 250 K – 340 K, i.e. also at temperatures where
only one component appears to be present according to the
spectrum. In contrast to the behaviour of the hard segments
the spin lattice relaxation of the soft segments , as exem-
plified for 260 K in Fig. 4, cannot be decomposed in two
components at any temperature even if the spectrum indicates
the presence of two components(s. Fig. 1).

DISCUSSION

If the two components are intimately intermixed in an interphase at the domain boundary the compositional gradient
should give rise (a) to a broad distribution of correlation

times of both components and (b) to a strong motional coupling. Whereas a broad distribution of T_1 values is in fact observed for the polyisoprene block, the two component T_1 behaviour of the styrene block cannot easily be understood by this model. If the component with long T_1 is assigned to the core of the hard domain the other component has to be associated with styrene segments in the interphase. The single T_1 value of this component indicates a constant composition of the interphase in contrast to the broad T_1 distribution of the isoprene block. This behaviour and the low degree of motional coupling between the components evident in the spectra point to a spatial separation of unlike segments which would be true in the case of a statistical boundary structure(5). The broad T_1 distribution of PI segments can then be explained by a dynamical gradient along mobile chains attached to a hard wall on one side and by the presence of soft segments embraced by hard material at the irregular boundary structure. The two T_1 components then correspond to the hard core and to the jagged parts of the coarse boundary structure of the hard microdomains.

REFERENCES

1) Siemann, U., Ruland, W., Colloid Polym.Sci., 1982, 260, 999

2) Annighöfer, F., Gronski, W., Makromol. Chem., 1984, 185, 2213

3) Morese-Seguela, St. Jacques, M., Renaud, J. M., Prud'homme, J., Macromolecules, 1980, 13, 100

4) Kraus, G., Rollmann, K. W., J. Polym. Sci., 1976, 14, 1133

5) Ruland, W., Macromolecules, 1987, 20, 87

SUPER-SLOW MOTIONS IN CRYSTALLINE POLYOXYMETHYLENE

A COMPLETE ELUCIDATION USING 2D SOLID STATE NMR

A.P.M. Kentgens, E. de Boer and W.S. Veeman

Department of Molecular Spectroscopy

University of Nijmegen, Toernooiveld, 6525 ED Nijmegen, The Netherlands

ABSTRACT

Super-slow motions in the crystalline parts of poly(oxymethylene) are detected via ^{13}C 2D exchange NMR. Determination of the orientation of the ^{13}C chemical shift tensor in the molecule via 2D correlation of dipolar interactions and chemical shift opens the way to a full characterization of the detected motions. It appears that the helical chains, in the crystalline part of POM, perform a screw-like rotation over $\pm200^\circ$. The activation energy for this motion is 20 ± 5 Kcal/mole.

INTRODUCTION

Molecular reorientations of macromolecules are very important because they are believed to influence the mechanical properties of polymers. One of the techniques used to study molecular motions is NMR. The slowest motions that can be studied via conventional NMR relaxation measurements are on a millisecond time scale. Spiess[1] used the Jeener-Broeckaert three-pulse technique to investigate even slower ($\tau_c \geqslant 1$ s) molecular reorientations in deuterated samples. We have shown that it is also possible to study such super-slow motions via natural abundance ^{13}C NMR using 2D exchange NMR[2,3].

At this moment most information about these slow molecular motions is obtained from mechanical and dielectrical relaxation measurements. Such experiments are, however, in general not very suitable to give an accurate description of the type of motion on a molecular level. The α-relaxation in poly(oxymethylene) (POM) for instance has been studied extensively, and is generally attributed to motions within the crystalline regions. Various models have been proposed, but no unique answer has emerged until this study.

RESULTS AND DISCUSSION

Poly(oxymethylene), with chain structure $(-CH_2-O-)_n$, is the polymerization product of formaldehyde. The stable form of POM has trigonal symmetry, with the chains arranged in a helical conformation. Within a single crystal only helices with the same handedness are found, with right or left handedness equally probable. Huggins[4] proposed the 9/5 model, meaning that the unit cell consists of chains with 9 CH_2-O monomer units in 5 turns of the helix. The local tetrahedral symmetry of the CH_2O_2 unit appears to be hardly distorted by the helix formation (O-C-O bond angle $\sim 111°$). The H-C-H angle is 109° and the C-H bond length is 1.09 Å. The H-C-H plane makes an angle of 55° with the helix axis[5]. This information of the proton positions allows us to find the orientation of the carbon chemical shift tensor within the molecule. Bulk-crystallized POM has a spherulitic structure, consisting of crystalline lamellae connected to each other by amorphous material. The lamellae, in which the molecular chains are folded, are approximately 100 Å thick[6].

Preliminary 2D exchange experiments on bulk crystallized POM (commercially available Hostaform-C from Hoechst) showed that there are indeed slow molecular motions, with a correlation time $\tau_c \approx 1$ sec at 50°C, present in the crystalline part of the polymer[3]. These results were obtained with Andrew-Beams type rotors at a spinning speed of 1500 Hz. At this speed the sidebands are very small, allowing no exact analysis of the pattern. The weak points of the Andrew-Beams type rotors for this kind of study is long-time instability of the rotor speed, and the fact that it will not allow rotation at speeds below $\sim$ 1200 Hz. Therefore the experiments were carried out once more in a double-bearing MAS probe with temperature control. A cylindrical piece of Hostaform-C was machined to fit exactly in the double-bearing rotor. Thus stable spinning could be achieved over a large interval of spinning frequencies and various temperatures.

Fig. 1 shows the result of the 2D exchange experiment, with a mixing time of 1 s, applied to POM at a temperature T = 316 K, and a spinning speed of 700 Hz. Besides the spinning sideband pattern of POM on the diagonal, some extremely weak cross-peaks show up indicating exchange between orientationally different sites. At a temperature of 336 K, the off-diagonal peaks are already very intense for $\tau_m = 1$ s. Increasing the mixing time (Fig. 2) or the temperature results in a further increase of the off-diagonal intensity. This temperature dependence shows that we are dealing with molecular motions and not with spin diffusion.

Analysis of these 2D spinning sideband patterns can yield the relative orientation of the CSA tensor of a group of spins in the detection period with respect to their orientation in the evolution period. In order to obtain information about the type of motion causing the off-diagonal peaks, it is essential that we know the orientation of the CSA tensor within the molecule and its principal values. The principal values were recovered from a static 1D powder pattern giving σ_{xx} = 67 ppm, σ_{yy} = 86 ppm and σ_{zz} = 111

166

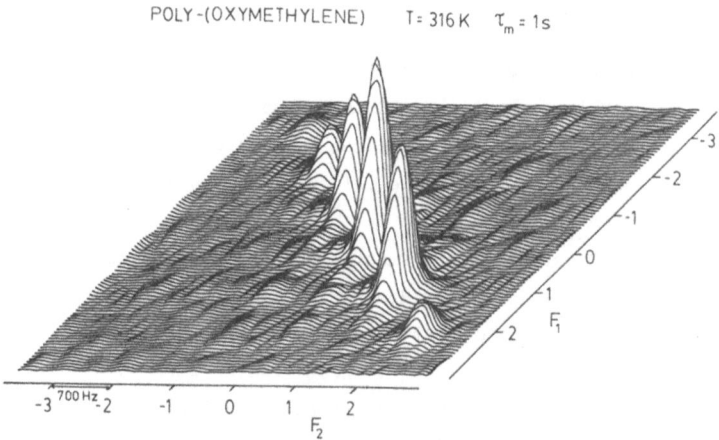

POLY-(OXYMETHYLENE)   T= 316 K τ_m = 1s

Figure 1. ^{13}C 2D exchange spectrum of poly(oxymethylene) at 316 Kelvin, with a rotor synchronous mixing time of 1s. The spinning speed was 700 Hz.

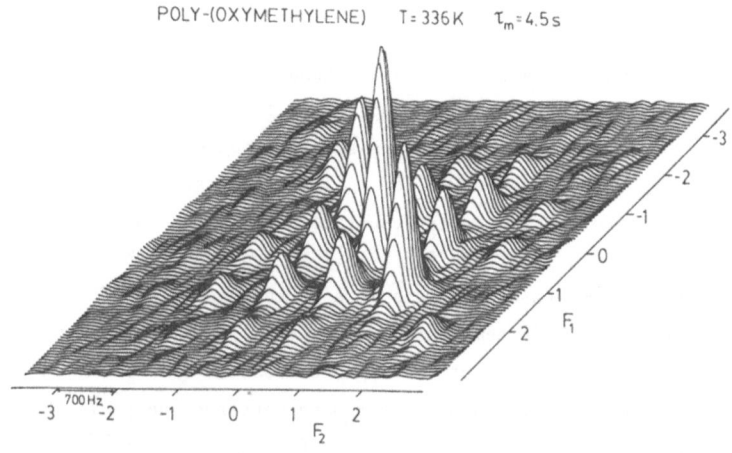

POLY-(OXYMETHYLENE)   T= 336 K τ_m= 4.5s

Figure 2. ^{13}C 2D exchange spectrum of POM at 336 Kelvin with a mixing time of 3150 rotor periods at a spinning speed of 700 Hz (τ_m = 4.5 s).

ppm. The orientation of the shielding tensor within the monomer unit was determined by correlating the chemical shielding anisotropy to the hetero-nuclear dipolar ^{13}C-^{1}H interaction, whose orientation is known. From this 2D dipolar correlation spectrum it appears that the orientation of the CSA tensor is determined by the local tetrahedral symmetry of the CH_2O_2 unit[7] i.e. the Z axis is oriented perpendicular to the O-C-O plane, and makes an angle of 55° with the helix axis. The Y axis lies within the O-C-O plane and bisects the O-C-O angle. The X axis is of course perpendicular to the Y and Z axis.

From the experimental 2D exchange spectra it is seen that the off-diagonal intensity is of considerable magnitude. This suggests that the motion involved cannot be restricted to the loops of the polymer chains at the surface of the crystal lamellae or to crystal defects. As the majority of polymer chains are involved in the motion a rotation of the chains, combined with a translation, seems the most plausible mechanism. When the helical polymer chains (9 monomer units in 5 turns) make a 200° jump, their starting and end position will be identical, and thus there is no elevation of the energy. So the energy required for this motion to occur, is the energy required to overcome the barrier to chain rotation. In a bulk-crystallized polymer, which is thought to consist of spherulites with crystalline lamellae connected to each other with amorphous material, the rotation of polymer chains in the crystals might be brought about by torsional forces exerted on the chains in the crystalline phase by chains in the highly mobile amorphous phase. Another explanation is that lattice defects move through the crystal, leaving behind a polymer chain that is rotated over 200°.

Hence, 2D spinning sideband intensity patterns were calculated for rotations over 0° for polymer chains that have not changed their position during the mixing time, ±200° for chains that have effectively undergone one jump during the mixing time, ±400° etc. and combined using the appro-priate statistics for a certain average jump time Ω. These simulations were then compared to the peak intensities in the experimental spectra. It appears that we are able to simulate the experimental spectra very well with this model.

In Fig. 3 $\log(\Omega)$, as obtained from the simulations, is plotted against the reciprocal temperature. The error intervals in this plot were obtained by varying Ω until simulations diverted to much from the experimental spectra, and are thus somewhat arbitrary. This plot obviously shows an

Arrhenius type of behaviour for the rotational motion. The activation energy is found to be 20 ± 5 kcal/mole, in agreement with values found from mechanical measurements[6,8], although also higher values have been reported[6].

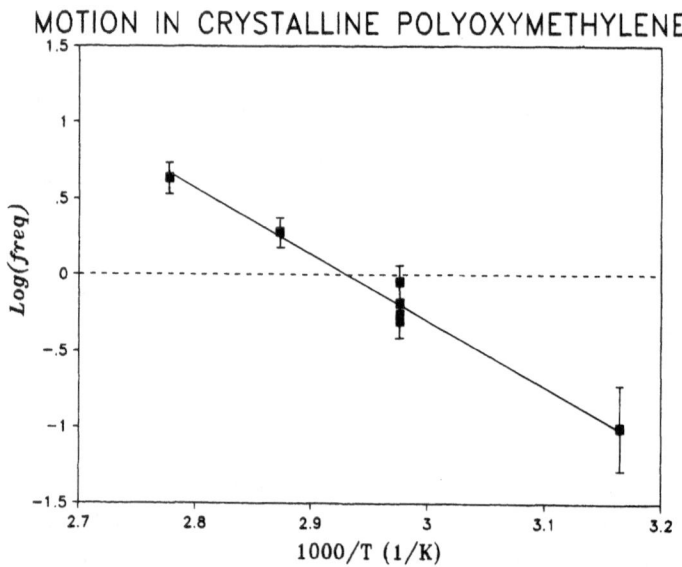

Figure 3. Arrhenius plot of the motion detected in poly(oxymethylene).

REFERENCES

1 H.W. Spiess, Advances in Polymer Science 66, 37, 1985.

2 A.F. de Jong, A.P.M. Kentgens and W.S. Veeman, Chem. Phys. Lett. 109, 337, 1984.

3 A.P.M. Kentgens, A.F. de Jong, E. de Boer and W.S. Veeman, Macromolecules 18, 1045, 1985.

4 M.L. Huggins, J. Chem. Phys. 13, 37, 1945.

5 T. Uchida and H. Tadokoro, J. Polym. Sci. A2 5, 63, 1967.

6 N.G. McCrum, B.E. Read and G. Williams, "Anelastic and Dielectric Effects in Polymeric Solids", Wiley, London, 1967.

7 A.P.M. Kentgens, Thesis 1987 (available on request).

8 H.W. Starkweather, Macromolecules 19, 2538, 1986.

MOLECULAR MOBILITY OF OLIGOMERIC METHYL-H-SILOXANES

R. Kosfeld, R. Krause, M. Heß
Universität -GH- Duisburg
FB 6/Physikalische Chemie

ABSTRACT

Linear and cyclic oligomeric Methly-H-siloxanes were prepared and investigated by ^{29}Si-NMR. Chemical shifts, T_1-Relaxation and NOE were measured isothermally for different chain length and ring size. In most cases it was possible to calculate the dipolar share of relaxation behaviour and thus to give some interpretation on ring geometry and chain flexibility at room temperature.

INTRODUCTION

Siliciumorganic polymers have been used frequently because of their outstanding thermic properties. Investigations of Polydimethylsiloxanes [1,2,3] show that this is due to the great mobility of the methyl side-chains and the increased mobility of the -Si-O-Si- back-bone compared with the -C-C-chain of pure hydrocarbons.

In contrast to the information gained by ^{1}H or ^{13}C spectroscopy, the chemical shifts of the ^{29}Si nuclei enables to distinguish between the construction units (M, D, T, Q) of the polyorganosiloxanes [4,5,6,7,8].

Beside the spectral parameters of the ^{29}Si spectroscopy like chemical shift, relativ intensity and coupling constants, further information about structure and mobility along the molecule chain (in quality and quantity) can be derived from nuclear magnetic relaxation studies [9,10].

Relaxation studies of the ^{29}Si nucleus are suitable to analyse the motion processes and structure of these compounds, because relaxation mechanisms are correlated with the distances and kind of neighbouring atoms and the translation and rotation of the molecules [11]. It was shown that different kinds of mechanisms are possible for the spinlattice relaxation process. The whole relaxation rate is given by [6,9]:

$$1/T_1 = 1/T_1^{DD} + 1/T_1^{SR} + 1/T_1^{CSA} + 1/T_1^{SC}$$

DD = dipol-dipol interaction; SR = spin-roation interaction
SC = scalar interaction; CSA = chemical shift anisotropy

The share of the dipol-dipol relaxation on the T_1-relaxation can be obtained by measuring the Nuclar Overhause Effect (NOE) which is quantified by the Overhauser factor [12,14]. In the present case:

$$\frac{I - I_o}{I_o} = \eta_{Si-H} = \frac{\gamma H \cdot T_1}{2\gamma_{Si} \cdot T_{1DD}}$$

is valid, with $\eta_{SI-H} \equiv$ NOE of the observed system; $\gamma H \equiv$ magnetogyric ratio of the proton; $\gamma_{Si} \equiv$ magnetogyric ratio ofthe proton the ^{29}Si-nucleus; I = signal intensity of the decoupled ^{29}Si-nucleus; I_o = signal intensity of the coupled ^{29}Si spectrum

Because of the complexity of ^1H coupled ^{29}Si spectra, the interpretation of the relaxation experiments is only possible for decoupled ^{29}Si spectra. To prevent the decrease of signal intensity by an increasing NOE the following conditions have to be regarded (9): AQ $\leq$ T /20, AQ = aquisition time; I/T $\geq$ 5; ; PI/AQ $\geq$ 100, PI = impulse interval The inversion recovery method was used.

Linear Methyl-H-Siloxanes

$$(CH_3)_3Si-O-\underset{\underset{CH_3}{|}}{\overset{\overset{H}{|}}{Si}}_a-O-\underset{\underset{CH_3}{|}}{\overset{\overset{H}{|}}{Si}}_b-O-\underset{\underset{CH_3}{|}}{\overset{\overset{H}{|}}{Si}}_c-O(-\underset{\underset{CH_3}{|}}{\overset{\overset{H}{|}}{Si}}-O-)_n-$$

Table 1

$\delta^{29}Si$/ppm [*]	T_1/s	η	T_{1DD}/s		
+10.20	54.6 ± 5.7	-0.40	344	(16 %)	
+10.18	58.1 ± 5.4	-0.50	293	(20 %)	
+10.14	57.8 ± 5.3	-0.40	364	(16 %)	M
+10.00	56.3 ± 4.7	-0.40	355	(16 %)	
-34.32	41.4 ± 4.6	-1.82	57.3	(72 %)	
-34.33	42.6 ± 3.5	-1.83	58.7	(73 %)	D_a'
-34.71	41.1 ± 4.0	-1.98	52.3	(79 %)	
-34.73	40.7 ± 3.8	-1.72	59.5	(68 %)	
-35.04	41.1 ± 6.1	-1.92	53.9	(76 %)	
-35.25	41.3 ± 4.2	-1.75	58.8	(70 %)	
-35.27	39.7 ± 4.2	-1.80	55.6	(71 %)	D_b'
-35.38	40.8 ± 6.2	-1.87	55.0	(74 %)	
-35.71	41.3 ± 4.2	-1.91	54.5	(76 %)	
-35.73	42.5 ± 3.8	-1.82	58.8	(72 %)	
-37.77	41.5 ± 4.8	-1.83	57.0	(73 %)	D_c'
-35.75	40.3 ± 2.7	-1.82	55.8	(72 %)	

Cyclic Methyl-H-Siloxanes

Asymmetrically substituted cycles have configuration-isomers, which could not be isolated up to now. Possible configuration isomers of the Tetramethyl-cyclotetrasiloxane are:

O = Si
o = O
Me = —CH₃

Results of the relaxation experiments at 298 K and 1m concentration in C_6D_6:

δ^{29}Si /ppm*)	T_1/s	η	T_{1DD}/s
-31.85	53.0 ± (3.0)	-2.18	61.7 (86 %)
-31.87	52.5 ± (0.8)	-2.21	59.3 (88 %)
-32.15	48.6 ± (1.0)	-2.20	55.8 (87 %)
-32.16	46.9 ± (0.5)	-2.26	52.3 (89 %)
-32.51	38.2 ± (0.8)	-2.24	43.0 (89 %)

$$D_5'$$

δ^{29}Si/ppm*)	T_1/s	η	T_{1DD}/s
-33.65	53.6 ± 1.8	-1.80	75 (71 %)
-33.76	57.7 ± 1.5	-1.82	80 (72 %)
-33.83	53.7 ± 1,3	-1.83	74 (72 %)
-33.89	56.8 ± 0.9		
-33.95	54.4 ± 0.6		
-33.97	54.5 ± 0.7	no values obtainable	
-33.99	53.7 ± 0.9		
-34.05	52.3 ± 1.7		
-34.07	53.9 ± 0.7		

$$D_6'$$

δ^{29}Si/ppm*)	T_1/s
-34.07	48.2 ± 1.4
-34.18	47.1 ± 0.9
-34.17	49.7 ± 0.7
-34.24	48.2 ± 0.7
-34.28	49.9 ± 1.0
-34.32	49.9 ± 0.4
-34.35	49.2 ± 0.6
-34.38	49.6 ± 0.5
-34.41	48.4 ± 0.5
-34.45	48.1 ± 0.5
-34.48	47.2 ± 0.9

*) TMS as reference

The terminal units of linear polymethyl-H-siloxanes dissipate round about 70% of their surplus energy via spin-rotation mechanism after nuclear magnetic exitation. However 70% of the relaxation of the D' into their state of equilibrium is goverened by the dipolar mechanism.
The share of the dipolar relaxation in the D'-units of linear polymethyl-H-siloxanes is about as high as those one of the D'_5-cycle.
The mobility of the D' units is comparable with in both the systems. But the share of the dipolar relaxation is larger in D'_4 and D'_6 cycles. Steric conditions lead to an increase of the number of protons close to the ^{29}Si nuclei.Thus the different relaxation times of the ^{29}Si nuclei of the D'_4 cycle isomeres plead against a planar constellation of the system. The renewed increase of the dipol relaxation in the D'_6 can be explained by the conformation of this cycle. The geometry of D'_6 cycle must be conditioned in a manner that not neighboured D' units close up with their protons that the Si nucleus under observation is enabled to relax by a dipolar mechanism.
All experiments were executed with a 300 MHz FT-NMR-Spectrometer, type CXP 300 (Fa. Bruker). The resonance frequency of the ^{29}Si nucleus is 59.63 MHz.

ACKNOWLEDGEMENT
The Deutsche Forschungsgemeinschaft is greatly indepted for her financial support.

Literature

[1] A.W. Henry, G.J. Saffert
    J. polym. Sci. A-2, 7, 433 (1969)
[2] V.R. Honnold, F. McCaffrey, B.A. Morowka
    J. Appl. Phys. 25, 1219 (1954)
[3] C.M. Huggins, L.E.St. Pierre, A.M. Bueche
    J. Phys. Chem. 64, 1304 (1960)
[4] A.L. Smith
    Chemical Analysis Vol. 41: Analysis of silicones,
    John Wiley & Sons, New York (1974)
[5] J. Schraml, J.M. Bellama, in:
    Determination of organic structures by physical methods"
    Vol. 6, Academic Press, New York
[6] H. Marsmann
    in "NMR, Basic Principles and Progress" Vo. 17, S.
    65-235, Springer Verlag, Berlin (1981)
[7] G. Engelhardt, H. Jancke
    J. Organometal. Chem. 28, 293-300 (1971)
[8] T. Axenrod, G.A. Webb (editors) in:
    "Nuclear magnetic resonance spectroscopy of nuclei other
    than protons",
    G.C. Levy, J.D. Cargioli
    ^{29}Si Fourier Transform NMR, S. 275 (1974)
[9] G.C. Levy, J.D. Cargioli, P.C. Juliano, T.D. Mitchell
    J.A.C.S. 95, 3445 (1973)
[10] T. Saluvere, J. Puskar, E. Lippmaa, G. Engelhardt
    Proc. XIXth Congress Ampere, Heidelberg 1976
[11] R. Gerhards
    Dissertation, Universität Bochum (1977)
[12] K.F. Kuhlmann, D.M. Grant, R. K. Harris
    J. Chem. Phys. 52, 3432 (1970)

CONFORMATIONAL CHANGES IN A SERIES
OF SOLUBLE POLYDIACETYLENES

D.J. Ando, D. Bloor, S.N. Bedford, J.S. Obhi
Department of Physics, Queen Mary College,
Mile End Road, London E1 4NS,
U.K.

and

S. Mann
GEC Research Limited, Marconi Research Centre,
West Hanningfield Road, Chelmsford, Essex CM2 8HN,
U.K.

ABSTRACT

New members of the n-BCMU class of soluble polydiacetylenes have been synthesised and their solvato-chromism studied in an attempt to resolve existing controversies regarding the nature of the polymer conformations in solution. Evidence is presented for the existence of relatively long-lived metastable intermediates of these polymers in solution. Preparative HPLC has also been used to isolate fractions of polydisperse 2-BCMU polymer, and analysis by spectroscopic methods has attempted to correlate chromism with molecular weight in this particular polymer system.

INTRODUCTION

A number of different series of soluble polydiacetylenes (PDAs) now exist, but of these, the class containing the symmetrically substituted butoxycarbonylmethylurethane (BCMU) side group, first reported by Patel in 1978 [1], has been most extensively studied.

$$R = -(CH_2)_n O.CO.NH.CH_2 CO.O(CH_2)_3 CH_3$$

n-BCMU, where n = 2,3,4,6,9.

The solvato- and thermo-chromism displayed by these soluble polymers, however, continues to attract a considerable amount of research effort and controversy. While it is clear that the chromism is a direct feature of the order-disorder transformation of the dissolved polymer, the nature of the polymer conformation in these phases continues to encourage active debate [2,3].

Within the n-BCMU series, investigations have centred on the 3- and 4-

BCMU compounds. Although some evidence exists to the contrary, it has generally been assumed that the transition between the ordered and disordered forms is a direct one. In this present work we have extended the solvatochromism studies to include other homologues in the series, by the synthesis and characterisation of the n=2,6 and 9 polymers, and present evidence to show the existence of relatively long-lived metastable intermediates of these polymers in solution. The 2-BCMU polymer has been reported previously [4], but little studied, while the n=6 and 9 species are new; the latter was conceived as an extension of the docosadiyne diol-acetate class of soluble PDAs, reported by Plachetta et al [5]. A preliminary account of the 9-BCMU polymer has recently appeared [6]. In addition, it appears that conformational analysis of the 3- and 4-BCMU species, as previously reported, is complicated by the polydispersity of the samples, and it is difficult to obtain well-defined molecular-weight fractions for examination. For the 2-BCMU derivative, polymerisation gives a product with a relatively low average molecular weight (compared to the 3- and 4- species, see Table 1), but with an extremely broad molecular weight distribution. Typical products, examined by gel permeation chromatography contain both oligomeric and high molecular weight species. Preparative High Performance Liquid Chromatography (HPLC) has been used in our work to separate fractions of this polymer, and preliminary qualitative results are reported for the dependence of the solvato-chromic behaviour as a function of molecular weight.

MATERIALS AND METHODS

All of the n-BCMU monomers were prepared by a slightly modified version of the method originally reported by Patel [1]. The corresponding polymers were obtained by γ-ray (^{60}Co) irradiation of the monomers, with dose rates ranging from 10-50 Mrads; unconverted monomer was extracted with acetone. The optical spectra were recorded with a Perkin-Elmer Lamda 9 UV-Vis-NIR spectrophotometer. Chromatography was carried out using a Philips HPLC system consisting of a PU4011 solvent delivery system, a model PU4700 auto-injector and a model PU4020 UV absorbance detector. Separations were achieved using a Lichrosorb (Merck) 25cm x 4.6mm column packed with silica gel (Si-60, 230 mesh grade), with dichloromethane and methanol as elution solvents.

RESULTS

Initial studies of 9-BCMU polymer dissolved in chloroform-hexane mixtures have indicated that the solvato-chromic transitions occur on a comparatively long time scale [6]. We have chosen, therefore, to record the evolution of the absorption spectra with time for systems containing an excess of hexane. These are shown in Figs. 1 and 2, for 9-BCMU and unfractionated 2-BCMU, respectively. Absorption data for all of the polymers under discussion, are collected together in Table 1, and where available, molecular weight measurements are also shown.

DISCUSSION

Unfractionated Polymers

Optical spectra of all of the n-BCMU polymers, when dissolved in a good solvent, such as chloroform, show great similarities in both their shape and peak absorption energy, which occurs at about 21,300cm^{-1} ("yellow" form) (see Table 1). This indicates a common morphology, i.e. a "worm-like" chain as reported by Wenz et al [3], Allegra et al [7] and Aime and Schott [8]. The side groups appear to have very little influence on the backbone, which suggests that side group hydrogen bonds are not formed, otherwise there would be significant differences between, for example, the 2- and the 9- polymers.

TABLE 1

Absorption and molecular weight data for n-BCMU polydiacetylenes

POLYDIACETYLENE	MOLECULAR WEIGHT, $\bar{M}_w$ (DP)	crystal λmax/nm (cm^{-1})	cast film λmax/nm (cm^{-1})	yellow sol^n λmax/nm (cm^{-1})	red sol^n λmax/nm (cm^{-1})	blue sol^n λmax/nm (cm^{-1})
2-BCMU	7.00×10^4 (150)	-	-	462 (21,650)	528 (18,950)	549 + 598 (18,215 + 16,730)
3-BCMU	1.36×10^6 (2830)	637 (15,700)	633 (15,800)	469 (21,300)	-	629 (15,900)
4-BCMU	1.6×10^6 (3150)	633 (15,800)	534 (18,700)	465 (21,500)	529 (18,900)	-
6-BCMU	-	-	546 (18,300)	468 (21,350)	526 (19,025)	-
9-BCMU	-	639 (15,640)	631 (15,850)	468 (21,350)	538 (18,600)	619 (16,151)

When a "poor" solvent, such as hexane, is added, two types of chromic behaviour are observed, viz:-

a) The development of both a long and an intermediate wavelength energy absorption ("blue" and "red" forms, respectively), which is seen for the 2-, 3-, and 9-BCMU polymers (see Figs. 1 and 2). In the case of the 3-BCMU species, however, the intermediate form is only very short-lived [9].

b) The appearance of an intermediate absorption peak only, i.e. the "red" form with the long wavelength signal being absent, which is found for the 4- and 6-BCMU polymers.

It should be noted here that there is no "odd-even" property alternation i.e. the 2-BCMU polymer does not behave as its 4- and 6- analogues.

Resonant Raman spectroscopy of the 9-BCMU polymer shows that the side-groups are disordered in the yellow solution, partially ordered in the red solution, and completely ordered in the blue solution. There is strong evidence to support the observation that the intermediate red form is meta-stable [6]. This is true, also, for both the 3- [9] and 2-BCMU materials [10] (see Fig. 2).

There is some sensitivity of the absorption maxima of the red forms to the side groups, i.e. varying between 18,600 and 19,025cm^{-1}, but this variation is only slightly larger than that observed for the yellow forms. In contrast, the blue forms show a much wider variation (Table 1), which can be likened to crystalline, i.e. more ordered, species, where internal strain can result in large differences in the energies of absorption maxima. Bitler and Wudl have recently reported the synthesis of polydiacetylene oligomers, endcapped by t-butyl groups, containing up to 38 carbon atoms; extrapolation of their electronic spectra data to an infinite chain length gives a value of 18,900cm^{-1} for the maximum absorption [11], corresponding to values found for the red and surprisingly not the yellow forms of the BCMU polymers. Indications are that these oligomers are more extended, possibly resulting from the intrinsic chain stiffness imposed by the pendant groups. In contrast, the soluble PDAs have large, disordered sidegroups in yellow solutions, which impose disorder on the backbone.

This sensitivity to changes in the side-groups, plus the very slow kinetics of changes after step reductions in solvent quality, as seen for the 9- and 2-BCMU polymers (Figs. 1 and 2) support previous studies of the 3-BCMU material [3,8], and suggests that this transformation to the blue form is consistent with a microcrystallisation phenomenon, aggregation processes

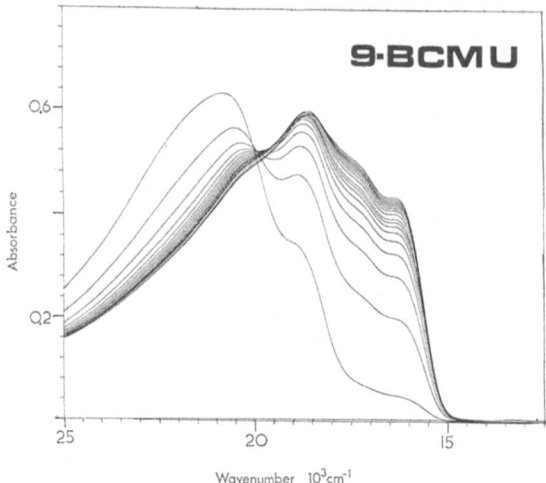

Figure 1. Evolution over a period of 2 hours of the spectra of 9-BCMU polymer in chloroform-hexane solution (40% : 60% by volume, respectively)

occuring even at low concentrations [6]. Further evidence is provided by
structure details in the spectrum of 2-BCMU (Fig. 2), where the double peaks
are similar to those seen for the low temperature phase of the toluene sulphonate (pTS) diacetylene [12]. The close relationship of the forms is
shown by a clear isobestic point in the spectra and is further supported by
fluorescent spectroscopy.

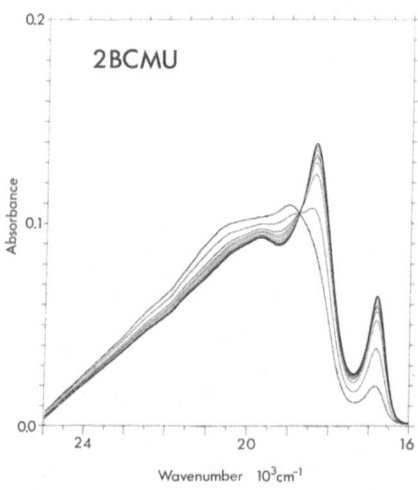

Figure 2. Evolution over a period of 2 hours of the spectra of unfractionated 2-BCMU polymer in chloroform-hexane solution, showing the transformation from the partially to the fully ordered forms.

It is instructive to consider why the 4- and 6-BCMU materials do not undergo further transformations from their intermediate forms. Although it remains a possibility that spectra could contain an overlap of intermediate and final absorptions, this seems unlikely because there are no sharp features present. A more reasonable explanation is that the intermediate form is thermodynamically more stable, a consequence of problems associated with the packing of the side groups.

Fractionated Polymers

Preliminary investigations of the spectra in chloroform of the fractionated 2-BCMU polymer, obtained by HPLC, show no striking changes in solvatochromism for the various fractions, when recorded as a function of increasing hexane composition. However, the trends do appear to be as expected and can be qualitatively assigned to a reduction in average molecular weight. In particular, a progressive increase in absorption at short wavelengths, with fraction examined, can be identified with an increasing contribution of a low molecular weight component. It is significant that a peak, occuring near $21,650cm^{-1}$, equivalent to the yellow, disordered form, is essentially independent of molecular weight. This is not unreasonable if the possibility of chains being disordered on the scale of 1-2 repeat units is considered. Excitation is confined within segments which are shorter than the chain length, for all chains other than the very shortest oligomers. Such strong localisation from weak disorder is surprising but is in fact consistent with recent neutron scattering studies reported for the 3-BCMU polymer [8]. It is worth mentioning that the spectra of one of our "lower" molecular weight fractions show a peak at $30,000cm^{-1}$, close to that reported by Bitler and Wudl, for a diacetylene "dimer" species [11].

Similar qualitative effects of changes in molecular weights are also observed for the corresponding time-evolved spectra, in this case working with a fixed concentration of 67% hexane : 33% chloroform (by volume), recorded after initial rapid mixing. Although the lower molecular weights show slower transformations to more ordered forms, some caution in interpretation must be exercised here, however, as these effects could be due either to concentration factors, or to differences in effective chain lengths, i.e. there is a smaller effective volume for shorter chain species so that they take longer to "find" one another and aggregate.

It is clear that better characterised samples are required before any quantitative interpretation of the effects of molecular weight dependence can be attempted. Corresponding thermochromic measurements also need to be made. Differences in thermal behaviour of unfractionated and separated "higher" molecular weight species have been observed in initial experiments from our laboratory, and it has also been possible to obtain well formed polymer spherulites from the fractionated sample, over a range of temperatures [13]. This work is continuing.

CONCLUSIONS

The spectroscopic results presented above indicate that the existence of an intermediate partially ordered conformation in solution is a general feature of the n-BCMU class of soluble PDAs. While further transformations, where observed, appear to be the result of a crystallisation process, the nature of the initial disordered form still remains uncertain. However, a model where disordered sidegroups are attached to a rigid backbone, is consistent with the proposed concept of a worm-like chain. Preliminary measurements carried out on fractionated polymer emphasise the need for well-characterised monodisperse samples to fully establish the effect of molecular weight on chromic behaviour.

178

This work was supported by funding from the Science and Engineering Research Council and the Department of Trade and Industry under the Joint Opto-Electronic Research Scheme. J.S.O. thanks the SERC for a studentship. We wish to thank Dr. P. Clay of Imperial College, London, for γ-irradiation of the monomers, and Mr. A.J. Etherton, at GEC-Marconi Research Centre for recording the absorption spectra of the fractionated samples.

REFERENCES

1. Patel, G.N., Soluble polydiacetylenes I. Synthesis and properties. Polym. Prepr., Am.Chem.Soc., Div.Polym.Chem., 1978, 19 154-9.

2. Lim, K.C., Sinclair, M., Casalnuovo, S.A., Fincher, C.R., Wudl, F. and Heeger, A.J., Ordered States for polymers : rod-to-coil transition of a conjugated polymer in solution. Mol.Cryst.Liq.Cryst., 1984, 105, 329-52.

3. Wenz, G., Muller, M.A., Schmidt, M. and Wegner, G., Structure of poly (diacetylenes) in solution. Macromols., 1984, 17, 837-50.

4. Patel, G.N. and Miller, G.G., Structure-property relationships of di-acetylenes and their polymers. J.Macromol.Sci.,Phys., 1981, B20, 111-31.

5. Plachetta, C., Rau, N.O., Hauck, A. and Schulz, R.C., Some soluble polydiacetylenes. Makromol.Chem.,Rapid Commun., 1982, 3, 249-54.

6. Bloor, D., Ando, D.J., Obhi, J.S., Mann, S. and Worboys, M.R., Meta-stable intermediate structures in polydiacetylene solutions. Makromol. Chem.Rapid Commun., 1986, 7, 665-70.

7. Allegra, G., Bruckner, S., Schmidt, M. and Wegner, G., Chain statistics of poly(diacetylenes) in solution. Macromols., 1986, 19, 399-405.

8. Aime, J.P. and Schott, M. Conformation of conjugated polymers and their relation to electron delocalisation. In Electronic Properties of Polymers : IWEPP 87, ed. H. Kuzmany, M. Mehring and S. Roth, Springer-Verlag, Heidelberg, 1987, to be published.

9. Chance, R.R., Washabaugh, M.W. and Hupe, D.J., Kinetics of a planar-nonplanar conformational transition in polydiacetylene solutions. Chemtronics, 1986, 1, 36-8.

10. Bloor, D., Structural and spectroscopic studies of polydiacetylenes. In Crystallographically Ordered Polymers, ed. D.J. Sandman, American Chemical Society, Washington, 1987, pp.128-39.

11. Bitler, S.P. and Wudl, F., Polydiacetylene oligomers. PMSE Preprints, Am.Chem.Soc.,Div. Polym.Mater.Sci.Eng., 1986, 54, 292-6.

12. Bloor, D., Fisher, D.A., Batchelder, D.N., Kennedy, R.J., Cottle, A.C., Lewis, W.F. and Hursthouse, M.B., A spectroscopic study of the second order phase transition in bis(p-toluene sulphonate) diacetylene poly-mer crystals. Mol.Cryst.Liq.Cryst., 1979, 52, 83-92.

13. Campbell, A.J., Private Communication.

TRANSIENT ELECTRIC BIREFRINGENCE STUDIES ON POLYELECTROLYTE SOLUTIONS

Wilhelm Oppermann
Institute of Physical Chemistry,
Technical University Clausthal,
Arnold-Sommerfeld-Str. 4, 3392 Clausthal-Zellerfeld,
West-Germany

ABSTRACT

The dynamic electric birefringence of dilute solutions of poly(styrene sulfonate) was studied with respect to the influence of concentration and the pattern of the electric field. The experiments revealed that the shape of the macroions is affected by the application of a static field, particularly in the case of very high molar mass polyelectrolytes. This change of molecular shape interferes with the transient rise and decay of the birefringence. The change of molecular shape can be avoided if the orientation of the macroions is induced by high-frequency AC pulses. This experimental method is therefore suitable to obtain the unperturbed dimensions of the macroions. It is shown that the macroions adopt a rod-like conformation at very low concentration. The rod-to-coil transition occurs approximately in the theoretically predicted concentration range.

INTRODUCTION

Solutions of polyelectrolytes show an extraordinarily strong birefringence on application of an electric field. The specific Kerr constant is generally several orders of magnitude larger than for non-ionic macromolecules [1]. There are two major reasons for this behaviour. First, the uncoiling of flexible macroions due to the electrostatic repulsion of ionic groups increases the persistence length of the chains, and the optical properties of the molecules may approach those of a rigid rod. Secondly, the orientation of the macroion in an electric field is caused by the polarization of the ionic atmosphere rather than by its dipolar and dielectric structure. This polarization arises from the displacement of counterions along the macroion and depends greatly on the ionic state of the solution [2].

The investigation of the electric birefringence is a powerful tool in studying polyelectrolyte solutions. It can be used to obtain information about the electrical and conformational properties of the macroions [3]. In the last years, some attempts have been made to predict theoretically the size and shape of a macroion as a function of concentration of polyelectrolyte and concentration of added low molecular salt [4,5]. The present study aims to come to an experimental verification of such theoretical predictions.

EXPERIMENTAL SECTION

Measurements of the electric birefringence were performed on solutions of sodium poly(styrene sulfonate) in double distilled water. Polyelectrolytes of various molar masses were obtained from Pressure Chemical Comp. and Polymer Standards Service. They had been made by sulfonation of nearly monodisperse polystyrene, $M_w/M_n \leqslant 1.06$. A comprehensive characterisation of the materials used has been given elsewhere [6].

The apparatus employed for the measurements of the electric birefringence is described in detail in ref. [6]. It offers the opportunity to apply rectangular DC pulses whose duration and field strength could be adjusted in wide ranges. It can also be used to produce electric fields whose polarity may be reversed once or repeatedly during the experiment, application of the latter method resulting in bursts of rectangular high-frequency AC fields of adjustable frequency. The birefringence of the solutions were determined as a function of time with a maximum resolution of 100 ns.

RESULTS AND DISCUSSION

Steady-State Birefringence with DC Pulses

Fig. 1 shows the dependence of the specific birefringence, $\Delta n_o/c$, at steady-state on the concentration of the solution. At very low concentrations, $\Delta n_o/c$ is independent of c. This indicates that within this regime the shape and size of the macroion does not change with concentration and that the birefringence is simply the sum of the contributions of separate molecules. The value of $\Delta n_o/c$ rises with increasing molar mass of the polyelectrolyte.

At higher concentrations, $\Delta n_o/c$ drops markedly with increasing concentration. This can be interpreted such that the conformation of the macroion changes with concentration. It agrees well with the results of viscosity measurements on polyelectrolytes where an increase of the reduced viscosity with decreasing concentration is generally observed.

A potential explanation for these findings is that the macroion adopts the shape of a statistical coil at high concentration. With decreasing concentration, a gradual uncoiling occurs until the macroion is fully extended. This is re-flected in an increase of $\Delta n_o/c$ with falling concentration. At the stage where the molecules are fully extended, no further change of size is possible and the horizontal course of $\Delta n_o/c$ is observed.

The theory developed by Odijk predicts the critical concentration, c*, where the rod-to-coil transition should occur. Such computed values for c* are represented as the arrows in Fig. 1. It is seen that there is a reasonable agreement between c* and those concentrations, where the curves change from the horizontal course into the steep, concentration dependent course.

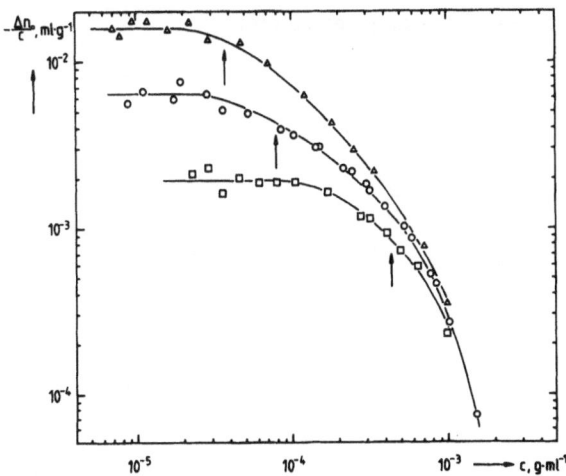

Figure 1. Specific birefringence at steady-state as function of concentration
$\triangle$: M=690000; o : M=390000; $\square$: M=70000 g mol^{-1};
E = 2.5·10^5 V m^{-1}; T = 20°C; arrows indicate calculated c*.

Difference between AC and DC Pulses

Fig. 2 shows two curves for the time dependent birefringence of a solu-tion of poly(styrene sulfonate) having M_n=1.1·10^6 g mol^{-1} at the same concen-tration. The upper curve has been obtained with an AC pulse of approx. 4.5 ms duration at a frequency of 50 kHz. The lower curve shows the bire-fringence generated by a DC pulse of the same field strength, the polarity being reversed after 1.7 ms.

It is striking that the AC pulse gives rise to about twice the steady-state birefringence although the field strengths are identical. It is further noticed that the rise and decay of the birefringence occur symmetrically to each other as expected for a purely induced dipole moment only when the AC field is applied. The time constant is much larger than in case of the DC field. A molecular model for the phenomenon observed has been proposed and discussed in detail elsewhere [6,7]. Basically, it is assumed that the AC pulse does not alter the shape of the macroion, and hence the time constants can be used to calculate the rotational diffusion coefficient of the unperturbed molecule. A DC field, however, causes a coiling of the molecule. The change of molecular shape interferes with the transient rise and decay, and the rotational diffusion coefficient becomes time dependent. As a result, the mean relaxation time for the rise curve is smaller than that of the field-free decay, and both are smaller than the one obtained with AC pulses. This effect is the larger the higher is the molar mass of the polyelectrolyte.

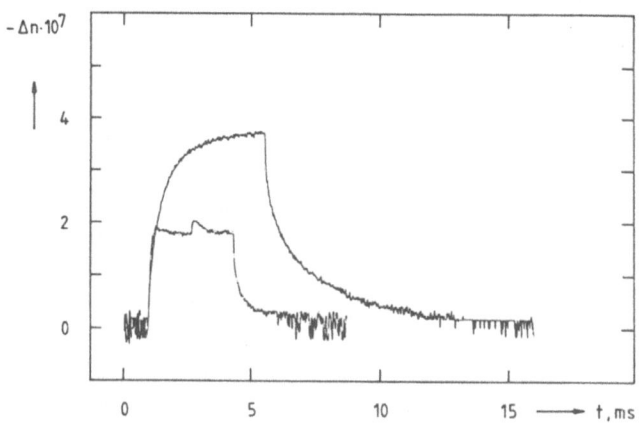

Figure 2. Birefringence curves for poly(styrene sulfonate), $M_n = 1.1 \cdot 10^6 \, g \, mol^{-1}$
c = $1.9 \cdot 10^{-5}$ g ml^{-1}; E = $1.3 \cdot 10^5$ V m^{-1};
upper curve: high-frequency AC field, 50 kHz;
lower curve: static field with reversal of polarity.

Analysis of the Relaxation Curves

When the orientation of the macroions in the very dilute regime is brought about by a high-frequency AC pulse, the field-free decay of the birefringence occurs essentially as a single exponential. Fig. 3 shows a log-log plot of the relaxation time, τ_1, versus the molar mass of the polyelectrolyte. The data points may be fitted to a straight line having a slope of 2.9. A slope close to 3 is expected for rigid rods.

In addition, it is possible to calculate the absolute values of τ_1 using the Broersma equation for the rotational diffusion coefficient, D_r, and the relation

$6 D_r \tau_1 = 1$. Such a calculation yields the dashed line in Fig. 3. (The diameter of the rod was assumed to be 1 nm, the results being rather insensitive to this choice.) The agreement between the experimental data and the calculated curve is surprisingly good. This is a second proof that the macroions adopt a rod-like conformation at very low concentrations.

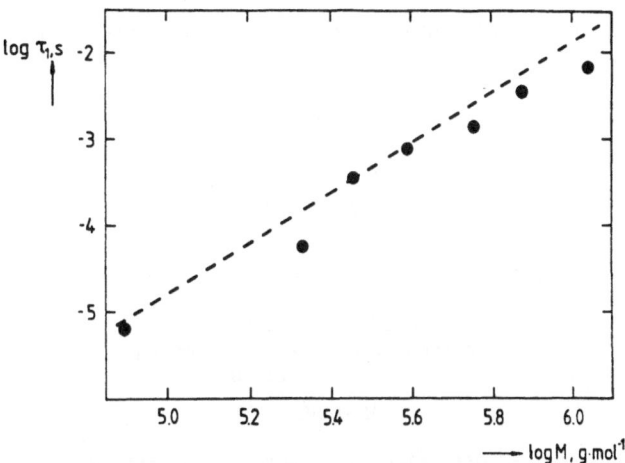

Figure 3. Relaxation time τ_1 (obtained by AC-pulse technique) as a function of the molar mass of the polyelectrolyte;
--- : calculated curve assuming rod-like molecules.

ACKNOWLEDGEMENT

Generous support for this work by the Deutsche Forschungsgemeinschaft is gratefully acknowledged.

REFERENCES

1. Tsvetkov, V.N. and Andreeva, L.N.; Adv. Polym. Sci. **39**, 95 (1981)

2. Mandel, M.; Mol. Phys. **4**, 489 (1961)

3. Wijmenga, S.S., van der Touw, F., Mandel, M.; Polym. Comm. **26**, 172 (1985)

4. Odijk, T.; J. Polym. Sci. Phys. **15**, 477 (1977); Macromolecules **12**, 688 (1979)

5. Skolnick, J. and Fixman, M.; Macromolecules **10**, 944 (1977)

6. Oppermann, W.; Habilitationsschrift, Clausthal (1986)

7. Oppermann, W.; Makromol. Chem. RC, submitted

ELECTRIC BIREFRINGENCE RELAXATION OF POLYELECTROLYTES WITHOUT ADDED SALT

S.S. Wijmenga
National Institutes of Health,
NIDDK,
Laboratory of Chemical Physics,
Bethesda, MD 20892
USA

present address: Katholieke Universiteit Nijmegen,
Nationale Hf-NMR Faciliteit (SON),
Biofysische Chemie,
Toernooiveld, 6525 ED Nijmegen
The Netherlands

ABSTRACT

We have investigated with electric birefringence the dynamic behaviour of sodium poly(styrene sulphonates)(NaPSS) in aqueous solutions
without added salt over a range of polyion concentrations encompassing the
dilute regime and two of the semi-dilute regimes predicted by Odijk [1].
We found that in the dilute regime the longest relaxation time τ_1 of the
multiexponential birefringence decay curves becomes concentration independent. Going to higher concentrations into the semi-dilute regimes τ_1 first
increases, goes through a maximum and then decreases. As τ_1 is related to
overall chain motion this concentration dependence can perhaps be understood qualitatively in terms of a competition between the influence of the
increasing hindrance by neighbouring chains and the decreasing chain
dimensions with increasing polyion concentration.

INTRODUCTION

Polyelectrolyte solutions are quite complicated systems as is
apparent from the scaling predictions made by Odijk [1], who extended the
scaling theories developed for polymer solutions to solutions of charged
macromolecules. A plethora of different regimes is predicted to exist.
Experimental support for the predicted existence of a dilute and semi-
dilute regime, in the case of added salt, has been obtained for example by
dynamic light scattering [2] and electric birefringence [3,4]. For polyelectrolytes without added salt a larger number of concentration regimes
is predicted, i.e. a dilute regime and a number of semi-dilute regimes. As
far as we are aware systematic studies covering all these concentration
regimes have not been done. With the electric birefringence technique it
is in principle possible to probe the various predicted concentration
regimes including the dilute regime which can only be reached at extremely
low polyion concentration (for NaPSS of $M=4.4\times10^5$ g/mol the polyion
concentration should be below 0.0048 g/l).

In this preliminary report we present some results of electric birefringence relaxation experiments on solutions of sodium poly(styrene sulphonate) (NaPSS) without added salt over a range of polyion concentrations which should encompass the dilute regime and two of the semi-dilute regimes according to the criteria given by Odijk |1|.

MATERIALS AND METHODS

The two NaPSS samples investigated ($M=4.4 \times 10^5$ and $M=1.03 \times 10^6$ g/mol) were of commercial origin (Pressure Chemical Co.). According to the manufacturer's specification $M_w/M_n < 1.1$. The materials as purchased were exhaustively dialized against deionized water and then freeze-dried. All solutions were prepared with deionized water. The polyelectrolyte concentration of every solution was determined spectrophotometrically at 261 nm |4|, except for the lowest concentrations which were determined from the dilution factor.

The electric birefringence instrument is of classical design and has been described elsewhere |4|. The rectangular electric field pulses to induce the birefringence in the NaPSS solutions were produced either by a Cober pulse generator or by a Systron Donner pulse generator. All the measurements were done with a Kerr cell with low intrinsic birefringence |5| kept at 21°C. The electric field pulses were taken long enough to ensure that the birefringence $\Delta n(t)$ reached a steady state Δn_0. After the electric field is cut off the birefringence $\Delta n(t)$ decays to zero. The moment at which the field is cut off is taken as t=0. The results are here presented in terms of the normalized birefringence defined as $\Delta n_n(t) = \Delta n(t)/\Delta n_0$. Electric birefringence decay curves consist most often of superposition of exponentials. The time constant τ_1 of the longest exponential is generally related to the overall motion of the polyelectrolyte chains |4| and is, therefore, of the main interest here. This time constant was found either from a multi-exponential fit of the decay curves as described elsewhere |4| or by fitting a linear least squares line to the straight tail part of a $\ln(\Delta n_n(t))$ versus t curve |6|.

RESULTS

For both systems investigated ($M=4.4 \times 10^5$ and 1.03×10^6) we found that at intermediate concentrations (between approximately 0.015 g/l and 4.5 g/l for $M=4.4 \times 10^5$ and 0.0027 g/l and 1.5 g/l for $M=1.03 \times 10^6$) the steady state birefringence Δn_0 showed a sign reversal at very low field strengths. (If the poly(styrene sulphonate) chains align on the average along the electric field a negative birefringence will be observed, whereas a positive birefringence will be observed if they align on the average perpendicular to the electric field). For field strengths significantly above and below the field strength E_c where the transition occurred, the birefringence decays $\Delta n_n(t)$ had the usual form, i.e. they were monotonically decreasing functions of time. However, in a narrow range of field strengths around E_c, $\Delta n_n(t)$ took on an abnormal form, e.g. at C=0.15 g/l for $M=4.4 \times 10^5$ $\Delta n_n(t)$ seemed to consist of a negative birefringence signal superimposed on a positive birefringence signal. Outside this concentration range Δn_0 was always negative and no abnormal signals were observed at any field strength investigated. The longest relaxation time τ_1 in the decays as found from either multi-exponential fits or from a linear least squares fit to the straight tail of the $\ln(\Delta n_n(t))$ versus time curves, is shown in figure 1. The relaxation times τ_1 were generally obtained from decays

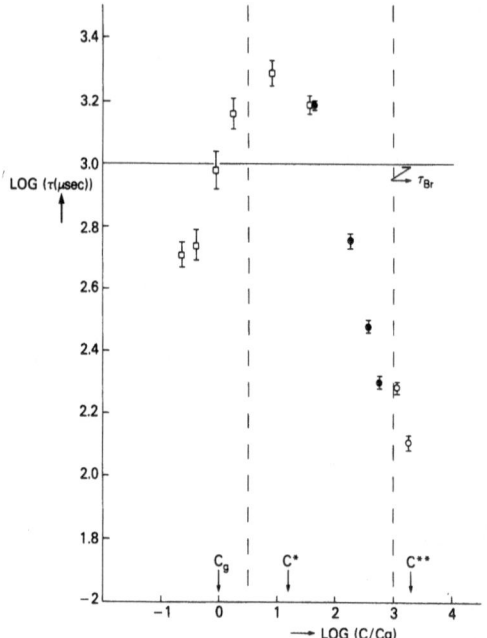

Figure 1: Concentration dependence of τ_1 for M=4.4x10^5 g/mol;
the critical concentrations C_g, C* and C** are defined
in Table 1; the vertical broken lines delineate the
concentration regime between which a sign reversal of
Δn_0 occurs; τ_1 at field strengths E where $\Delta n_0 < 0(\square, \circ)$,
$\Delta n_0 > 0(\bullet)$; τ_{Br}: see text.

measured at the lower field strength range where measurements were done.
At these field strengths this exponential constituted the main part of the
decay. As can be seen from figure 1, τ_1 seems to become independent of
concentration in the dilute regime, i.e. for $C < C_G$. Towards higher concentrations τ_1 increases with concentration, goes through a maximum and then
decreases.

DISCUSSION

In the table 1 the concentration regimes as proposed by Odijk |1|
are summarized as well as the critical concentrations separating them.
Note that according to the relations derived by Odijk, the electrostatic
persistence length L_e of the polyelectrolyte is inversely proportional to
the polyion concentration and that for the NaPSS concentrations investigated here, L_e is much larger than the bare persistence length L_p, so that
the total persistence length $L_t = L_e + L_p = L_e = (1/16 \pi AC)$. A calculation of L_e
according to this relation shows that the chains of the two NaPSS samples
are predicted to be fully extended in the dilute regime $(C < C_G)$. At $C = C_G$ the
fully extended chains start to overlap and it is only at $C = C^*$ that the
chains start to bend appreciably. Above C*, but below C**, the average
distance between overlap points d is, however, still smaller than $2L_e$. It

TABLE 1

Concentration regimes of salt-free polyelectrolytes solutions
according to the criteria of Odijk |1|

$C<C_G=(1/l^2A)$	dilute regime	
$C_G<C<C^*=(1/16\pi QAl)$	semi-dilute I	overlapping rigid chains; at C^* the persistence length equals the contour length l.
$C^*<C<C^{**}=0.04(1/4\pi Q^2A)$	semi-dilute II	the overlapping chains start to bend appreciably.
$C>C^{**}$	semi-dilute III	the average distance between overlap points is larger than $2 L_e$.

l = chain contour length; A = length of a monomer unit; Q = the
Bjerrum length equal to about 7.1 Å in H_2O at 21^oC; for NaPSS A =
2.5 Å and L_p = 10 Å |5|; the contour length follows from $l = (M/M_0) \times A$
where M is the molar mass of the polyion and M_0 is the molar mass of
the monomer unit, which is 206 for NaPSS.

is only at $C=C^{**}$ that d becomes larger than $2L_e$, so that above C^{**} this
semi-dilute regime becomes comparable to the one found in the case of
added salt.

Comparing these predictions with our experimental results, presented
in figure 1, it can be seen that a clear transition can be observed between
the dilute and the semi-dilute regimes, which occurs approximately at C_G.
Below C_G, τ_1 seems to become concentration independent as is expected for
unhindered rotation. However, this τ_1 is smaller than τ_{Br}, the rotation
time of a fully extended chain calculated with Broersma's equation |7|,
indicating that the chains are not yet fully extended. Between C_G and C^*,
τ_1 increases strongly with increasing concentration as is expected for an
extended chain that experiences more and more hindrance in its rotation
|8|. The decrease of τ_1 above C^* can possibly be correlated with the
decreasing chain dimensions predicted to take place above C^*. At the
highest polyion concentrations the decrease of τ_1 can perhaps be understood
in terms of the reptation of a chain through a network |4,9|. A more
detailed analysis of the concentration dependence of τ_1 in terms of the
various models for hindered rotation |8,9,10| will be presented elsewhere.

CONCLUSIONS

From the concentration dependence of τ_1 we can at least conclude that
of the various concentration regimes predicted to exist for polyelectrolytes without added salt clearly a separation can be seen between the
dilute and the semi-dilute regimes at approximately the predicted
concentration.

ACKNOWLEDGEMENT

This investigation has been carried out under auspices of The Netherlands Foundation for Chemical Research (SON) and with financial aid from The Netherlands Organization for the Advancement of Pure Research (ZWO). The experimental work was done in the Department of Physical and Macromolecular Chemistry, Gorlaeus Laboratories, University of Leiden, of Prof. M. Mandel. The author wants to thank J.van der Ploeg for his excellent technical assistance.

REFERENCES

1. Odijk, T., Possible scaling relations for semidilute polyelectrolyte Solutions. Macromolecules, 1979, 12, 688-693.

2. Koene, R.S., and Mandel, M., Scaling relations for aqueous polyelectro-lyte-salt solutions. 1. Quasi-elastic light scattering as a function of polyelectrolyte concentration and molar mass. Macromolecules, 1983, 16, 220-229.

3. Wijmenga, S.S., van der Touw, F. and Mandel, M., Dynamics of poly-electrolytes in aqueous solutions probed by electro-optical relaxation. In Physical Optics of Dynamic Phenomena and Process in Macromolecular Systems, ed. B. Sedlacek, Walter de Gruyter & Co., Berlin, 1985, pp. 87-105.

4. Wijmenga, S.S., van der Touw, F. and Mandel, M., Scaling relations for aqueous polyelectrolyte-salt solutions. 4. Electric birefringence decay as a function of molar mass and concentration. Macromolecules, 1986, 19, 1760-1768.

5. Wijmenga, S.S., van der Touw, F. and Mandel, M., An electric biref-ringence cell with small intrinsic optical retardation. J. Phys. E.: Sci. Instrum., 1985, 18, 673-675.

6. Wijmenga, S.S. and Maxwell, A., Rotational Diffusion of short DNA fragments in polyacrylamide gels: An electric birefringence study. Biopolymers, 1986, 25, 2173-2186.

7. Broersma, S., Rotational diffusion constant of a cylindrical particle. J. Chem. Phys., 1960, 32, 1622-1631.

8. Doi, M. and Edwards, S.F., Dynamics of rod-like macromolecules in concentrated solution. J. Chem. Soc., Far. Trans. II, 1978, 74, 560-570.

9. de Gennes, P.G., Reptation of a polymer chain in the presence of fixed obstacles. J. Chem. Phys., 1971, 55, 572-579.

10. Odijk, T., On the statistics and dynamics of confined or entangled stiff polymers. Macromolecules, 1983, 16, 1340-1344.

SINGLE AND DOUBLE STRUCTURE OF XANTHAN

Guy Muller
UA500 du CNRS "Polymères,biopolymères,membranes"
Faculté des Sciences de l'Université de Rouen
76130 Mont-Saint-Aignan,France

Jacqueline Lecourtier and Guy Chauveteau
Institut Français du Pétrole
BP 311,92500 Rueil-Malmaison,France

ABSTRACT

Two highly pyruvated xanthan samples supplied as fermentation broths have been investigated with regard to their conformational changes associated with the salinity driven order-disorder transition. Analysis and interpretation of the intrinsic viscosity,light scattering and optical activity data as a function of the treatment the samples underwent after the fermentation step show that xanthan can exist under a single or a double strand. This suggests that apparent discrepancies concerning the xanthan structure and solution properties may originate from differences in salinities and temperatures of the treatments that the samples underwent.

INTRODUCTION

Xanthan produced by the bacterium Xanthomonas campestris displays unusual and interesting solution properties which result in its use as a viscosity enhancing agent in the food industry and for controlling water mobility in enhanced oil recovery. The primary structure consists of a linear chain of 1,4-linked β-D-glucopyranosyl residues with charged trisaccharide side chains attached to alternate residues of the main chain to give a pentasaccharide repeat unit (1). The internal mannosyl residues are substituted generally stoechiometrically with an acetyl group whereas the terminal mannose of the side chains are substituted with a pyruvate ketal,generally in less than stoechimetric (Figure 1). The actual pyruvyl and acetyl contents are affected by culture conditions (2-4). As they have an effect on solution properties of xanthan this means that the viscosity and the molecular weight could also be influenced by growth conditions. Xanthan can assume a rod-like helical ordered conformation in solution which melts out into a disordered one under conditions of elevated temperature and low salinity (5-8).

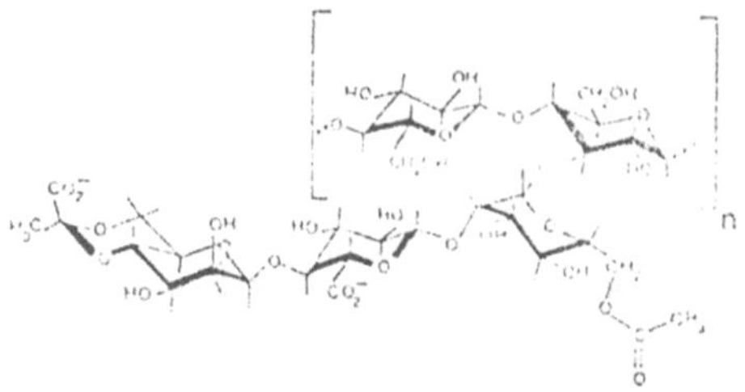

Figure 1. Primary structure of xanthan

The exact nature of this ordered structure in the solid state
is uncertain and two possible conformations are equally consistent with X-ray diffraction patterns namely an extended isolated
helix stabilized by binding of the side chains to the backbone
and a coaxial double-helix structure (8,9),the latter being
suggested by electron microscopy (10). Although the molecular
organization existing in the solid state is not necessarily
prevailing in solution,the same controversy exists concerning
the question of whether the ordered xanthan in solution is
a single or a double strand. The debate is still persisting
despite extensive investigations (7,8,10-16).

The reasons for such apparently contradictory conclusions could
originate from the difficulty to obtain true solutions free of
microgels and also from the fact that comparison generally is
made on xanthan samples the history of which is not comparable
with regard to the growth conditions and the nature of post-
treatments. On the other hand xanthan samples can also contain
different extents of contaminants such as proteins and nucleic
acids (17),the proportion of which can be variable.Unfortunately
exact and precise data concerning all these factors are rather
scarce.
The aim of this paper is to present evidences deduced from
macromolecular probes from which it can be argued that the
ordered rigid conformation of xanthan can be either a single or
a double-stranded helix depending on the treatment the biopolymer underwent after isolation from the fermentation broth.

MATERIALS AND METHODS

Two highly pyruvated xanthan samples manufactured by Rhone-
Poulenc and supplied as fermentation broths were investigated
with regard to their conformational changes (as monitored by
optical activity,low shear viscosity and low angle laser light
scattering) associated with the salinity driven order-disorder
transition. One (X_A) comes from a culture broth which was
pasteurized,the other (X_B) is a truly native product. The initial
ordered conformation (0.1M NaCl) was denatured at room temperature by decreasing the salinity (ultrafiltration). For both
samples the transition is initiated in the NaCl range 10^{-2}-10^{-3}M
and the disordered conformation is found prevailing for lower
salinity as indicated by the optical activity. Restoration of
the ordered conformation was achieved after adjusting the

salinity of xanthan solutions which have been desalted at various levels (10^{-2}-10^{-5}M) at its initial value (0.1M NaCl). The change in molecular characteristics of both samples as a result of such a salinity driven order-disorder-order cycle were analyzed by light scattering and viscosity.

RESULTS

From the comparison between the measured viscosities for fractions obtained by surface exclusion chromatography and the theoretical values calculated from Yamakawa-Yoshizaki's theory (18),X_A has been evidenced to exist as a single helix (q=50 nm,M_L=1000 nm^{-1}) at salinity where the ordered conformation prevails. We observed that the transition of X_A occurs without an associated change in $\bar{M}_w$.In the same way,the viscosity of the renatured ordered form has the same value than that of the initial ordered one (16). Therefore the conformational ordering is only the result of a reorganization of the secondary structure.This agrees with some reported data claiming that xanthan exists as a single isolated helix.As this sample was isolated from a broth which has been heat treated it can be thought that it probably existed in a renatured ordered form before it underwent the above cycle.

Xanthan X_B which can be considered as a truly native product behaves quite differently with regard to its viscosity and light scattering behaviour associated with the salt driven transition process as indicated from data reported in Figure 2.

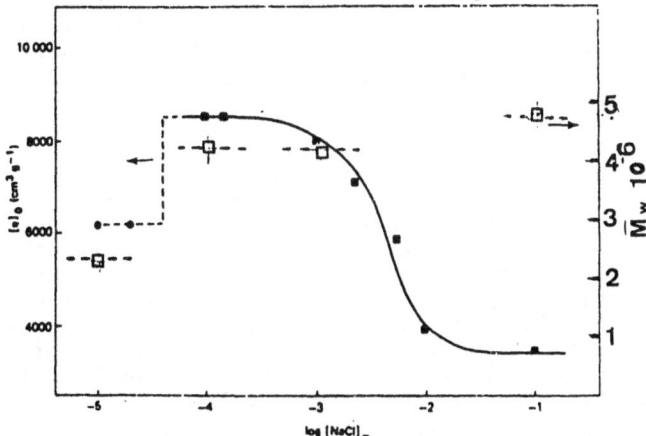

Figure 2. Intrinsic viscosity at zero shear rate (left scale) and molecular weight (right scale) of the native X_B xanthan in 0.1M NaCl after treatment at different salinities $(NaCl)_m$ (in mole/l).

After X_B has been treated at low salinity (10^{-3}-10^{-4}NaCl) the viscosity is increased from 3400 to 8500 cm^3/g whereas in the same time M_w is only slightly affected.This suggest an intra-molecular conformational change as a result of an extension of the native compact structure.After extensive salt removal (below 10^{-5}M NaCl),the viscosity measured in 0.1 NaCl reaches a value near 6100 cm^3/g; in the same time the molecular weight is roughly halved thus indicating that dissociation of the native structure occured (19,20).A persistence length close to 150 nm is found for the extended double strand in good agreement with that reported by Sato (14) for a Kelco sample which was shown to be a double helix.The dissociation at lower salinity also agrees with the observation of Sato (15) who reported dissociation of the native form in cadoxen or after heating in water at high temperature.

DISCUSSION AND CONCLUSION

The above reported data suggest that different ordered structures can exist and that care should be taken when comparing the solution properties of xanthan samples the history of which is not comparable or not precisely known.However it is a fact that the ordered conformation melts out in response to a change in salinity or temperature.We have shown that melting out of x_B could take place without or with dissociation of the native dimeric structure depending at which extent the salt was removed. The same probably holds when heating a native xanthan beyond the melting temperature;the nature of the renatured form obtained by cooling should depend on both the temperature at which xanthan was heated and the external salinity at which heat treatment was performed.The unknown conditions of post-treatment are thought to be at the origin of the debate concerning the nature of the secondary ordered structure of xanthan. In this regard it should be recalled that the xanthan samples (from Kelco) shown by some authors (7,15,21) to be a single helix had been heat treated in pure water.On the other hand Sato(14) has clearly given evidence that the same sample was under a double strand before heat treatment. More recently Rinaudo and Milas (22,23) have reported the changes in the molecular characteristics of a xanthan obtained from a culture broth manufactured by Shell after heating beyond the transition temperature and then cooling at ambient temperature.From the measured viscosity increase and the nearly constancy in molecular weight they concluded that this sample was a single strand with two different ordered conformations only differing by the stiffness of the chain.However in our opinion their results are also compatible with the existence of a double strand for both conformations which is not dissociated probably because the treatment conditions (both in temperature and salinity) were not severe enough.

As a conclusion the origin of the much debated question concerning the existence of the ordered xanthan molecule as a double or a single strand is probably due to the nature of post-fermentation treatments.Both ordered structures should probably exist and it is clear from many observations that a double strand can effectively be dissociated into a single one with the help of various treating processes including extensive desalting at room temperature (19),heating at low salinity (7,14) or change in solvent conditions (15).Three parameters seem to be very important for obtaining dissociation of the native structure namely salinity,temperature and heating time at a given salinity. If the treatment is performed at a salinity not sufficiently low or at a temperature not high enough or during a too short time for given salinity conditions,the transition may occur without dissociation (19,23).

REFERENCES

1. Jansson,P.E.,Kenne,L. and Lindberg,B.,Carbohydr.Res.,1975, 45,275.
2. Cadmus,M.C.,Knutson,C.A.,Lagoda,A.A.,Pittsley,J.E. and Burton,K.A.,Biotechnology and Bioengineering 1978,20,1003.
3. Sutherland,I.W.,Carbohydr.Polymers,1981,1,107
4. Davidson,I.W.,FEBS Microbiology Letters,1978,3,347
5. Holzwarth,G.,Biochemistry,1976,15,4333.
6. Milas,M. and Rinaudo,M., Carbohydr.Res.,1979,76,189.
7. Norton,I.T.,Goodall,D.M.,Morris,E.R. and Rees,D.A.,J.Chem. Soc.,Chem.Commun.,1980,545.
8. Moorhouse,R.,Walkinshaw,M.D.,and Arnott,S.,Am.Chem.Soc.Symp. Ser.,1977,45,90.
9. Ukayama,K.,Arnott,S.,Moorhouse,R.,Walkinshaw,M.D.,Atkins,E. D.T. and Wolf-Ullish,C.,ACS Symp.Ser.,1980,141,411.
10. Holzwarth,G. and Prestridge,E.B.,Science,1977,197,757.
11. Frangou,S.A.,Morris,E.R.,Rees,D.A.,Richardson,R.K. and Ross Murphy,S.B.,J.Polym.Sci.,Polym.Letters Ed.,1982,20,531.
12. Muller,G.,Lecourtier,J.,Chauveteau,G. and Allain,C., Makromol.Chem.,Rapid Commun.,1984,5,203.
13. Paradossi,G. and Brant,D.A.,Macromolecules,1982,15,874.
14. Sato,T.,Kojima,S.,Norisuye,T. and Fujita,H.,Polymer J.,1984, 16(5),427.
15. Sato,T.,Norisuye,T. and Fujita,H.,Polymer J.,1984,16(4),341.
16. Muller,G.,Anrhourrache,M.,Lecourtier,J. and Chauveteau G., Int.J.Biol.Macromol.,1986,8,167.
17. Tait,M.I.,Sutherland,I.W.and Clarke-Sturman,A.J.,J.General Microbiology,1986,132,1483.
18. Yoshizaki,T. and Yamakawa,H.,J.Chem.Phys.,1980,72,57.
19. Lecourtier,J.,Chauveteau,G. and Muller,G.,Int.J.Biol.Macro-mol.,1986,8,306.
20. Muller,G.,Lecourtier,J. and Chauveteau,G., Macromolecules 86, Functional Polymers and Biopolymers,Oxford,1986,pp.63.
21. Morris,E.R.,Rees,D.A.,Young,G. and Walkinshaw,M.M.,J.Mol. Biol.,1977,110,1.
22. Milas,M. and Rinaudo,M.,Polym.Bull.,1984,12,507.
23. Milas,M. and Rinaudo,M.,Third European Symp. on Carbohydrates Grenoble,1986,sept.16-20.
24. Seeger,B.,Die Nahrung,1981,25,7,655.

Part 3
THERMODYNAMICS/BLENDS

MEAN FIELD LATTICE EQUATIONS OF STATE

Eric J. Beckman, Roger S. Porter, Ronald Koningsveld

(Polymer Sci.& Eng., Univ. of Massachusetts, Amherst, MA 01003, USA),

and

Ludo A. Kleintjens

(DSM Research, PO Box 18, 6160 MD Geleen, Netherlands)

Synopsis

Following the work of Schottky, many have used a lattice of filled and vacant sites to model compressible fluids. The lattice models of Kanig, Kilian, Sanchez & Lacombe, Okada & Nose, Kleintjens et al. (KK), Vera & Panayioutou and similar treatments, can all be summarized by a single equation of state, the various versions representing special cases. Liquid/vapor coexistence data and p(V) isotherms on small molecules, as well as p(V) data on polystyrene, were used to test the variations of the general model. The descriptions of the experimental data improve in the sequence Kanig-Kilian-Sanchez-Vera-KK. The KK model contains two empirical parameters, the molecular origin of which can be explained with Staverman's contact number statistics. Differences in molecular size and, hence, in number of nearest neighbors come to the fore as important factors that should not be ignored in molecular modelling. Thus, combinatorial entropy contributions are probably more complex than is usually assumed, and may, for instance, by themselves cause lower and upper critical miscibility to occur in the same polymer blend.

Introduction

Equations of state (EoS) based on the model of a compressible lattice containing randomly distributed vacant sites have been suggested quite frequently [1-11]. Most of these treatments were concerned with molecular considerations rather than accurate descriptions of pVT data. Kleintjens et al. [12-18] found that the introduction of empirical parameters can hardly be avoided if a better than qualitative agreement between model and experiment is to be obtained. In their formulation the mean-field lattice-gas EoS assumes the following form:

$$- pv_0/RT = ln(1 - \phi_1) + (1 - 1/m_1)\phi_1 + \chi\phi_1 \tag{1}$$

where $\phi_1 = v_0 m_1/\hat{V}_1$, v_0 = the molar volume of the empty sites, and m_1 = the number of sites occupied by the molecules. The molar volume of the system is $\hat{V}_1$, p is the pressure, R and T are gas constant and absolute temperature, and χ is an interaction parameter. With small-molecule substances we expect m_1 to be close to unity.

We note that eq.(1), a) reduces to Boyle's law for $\hat{V}_1 \rightarrow \infty$ and b) is identical to the rigid lattice expression for the osmotic pressure of a polymer solution [19] if p and v_0 are replaced by osmotic pressure and molar volume of the solvent. The solvent molecules fill the vacant sites and the lattice is then incompressible. The empty sites may be understood to play the role of solvent molecules and the MFLG model treats a single-component substance as a binary mixture.

About half a century ago Staverman pointed out that molecular models should account for the physical fact that virtually no molecular species exist that do not differ in size and shape. Hence, they differ also in molecular surface area and number of nearest-neighbor contacts [20,21]. Therefore, the usual lattice treatments, assigning the same coordination number to all molecular species, seem to start from an unfavorable position. Staverman found the concentration dependence of heats of mixing, then available in the literature, to be well covered with a single parameter representing the ratio of molecular surface areas.

It has been shown for a wide variety of systems that Staverman's procedure provides an essential step toward a quantitative description of thermodynamic data. The examples range from ^{4}He close to its vapor/liquid critical point [22], gas/gas demixing in noble gas mixtures [23], liquid/liquid phase equilibria and spinodals in polymer solutions [24,25], partial miscibility in polymer blends [26], equilibrium swelling of polymer networks [27], to solid/near-critical fluid phase equilibria at a second critical end point [28]. Not seldom do values for γ, derived from fits to data with expressions like eq. (1), agree quite well with those estimated independently with Bondi's method [29].

In terms of Staverman's concept we may define χ as [12,18]

$$\chi = - (z_{11}w_{11}/2RT)(1 - \gamma)/(1 - \gamma\phi_1)^2 \equiv \beta_1(1 - \gamma)/T(1 - \gamma\phi_1)^2 \qquad (2)$$

where z_{11} = the number of nearest neighbors of an occupied site, w_{11} = the contribution to the internal energy per nearest-neighbor contact, and $\gamma = 1 - \sigma_1/\sigma_0$. The ratio z_{11}/z_{00} of the contact numbers z_{11} and z_{00} of occupied and vacant sites is set equal to the ratio σ_1/σ_0 of the molecular surface areas σ_1 and σ_0.

Application of eqs (1) and (2) to a large variety of pVT data has indicated that better than qualitative descriptions call for the introduction of two empirical parameters α_0 and β_0 in such a manner that χ is given by [12,18]

$$\chi = \alpha_0 + (\beta_0 + \beta_1/T)(1 - \gamma)/(1 \; \gamma\phi_1) \qquad (3)$$

It is shown below that eqs (1) and (3) include most other mean-field lattice EoS as special cases. It is further demonstrated that the two empirical parameters α_0 and β_0, that have in many instances proven to be indispensible, can be given a molecular significance with the aid of Staverman's contact number statistics [18]. Many years ago Staverman said that, once the internal energy has been calculated on the basis of numbers of nearest-neighbor contacts rather than numbers of molecules, the same ought to be done for the entropy [21]. This approach leads, in a rough approximat-

already, to a possible molecular explanation for α_0 and β_0 [18].

Objections may be raised against the use of the random mixing approach, and that for nearest neighbor contacts only. We must also realise that the introduction of different contact (coordination) numbers turns the lattice into a physically unrealistic abstraction. However, no fluid or liquid conforms exactly to any model and the latter will always represent an abstraction of sorts. Thus, there is nothing against using the lattice merely as a tool in the enumeration of possible contacts and conformations as long as the procedure finds its justification in the adequacy of the descriptions it provides. We employ the simplest approach, viz. the *strictly-regular* approximation [30] and find in the present state of development no need for more involved treatments, such as the quasi-chemical approximation [9-11,30].

Comparison with other mean-field lattice equations of state

Several of the lattice equations of state developed in the past can be shown to be special cases of the zero-order MFLG EoS. Making the substitutions

$$V^* = m_1 v_0, \quad p^* = \beta_1 R / v_0, \quad T^* = \beta_1 \tag{4}$$

we transform eqs (1) and (3) to

$$- \tilde{P}/\tilde{T} = ln(1 - 1/\tilde{V}) + (1 - 1/m_1)(1/\tilde{V}) + \{\alpha_0 + (\beta_0 + 1/\tilde{T})(1 - \gamma/\tilde{V})^{-2}\}(1/\tilde{V}^2) \tag{5}$$

where $\tilde{P} = p/P^*$, $\tilde{T} = T/T^*$ and $\tilde{V} = \tilde{V}_1/V^*$. If we neglect the effect of molecular surface areas on entropy ($\beta_0 = 0$) and internal energy ($\gamma = 0$), and further assume the empirical correction factor α_0 to be negligible, we have

$$- \tilde{P}/\tilde{T} = ln(1 - 1/\tilde{V}) + (1 - 1/m_1)(1/\tilde{V}) + 1/\tilde{T}\tilde{V}^2 \tag{6}$$

which is the Sanchez-Lacombe EoS for chain molecule fluids [8,31]. Unlike the MFLG, the Sanchez-Lacombe model treats v_0, the molar volume per lattice site, as a material parameter. This may compensate for the lack of an explicit surface area dependence, but it leads to complications in that a combination rule must then be devised for v_0 in order to model fluid mixtures. We prefer using a constant value for v_0 for all systems, which is a mixing rule in itself, but one whose simplicity avoids complications in mixtures.

Vera and Panayloutou [10] derived a lattice EoS which uses a constant volume per lattice site but does not explicitly recognize the contribution of different contact numbers to the free energy. Both a random and non-random mixing EoS were derived, here we discuss only the random mixing version:

$$\tilde{P}/\tilde{T} = ln\{\tilde{V}/(\tilde{V} - 1)\} + (z/2)ln\{(\tilde{V} + q/r - 1)/\tilde{V}\} - \Theta^2/\tilde{T} \tag{7}$$

where z = the lattice coordination number, assumed to be a universal constant, r = the number of segments per molecule, and q = the number of non-bonded contacts per molecule, given by [30]

$$q = (z - 2)r/z + 2/z \tag{8a}$$

and

$$\Theta = q\phi_1 / \{(1 - \phi_1)r + \phi_1 q\} \tag{8b}$$

The definitions of P^* and T^* differ from those in eqs (4) but can be shown to be related to our parameters β_0 and β_1. Further, expanding the second term on the right-hand side of eq. (7) in powers of ϕ_1 and retaining the leading term, we can transform Vera and Panayioutou's EoS to [31]

$$- pv_0/RT = ln(1 - \phi_1) + (1 - 1/m_1)\phi_1 + (\beta_0 + \beta_1/T)(1 - \gamma)(1 - \gamma\phi_1)^{-2}\phi_1^2 \tag{9}$$

which is the zero order MFLG EoS (eqs (1) and (3)) under the constraint $\alpha_0 = 0$, except for a different definition of γ, which is now given by

$$\gamma = 1 - q/r \tag{9a}$$

While the form of the two EoS are similar, rigorous application of Vera's model couples γ to the number of segments per molecule via eqs (8a) and (9a), since the coordination number is taken to be a universal constant. All molecular segments are therefore assumed to have similar shapes and, when the number of segments becomes very large, as in the case of polymers, γ approaches a constant value. In a later version of his theory Vera explicitly introduces molecular surface areas to calculate the number of segment/segment contacts, while retaining the form of the EoS. The number of contacts per segment (q/r above) then becomes the ratio of surface areas, σ_1/σ_0, as in the zero-order MFLG model. Rather than treating γ as a material parameter, Vera uses Bondi's group contribution method to estimate γ. In this study, γ is considered an adjustable parameter and $\alpha_0 = 0$ when dealing with Vera's model.

Manipulation of the Okada-Nose random-mixing lattice EoS [9] in a similar fashion leads to the zero-order MFLG with both α_0 and β_0 equal to zero. This model, like Vera's earlier version, employs constant values for both v_0 and the lattice coordination number and defines the parameter q/r to represent the number of contacts per segment.

The lattice EoS developed by Kanig [4], though similar in form to the Okada-Nose expression, includes molecular surface areas and thus avoids the coupling between γ and m_1 through q/r. Kanig's treatment differs from the zero-order MFLG in that the surface areas are taken temperature dependent, which makes the lattice one of holes and 'cells'. Such a model of fluids has been pursued in depth by Simha et al. [32-34]. In the present comparison we assume γ to be independent of temperature when dealing with Kanig's model. This will naturally decrease the accuracy of description but allow easier comparison with the other models.

Finally, Kilian's version of the mean-field lattice EoS, derived for modelling polymer pVT behavior, is similar to the zero-order MFLG with γ and β_0 set equal to zero [5].

Table 1 illustrates the relationship between the various lattice EoS and the zero-

Table 1

Relationship of various lattice equations of state

Author	v_0	γ	α_0	β_0
Sanchez	Floating	0	0	0
Kanig	Fixed	Non-zero	0	0
Kilian	Fixed	0	Non-zero	0
Vera	Fixed	Non-zero	0	Non-zero
KK	Fixed	Non-zero	Non-zero	Non-zero

order MFLG, defined by eqs (1) and (3). The models shown in the table have been tested against literature data on CO_2, SO_2, CF_3H, and on polystyrenes covering a large range of chain lengths [31]. Here we discuss the results for CO_2 and for poly-styrene.

Carbon dioxide

The material parameters were found by fitting saturation density, pVT and critical point data to eqs (1) and (3) for the various parameter combinations indicated in table 1. At a vapor/liquid critical point three conditions are valid [12,35]: a) the EoS for the critical values of p, T and $\hat{V}_1$, b) the spinodal condition given by

$$1/(1 - \phi_1) + 1/m_1\phi_1 = 2\alpha_0 + 2(\beta_0 + \beta_1/T)(1 - \gamma)/(1 - \gamma\phi_1)^3 \tag{10}$$

and c) the critical condition, defined for the zero-order MFLG by

$$1/(1 - \phi_1)^2 - 1/m_1\phi_1^2 = 6(\beta_0 + \beta_1/T)\gamma(1 - \gamma)/(1 - \gamma\phi_1)^4 \tag{11}$$

It appears that the quality of the description shows little sensitivity to the values chosen for v_0, the molar volume per lattice site, and z_{00}, the hole-to-hole coordination number. It is thus possible to adopt physically meaningful values for these constants, we used $z_{00} = 10$, and $v_0 = 25$ cm^3/mole. Frenkel estimates the equilibrium size of a hole to have a value comparable to that of a small molecule [2], which means that values of v_0 in the range 10 - 30 cm^3/mole would be acceptable.

Vapor/liquid equilibrium data and the descriptions by the various models are shown in figure 1, representative EoS results are given in figure 2. Parameter values are given in ref. 31.

The figures illustrate the importance of the parameters containing the contributions of the molecular surface areas to entropy and internal energy. The zero order MFLG and Vera models come closest to the data. The β_0 parameter is crucial to obtaining the proper temperature dependence, as is evident by comparing the results for the Kanig and Vera versions. In conjunction with a non-zero γ a good overall

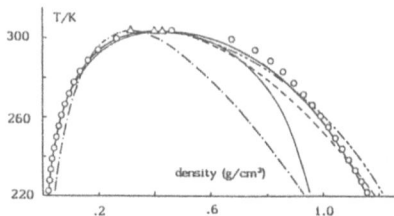

Fig. 1.   Vapor-liquid coexistence data for CO_2 (o) vs model calculations. Zero-order
MFLG,━━━ ; Vera,━ ━ ━; Kilian,━━━ ; Sanchez, ━━ ━ ━ ;
Kanig, ━ • ━ • ━. Calculated critical points: Δ.

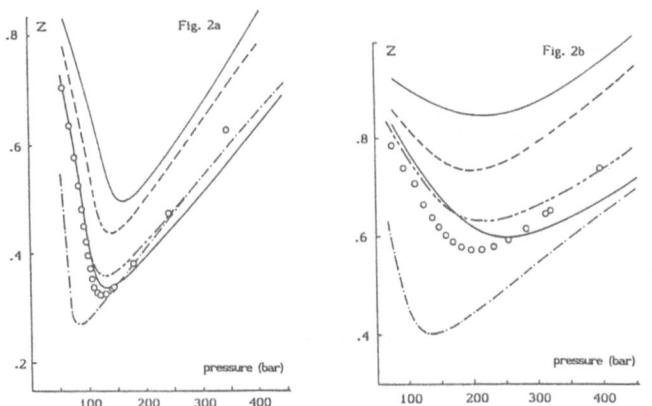

Fig. 2.   Compressibility factor Z (= pV/RT) vs pressure for CO_2. Data, o; zero-
order MFLG,━━━ ; Vera,━━ • ━ ━; Kilian, ━━━ ;Sanchez, ━━ ━ ━ ;
Kanig,━ • ━ • ━ ; a, 323 K; b, 373 K.

description is provided (compare Kilian and Vera results). The Sanchez model
performs reasonably well at the expense of unrealistic values of m_1 (9.75) and v_0
(2.73 cm³/mole), implying holes smaller than hydrogen atoms. With other
substances quite similar results are obtained [31].

Polystyrene

A polymer melt differs from the small molecule substances dealt with so far in that
only states of high density exist. Critical conditions cannot be realized and the information supplied by a horizontal point of inflexion on the critical isotherm is
therefore not available. On the other hand, the chain length now offers an extra
variable at constant values of the other parameters. Polydispersity in chain length
affects the analysis only in that m_1 in eq. (1) must be replaced by its number-average
value. Polystyrene is particularly suited for the present purpose due to the availability of pVT data on a broad range of molar mass, temperature and pressure [31].

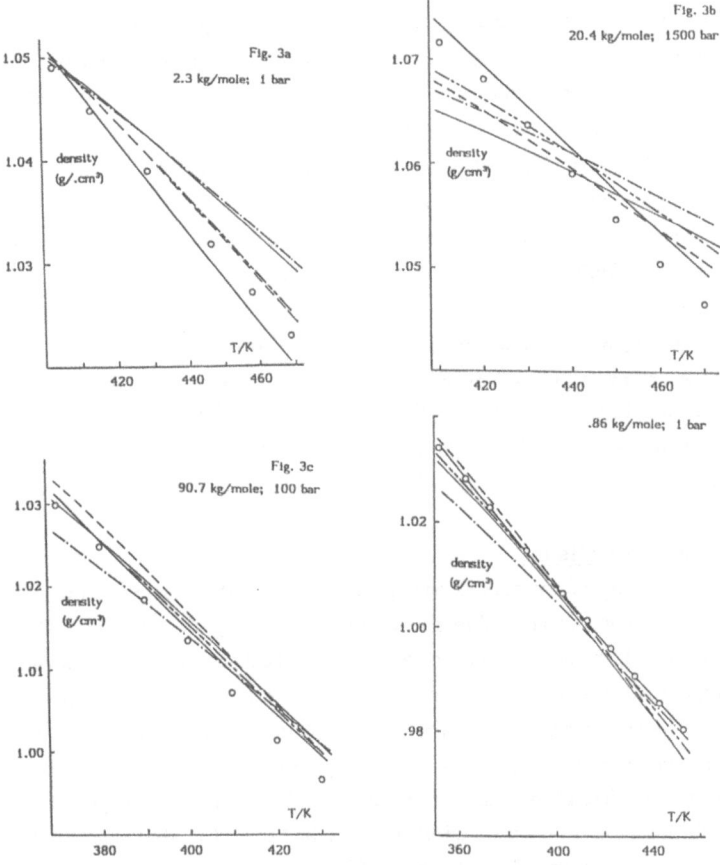

Fig. 3. Density-temperature data (o) for polystyrene for indicated molar mass and pressure. Zero-order MFLG,━━━ ; Vera,━ ━ ━ ; Kilian, ───── ; Sanchez,━ ━ ━ ; Kanig,━ · ━ · ━ .

Fig. 4. Density-temperature data (o) for an oligomeric polystyrene and predictions by various lattice-gas models indicated as in fig. 3.

Variations of v_0 and z_{00} within their constraints again proved not to influence the quality of the description and we used the same values as above. Data on samples with molar mass 2.3, 3.7, 20.7, 90.7 and 125 kg/mole were used to obtain values of the parameters for the various versions of the model. The latter are given in ref. 31, figures 3 show representative examples.

There is a notable difference in the relative performance of the models compared to small-molecule systems. Comparing the Vera and zero-order MFLG descriptions, as well as those of Kanig vs Vera, we see that α_0 is far more important than β_0 in achieving a good fit. This is not surprising since the packing constraints of segments and, therefore, deviations from the combinatorial entropy, as expressed in the first two terms on the right-hand side of eq. (1), should be more severe for a polymer with ϕ_1 approaching unity than for a gas. Similarities to small-molecule systems include the importance of a surface area contribution to the internal energy (see Kilian vs Vera), and the variable lattice cell volume of the Sanchez-Lacombe model approximating the effect of surface-area terms, since the descriptions are of the same quality as those of the Kanig model.

The zero-order MFLG model shows the best performance. In the molecular mass range of 2.3 - 125 kg/mole at pressures up to 2000 bar the deviation between cal-culated and experimental densities is $< 0.1\%$. In addition, the predicted temperature dependence of the density is more accurate than with the other models.

To test the capability of lattice EoS to cover the oligomeric chain length range, we used the parameter values obtained above to *predict* the V(T) behavior at 1 bar for a .86 kg/mole polystyrene. Figure 4 shows that the predictions are quite adequate for all models, and very accurate with the zero-order MFLG model.

Higher order contact statistics

The greater versatility of the zero-order MFLG model in the description of pVT data is probably due to the two adaptable parameters α_0 and β_0. For this situation to be acceptable, we should be able to present a molecular background to these parameters, if only in a qualitative sense. To this end we use Staverman's contact number statis-tics to check whether corrections to the combinatorial entropy might arise from the differences in numbers of contacts due to disparities in molecular size and shape.

Eq.(1) originates from the usual enumeration of Ω_0, the number of possible arran-gements of N_0 vacant and N_1 occupied sites on N (= $N_0 + N_1$) sites sharing the same coordination number. We use Silberberg's [36] approximation to account for the fact that species 1 might occupy more than one site:

$$\Omega_0 = (N!/N_0!\nu_1!)(V^\ddagger/V)^{\nu_1(m_1 - 1)}$$

(12)

where $N_1 = \nu_1 m_1$, V is the total volume and $V^\ddagger$ is a parametric volume accounting for limitations of placement after the first site of a molecule of species 1 has been allocated. In the calculation of the entropy of mixing $V^\ddagger$ vanishes if, as we do here, it is assumed not to depend on ϕ_1.

Applying Silberberg's method also to account for the different numbers of nearest neighbors z_{00} and z_{11} we may write [18]

$$\Omega_1 = \Omega_0 (z_{01}/\bar{z})^{P_{01}} (z_{10}/\bar{z})^{P_{10}} = \Omega_0 (z_{01} z_{10}/\bar{z}^2)^{P_{10}}$$

(13)

since Ω_0 needs correction only for the P_{01} and P_{10} neighbors that involve different z. For those contacts we rather arbitrarily assume Ω_0 to be wrong by a factor $z_{ij}/\bar{z}$, where $\bar{z}$ is the average number of contacts per site at a given composition of the system. We have

$$\bar{z} = (2P_{00} + 2P_{11} + P_{01} + P_{10})/N = z_{00}Q \tag{14}$$

and, using standard procedures, we find α_0 and β_0 in eq. (3) to be defined by

$$\alpha_0 = 2z_{11}(lnQ)/Q; \quad \beta_0 = - z_{11}lnz \tag{15}$$

where $z = z_{01}z_{10}/z_{00}^2$ and $Q = 1 - \gamma\phi_1$.

In a rough approximation the number of nearest neighbors z_{ij} may be taken proportional to σ_i and inversely proportional to σ_j,

$$z_{ij} = b_i\sigma_i/\sigma_j \tag{16}$$

Thus, $b_i = z_{ii}$ and $z = b_1/b_0 = z_{11}/z_{00}$.

We note that this **first order** approximation assigns a clear physical significance to the hitherto unexplained empirical parameters α_0 and β_0 in the zeroth order approximation (eqs (1) and (3)). In stead of α_0 and β_0 we have z_{11}, a coordination number that must be expected to be close to 10, and z, a ratio of coordination numbers that will not be far from unity. Thus we have six physically meaningful parameters in the first order MFLG EoS, v_0, σ_1/σ_0, m_1, β_1, z_{11} and z, each of which can be judged as to the acceptability of its value (β_1 values should be consistent with Van der Waals interactions).

We further observe that the need for a term β_0 in the zeroth order approximation is not surprising after all, eq. (17) indeed defines it as a constant for a given system. The first order treatment makes α_0 dependent on ϕ_1, a dependence we missed when we introduced α_0 into the zeroth order approximation. It is of interest to note here that the second term in Vera's EoS (eq. (7)) can be rewritten as $(z/2)lnQ$, which expression covers part of the density dependence of α_0 revealed by the first order approximation.

The latter still has an unsatisfactory feature in that it treats the two coordination numbers z_{00} and z_{11} as average numbers, valid throughout the entire density range. Such a treatment must necessarily be very crude since though molecules 1 are, at high density (ϕ_1 close to 1), mainly surrounded by their own kind, at low density most nearest neighbor sites of 1 will be vacant sites. Hence, the coordination numbers must vary with density.

In a **second order** approximation [18] we allow the coordination numbers to depend on density so as to at least give correct numbers at high and low density ($\phi_1 = 1$ and $\phi_1 = 1 - \phi_0 = 0$). We can do so without adding new parameters, writing

$$2P_{00} + P_{01} = z_{00}N_0(\phi_0 + s_{01}\phi_1) = z_{00}N_0Q_0$$
$$2P_{11} + P_{10} = z_{11}N_1m_1(\phi_1 + s_{10}\phi_0) = z_{11}N_1m_1Q_1 \tag{17}$$

where P_{ij} = the number of contact pairs i-j, and $s_{10} = 1/s_{01} = \sigma_1/\sigma_0$.
The second order EoS reads

$$- pv_0/RT = ln(1 - \phi_1) + (1 - 1/m_1)\phi_1 + \{Q^*\Lambda - (1 - \phi_1)\partial(Q^*\Lambda)/\partial\phi_1\}\phi_1^2 \tag{18}$$

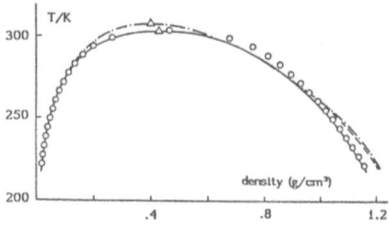

Fig. 5.   Vapor/liquid coexistence data for CO_2 (o) vs model calculations. Zero-order
MFLG, ━━━ ; first order, ━•━• ; second order, ━ ━ ━ . Calculated
critical points, Δ.

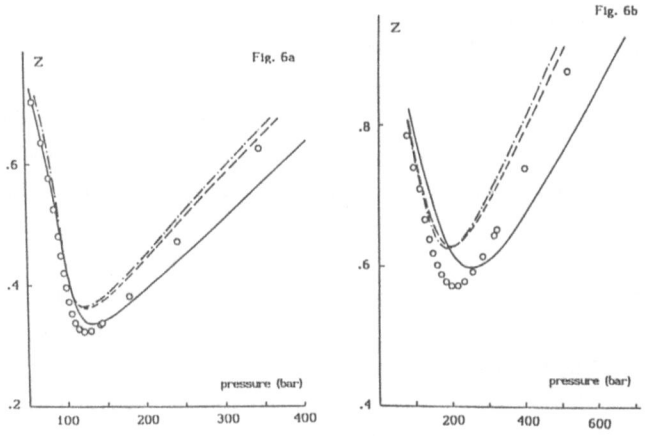

Fig. 6   Compressibility factor Z (= pV/RT) vs pressure for CO_2 (o). Model calculations indicated as in fig. 5; a, 323 K; b, 373 K.

where $Q^* = Q_0 Q_1/Q_2$, $\Lambda = - z_{11}\{ln(z/Q_2^2) + w_{11}/2RT\}$ and $Q_2 = \phi_0 Q_0 + z\phi_1 Q_1$.

Expressions for spinodal and critical point become quite complex. We shall not
cite them here but refer to previous, more elaborate reports [18,31]. The latter also
discuss applications of first and second order MFLG models from which we select
CO_2 as an example. Figures 5 and 6 reveal that, though the higher order approximations offer an explanation for the empirical terms needed in the zeroth-order MFLG
model, the descriptions of pVT behavior they provide do not quite equal the quality of
the zeroth order approximation. They might therefore be discarded if it were not for
the fact that the γ values obtained from fits to first and second order EoS come
considerably closer to Bondi's estimations than those from the zero-order model.
This is seen in Table 2 where we show Bondi's estimations of ratios of surface areas
for pairs of substances a and b compared with the values $(1 - \gamma)_a/(1 - \gamma)_b$ from fits
of pVT data to zeroth, first and second order MFLG EoS.

It should be mentioned that, though the first order model contains only parameters

Table 2

Molecular surface area ratios by Bondi's method
and from fitting of lattice equations of state

pairs of molecules a/b	Bondi	σ_a/σ_b		
		mean-field lattice-gas model (KK version)		
		zero-order	first order	second order
CO_2/SO_2	.87	1.28	.81	.82
CO_2/CF_3H	.54	1.46	.77	.72
CO_2/PSt^*	.30	.38	.44	.30

* repeat unit

having a molecular significance, the second order model needs reintroduction of β_0 in order to function properly. Application of the higher-order models to polystyrene has not been found to improve the description already obtained with the zeroth order EoS. This is probably due to the high densities involved.

Conclusions

The Kanig, Kilian, Okada-Nose, Vera-Panayioutou, and Sanchez-Lacombe equations of state have been shown to be special cases of the zero-order MFLG model. Each of the above EoS can be derived from the MFLG EoS by removing the effect of disparity in numbers of nearest neighbor contacts from the entropy and/or the internal energy of mixing vacant and occupied sites. Comparison of the fit of these EoS to pVT data on various substances, including a polymer, demonstrates the value of Staverman's suggestion that differences in size should primarily reveal themselves in thermodynamic properties via the ratio of molecular surface areas.

The relative accuracies of the zeroth and higher-order MFLG models imply that, while the parameter α_0 is derived from the effect of molecular surfaces on the number of conformations, the origins of β_0 lie at least partly elsewhere, possibly with the temperature dependence of the segmental vibrations [31]. The relatively poor performance of the second-order model suggests that future work should concentrate upon improving the temperature dependence in the MFLG model rather than further elaboration of the density dependence.

The importance of combinatorial entropy corrections due to disparities in molecular surface areas and, hence, contact numbers has also been illustrated recently in an application of the *second-order* model to a rigid lattice containing two kinds of polymer chains (polymer blend) [25]. It could be shown that combinatorial reasons alone may be responsible for the simultaneous occurrence of a lower and an upper

208

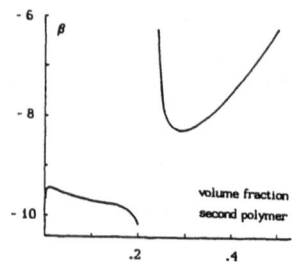

Fig. 7   Second-order contact statistical calculation of a spinodal for a polymer blend
with $m_1 = 1500$, $m_2 = 3000$, $\sigma_1/\sigma_2 = 2$, $z_{22} = 6$, $z = .8$ ($\beta = \beta_0 + \beta_1/T$).

miscibility gap in a polymer blend if the two types of repeat units in the chains differ in numbers of nearest neighbors, and the coordination number of a repeat unit is assumed to vary linearly with concentration (as indicated in eqs (15)). Figure 7 gives an example.

References

1. Schottky, W.; Ulich, H.; Wagner, C. *Thermodynamik*, Springer, Berlin, **1929**.
2. Frenkel, J. *Kinetic Theory of Liquids*, Oxford University Press, London, **1946**; Dover Reprint, New York, **1955**.
3. Cernuschi, F.; Eyring, H. *J.Chem.Phys.*, **1939**, 7, 547.
4. Kanig, G. *Kolloid Z.&Z.Polym.* **1963**, *190*, 1; *J.Polym.Sci.,Part C*, **1967**, No. *16*, 1957.
5. Kilian, H.-G. *Kolloid Z.&Z.Polym.*, **1974**, *252*, 353; *Progr.Colloid&Polym.Sci.* **1975**, *58*, 53.
6. Trappeniers, N.J.; Schouten, J.A.; Ten Seldam, C.A. *Chem.Phys.Lett.*, **1970**, *5*, 541.
7. Schouten, J.A.; Ten Seldam, C.A.; Trappeniers, N.J. *Physica*, **1974**, *73*, 556.
8. Sanchez, I.C.; Lacombe, R.H. *J.Phys.Chem.*, **1976**, *80*, 2352, 2568; *J.Polym. Sci.,Polym.Lett.Ed.*, **1977**, *15*, 71.
9. Okada, M.; Nose, T. *Polym.J.*, **1981**, *13*, 399, 591.
10. Panayiotou, C.; Vera, J.H. *Polym.J.*, **1982**, *14*, 681; *Polym.Eng.Sci.*, **1982**, *22*, 345.
11. Kumar, S.K.; Suter, U.W.; Reid, R.C. *Fluid Phase Eq.*, in press; *J.Chem.Phys.*, in press.
12. Kleintjens, L.A. *PhD Thesis*, Essex University, UK, **1979**.
13. Kleintjens, L.A.; Koningsveld, R. *Colloid&Polym.Sci.*, **1980**, *258*, 711.
14. Kleintjens, L.A.; Koningsveld, R. *J.Electrochem.Soc.*, **1980**, *127*, 2352.
15. Kleintjens, L.A.; Koningsveld, R. *Sep.Sci.Technol.*, **1982**, *17*, 215.
16. Kleintjens, L.A. *Fluid Phase Eq.*, **1983**, *10*, 183.
17. Nies, E.; Kleintjens, L.A.; Koningsveld, R.; Simha, R.; Jain, R.K. *Fluid Phase Eq.* **1983**, *12*, 11.
18. Koningsveld, R.; Kleintjens, L.A.; Leblans-Vinck, A.M., subm. to *J.Phys.Chem.*
19. Flory, P.J., *Principles of Polymer Chemistry*, Cornell Univ. Press, Ithaca, **1953**
20. Staverman, A.J. *Rec.Trav.Chim.*, **1937**, *56*, 885.
21. Staverman, A.J. *PhD Thesis*, Leiden University, Netherlands, **1938**.

22. Koningsveld, R.; Kleintjens, L.A.; Swenker, A.G.; Vanmeulebrouk, M.G.G. *IUPAC Conf.Chem.Thermod.*, **1982**, London.

23. Keller, P.; Kleintjens, L.A.; Koningsveld, R. *IUPAC Conf.Chem.Thermod.*,**1982**, London.

24. Nies, E.; Koningsveld, R.; Kleintjens, L.A. *Progr.Coll.Polym.Sci.*, **1985**, *71*, 2.

25. Koningsveld, R.; Kleintjens, L.A.; Nies, E. *Croat.Chim.Acta*, in press.

26. Voigt-Martin, I.G.; Leister, K.-H.; Rosenau, R.; Koningsveld, R. *J.Polym.Sci.*, Part B: *Polym.Phys.*, **1986**, *24*, 723.

27. Petrović, Z.S.; MacKnight, W.J.; Koningsveld, R.; Dušek, K. *Macromol.*, in print

28. Leblans-Vinck, A.M.; Koningsveld, R.; Kleintjens, L.A.; Diepen, G.A.M. *Fluid Phase Eq.*, **1985**, *20*, 347.

29. Bondi, A. *J.Phys.Chem.*, **1964**, *68*, 441.

30. Guggenheim, E.A. *Mixtures*, Clarendon Press, Oxford, **1952**.

31. Beckman, E.J.; Porter, R.S.; Koningsveld, R., subm. to *J.Phys.Chem.*

32. Nanda, V.S.; Simha, R.; Somcynsky, T. *J.Polym.Sci.*,Part C, **1966**, *12*, 277.

33. Simha, R.; Somcyncky, T. *Macromolecules*, **1968**, *2*, 342.

34. Simha, R. *Macromolecules*, **1977**, *10*, 1025.

35. Gibbs, J.W. *Scientific Papers*, Vol. I., Dover Reprint, New York, **1961**.

36. Silberberg, A. *J.Chem.Phys.*, **1968**, *48*, 2835.

LIGHT SCATTERING CHARACTERIZATION
OF POLYMER BLENDS IN SOLUTION

Takeshi Fukuda and Hiroshi Inagaki
Institute for Chemical Research,
Kyoto University,
Uji, Kyoto 611,
Japan

ABSTRACT

Experimental and theoretical problems associated with the light scattering characterization of polymer-polymer interactions in solution are discussed. A series of light scattering studies on several blend systems carried out under the special condition termed "optical θ" reveal how polymer-polymer interactions manifest themselves in dilute and nondilute solutions. The apparent interaction observed in solution is, if corrected for the excluded volume effect, seemingly similar to that in bulk, though they may be different in details.

INTRODUCTION

Interactions between different polymers are, in many cases, difficult to characterize by using dry blends, especially when they are immiscible under usual conditions. Introduction of a suitable solvent brings about miscibility between otherwise immiscible polymers, thus rendering experimental approaches feasible. However, it often brings about serious complexities, too, arising from solvent-polymer interactions [1]. For Example, two polymers miscible in bulk become immiscible in solution in certain cases [2]. This implies the dificulty of characterizing polymer-polymer interactions in solution. In these days, even the pessimistic view seems to be prevailing that solution approaches to blend compatibility are unreliable even qualitatively. In this contribution, we wish to show that reasonably reliable light scattering analyses are possible, if experimental conditions are suitably chosen.

In dilute solution with a good solvent, the effect of excluded volume appears most strongly, making the polymer-polymer interaction to appear as if very small in magnitude. Krigbaum and Flory [3] were the first to point this out. Correct description of such a solution requires precision in both experiment and theory. We have attempted to improve precision in the classical light scattering technique originating from the work of Stockmayer and Stanley [4] as well as in the excluded-volume theory to achieve considerable success. Well-defined knowledge about dilute solutions has helped us understand, to some extent at least, properties of nondilute solutions as well. The following is a brief summary of the recent light scattering work [5-11].

DEFINITIONS AND METHODS

We give the free energy of mixing ΔG of two polymers 1 and 2 and one solvent 0 by the following form (cf. Koningsveld et al. [1]):

$$Z = \Delta G/NRT = \phi_0 \ln \phi_0 + (\phi_1/m_1)\ln \phi_1 + (\phi_2/m_2)\ln \phi_2$$
$$+ \phi_0\phi_1 g_{01}(\phi) + \phi_0\phi_2 g_{02}(\phi) + \phi_1\phi_2 g_{12}(\phi_1,\phi_2) \qquad (1)$$

where ϕ_i is the volume fraction, m_i is the degree of chain length, and other symbols have usual significances [6]. We assume that g_{0i}'s depend only on the total polymer fraction $\phi = \phi_1 + \phi_2$, and thus g_{12} is a function of both ϕ_1 and ϕ_2 in general. We write the forward light scattering R_0 in the general form (cf. Stockmayer [12] and ref 6)

$$R_0 = \text{const.} \times \frac{\psi_2{}^2 a_{11} + \psi_1{}^2 a_{22} - 2\psi_1\psi_2 a_{12}}{a_{11}a_{22} - a_{12}{}^2} \qquad (2)$$

$$a_{ii} = (1 - \phi)^{-1} + (m_i x_i \phi)^{-1} - 2\chi_{0i} \qquad (3)$$

$$a_{12} = (1 - \phi)^{-1} - \chi_{01} - \chi_{02} + \chi_{12} \qquad (4)$$

where $\psi_i = \partial n/\partial \phi_i$ is the refractive index increment, and $x_i = \phi_i/\phi$ is the blend composition. Equations 3 and 4 define the "light scattering interaction parameter" χ_{ij}'s. They are related to g_{ij}'s through

$$a_{ij} = \partial^2 Z/\partial\phi_i\partial\phi_j, \quad i, j = 1 \text{ or } 2 \qquad (5)$$

The expression for χ_{12} is relatively simple, i.e.

$$\chi_{12} = (1/2)[\partial^2(x_1 x_2 g_{12})/\partial x_1 \partial x_2] \qquad (6)$$

which, if g_{12} is independent of x, is identical with g_{12}. However, χ_{01} and χ_{02} depend on subtle details of the composition dependence of all g_{01}, g_{02} and g_{12} [6]. For this reason, a rigorous light scattering analysis is difficult and impractical except for two cases, i.e., infinitely dilute systems and symmetrical systems (see below).

For $\phi \to 0$, eq 2-4 give (cf. Stockmayer and Stanley [4])

$$R_0/K'\phi = \psi_1{}^2 m_1 x_1 + \psi_2{}^2 m_2 x_2 - [\psi_1{}^2 m_1{}^2 x_1{}^2(1 - 2\chi_{01}) + \psi_2{}^2 m_2{}^2 x_2{}^2(1 - 2\chi_{02})$$
$$+ 2\psi_1\psi_2 m_1 m_2 x_1 x_2(1 - \chi_{01} - \chi_{02} + \chi_{12})]\phi + \cdots \qquad (7)$$

where K' is a constant, and the following relations hold for $\phi = 0$:

$$\chi_{0i} = g_{0i} - (\partial g_{0i}/\partial\phi)_{\phi=0} \qquad (8)$$

$$\chi_{12} = g_{12} \qquad (9)$$

Thus independent measurements on the two polymer/solvent binaries and one polymer/polymer/solvent ternary solution allow determination of χ_{12}. This general method, however, is often unreliable, because χ_{12}, as compared with $1 - 2\chi_{01}$ or $1 - 2\chi_{02}$, is in many cases too small to be determined with meaningful accuracy. We have proposed that measurements be made under the condition represented by eq 10, which transforms eq 7 to eq 11:

$$\psi_1 m_1 x_1 + \psi_2 m_2 x_2 = 0 \qquad (10)$$

$$K\phi/R_0 = (m_1 x_1)^{-1} + (m_2 x_2)^{-1} - 2\chi_{12}\phi + \ldots \qquad (11)$$

where $K = K'(\psi_1 - \psi_2)^2$. Therefore, if one meets the "optical Θ" condition in eq 10, one can determine χ_{12} without knowing χ_{01} and χ_{02}. This not only simplifies experiment but also enhances accuracy [5].

The spinodal point, at which R_0^{-1} vanishes, is represented by

$$a_{11}a_{22} - a_{12}{}^2 = 0 \qquad (12)$$

This equation with eq 3 and 4 is solved for χ_{12} to give

$$\chi_{12sp} = \chi_{12sym} - (1/2)(a_{11}{}^{1/2} - a_{22}{}^{1/2})^2 \qquad (13)$$

where the subscript "sp" denotes spinodal, and the new parameter χ_{12sym} is defined in terms of the spinodal concentration ϕ_{sp} by

$$\chi_{12sym} = [(m_1 x_1)^{-1} + (m_2 x_2)^{-1}]/(2\phi_{sp}) \qquad (14)$$

Clearly, χ_{12sp} is equal to χ_{12sym} only when $a_{11} = a_{22}$, but otherwise smaller than the latter. In order for the equality $a_{11} = a_{22}$ to hold, the system has to be thermodynamically symmetrical with respect to the two polymers, i.e., $m_1 = m_2$, $\chi_{01} = \chi_{02}$, and $x_1 = x_2$.

It may be more infomative to express eq 12 or 13 with eq 3 and 4 in the following form:

$$\chi_{12sp} = \chi_{12sym} - \frac{[(\chi_{01} - \chi_{02}) + \{(m_1 x_1)^{-1} - (m_2 x_2)^{-1}\}/(2\phi_{sp})]^2}{2[(1 - \phi_{sp})^{-1} - \chi_{01} - \chi_{02} + (\chi_{12sp} + \chi_{12sym})/2]} \qquad (15)$$

This equation specifically shows how much difference between χ_{12sp} and χ_{12sym} is expected, when the system is asymmetrical. For the practically interesting case with $m_1 x_1 \cong m_2 x_2$, eq 15 may be approximated by [13]

$$\chi_{12sp} \cong \chi_{12sym} - \frac{(\chi_{01} - \chi_{02})^2}{2[(1 - \phi_{sp})^{-1} - \chi_{01} - \chi_{02} + \chi_{12sym}]} \qquad (16)$$

Thus the second term in the right hand side of eq 16 may be neglected, when χ_{01} and χ_{02} are sufficiently, if not perfectly, close to each other. Clearly, when the optical and thermodynamic symmetry conditions are simultaneously met, one can determine χ_{12} at any concentrations between $\phi = 0$ and ϕ_{sp} without detailed knowledge of the solvent-polymer interactions.

EXPERIMENTAL RESULTS

Dilute Solution

Blend systems thus far examined by the optical Θ method include polystyrene (PS)/poly(methyl methacrylate) (PMMA), PS/polyisoprene (PIP), PS/poly(ethylen oxide) (PEO), PS/polyisobutylene (PIB), and PS/poly(vinyl methyl ether) (PVME). The last system was studied by Klotz et al. [11]. Bromobenzene (BB), 1,4-dioxane/1-bromonaphthalene mixtures (DO/BN), and chlorobenzene (CB) were used as solvent. In all cases, the refractive index increment ψ of PS is positive, while those of all other polymers are negative,

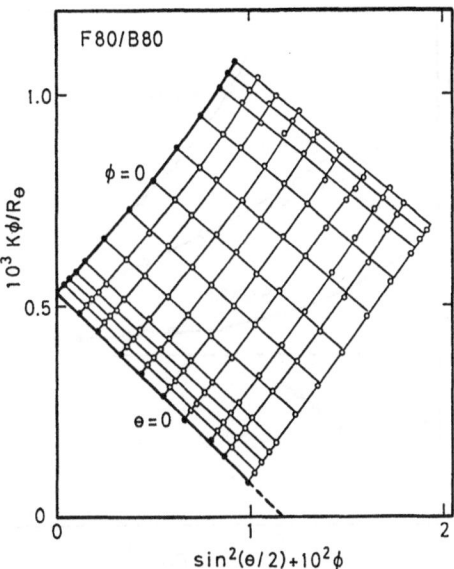

Figure 1. Zimm's plot for a PS (M_w = 7.8 × 10⁵)/PIB (7.6 × 10⁵)/BB system under an optical Θ condition (30°C)

so that the condition in eq 10 could be exactly met by adjusting the blend composition x. To avoid theoretical ambiguity as much as possible (see the following), two polymers of similar molecular weights were paired in most cases. An example of the Zimm plot is given in Figure 1. The ordinate intercept of the plot well agrees to the theoretical value, the sum of the first two terms in the right hand side of eq 11, confirming linearity of the plot in a very dilute region. We obtain the χ_{12} value for the dilute limit directly from the initial slope of the θ = 0 curve, where θ is the scattering angle.

In Figure 2, thus determined values of χ_{12} are plotted as a function of an average molecular weight $\bar{M} \equiv (M_{1w}M_{2w})^{1/2}$, where the subscript "w" denotes weight average. In all the systems excepting the PS/PVME, χ_{12} is positive. From this result, one would expect that a PS/PVME blend is compatible, while all others are incompatible, in the bulk. This in fact is the case, as is known. However, such an argument is not always correct, because χ_{12} reflects not only the interaction between unlike segments but also the excluded-volume effect between the two molecules. Figure 2 clearly shows that the absolute χ_{12} decreases with increasing $\bar{M}$ in all cases, indicating that the excluded volume effect becomes more and more significant as the chain size becomes larger.

This problem was firstly discussed by Krigbaum and Flory [3]. According-ing to them, the present χ_{12} defined for an infinitely dilute solution may be represented by

$$\chi_{12} = (1 - \chi_{01}{}^0 - \chi_{02}{}^0 + \chi_{12}{}^0)h_0(\bar{Z}_{12})$$

$$- (1/2)[(1 - 2\chi_{01}{}^0)h_0(\bar{Z}_{11}) + (1 - 2\chi_{02}{}^0)h_0(\bar{Z}_{22})] \quad (15)$$

where $h_0(\bar{Z}_{ij})$ is the function, appearing in the theory of the second virial

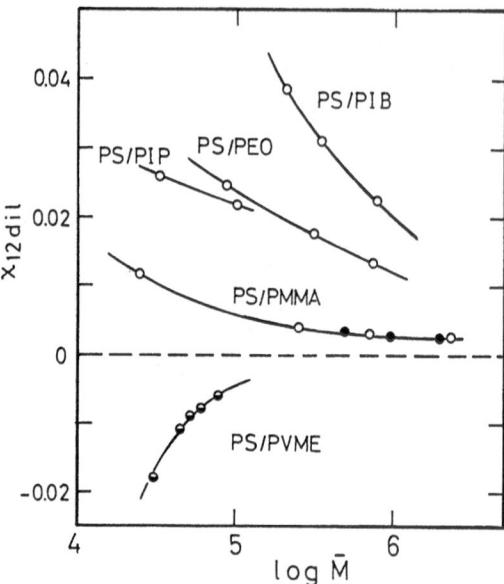

Figure 2. Values of χ_{12} at the dilute limit plotted against $\bar{M} = (M_1 M_2)^{1/2}$: the solvent and temperature are BB/30°C (open circles), DO/BN/30°C (filled circles, and CB/20°C (half-filled circles, from ref 11).

coefficient, that describes the excluded volume effect between molecules i and j, and $\chi_{ij}{}^0$ describes the intearction between segments i and j. Practically, $\chi_{ij}{}^0$ may be understood as being equal to χ_{ij} for M → 0. Because the Flory-Krigbaum theory [3,14] was not always satisfactory, we attempted to revise it in two respects. First, we determined the parameter $\bar{Z}_{12}$ so as to be consistent with the perturbation expansion of the second virial coefficient to obtain

$$\bar{Z}_{12} = 0.3723[\sigma^{5/2} + \sigma^{-5/2} - (\sigma + \sigma^{-1})^{5/2} + (5/2)(\sigma^{1/2} + \sigma^{-1/2})]$$

$$\times (\varepsilon\sigma^{-3/2}\bar{Z}_{11} + \varepsilon^{-1}\sigma^{3/2}\bar{Z}_{22})(1 + \kappa) \qquad (16)$$

$$\varepsilon = m_2/m_1, \quad \sigma = [<S_2{}^2>/<S_1{}^2>]^{1/2}, \quad \kappa = \chi_{12}{}^0/(1 - \chi_{01}{}^0 - \chi_{02}{}^0)$$

where $<S_i{}^2>$ is the mean-square radius of polymer i. Second, we adopted the modified Kurata-Yamakawa equation 17 for h_0 [15] instead of the modified Flory-Krigbaum-Orofino function [14].

$$h_0(\bar{Z}) = (2.193)^{-1}[1 - (1 + 3.537\bar{Z})^{-0.620}] \qquad (17)$$

We have shown that precision is significantly enhanced by these treatments [8].

Provided that the properties of the two polymers in the respective polymer/solvent binaries, i.e., $<S_i{}^2>$, χ_{0i} and $\bar{Z}_{ii}$, as well as χ_{12} are known, eq 15 with eq 16 and 17 can be solved for the parameter $\chi_{12}{}^0$. This parameter should be independent of m_i's and x_i, and, at least to a first approximation, of solvent also. Figure 3 shows results of this analysis.

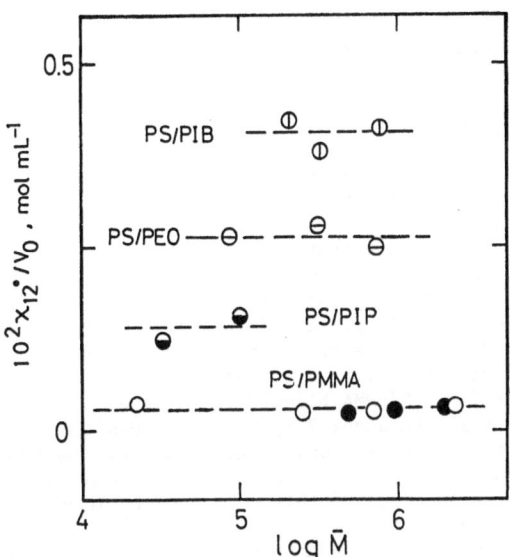

Figure 3. Values of $\chi_{12}{}^0/V_0$ plotted against $\bar{M}$: V_0 is the solvent molar volume in mL/mol (30°C)

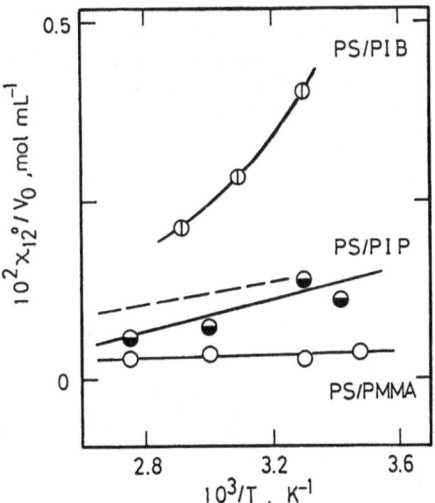

Figure 4. Values of $\chi_{12}{}^0/V_0$ as a function of inverse temperature: V_0 is the molar volume of solvent (or segment), and the broken line shows the χ_{12} data for a dry PS-PIP block copolymer from ref 16.

We could not analyze the PS/PVME data due to lack of information. Figure 3 shows that the $\chi_{12}{}^0$ thus estimated is independent of $\bar{M}$ within ± 20 % for PS/PMMA and within ± 10 % for other systems.

The temperature dependence of $\chi_{12}{}^0$ has been examined for some systems [10]. Preliminary results are given in Figure 4. In each system, $\chi_{12}{}^0$

decreases with increasing temperature. The temperature dependence is the smallest for PS/PMMA and the largest for PS/PIB. Some discussion on these results will be given later.

Nondilute Symmetrical Solutions

We have examined several PS/PMMA/BB solutions, meeting the optical Θ condition at the dilute limit, continuously from a dilute- to a spinodal region, while keeping x and T constant. Each system met the thermodynamic symmetry conditions at all concentrations to such a high degree that we could evaluate χ_{12} as a function of concentration with little theoretical ambiguity [6]. More specifically, we solved eq 2-4 for given values of R_0 and ϕ and approximate values (e.g., the dilute-limit values) of χ_{0i}'s. It should be noted that BB is a moderately good solvent for both PS and PMMA.

Figure 5 shows values of χ_{12} as a function of ϕ. The figure collects the data obtained for four PS/PMMA blends with average molecular weight $\bar{M}$ ranging from about 2×10^4 to about 2×10^6 and composition x about 0.5. Apparently, there exists in each blend a characteristic concentration ϕ^* below which χ_{12} is approximately constant, but above which it is an increasing function of ϕ. The location of this ϕ^* for each blend is indicated by a vertical arrow head in the figure. The constant value of χ_{12} in the region $\phi < \phi^*$, shown by a horizontal broken line, is equal to the dilute-limit value already discussed. The data points for ϕ higher than the ϕ^* of each system seem to form a single composite curve. This brings to mind the notion of "dilute" and "semidilute" solutions [17]. In these terms, χ_{12} in the dilute regime is approximately constant and a function of molecular weight, whereas that in the semidilute regime is a function of ϕ alone, independent of molecular weight. We have estimated the "overlap" concentration ϕ^*_c, which is supposed to characterize the dilute-to-semidilute crossover, according to the following relation:

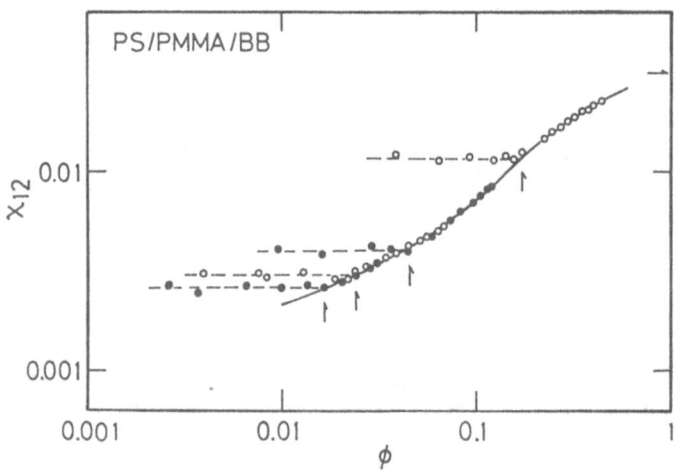

Figure 5. Values of χ_{12} for four PS/PMMA/BB systems plotted against polymer volume fraction ϕ (30°C): the blend composition is approximately of equivolume in all cases. The vertical arrow heads show concentrations ϕ^*, and the horizontal one shows the χ_{12} value extrapolated to bulk, i.e., ~ 0.03_0.

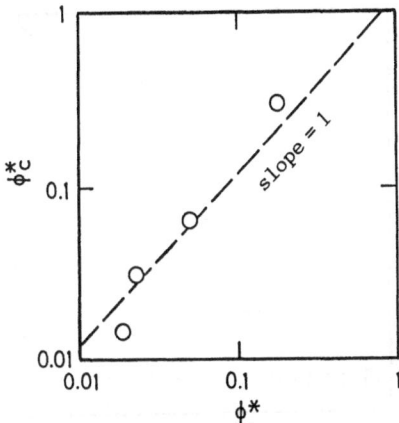

Figure 6. Comparison of the concentration ϕ^* from Figure 5 with the calculated "crossover" concentration ϕ^*_c for PS/PMMA/BB systems (30°C).

$$\phi^*_c = (\text{average molecular volume})/(\text{average radius of gyration})^3 \quad (18)$$

There is a strong correlation between ϕ^* and ϕ^*_c, as Figure 6 shows.

According to Figure 5, the slope of the semidilute branch of the
log χ_{12} vs. log ϕ curve may not be constant, being larger than about 0.5 in
the entire range of ϕ studied. This slope should be zero, accoding to the
Flory-Huggins mean-field approximation [18], and 1/4, according to the
scaling arguments of de Gennes [17]. We also note that recent renormalization calculations predict it to be some 0.53 [19].

DISCUSSION AND CONCLUSIONS

Comparison of the Light Scattering and Other Data

Figure 7 compares the χ_{12}^0 derived from the dilute solution data with
the similar parameter calculated by the solubility parameter method:

$$\chi_{12}^0{}_{cal} = (V_0/RT)(\delta_1 - \delta_2)^2 \quad (19)$$

where V_0 is the molar volume of the (arbitrarily defined) segment. We calculated values of the solubility parameter δ using Hoy's table [20]. The
experimental and calculated values agree unexpectedly well for PS/PMMA, PS/
PIP and PS/PIB, but very poorly for PS/PEO. The poor agreement in the last
system may imply a specific interaction which cannot be taken into account
in this calculation scheme. Experimentally, PEO exhibits several different
values of δ dependeing on the type of solvents employed for its determination [21].

Table 1 lists values of χ_{12}^0 (or similar parameter) estimated by
various methods. As to PS/PMMA blends, the dilute ternary solution values
obtained in the two different solvent systems well agree to each other and
also to the "intrinsic viscosity value", which was obtained by analyzing
the intrinsic viscosities of a series of styrene-methyl methacrylate random
and alternating copolymers in _various_ solvents [22]. This suggests that

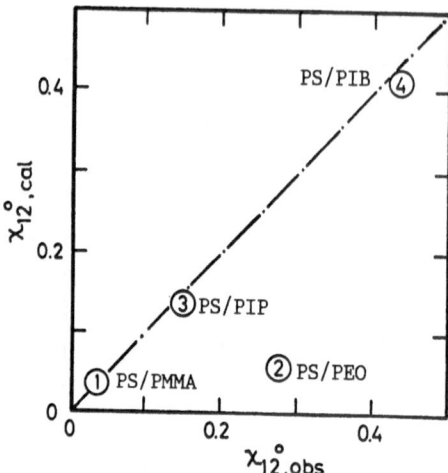

Figure 7. Comparison of the observed values of χ_{12}^0 with those calculated by the solubility-parameter method (around 30°C, V_0 = 106 mL/mol).

solvent effects on χ_{12}^0, if any, may be insignificant. This must be so especially for highly incompatible (i.e., large-χ_{12}^0) systems. This view may be confirmed by the above-noted, generally good agreement between the ternary solution- and the solubility parameter values of χ_{12}^0.

Also important may be the fact that the PS/PMMA/BB concentrated solution data can be extrapolated to $\phi = 1$ to give a χ_{12} value similar to the χ_{12}^0's just discussed (Figure 5; Table 1). This implies a close correlation between solution- and bulk properties. As to PS/PIP systems, Mori et al. recently studied the X-ray scattering from a disordered PS-PIP block copolymer to determine χ_{12} as a function of temperature [16]. The broken curve in Figure 4 shows part of their results, to which our χ_{12}^0 data are rather

TABLE 1

Values of $10^2\chi_{12}^0/V_0$ (or the like) estimated by various methods (30°C)[a]

Blend	Dilute ternary solution[b]	Solubility parameter	Other method	Bulk
PS/PMMA	0.029, 0.026[c]	0.036	0.032[d]	0.028[e]
PS/PIP	0.14	0.13		0.16[f], 0.15[g]
PS/PEO	0.26	0.055		
PS/PIB	0.40	0.39		(0.4-0.5)[h]

[a] V_0 is the molar volume of solvent (or segment) in mL mol^{-1}. [b] Average value in BB except for (c). [c] In DO/BN. [d] Intrinsic viscosity analysis on random and alternating copolymers [22]. [e] Concentrated ternary solution data extrapolated to the bulk [6]. [f] Rounds and McIntyre [24]. [g] Mori, Hasegawa and Hashimoto [16]. [h] Estimated with the Koningsveld and Kleintjens data [23] (see ref 9 for detail).

close.   Koningsveld and Kleintjens have studied the phase behavior of a low-molecular weight blend of PS/PIB [23]. They observed strong composition dependence of the interaction parameter g_{12} in their system, and for this reason, comparison of the solution- and the bulk data would make no strict sense. Nevertheless, it may be noteworthy that our $\chi_{12}{}^0$ are comparable with their g_{12} in the order of magnitude [9]. In the light of thieir result, however, the temperature dependence observed for our $\chi_{12}{}^0$ may be somewhat unreasonably large (Figure 4). The reason is not clear at this time.

<u>Spinodal Composition for Highly Incompatible Systems</u>

It is generally difficult to derive well-defined information from non-dilute, asymmetrical ternary solutions, as has been pointed out. However, we wish here to emphasize that a study of the spinodal composition provides at least semi-quantitative information, if the system is highly incompatible. Because the spinodal concentration ϕ_{sp} can usually be determined with high precision, the ambiguity in an estimated value of χ_{12} arises mainly from that in χ_{0i}'s. Equation 15 or 16 clearly shows that this ambiguity becomes smaller and smaller, as $\chi_{12\,sp}$ or $\chi_{12\,sym}$ increases   (or ϕ_{sp} decreases). Another point we wish to make is that when $\chi_{12\,sp}$ is sufficiently large, the spinodal can exist in the dilute regime in which χ_{12} remains nearly constant. In such a case, we may expect the approximate equalities, $\chi_{12\,sym} \cong \chi_{12\,sp} \cong \chi_{12\,dil}$, where the subscript "dil" denotes the dilute limit.

Figure 8 shows $K\phi/R_0$ vs. ϕ plots for a PS/PMMA/BB, a PS/PEO/BB and a PS/PIB/BB. Each system has a PS in common and a second polymer of similar size, blended in a roughly similar ratio ($x \cong 0.5$). The plots for the PS/PEO and PS/PIB are nearly linear up to the spinodal, whereas that for the PS/ PMMA has a significant curvature at high concentrations. The spinodal concentration decreases in the order PS/PMMA > PS/PEO > PS/PIB, essentially reflecting the increasing order of $\chi_{12}{}^0$. As has been noted, the spinodal of the PS/PMMA exists in the semidilute regime. The crossover concentration ϕ^* for this system was found to be about 0.023, and this value should approximately apply to the other systems. Thus the PS/PEO solution may have its spinodal in the crossover region, whereas the PS/PIB solution may have its spinodal in the dilute regime. We below compare the values of $\chi_{12\,dil}$ and

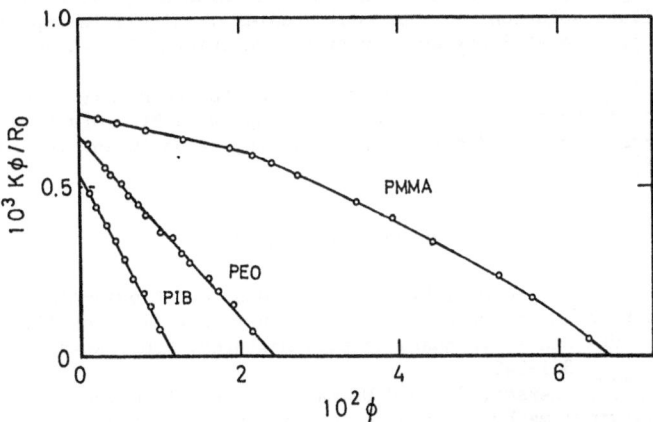

Figure 8. Comparison of the $K\phi/R_0$ vs. ϕ curves for PS/PMMA/BB, PS/PEO/BB, and PS/PIB/BB systems including a common PS (30°C): $10^{-5}M_w = 7.8$ for PS, 6.3 for PMMA, 6.8 for PEO, and 7.6 for PIB.

χ_{12sym} for the three systems:

	PS/PMMA	PE/PEO	PS/PIB
$\chi_{12}dil$	0.0030	0.0133	0.0225
$\chi_{12}sym$	0.0053	0.0133	0.0239

In the PS/PMMA, χ_{12sym} is considerably larger than χ_{12dil}. As we have noted, this is essentially due to an increase of χ_{12} itself in the semidilute regime, and possibly reflects an increased contact number of unlike (as well as like) segments in this regime. In the other two systems, the two parameters are nearly the same. According to the argument given above, we consider that these systems, especially the PS/PIB, are typical examples of those highly incompatible systems in which bad balance of the solvent properties for the two polymers makes no serious contribution to the location of the spinodal, and in which ϕ_{sp} is so small that the ϕ-dependence of χ_{12} is trivial. We should add that solvent BB is moderate for PS, fairly good for PIP, and nearly Θ for PIB [9].

Concluding Remarks

The light scattering observation of χ_{12} has revealed that the structure of a symmetrical ternary solution is "nonrandom" throughout the studied concentration range ($0 < \phi < \sim 0.5$): For each pair of polymers, there exists a characteristic concentration ϕ^* below which χ_{12} is approximately constant and equalt to χ_{12dil}, but above which it is an increasing function of ϕ only, independent of molecular weight. A ternary solution of a highly incompatible blend can have its spinodal in the region $\phi < \phi^*$ so that the approximate equalities $\chi_{12dil} \sim \chi_{12sp} \sim \chi_{12sysm}$ may hold. These remarks should be essentially correct also for assymetric systems.

On the basis of the phenomenological parameter χ_{12dil} and other information, the "molecular" parameter $\chi_{12}{}^0$ can be estimated by the aid of the excluded volume theory. This parameter, basically independent of molecular weight, blend composition, and solvent, is similar in value to the corresponding parameter for the bulk.

Selection of a suitable optical condition is essential to determine the parameter χ_{12}, especially χ_{12dil}, with sufficient accuracy. Thermodynamical symmetry is another requirement for obtaining unequivocal information out of nondilute solutions. Spinodal studies on highly incompatible systems may provide at least semi-quantitative information in any case.

Acknowledgment. We thank Messrs. M. Nagata and H. Miyashita for collaboration in this work. This work was supported by a Grant-in-Aid for Scientific Research, the Ministry of Education, Japan (Grant-in-Aid 5670081 and 59550613).

REFERENCES AND NOTES

1. Koningsveld, R., Chermin, H. A., and Gordon, M., Liquid-liquid phase separation in multicomponent polymer solutions VIII. Stability limits and consolute states in quasi-ternary mixtures. Proc. Royal Soc. London, A, 1970, 319, 331-349.
2. Robard, A., Patterson, D., and Delmas, G., The "$\Delta\chi$ effect" and polystyrene-poly(vinyl methyl ether) compatibility in solution. Macromolecules, 1977, 10, 706-707.
3. Krigbaum, W. R. and Flory, P. J., Statistical mechanics of dilute polymer solutions. III. Ternary mixtures of two polymers and a solvent.

J.Chem. Phys. 1952, 20, 873-876.

4. Stockmayer, W. H. and Stanley, H.E., Light scattering measurement of interactions between unlike polymers. J. Chem. Phys., 1950, 18, 143-154.

5. Fukuda, T. Nagata, M. and Inagaki, H., Light scattering from ternary solutions. 1. Dilute solutions of polystyrene and poly(methyl methacrylate). Macromolecules, 1984, 17, 548-553.

6. Fukuda, T., Nagata, M. and Inagaki, H., Light scattering from polymer blends in solution. 2. Nondilute solutions of polystyrene and poly(methylmethacrylate). Macromolecules, 1986, 19, 1411-1416.

7. Nagata, M., Fukuda, T. and Inagaki, H., Light scattering from polymer blends in solution. 3. Accurate determination of polymer-polymer interactions by using mixed solvents. Makromol. Chem. Rapid Commun., 1986, 7, 127-131.

8. Fukuda, T., Nagata, M. and Inagaki, H., Light scattering from polymer blend solutions. 4. Data analysis for asymmetrical dilute solutions. Macromolecules, 1987, 20, to appear in No. 3.

9. Nagata, M., Fukuda, T. and Inagaki, H., Light scattering from polymer blend solutions. 5. Characterization of systems of relatively high incompatibility. Macromolecules, in press.

10. Fukuda, T., Miyashita, H., Nagata, M. and Inagaki, H., Light scattering from polymer blend solutions. 6. Temperature-dependence of interaction parameter. Bull. Inst. Chem. Res., Kyoto Univ., in press.

11. Klotz, S., Cantow, H.-J. and Kögler, G., Interaction in polymer blend via dilute solution light-scattering under optical Θ-conditions. Polymer Bull., 1985, 14, 143-146.

12. Stockmeyer, W. H., Light scattering in multicomponent system. J. Phys. Chem., 1950, 18, 58-61.

13. There is an error in eq 22 in ref 6, which should read as eq 16 in this paper. The discussion in ref 6 need not in the least be modified, however.

14. Stockmayer, W. H., Problems of the statisatical thermodynamics of dilute polymer solutions. Makromol. Chem., 1960, 35, 54-74.

15. Tanaka, G. and Šolc, K., Second virial coefficient of polydisperse polymers. Macromolecules, 1982, 15, 791-800.

16. Mori, K., Hasegawa, H. and Hashimoto, T., Small-angle X-ray scattering from bulk block copolymers in disordered state. Estimation of χ-values from accidental thermal fluctuations. Polymer J., 1985, 17, 799-806.

17. de Gennes, P. G., Scaling Concepts in Polymer Physics, Cornell University, Ithaca, NY, 1979, Chapter 4.

18. Flory, P. J., Principles of Polymer Chemistry, Cornell University, Ithaca, NY, 1953, Chapters 12 and 13.

19. Broseta, D., Leibler, L. and Joanny, J.-F., private communication, 1987.

20. Hoy, K. L., New values of the solubility parameters from vapor pressure data. J. Paint Technol., 1970, 42, 76.

21. Graham, N. B., Nwachuku, N. E. and Walsh, D. J., Interaction of poly(ethylene oxide) with solvents: 1. Preparation and swelling of a crosslinked poly(ethylene oxide) hydrogel. Polymer, 1982, 23, 1345-1349.

22. Fukuda, T. and Inagaki, H., Interactions between unlike polymers versus dilute solution properties of copolymers. Pure Appl. Chem., 1983, 55, 1541-1551.

23. Koningsveld, R. and Kleintjens, L., Thermodynamics of polymer mixtures. J. Polym. Sci., Polym. Symp., 1977, 61, 221-249.

24. Rounds, N. A. and McIntyre, D., cited in Helfand, E. and Wasserman, Z. R., Macromolecules, 1976, 9, 879-888.

ON THE INTERPENETRATION EFFECT OF TWO HOMOGENEOUS POLYMERS DIFFERING IN MOLECULAR WEIGHT IN A GOOD SOLVENT

Marcel Mennen, Jan Smit and Michel Mandel
Department of Physical and Macromolecular Chemistry,
Leiden University, P.O.Box 9502, 2300 Leiden,
The Netherlands.

ABSTRACT

A study of the osmotic second virial coefficient A of mixtures of two chemically identical polymers differing in molecular weight in a good solvent has been performed. We have measured A as a function of the relative composition of the dissolved polymers for ten different combinations with molecular weight ratios varying from 1.8 to about 100. The experimental results were compared to the predictions of more than ten different theories. None of them can satisfyingly describe the behaviour of A in the whole region of molecular weight ratios. Our results show that when the two polymers have about the same size they behave like quasi-hard spheres and in the other limit of large size differences an interpenetration of the smaller molecule into the larger one is probable.

INTRODUCTION

The second virial coefficient A of mixtures of two chemically identical polymers differing in molecular weight in a good solvent has been subject of many studies, either experimentally [1-6] or theoretically [7-11]. Until now experiments failed to confirm the current theories, because the measurements were neither accurate enough nor sufficient in number. Hence we decided to set up a large number of accurate measurements comprising different combinations of two polymers with molecular weight ratios varying from about 1 to 100.

THEORY

The second virial coefficient of a mixture of two polymers is given by [7]

$$A = A_S x_S^2 + 2A_{SL} x_S x_L + A_L x_L^2 \qquad (1)$$

in which x is the mass fraction of the polymers and the subscripts S and L refer to the smaller and larger polymer respectively. A_S is a measure for the interaction between two small polymer molecules, A_L between two large ones and A_{SL} between a small and a large one.

This paper is concerned with the coefficient A_{SL}. At least fifteen papers [7-11] have dealt with a theoretical calculation of this quantity as a function of M_L, M_S, A_L and/or A_S. In some treatments other parameters such as the expansion factor α and the excluded volume parameter z [12] have been used. Here it would need too much space to consider them all with regard to their expressions for A_{SL}. Therefore we have calculated values of A_{SL}/A_S as a function of M_L/M_S for each theory. Some of them are shown together with our results in figure 3. Their physical meaning will be discussed later.

EXPERIMENTAL

Nearly monodisperse polystyrenes from the Pressure Chemical Company with $1.04 \leq \overline{M}_w/\overline{M}_n \leq 1.12$ were used as the solutes and toluene was used as the (good) solvent. Ten different combinations of two polystyrenes were studied (see table 1 for their molecular weights). For each of them we prepared nine stock solutions containing a weight fraction of both polymers in the range $0 \leq x_L \leq 1$ with a polymer concentration far below

TABLE 1
Number averaged molecular weight data of the used polystyrene sample, determined by osmotic pressure measurements.

no	M_S (kg/mole)			M_L (kg/mole)			M_L/M_S		
1	87.5	±	0.7	161.1	±	1.6	1.84	±	0.02
2	161.1	±	1.6	550	±	24	3.41	±	0.15
3	87.5	±	0.7	550	±	24	6.3	±	0.3
4	29.45	±	0.08	550	±	24	11.1	±	0.2
5	47.7	±	0.2	327	±	5	11.5	±	0.5
6	29.45	±	0.08	550	±	24	18.7	±	0.8
7	24.45	±	0.08	875	±	30[*]	29.7	±	1.0
8	17.02	±	0.08	550	±	24	32.3	±	1.4
9	17.02	±	0.08	875	±	30[*]	51.4	±	1.8
10	17.02	±	0.08	1660	±	30[*]	97.5	±	1.8

[*] calculated from $\overline{M}_w$ determined by light scattering and $\overline{M}_w/\overline{M}_n$ as given by the PCP.

c^*. Each stock solution was diluted seven times, after which the osmotic pressures of these solutions were measured in a Wescan osmometer (model 230) provided with a cellulose RC51 membrane supplied by Schleicher & Schüll. This membrane is impermeable for molecules with M > 10.000 g/mole. The temperature was kept at 25°C.

Following the virial expansion of van 't Hoff's law

$$\Pi/RTc = \bar{M}_n^{-1} + Ac + \dots \qquad (2)$$

(in which Π is the osmotic pressure, T the temperature, c the concentration, R the gas constant and $\bar{M}_n$ the number averaged molecular weight) we calculated $\bar{M}_n$ and A of each mixture by a linear plot of Π/RTc against c.

RESULTS AND DISCUSSION

Expression (1) implies that for each combination the second virial coefficient A can be represented by a second order polynomal in x_L or x_S (because $x_L+x_S=1$). By using a quadratic least squares fit we have determined A_S, A_L and A_{SL} from our measurements (see table 2). These calculated values of A_S and A_L were, within the accuracy of the measurements, always the same as their values measured at x_S or $x_L=0$ respectively. An example of the fit is represented in figure 1a.

Figure 1b shows the dependence of $\bar{M}_n^{-1}$ on x_L, which ought to be linear according to the definition of $\bar{M}_n$.

TABLE 2

Measured values of A_S (at $X_L=0$), A_L (at $x_S=0$) and A_{SL} (by the least squares method), all given in 10^{-10} mole m^3/g^2

no.	M_L/M_S	AS	A_L	A_{SL}	A_{SL}/A_S
1	1.84 ± 0.02	6.22 ± 0.18	5.60 ± 0.15	6.18 ± 0.13	0.99 ± 0.04
2	3.41 ± 0.15	5.60 ± 0.15	4.28 ± 0.30	5.55 ± 0.21	0.99 ± 0.05
3	6.3 ± 0.3	6.22 ± 0.18	4.28 ± 0.30	6.31 ± 0.21	1.01 ± 0.04
4	11.1 ± 0.2	7.93 ± 0.15	4.84 ± 0.12	8.30 ± 0.33	1.05 ± 0.04
5	11.5 ± 0.5	7.25 ± 0.16	4.28 ± 0.30	7.25 ± 0.27	1.00 ± 0.04
6	18.7 ± 0.8	7.93 ± 0.15	4.28 ± 0.30	8.88 ± 0.43	1.12 ± 0.06
7	29.7 ± 1.0	7.93 ± 0.15	3.86 ± 0.80	8.89 ± 0.47	1.12 ± 0.06
8	32.3 ± 1.4	9.30 ± 0.30	4.28 ± 0.30	10.08± 0.35	1.08 ± 0.05
9	51.4 ± 1.8	9.30 ± 0.30	3.86 ± 0.80	10.54± 0.37	1.13 ± 0.05
10	97.5 ± 1.8	9.30 ± 0.30	3.55 ± 1.30	10.55± 0.62	1.13 ± 0.07

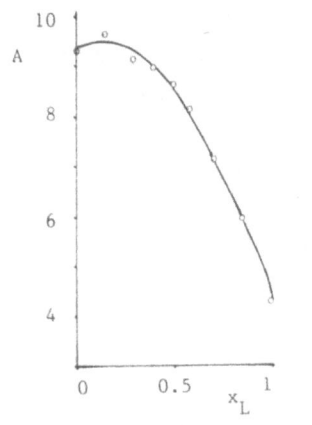 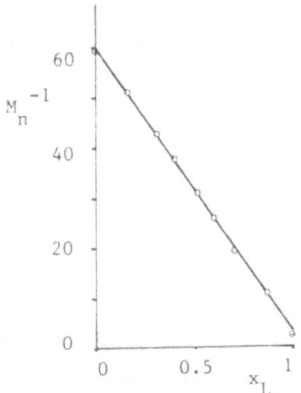

(a)   (b)

Figure 1. Measured values of A (a) and $\bar{M}_n^{-1}$ (b) as a function of x_L for combination 8. A is given in 10^{-10} mole m^3/g^2 and $\bar{M}_n^{-1}$ in 10^{-6} mole/g.

A linear plot of ln A against ln M (figure 2a) gives us a slope of 0.22±0.01, in agreement with theoretical predictions [12-13].

Let us now focus on the main problem. A good comparison between theories and experiment can be made in a plot of A_{SL}/A_S against M_L/M_S (figure 2b). Of course $A_{SL}/A_S \rightarrow 1$ for $M_L/M_S \rightarrow 1$. Witten and Prentis [10]

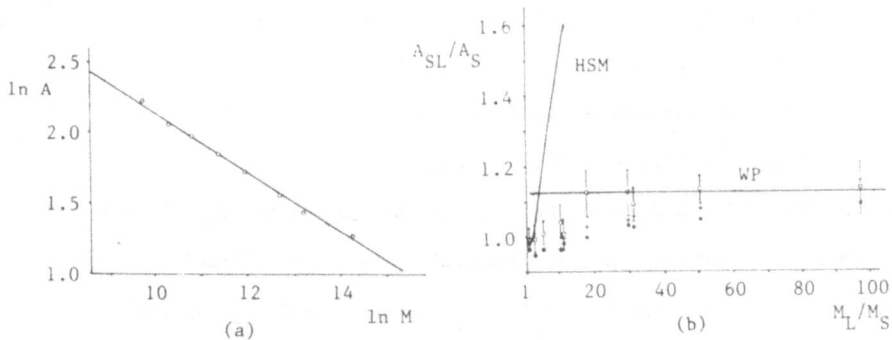

(a)   (b)

Figure 2. (a) Double-logarithmic plot of A (in 10^{-10} mole m^3/g^2) against M (number-averaged; in g/mole). The straight line has a slope of -0.22±0.01; (b) A_{SL}/A_S as a function of M_L/M_S for our experiments (0) and for the theories of Casassa (Hard Sphere Model, full curve denoted by HSM), Witten and Prentis (full line, denoted by WP), Casassa and Markovitz (+) and Tanaka and Solc with their A_{ij}-averaging (*). The two latter sets of points have been calculated using measured values of the radius of gyration of the large molecule. To keep the figure comprehensive we have presented only the theories which give values comparable to our experimental results.

have predicted a slightly larger value, as they claim their result to be independent of the ratio M_L/M_S. This value seems to be rather good only for large M_L/M_S, as our results show. Their calculation is based on the assumption that the smaller molecule interpenetrates into the larger one. The hard sphere approximation of Casassa [7], deviates strongly from our results except for $M_L/M_S \approx 1$.

The theories of Casassa and Markovitz [8] and of Tanaka and Solc [9] give values which are somewhat but systematically lower than the experimental ones. Moreover, in these two theories the problem is treated with the perturbation expansion method, which have been proved to be unvalid for large excluded volume effects.

We feel that from our results it may be concluded that two chains of comparable size behave like quasi-hard spheres, but with increasing size differences the smaller molecule tends to interpenetrate into the larger one.

REFERENCES

1. Noda, I., Kitano, T. and Nagasawa, M., J.Polym.Sci., Polym.Phys.Ed., 1977, 15, 1129-42.

2. Krigbaum, W.R. and Flory, P.J., J.Amer.Chem.Soc., 1953, 75, 1775-83.

3. Chien, J.Y., Shih, L.-H. and Yu, S.-C., J.Polym.Sci., 1958, 29, 117-25.

4. Varadaiah, V.V. and Rao, V.S., J.Polym.Sci., 1961, 50, 31-34.

5. Kok, C.M. and Rudin, A., Eur.Polym.J., 1986, 22, 107-9.

6. Wallace, T.P. and Casassa, E.F., Polym.Prepr., 1970, 11, 136-41.

7. Casassa, E.F., Polymer, 1962, 3, 625-38.

8. Casassa, E.F., and Markovitz, H., J.Chem.Phys., 1958, 29, 493-503.

9. Tanaka, G. and Solc, K., Macromolecules, 1982, 15, 791-800.

10. Witten, T.A. and Prentis, J.J., J.Chem.Phys., 1982, 77, 4247-53.

11. The list with theoretical papers on this subject could be extended with at least ten other ones, but because of the small length of this paper we only refer to the best fitting theories.

12. Yamakawa, H., Modern theory of polymer solution, Harper & Row, New York, 1971.

13. Gennes, P.G. de, Scaling concepts in polymer physics, Cornell University Press, Ithaca/London 1979.

DETERMINATION OF BINODAL AND SPINODAL POINTS BY MEANS OF THE PRESSURE PULSE INDUCED CRITICAL SCATTERING METHOD

F. Kiepen, W. Borchard
Angewandte Physikalische Chemie der
Universität -GH- Duisburg, FRG

ABSTRACT

Thermodynamic properties of polymer-solvent systems can be determined near the miscibility gap in the temperature(T)-pressure(P)-concentration(y) phase diagram with the pressure pulse induced critical scattering method, which was developed by the authors. It is shown that turbidity corrections of the experimental determined light scattering intensities in the critical region of the binodal plane are necessary, because only the light scattering intensities without attenuation are of thermodynamic relevance.

INTRODUCTION

The thermodynamics of polymer solutions has taken much attention in the physics of polymers in the last decade. The development is partially based on scattering methods, which makes it possible to determine binodals and spinodals of these multicomponent systems. In our investigations we are interested in the pressure dependence of the separation phenomena of polystyrene-solvent systems at upper and lower critical solution temperatures. For this reason we have developed the pressure pulse induced critical scattering (PPICS) method [1], which is a continued development of the thermally induced light scattering investigated by Scholte [2,3] and Gordon [4].

THEORY

There are high turbidities near the binodal curves of polymer-solvent systems especially in the critical region. This means that the scattered light will be attenuated before having reached the scattering volume and after having left it. For thermodynamic calculations the scattered light

$I_{,0}^{corr}$ at the scattering angle of 90° without any attenuation has to be known. $I_{,0}^{corr}$ can be calculated from transmission light intensities I_{tr} and the directly measured light scattering intensity $I_{,0}^{exp}$ after attenuation [1]. $I_{,0}^{corr}$ is named the corrected light scattering intensity at a scattering angle of 90°.

In fig. 1 the extrapolation procedure close to the binodal curve in the $P-y_2$ phase-diagram is shown schematically. The binodal curve has an upper critical solution pressure at con-

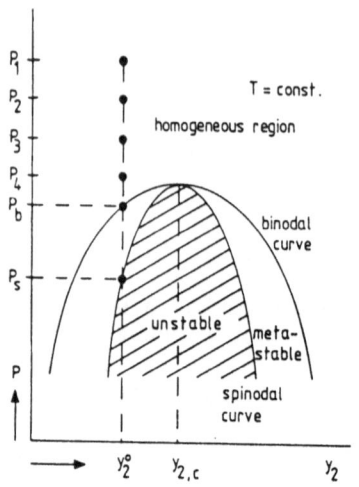

Fig. 1: Schematic $P-y_2$ phase diagram for the explanation of the PPICS method

stant temperature T. The binodal curve separates the homogeneous from the heterogeneous region. Inside the heterogeneous region the spinodal curve is the limit between the metastable and the absolutely unstable area. At the total mass fraction y_2^0 of the polymer in the mixture, I_{tr} and $I_{,0}^{exp}$ are registrated for the different pressures P_1 up to P_4 in the homogeneous region. $I_{,0}^{corr}$ is calculated from the corresponding directly measured values of $I_{,0}^{exp}$ and I_{tr}. The binodal pressure P_s can be determined from anomalies in I_{tr} and $I_{,0}^{corr}$ in reversing the pressure changes, this is typical for the occurence of new phases. The spinodal pressure P_s is obtained from the linear extrapolation of $(I_{,0}^{corr})^{-1}$ to zero against the pressure [5]. At the critical concentration $y_{2,c}$ of a binary system the binodal and spinodal curves have a common tangent, which means $P_b = P_s$ [6]. The extrapolation procedure can be carried out by means of an isothermal and an adiabatic method [1].

MATERIALS AND METHODS

The experimental equipment and the method of operation have been described previously [1]. The oligostyrene sample PSA 35 has a number average of molar mass of $\bar{M}_n = 1000$ gmole^{-1} with a ratio of the weight to the number average of the molar masses $\bar{M}_w/\bar{M}_n = 1.1$. The n-pentane was a Fluka product of 99.5 %.

RESULTS

The extrapolation procedure will be explained for a solution with $y_{2,c} = 0.400$, which is the critical mass fraction of the oligostyrene in n-pentane. The measurements were carried out with the isothermal method at the temperature T = 13.78 °C. Fig. 2 shows the reciprocal directly measured scattering intensities $(I_{90}^{exp})^{-1}$ at a scattering angle of 90°, the transmission light intensities I_{tr} and the reciprocal corrected scattering intensities $(I_{90}^{corr})^{-1}$ against the pressure P.
Starting at high pressures the $(I_{90}^{exp})^{-1}$-curve begins with a decrease with decreasing P, passes through a minimum and increases up to the binodal pressure P_b. At this critical concentration the binodal pressure is identical to the spinodal pressure P_s. The I_{tr}-curve is decreasing monotonously with decreasing pressure and becomes zero at

$P_b = P_s = 127$ bar. Similar behaviour has been found by B.A. Wolf et al. in the system polystyrene-t-decalin near the critical composition [7]. The $(I_{90}^{corr})^{-1}$-curve is completely linear and becomes zero at $P_b = P_s = 127$ bar as has been predicted by the theory [5].

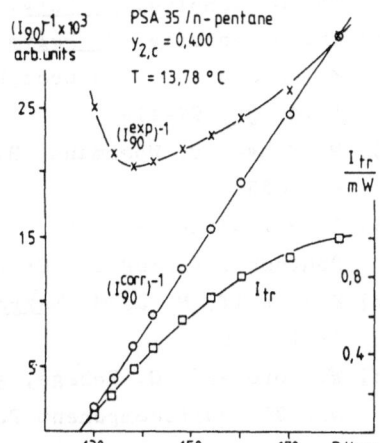

Fig. 2: Pressure dependence of transmitted and scattered light intensities in the system PS-n-pentane

DISCUSSION

By comparison of the $(I_{,0}^{exp})^{-1}$-P-curve and the $(I_{,0}^{corr})^{-1}$ -P-curve qualitative and quantitative differences are shown. The $(I_{,0}^{corr})^{-1}$-P-plot is totally linear according to the theory of Debye [6]. The $(I_{,0}^{exp})^{-1}$-P-curve has a minimum which does not make a thermodynamic sense. The minimum is related to the high turbidity of the solution and the relative large distance which the scattered light beam has to pass before leaving the high pressure cell through a window. Minima of this kind have also been reported studying the critical opalescence of polymer solutions [8]. In further experiments we have shortened the optical pathway in the high pressure cell, which gives a $(I_{,0}^{exp})^{-1}$ vs. P-curve without a minimum, but still with large quantitative differences with respect to the above mentioned $(I_{,0}^{corr})^{-1}$ vs. P-curve, which belongs to the large optical pathway.

ACKNOWLEDGEMENT

Financial support by the "Deutsche Forschungsgemeinschaft" is greatfully acknowledged.

REFERENCES

[1] F. Kiepen, W. Borchard, presented for publication,
Ber. Bunsenges. 1987

[2] Th. G. Scholte, J. Polym. Sci. A-2, 1971, 9, 1553-1577

[3] Th. G. Scholte, J. Polym. Sci. C, 1972, 39, 281-291

[4] M. Gordon, J. Goldsbrough, K. Derham, Pure Appl. Chem.,
1974, 38, 97-116

[5] P. Debye, D. Woermann, B. Chu, J. Chem. Phys., 1962, 36,
851-855

[6] I. Prigogine, R. Defay, "Chemische Thermodynamik", VEB
Deutscher Verlag für Grundstoffindustrie, Leipzig 1962

[7] B.A. Wolf, R. Jend, Makromol. Chem., 1977, 178,
1811-1822

[8] W. Borchard. G. Rehage, Advances in Chemistry Series,
Nr. 99 "Multicomponent Polymer Systems", 1971, 4, 42-52.

The Simha-Somcynsky Hole Theory: Thermodynamic Properties
Based on Compositional Derivatives

Erik Nies and Alexander Stroeks
Laboratory of Polymer Technology
Eindhoven University of Technology
P.O. Box 513
5600 MB Eindhoven
The Netherlands

Abstract

The compositional derivatives, e.g. binodal, spinodal, and critical
conditions represent an important set of thermodynamical properties both
from an experimental and theoretical point of view. Compositional
derivatives of the Simha-Somcynsky expression for the Helmholtz free
energy are computed. Phase diagrams for a binary mixture of hypothetical
low molar mass components as well as a mixture of a low and a high molar
mass component are shown. These results are compared with those based on
the classical Flory-Huggins rigid lattice theory. Finally, the phase
diagram of the system polyethylene/n-hexane is constructed and compared
with the binodal and critical conditions computed by Jain and Simha. The
present results do not intend to give a systematic survey of the
thermodynamic properties considered, but merely serve to demonstrate the
feasibility of these computations.

Introduction

Most of the applications of the Simha-Somcynsky (SS) theory, [1]
presented so far, involved volume dependent derivatives of the free
energy, e.g. equation of state, expansion coefficient, etc.. Recently, a
study of the possibilities of the SS theory in the prediction of
compositional derivatives of the free energy was initiated. In one of
these applications, the free energy of mixing for the system polyethylene
/n-hexane (PE/nH), considered as a monodispers polymer solution was
approximated by quartics in the mole fraction of polymer [2]. These
polynomials were then used to calculate binodal and critical conditions.

Theoretical considerations

1. The Simha Somcynsky hole theory

The Simha Somcynsky theory has been reviewed in some detail [3].
Therefore, only a brief recapitulation of the most important aspects
pertinent to pure components and binary mixtures thereof is presented
here.

1.a. Pure components

A system in the liquid state is modelled on a quasi-lattice with sites either vacant or occupied by segments of chain molecules (s-mers). The configurational partition function Z reads

$$Z = g \ v_f{}^{cN} \exp \ (-E_0/kT) \tag{1}$$

For the combinatorial entropy g, derived from the mixing of vacant and occupied sites, the expression of Flory is adapted [4]

$$g = [(z-1)/e]^{(s-1)N} (y/s)^{-N} (1-y)^{-Ns(1-y)/y} \tag{2}$$

where y is the fraction of occupied sites, N the number of s-mers and z the lattice coordination number.

The internal energy E_0 is equated to the lattice energy, all segments placed in rest position on the lattice. Simha and Somcynsky adapt an equation due to De Boer and Prigogine [5].

$$E_0 = yNqz\varepsilon^*/2 \ (1.011 \ \tilde{w}^{-4} - 2.409 \ \tilde{w}^{-2}) \tag{3}$$

where $\tilde{w}$ (=yv/v^*) is the reduced cell volume, ε^* and v^* are the characteristic energy and volume of a segment, qz (= s(z-2)+2) is the number of external contact sites of the s-mer.

The cell partition function v_f accounts for the motion of the molecules in their respective cells. Following Prigogine [6], the concept of volume dependent (external) degrees of freedom ('c' parameter) is introduced. For a single external degree of freedom Simha and Somcynsky write [1]

$$v_f{}^{1/3} = [y(\tilde{w}^{1/3}-2^{-1/6}) + (1-y)\tilde{w}^{1/3}](v^*)^{1/3} \tag{4}$$

A chain molecule is thus characterised by ε^*, v^*, a geometric (s) and dynamic (c) parameter. This last parameter is related to the flexibility of the molecule.

The fraction of occupied sites y, is determined by the minimalisation of the free energy, or alternatively

$$(\partial Z/\partial y)_{V,T} = 0 \tag{5}$$

1.b. Binary mixture

For a mixture of N_1 and N_2 molecules having respectively s_1 and s_2 segments the form of the expression for Z is retained. However, the molecular parameters become dependent on composition through appropriate averages [7].

$$\langle s \rangle = x_1 s_1 + x_2 s_2 \qquad \langle c \rangle = x_1 c_1 + x_2 c_2$$

$$\langle \varepsilon^* \rangle \langle v^* \rangle^2 = \varepsilon^*_{11} v^*_{11}{}^2 \ x_1{}^2 + 2\varepsilon^*_{12} v^*_{12}{}^2 \ x_1 x_2$$

$$+ \ \varepsilon^*_{22} v^*_{22}{}^2 \ x_2{}^2 \tag{6}$$

$$\langle\varepsilon^*\rangle\langle v^*\rangle^4 = \varepsilon^*_{11}v^{*4}_{11}X_1^2 + 2\varepsilon^*_{12}v^{*4}_{12}X_1X_2$$
$$+ \varepsilon^*_{22}v^{*4}_{22}X_2^2$$

$$X_1 \ (= 1-X_2) = x_1q_1z/(x_1q_1z + x_2q_2z)$$

with $x_1(x_2)$ the mole fraction of component 1(2).

Two mixing parameters are introduced, i.e. ε^*_{12} and v^*_{12} the characteristic parameters of intersegmental contacts. The combinatorial factor for the binary mixture becomes

$$g' = (x_1^{-N1}x_2^{-N2})\ g \tag{7}$$

2. Rigid lattice model

For comparison, we will use an expression for the free energy of mixing based on the rigid lattice theory, i.e. the classical Flory-Huggins equation [4,8]. In this theory the energy of mixing ΔE_0 can be written as

$$\Delta E_0 = N_1s_1\phi_2 \ (b_0T + b_1)$$

where ϕ_2 (= $N_2s_2/(N_1s_1+N_2s_2)$) is the volume fraction of component 2, b_0 and b_1 are adjustable parameters, describing the cooperative free energy [9].

Thermodynamic equations of interest

The partition function can be used to calculate thermodynamic properties. Those of interest to us are summarized in Table 1.

Table 1. Thermodynamic relationships

Helmholtz free energy $F = -kT \ln(Z)$
Equation of state $\quad p = -(\partial F/\partial V)_T$
Gibbs free energy $\quad G = F + pV$
Binodal condition $\quad \mu_i' = \mu_i''$, $i = 1,2$
Spinodal condition $\quad I_{sp} = (\partial^2 G/\partial x_1^2)_{p,T} = 0$
Critical condition $\quad I_{sp} = 0$ and
$\quad\quad\quad\quad\quad\quad\quad J_{cr} = (\partial^3 G/\partial x_1^3)_{p,T} = 0$

Results

For the SS theory binodal, spinodal and critical conditions result in a complicated set of non-linear equations. For example, the spinodal conditions are determined by a set of seven implicit equations. Consisting of the spinodal condition and six extra equations, i.e. the equation of state, the minimalisation condition (5) and the first and second derivatives of both equations with respect to the mole fraction x_1. The resulting sets of implicit equations defining the different properties, are solved in an iterative process without having recourse to approximations (e.g. approximation of the free energy by quartics). Here we present the first results of these computations.

Mixture of two low molar mass components

Figure 1 represents binodal, critical and spinodal conditions for a hypothetical mixture of two low molar mass components. Hypothetical values for the parameters of the pure components are adapted. The mixing parameter ε_{12}^* is adjusted to give a two phase region in a convenient temperature range. Binodal and spinodal make contact and have a common tangent in the maximum of both curves; i.e. the critical point. Both curves are symmetrical around the critical composition ($x_1 = 0.5$). These results are as expected for a binary mixture of spherical molecules and make us feel confident in the present computational procedures.

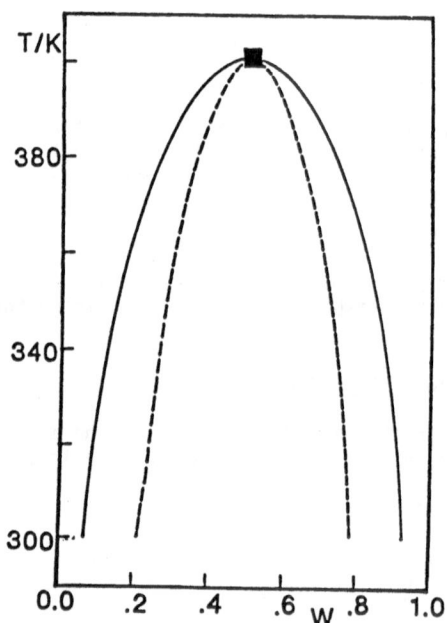

Figure 1. Binary mixture of low molar mass components. Binodal (——), spinodal (--) and critical conditions (■) as function of mass fraction w_2

Polymer solutions: UCST behaviour

Spinodal and critical conditions [10-14] for the well documented system polystyrene/cyclohexane (PS/CH) are presented in Figure 2. For comparison the computations based on the Flory-Huggins expression are drawn also [15].

- The FH treatment (-.-., ■) gives the familiar picture, i.e. a good description of the critical temperatures (used in the estimation of the parameters b_0 and b_1) but a less satisfactory prediction of the critical concentrations and spinodal conditions.

— In the computation with the SS theory (—, ■) the parameters for the pure components were taken from literature [16,17]. We only adjusted the characteristic energy ε^*_{12} to give good agreement between the computed and experimental critical temperature for the polymer sample M_1. The characteristic volume v^*_{12} is fixed at the value given in Table 2. With this single adjustable parameter, all the remaining critical and spinodal conditions are predicted.

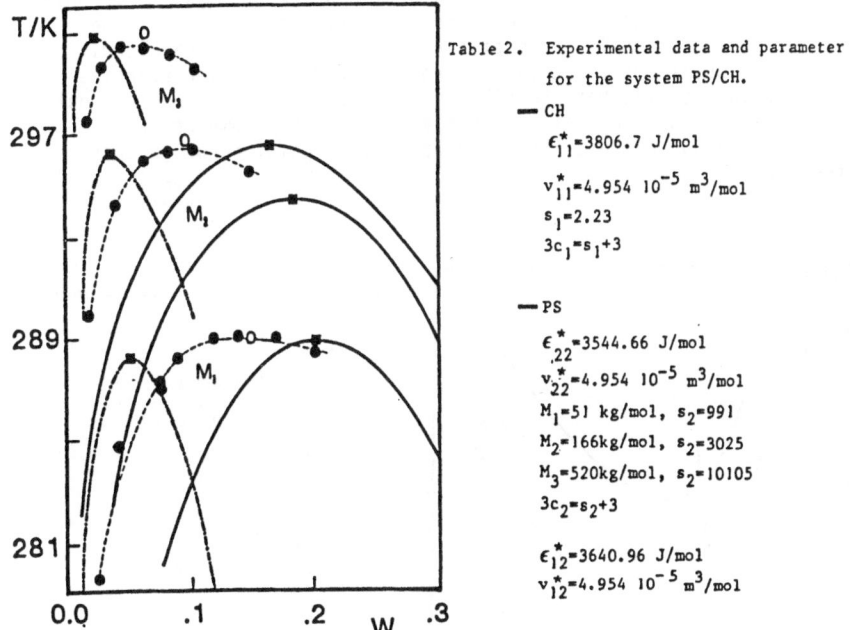

Table 2. Experimental data and parameter for the system PS/CH.

— CH

$\varepsilon^*_{11} = 3806.7$ J/mol

$v^*_{11} = 4.954 \ 10^{-5}$ m^3/mol

$s_1 = 2.23$

$3c_1 = s_1 + 3$

— PS

$\varepsilon^*_{22} = 3544.66$ J/mol

$v^*_{22} = 4.954 \ 10^{-5}$ m^3/mol

$M_1 = 51$ kg/mol, $s_2 = 991$

$M_2 = 166$ kg/mol, $s_2 = 3025$

$M_3 = 520$ kg/mol, $s_2 = 10105$

$3c_2 = s_2 + 3$

$\varepsilon^*_{12} = 3640.96$ J/mol

$v^*_{12} = 4.954 \ 10^{-5}$ m^3/mol

Figure 2. Spinodal (●,—,---) and critical conditions (■,O) for the system PS/CH as a function of mass fraction w_{PS}. Experimental data from different references [10-14].

One can observe that the predicted chain length dependence of the critical temperatures doesn't agree quantitatively with the experimental data. Furthermore, the critical concentrations are predicted to be too large. The correspondence between experimental and predicted critical concentrations certainly can be improved by adjusting the mixing parameter v^*_{12}. Compared to the FH treatment, the predicted spinodal curves are in much better agreement (much broader) with experimental data and are comparable to the descriptive quality of spinodal conditions obtained with the Flory-Huggins-Staverman-Guggenheim expression for the free energy of mixing studied by Koningsveld and coworkers [15].

Polymer solutions: LCST behaviour

In Figure 3 the computation of the LCST behaviour for the system PE/nH
are presented for the parameters (Table 3) used by Simha and Jain [2].
Because of numerical difficulties, probably due to the use of mole
fractions, we were not able yet to compute binodals. It is hoped that
the reformulation of the appropriate equations in volume or mass
fraction will remedy this problem. For comparison, the binodals and
critical conditions computed by Simha and Jain are drawn also. If both
computations were consistent, one would expect the spinodals to be
confined by the binodals and meet each other in the critical conditions.
This is obviously not the case. The reasons for these discrepancies
haven't been resolved yet. Possibly, it is due to the approximate nature
of the quartics used by Jain and Simha or, possibly some errors are
still left in our computational software. This point needs further
attention.

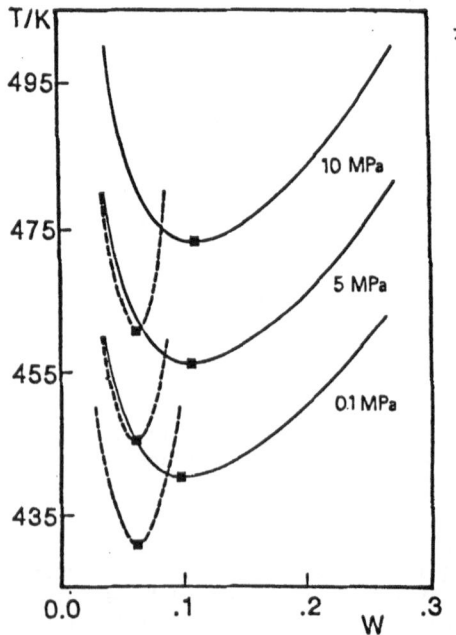

Table 3. Parameter values for the system PE/nH

— nH

$\epsilon_{11}^{*} = 2146.9$ J/mol

$v_{11}^{*} = 3.863 \ 10^{-5}$ m^3/mol

$s_1 = 2.5$

— PE

$\epsilon_{22}^{*} = 3316.5$ J/mol

$v_{22}^{*} = 4.948 \ 10^{-5}$ m^3/mol

$s_2 = 252.2$

$\epsilon_{12}^{*} = 2146.9$ J/mol

$v_{12}^{*} = 4.4156 \ 10^{-5}$ m^3/mol

Figure 3. Predicted spinodal (—), binodal (—) and critical (■) conditions
for the system PE/nH as a function of mass fraction w_{PE}.

Conclusions and future research

It is shown that exact calculations of compositional derivatives with the SS theory are possible. However, some discrepancies between our calculations and those of Simha and Jain still have to be resolved.

The computations presented here, suggest that the SS theory can be useful in the interpretation of the phase behaviour of polymer systems. Furthermore, the presented computations can be extended easily to deal with polydispersity.

The performance of the SS theory with respect to compositional derivatives will be elaborated in the near future.

Literature

1. R. Simha, T. Somcynsky, Macromolecules, 2, 342 (1969)
2. R.K. Jain, R. Simha, Macromolecules, 17, 2663 (1984)
3. R. Simha, Macromolecules, 10, 1025 (1977)
4. P. Flory, J. Chem. Phys, 10, 51 (1942)
5. V.S. Nanda, R. Simha, T. Somcynsky, J. of Pol. Sci., part C, 12, 277 (1966)
6. I. Prigogine, "The Molecular Theory of Solutions", North Holland Publishing Co., Amsterdam (1957)
7. R.K. Jain, R. Simha, Macromolecules, 13, 1501 (1980)
8. M.L. Huggins, Ann. N.Y. Acad. Sci., 43, 1 (1942)
9. E.A. Guggenheim, "Mixtures", The Glarendos Press, Oxford (1952)
10. J. Goldsbrough, Science Progress (Oxford), 60, 281 (1972)
11. K.W. Derham, J. Goldsbrough, M. Gordon, Pure Appl. Chem., 38, 97 (1974)
12. Th.G. Scholte, J. Pol. Sci., A2, 9, 1553 (1971)
13. P.A. Irvine, M. Gordon, Macromolecules, 13, 761 (1980)
14. R. Koningsveld, L.A. Kleintjens, A.R. Schultz, J. Polym. Sci., A2, 6, 1261 (1970)
15. E. Nies, R. Koningsveld, L.A. Kleintjens, Progr. Colloid & Polym. Sci., 71, 2 (1985)
16. R. Simha, R.K. Jain, Colloid and Polymer Sci., 263, 905 (1985)
17. R.K. Jain, R. Simha, P. Zoller, J. Polym. Sci., Polym. Phys. Ed., 20, 1399 (1982)

MISCIBILITY AND MELT MORPHOLOGY OF BINARY POLYMER BLENDS OF SOLUTION CHLORINATED POLYETHYLENE AND POLYCAPROLACTONE

G. Defieuw, G. Groeninckx and H. Reynaers
Catholic University of Leuven
Laboratory for macromolecular structural chemistry
Celestijnenlaan, 200 F, 3030 Heverlee
Belgium.

ABSTRACT

Last decade, the system polyvinylchloride (PVC) - Polycaprolactone (PCL) has been studied by various authors (1-4). It is miscible over the entire composition and temperature range and exhibits no lower critical solution temperature behavior (LCST). Decreasing the chlorine content allows to pass into a system which exhibits LCST behavior or which is partially miscible.

EXPERIMENTAL

Randomly chlorinated polyethylene (CPE) samples with different chlorine contents were obtained by bubbling chlorine-gas through a 1% solution of HDPE in 1,1',2,2' tetrachloroethane at 110°C. A recent study in the literature (5) reveals that CPE produced by solution chlorination exhibits a random distribution of the chlorine atoms along the backbone. The chlorine contents were determined by elemental analysis (Schoninger method) and are given in weight percent (mentioned after the name CPE). DSC measurements show that CPE30.1 is already amorphous. In this study, only amorphous CPE will be used.
Molecular weight characteristics were determined by GPC analysis relative to polystyrene standards and are given in Table I.

SAMPLE	$\overline{MW}$	$\overline{MN}$	$\overline{MW}/\overline{MN}$
CPE35.6	198000	59000	3.35
CPE42.1	193000	50000	3.84
CPE49.1	293000	53000	4.64
PCL	22500	14000	1.60

TABLE I. Molecular parameters of the samples

The polymer blends were prepared by coprecipitation of a 3 % tetrahydrofuran solution in hexane. This method of preparation of the polymer blends was preferred because it allows to obtain a blend which is mixed on a molecular level and the choice of the solvent and the non-solvent does not affect the state of miscibility as it is the case in solvent casted films.

DSC scans were run at 5°C/min on a Perkin Elmer DSC 2C equiped with a Thermal Analysis Data System (TADS).
Prior to Dynamic Mechanical Analysis (DMA), the powdered blends were compressed at 120°C. These samples were crystallized at 25°C. DMA experiments were performed on a Dupont 1090 instrument at a scanning rate of 5°C/min.

RESULTS AND DISCUSSION

Miscible systems

Optical microscopy gives a first indication for the degree of miscibility in these blends. CPE49.1 seems to be miscible with PCL over the entire composition and temperature range, while CPE42.1 exhibits a temperature and composition dependent miscibility (fig. I). The cloud point curve is located at the PCL-side due to the difference in molecular weight between both polymers (6). The reversibility of this phase separation process is still a matter of further investigations.

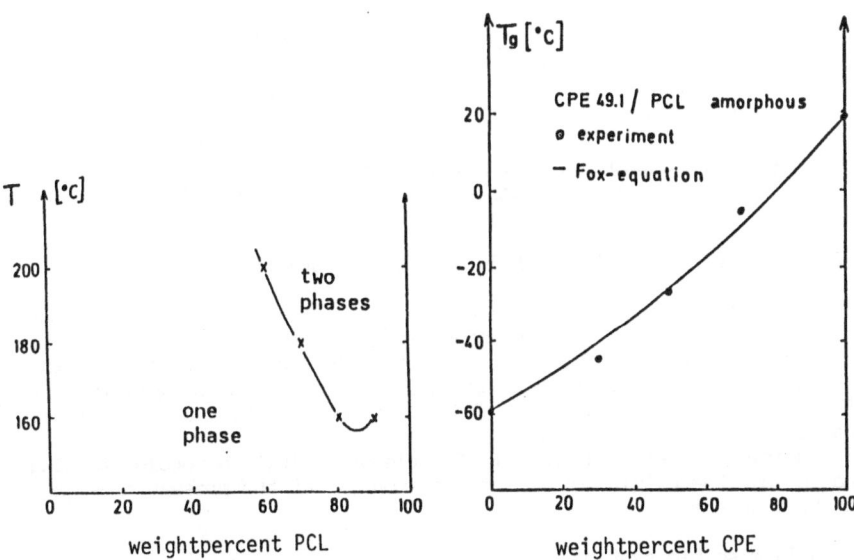

Fig. I. Cloud point curve of   Fig. II. Glass-transition temperatures
        CPE42.1/PCL   of amorphous CPE49.1/PCL blends

Since miscible polymer blends exhibit one glass transition temperature between those of the pure components, DSC measurements can give information about the state of miscibility. In order to avoid difficulties in calculating the composition of the amorphous phase in a semicrystalline blend, samples were heated to 100°C and quenched in liquid nitrogen. Only samples containing 90% PCL or more were not amorphous after quenching. No broadening of the glass-transition is observed and the data for these miscible systems fit rather well with the Fox-equation (fig. II.).

Since quenching of samples for DMA experiments in the amorphous state is very difficult, semicrystalline samples were used to determine the glass-transition behavior (fig. III). The loss modulus G" peak is shifted to higher temperature due to the change in composition of the amorphous phase after crystallization.

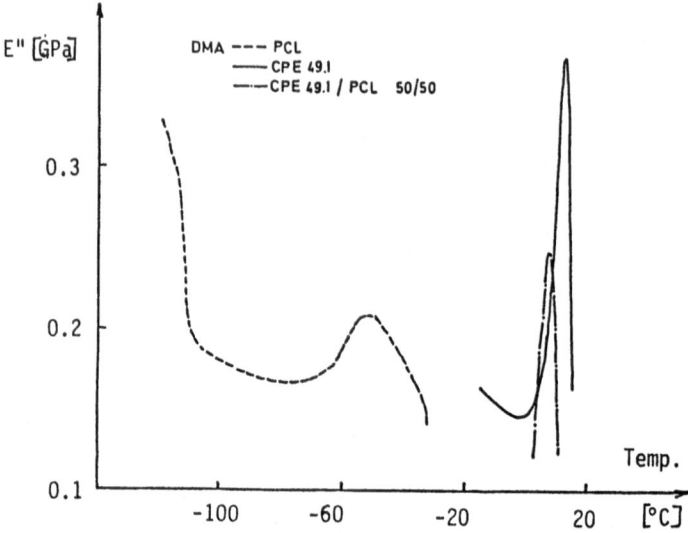

Fig. III. DMA of CPE49.1/PCL 50/50 after crystallization of PCL

Crystallization kinetics can also give an indication for the presence of a miscible melt (fig. IV). The spherulitic growth rate of PCL in the blends depends on the CPE content and the chlorine content of CPE. The spherulitic growth rate of PCL is much more affected by addition of CPE49.1 than by addition of CPE42.1 since CPE49.1 has a higher glass-transition temperature.
All miscible semicrystalline CPE/PCL blends exhibit a double melting behavior but the origin of this phenomenon is at the present time not yet fully understood.

Partially miscible systems

If the chlorine content of CPE is decreased below 42.1 wt.%, the cloud point curve is shifted to lower temperatures ; a partially miscible blend results when the minimum of this curve is located below room temperature. At 100°C, two molten phases are present in CPE35.6/PCL 50/50 (fig. V.a) and the small domains seem to fuse at higher temperatures (fig. V.b). All compositions of the CPE35.6/PCL blend crystallize during the quenching process and the measurement of the glass-transition temperatures with DSC is rather difficult.
Only one loss modulus peak is observed in DMA measurements corresponding to the glass-transition of the CPE-rich phase (fig. VI) ; the glass

transition of the PCL-rich phase is not detectable due to the high degree of crystallinity of this phase.

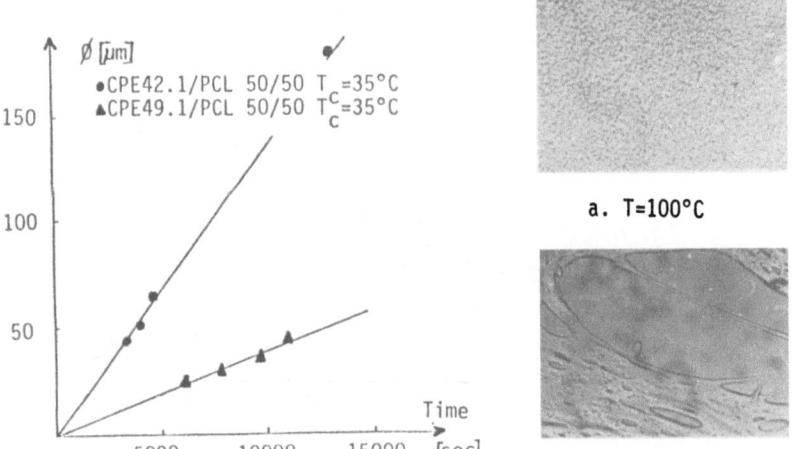

a. T=100°C

b. T=200°C   50 μm

Fig. IV. Spherulitic growth rate of CPE/PCL 50/50 blends

Fig. V. Melt morphology of CPE35.6/PCL 50/50

The presence of CPE in the PCL-rich phase can be proved by measuring the spherulitic growth rate of PCL in this phase (fig. VII). A small decrease is observed for all compositions.

Since CPE35.6/PCL 90/10 forms a one phase system at 100°C (from optical microscopy) while this is not the case for the 10/90 blend, one can conclude that the solubility of PCL in CPE is higher then the reverse case. This is probably caused by the pronounced difference in molecular weight between both polymers.

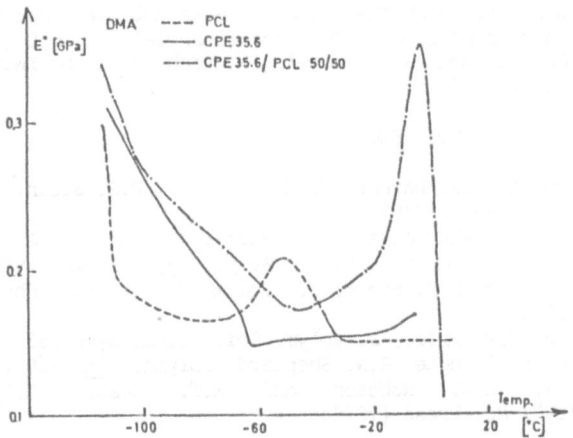

Fig. VI. DMA of CPE35.6/PCL 50/50 after crystallization of PCL

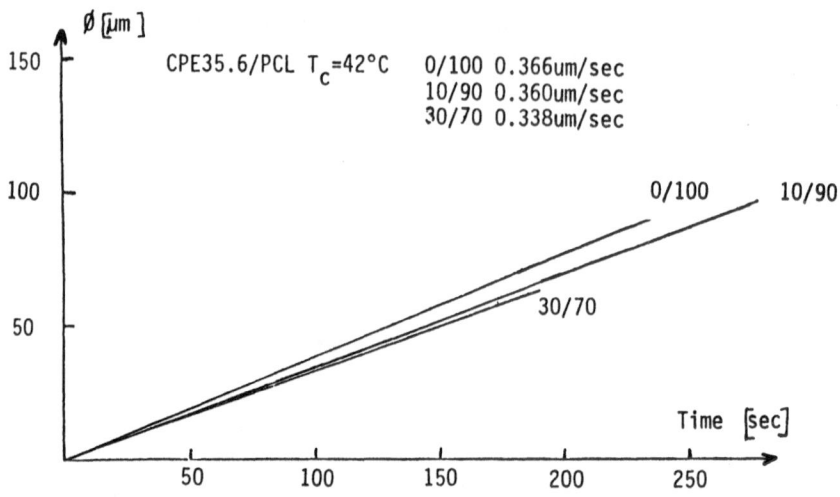

Fig. VII. Spherulitic growth rate of the PCL-rich
phase in CPE35.6/PCL blends

CONCLUSIONS

The present state of the study on CPE/PCL blends reveals that CPE containing 49.1% chlorine is miscible with PCL over the entire composition and temperature range, while CPE42.1/PCL blends exhibit a composition dependent LCST behavior. A further decrease of the chlorine content of CPE to 35.6% results in a partially miscible CPE/PCL blend.

AKNOWLEDGEMENTS

The authors wish to thank the Research Fund KULeuven and the Belgian National Science Foundation (NFWO) for financial support given to the laboratory. One of them (G. Defieuw) is indebted to the IWONL Brussels for a research grant.

REFERENCES

1. F.H. Khambatta, F.P. Warner, T. Russell, and R.S. Stein, J. Polym. Sci., A2, 14, 1391 (1976)
2. R.S. Stein, F.B. Khambatta, F.P. Warner, T. Russell, A. Escala, and E. Balizer, J. Polym. Sci. Polym. Symp., 63, 313-328 (1978)
3. T.P. Russell and R.S. Stein, J. Polym. Sci. Polym. Phys. Ed., 21, 999-1010 (1983)
4. C.J. Ong and F.P. Price, J. Polym. Sci. Polym. Symp.,63, 45-75 (1978)
5. C. Zhikuan, S. Lianghe, R.N. Sheppard, Polymer, 25, 369 (1984)
6. O. Olabisi, L.M. Robeson and M.T. Shaw, "Polymer-polymer miscibility", Ac. Press (1979)

A UNIFORM CHARACTERISATION OF THE PARTICLE SIZE AND PARTICLE SIZE DISTRIBUTION IN POLYMER MIXTURES

J.L. Cohen
Enka bv Research Institute Arnhem,
Engineering Plastics Department,
Velperweg 76, 6824 BM Arnhem,
The Netherlands

ABSTRACT

Polymer mixtures or blends are playing an increasingly important role in the field of engineering plastics. The way the polymers are intermixed is of extraordinary and dominating influence on the properties of these mixtures.

Generally one of the polymers serves as the matrix into which a second minor proportion of polymer is dispersed in the form of fine particles. The nature of the dispersion and consequently the resulting properties can be influenced by the various processing steps.

It is therefore the relationship between on the one hand particle size or particle size distribution, processing conditions and on the other the properties which should be given the utmost consideration.

To simplify discussions on these phenomena a uniform method has been developed for the determination of the particle size and particle size distribution.

The method essentially consists in measuring scanning electron micrographs of cut surfaces, with a contrast having been introduced to distinguish the various polymer phases with a Quantimet. The two-dimensional information thus obtained can be converted into a particle size distribution by means of a statistical mathematical method.

INTRODUCTION

In addition to the wide range of neat polymers with specific properties and all kinds of compounds derived from them manufacturers of engineering plastics tend to produce more and more polymer mixtures or blends.

The mixtures are composed in a special way with a view to overcoming particular drawbacks to one of the polymer components. Thus a second polymer is added which compensates for the weak points of the first polymer and preferably adds some strong points.

In all cases where product development aims at polymer blends with optimum properties, the morphology of the mixture is of the utmost importance. The morphology is developed during the production of the mixture and may be subject to changes in subsequent processing steps.

Important factors are the mixing equipment to be used, the processing conditions, the ratio of melt viscosities at the processing temperature and the shear sensitivity of the respective polymer melts as well as the mutual chemical affinity of the constituent polymers.

These various factors can reasonably well be determined, individually, i.e. they can be measured experimentally, but the overall result of the combination of them being the morphology of the produced mixture poses great problems if a uniform characterisation is desired.

In this paper a method is described which we think gives a clear picture of the particle size distribution in polymer mixtures.

DESCRIPTION OF THE METHOD

In principle the method simply consists of the evaluation of photomicrographs of surfaces.

In order that these photographs should provide useful information the surface of the object must be very regular, i.e. very flat. This can be achieved with the use of a microtome. To prevent blurring of the different phases during the cutting operation it may be necessary - depending on the properties of the polymer mixture - for the sample and the knife to be cooled to a lower temperature.

The next step is the introduction of a sufficient contrast between phases to make them visible in the micrograph. Among the techniques available the best known are staining or etching of the surface. The technique to be preferred depends on the chemical nature of the polymer blend.

The examples of micrographs shown below were obtained from etched cut surfaces. Especially in that case the use of a Scanning Electron Microscope (SEM) has the advantage of a large depth of field.

The photograph is further analysed with a Quantimet particle size analyser. This unfortunately can not be done directly since in a number of cases, viz. because of the shallow holes the contrast obtained is insufficient to activate the apparatus. Obviously the human eye is the most sensitive organ capable of observing such small differences. To enhance the contrast the visible circular spots are encircled with black ink.

The data provided by the Quantimet are used to calculate the total weight fraction of particles having a diameter smaller than a predetermined value δ. The mathematical treatment [1] is based on the well-known Wicksell equation (1925):

$$H(y) = 1 - \frac{1}{\mu} \int_0^\infty \sqrt{(x^2-y^2)} \ dG(x)$$

This equation represents the two-dimensional distribution H(y) of circular section diameters as a function of the three-dimensional distribution G(x) of spherical particles with radius x.

Ultimately, the result can be expressed in a so-called weight undersize distribution (W.U.D.) W(δ) as a function of the two-dimensional distribution:

$$W(\delta) = \frac{\int_0^\delta x^3 \ dG(x)}{\int_0^\infty x^3 \ dG(x)}$$

The three different samples shown were evaluated in accordance with the method described above and some W(δ) values for δ will be presented at the conference.

It is evident that the clear visible differences can now be expressed in numerical data permitting correlation of a given particle size distribution either to properties or to process or to both.

ACKNOWLEDGEMENT

This procedure is the result of cooperation in the present matter among Ms. Stigter and Mr. Schulenberg of Enka bv Research Institute Arnhem Engineering Plastics Department CDK who supplied the samples, Messrs. van Blokland and Mos of Akzo Corporate Research Applied Physics Department CRT who performed the microscopic experiments and the measurements and Mr. Kuiper of the Akzo Corporate Research Applied Mathematics Department CRS who developed the mathematical tools and performed the calculations.

REFERENCE

1. Paper by Mr. Engels to be presented at The International Stereological Conference at Caen, Normandie, France; september 1987.

A uniform characterisation of the particle size and particle size distribution in polymer mixtures

J.L. Cohen, Enka bv, Research Institute Arnhem, Velperweg 76, 6824 BM Arnhem, The Netherlands

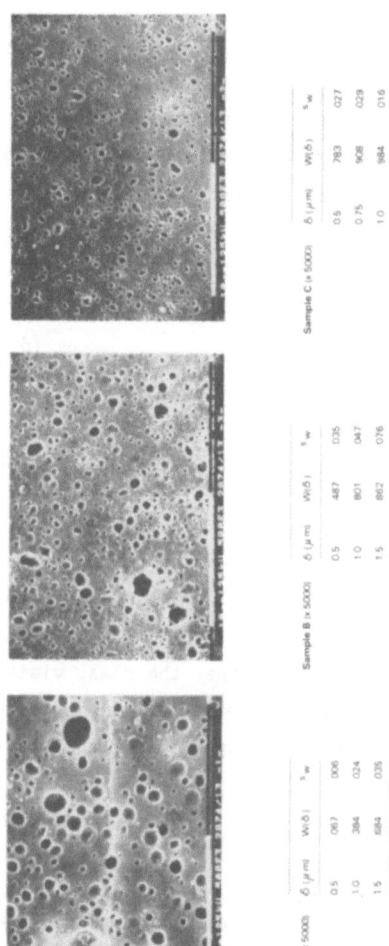

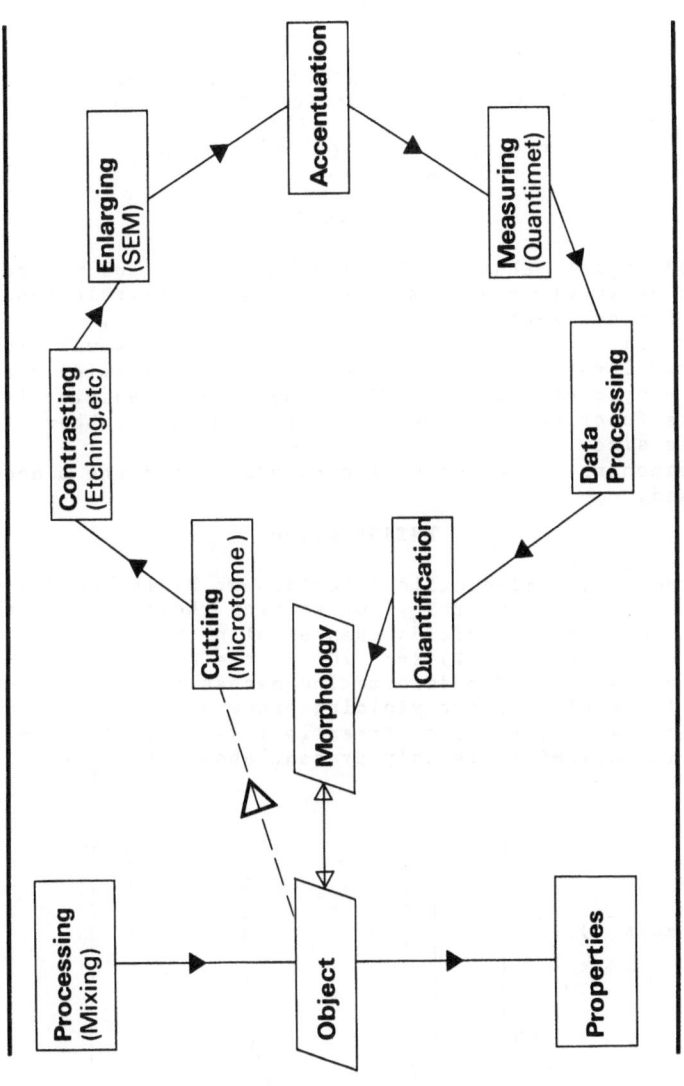

NYLON-RUBBER BLENDS

R.J. Gaymans and R.J.M. Borggreve
Department of Chemical Technology
University of Twente
P.O.Box 217, 7500 AE Enschede,
The Netherlands

ABSTRACT

The Izod impact behavior of polyamide 6, dry and wetted was studied as function of temperature and the brittle-tough transition correlated with the glass transition.
The Izod impact behavior of nylon-rubber blends was also studied. These blends were prepared from polyamide 6 and a maleic anhydride modified EPDM. The impact behavior is given as function of rubber concentration and rubber particle size.
The toughening mechanism of the polyamide and the blends are discussed.

INTRODUCTION

Polyamides are semi-ductile materials at RT in the dry state and ductile above their T_g. On wetting these polyamides the brittle to tough transition changes with the glass transition (fig 1 and 2) from 70° in the dry state to -30°C in the wet state. The deformation mechanism in the ductile region is mainly a shear yielding process and is allowed to take place when the yield stress is low. This low yield stress in polyamides is only present above the T_g (1,2).

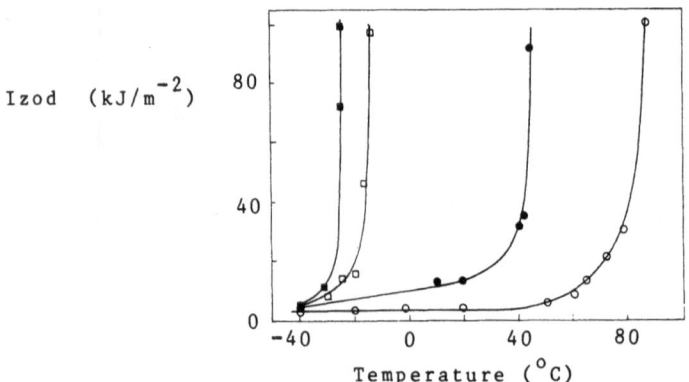

Figure 1. Notched Izod v.s. temperature of nylon 6, influence of waterconentration: dry ○, 2.3% ●, 7.6% □, 11.6% ■

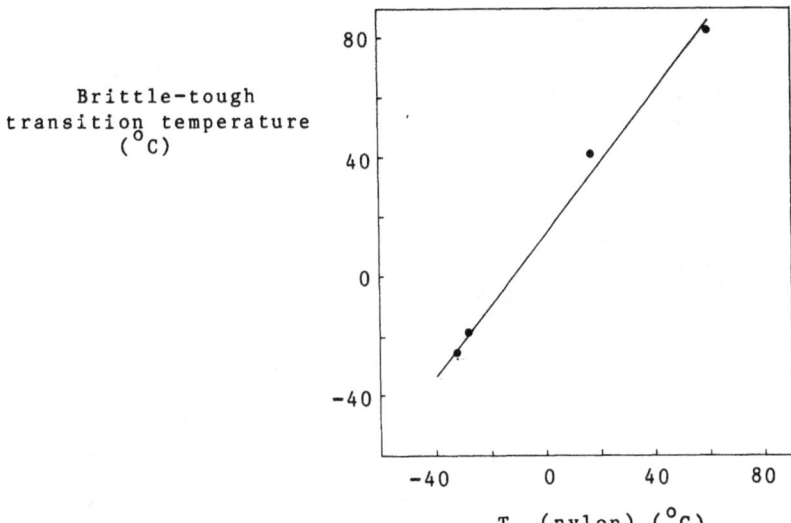

Figure 2. Brittle-tough transition v.s. glass transition of wetted-nylons

Rubber modified polyamides as obtained by reaction blending have a ductile behavior (notched Izod) well below the T_g of the polyamides (fig 3) (3-7). The brittle-tough transition is sharp and in the brittle region the deformed sample has a stress whitening at the notch tip. A tough deformed sample shows whitening over the whole fracture surface.

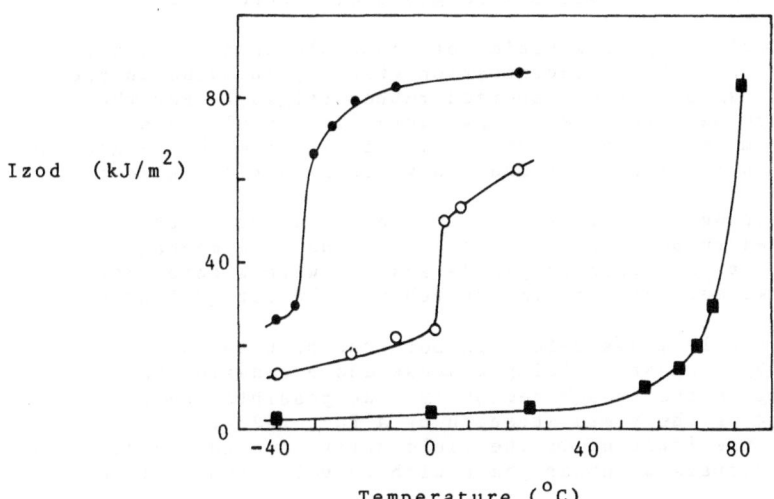

Figure 3. Notched Izod v.s temperature of nylon-blends, influence of rubber con.: 0% ■ , 13% o, 26% ● . (particle diameter $d_w = 0.32 \mu m$)

In figure 4 the effect of rubber concentration on the brittle-tough transition is given (7). As can be seen with only 13 vol % rubber the polyamide is tough down to -5°C. At higher concentrations the brittle-tough transition approaches the T_g of the rubber which is apparently a limiting value.

Transition temperatures ($^{\circ}$C)

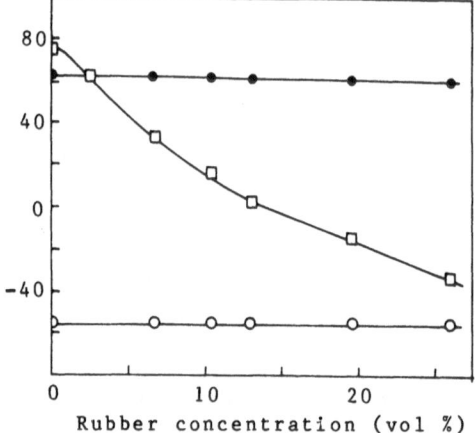

Figure 4. Brittle-tough transition □, T_g nylon ● and T_g rubber○as function of rubber conc. ($d_w=0.32 \mu$m)

Nylon rubber blends are prepared by reaction blending a polyamide 6 with EPDM rubber (Keltan 740) modified with maleic anhydride (0.4 wt%). The blends are made on a 40 mm single srew extruder fitted with a cavity transfer mixing head. The methods of analysis of the blend structure and the mechanical properties are given elsewhere (7).

The brittle-tough transition of these blends depends not only on the rubber concentration (fig. 4) but also on the particle size of the dispersed rubber (fig.5). For the rubber to be effective the particle size should be smaller than 1 m and as the particle size is lowered the toughening effect is increased (at least down to a particle size of 0.1μm).
The question is whether this toughening effect with dispersed rubber particles is due to the same mechanism as with the plasticized polyamide and can with a combination of these two methodes the impact behavior be futher improved

The neat polyamides deform in both the brittle and tough region by a shear yielding process and excessive shear yielding in the though region is made possible due to a lowering of the yield strength by a factor 2-4.
However the lowering of the yield stress due to the presence of the dispersed rubber phase with 26 vol % rubber is only 26 %.
If the blends are wetted the toughening behavior only changes at very high water concentrations and is at these very high water concentrations not different from the wetted

neat polyamide. The toughening behavior of the blends must therefore have another mechanism than the neat polyamide and might be either a particle induced or an interparticle distance process (6,7).

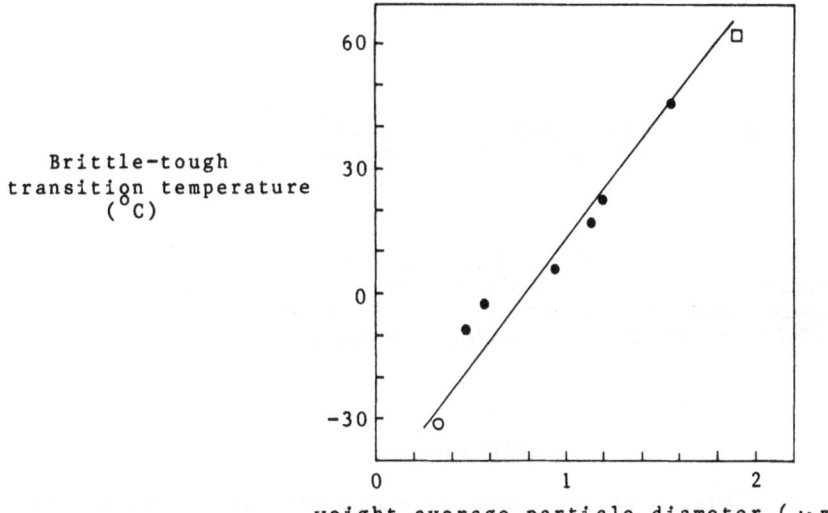

weight average particle diameter (μm)

Figure 5. Influence of particle diameter (d_w) on tough-brittle transition (26% EPDM) with different malic anhydride contents: 0.1% □ ,0.25% ● and 0.4% ○ .

Acknowledgement

This work is part of the research program of the University of Twente and was financially supported by the Netherlands Technology Foundation (STW) and DSM Research, Geleen.

References

1. Kohan, M.L., Nylon Plastics, SPE Monograph, John Wiley & Sons, New York, 1973.

2. Ogorkiewicz, R.M., Engineering Properties of Plastics, John Wiley & Sons, New York, 1970.

3. Flexman, E.A., Kunststoffe, 1979, 69, 172.

4. Neuray, D. and Ott, K.H., Angew. Makromol. Chem., 1981, 98, 213.

5. Fahler, F. and Merten, J., Kunststoffe, 1985, 75, 157.

6. Wu, S. Polymer, 1985, 26, 1855.

7. Borggreve, R.J.M., Gaymans, R.J., Schuijer, J. and Ingen Housz, J.F. Polymer, submitted for publication.

SPECIFIC INTERACTIONS AND INTERDIFFUSION IN POLYMER BLENDS BY FT-IR SPECTROSCOPY

S. Klotz, H.-J. Cantow and M. Knopf
Institut für Makromolekulare Chemie der Universität
Freiburg, Hermann-Staudinger-Haus, Stefan-Meier-Str.31,
D-7800 Freiburg i. Br.,Federal Republic of Germany

ABSTRACT

The interdiffusion of two compatible polymer pairs was measured using two different FT-IR spectroscopic techniques. In the system polystyrene (PS) and polyvinylmethylether (PVME) the interdiffusion was observed by recording the arising interaction spectrum as a function of time. The mixing of the two polymers was found to be a diffusion controlled process. For the other polymer pair, polystyrene (PS) and polyphenyleneoxide (PPE), a new technique, the scanning FT-IR microscopy (SIRM), was applied. With this technique the concentration profile along the phase boundary was evaluated.

INTRODUCTION

Diffusion measurements of bulk polymers are interesting for practical purposes (i.e. coextrusion for polymer blending) as well as for theoretical interest (i.e. diffusion mechanism, molecular weight dependence). In this context we present two different techniques for measurements of the interdiffusion of two initially phase separated polymers.

The first method records the differences in the IR transmission spectra during the diffusion in a sandwich-like film. These differences are evaluated as an arising interaction spectrum, which may be compared with the interaction spectrum of the completely mixed system.

In the second procedure (SIRM) a microcut of the sample is scanned along the diffusion axis in 10 μm steps with a computer controlled mouvable table. The IR spectra were measured with an IR microscope with spot size of 18 μm† With the sample coordinates of the IR spectra the concentration profile of the components along the x-axis may be evaluated.

RESULTS

Fig.1 shows the IR spectra of a homogenous blend of PS/PVME and of the pure components respectively. In order to visualize the interaction between the components in the blend the pure polymer spectra are subtracted to get the so-called interaction spectrum (fig.2). Fig.2 also shows the differences between the phase separated blend (at the top) and the homogenous blend (at the bottom). The changes in the C-O-C stretching vibration (1150 cm^{-1}) of PVME and the out of plane ring vibration (700 cm^{-1}) of PS are evident.

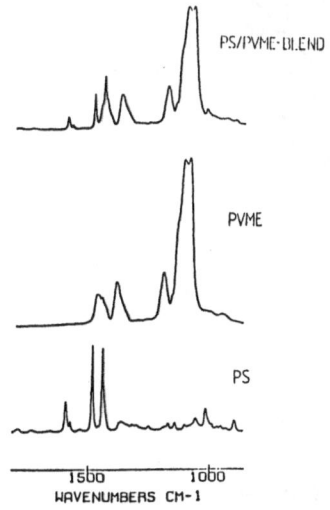

Fig.1:Spectra of a homo-
genous PS/PVME blend
and the pure compo-
nents

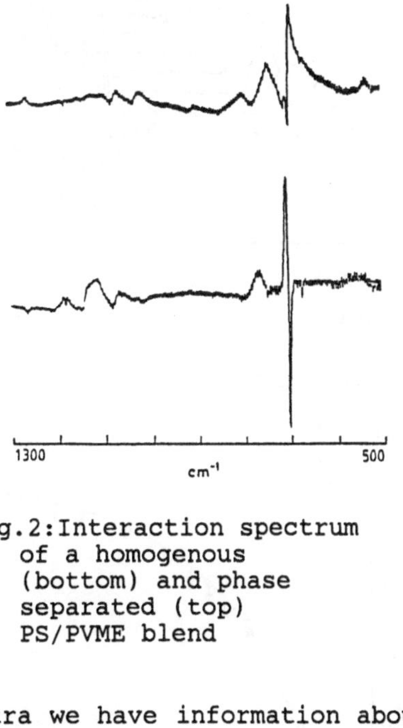

Fig.2:Interaction spectrum
of a homogenous
(bottom) and phase
separated (top)
PS/PVME blend

With these interaction spectra we have information about the degree of mixing in the polymer blend.

In the first diffusion experiment a sandwich was made of a thin PS film and a thin PVME film, cast from different solvents. After completion of the sample preparation the transmission spectrum of the sandwich was measured as a function of time. From these spectra the pure component spectra were subtracted and we got the interaction spectrum as a function of time (fig.3). The increasing bands in the ether stretching and the out of plane ring vibration region are evident. The interaction spectrum, marked with a star, is taken after three months. For a more quantitative evaluation of the measured effect, the band in the ether stretching region was integrated,

and in fig.4 the respective intensities are plotted as a function of the square root of time.

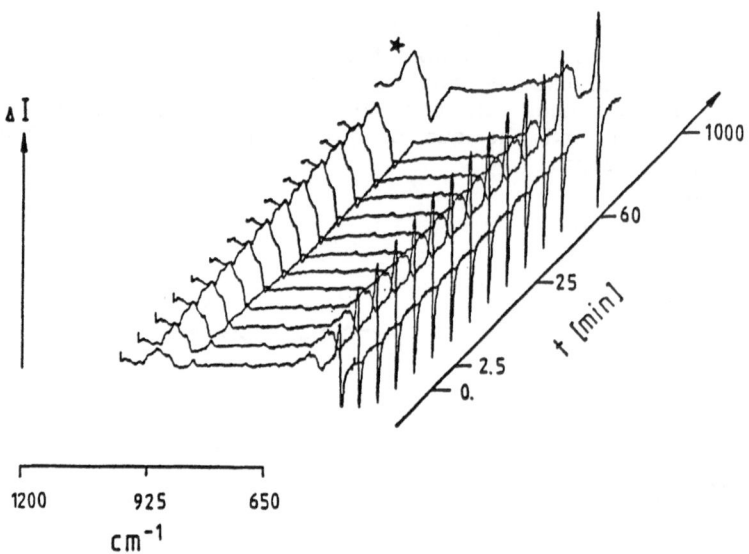

Fig.3:Interaction spectrum as function of time

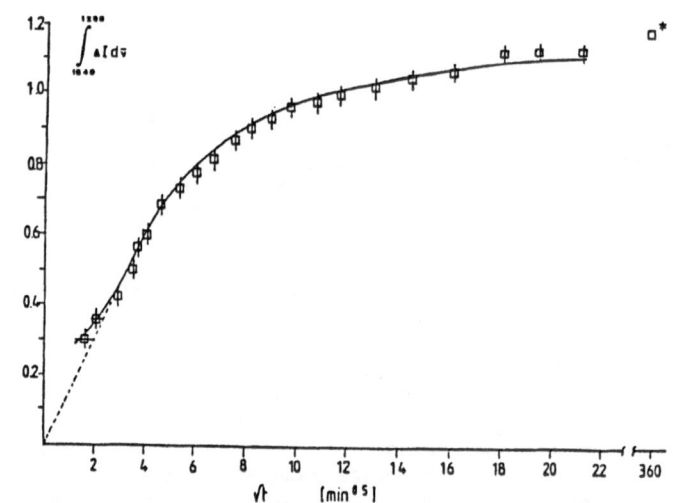

Fig.4:Intensities of the interaction spectrum between
1200 cm^{-1} and 1040 cm^{-1} as a function of time

Again the point with the star is taken after three months.
The trace of the curve strongly suggests a diffusion controlled
process and may be described as a Fickian diffusion. With an

empirical time-concentration-intensity relationship we have calculated a diffusion coefficient in the order of 1.1E-8 cm^2/sec. Reminding the low molecular weights of the components, $M_w(PS)=800$ and $M_w(PVME)=3200$, the calculated diffusion coefficient is supposed to be a quite reasonable value.

The idea of the second procedure is based on measurements of Klein and Briscoe[1]. We have combined their method with the IR microscope technique (minimum spot size 8 μm) connected with a computer controlled mouvable table (step size 5 μm). Fig.5 shows scematically the sample preparation and the measurement procedure.

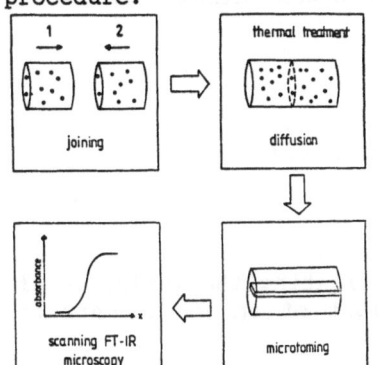

We have applied this technique to the compatible polymer pair PS/PPE. After joining the polymers, the sample was heated up to 200°C, then cooled down and a microcut of 4μm thickness was taken in the x-axis of the sample. The microcut was scanned under the IR microscop along the x-axis in 10μm steps.

Fig.5:Sample preparation and measurement procedure

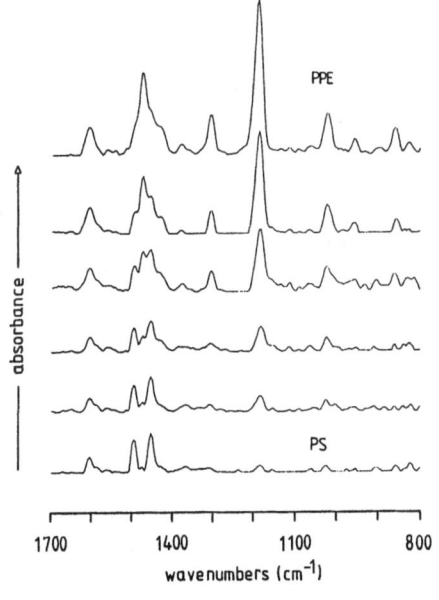

Fig.6 shows some spectra as well from the PPE rich part (at the top) as from the PS rich part (at the bottom) and from the phase boundary (in between). Especially the vanishing of the ether stretching vibration of PPE at 1185 cm^{-1} and the arising of the C-C stretching ring modes of PS at 1452 $cm^{-1}$and 1495cm^{-1} are evident.

Fig.6:Spectra of PPE (at the top), PS (at the bottom) and the phase boundary region (in between)

Normalizing the absorbance of the ether stretching vibration of pure PPE to unity and plotting these intensities versus their x-axis coordinates, we got the concentration profile in the system PS/PPE at a given time t_i (fig.7).

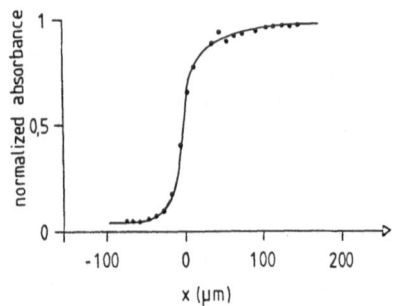

From the concentration profile it is possible to evaluate the diffusion coefficient [1]. A value for the diffusion coefficient of at about 1.2E-9 cm^2/sec at 200^{0}C was calculated.

Fig.7: Concentration profile of the system PS/PPE at the phase boundary measured at 200^0 C

In our current studies we are investigating the influence of the temperature and the molecular weight of the components on the diffusion coefficient.

Aknowledgements

The authors would like to thank the Bundesministerium für Forschung und Technologie (BMFT) for financial support and Dr. B. J. Schmitt, BASF Ludwigshafen, for the supplying the samples.

Reference

1. J. Klein, B. J. Briscoe, Proc. R. Soc. Lond. 1979, A365,53

MISCIBILITY OF POLY(CARBONATE) AND COPOLYESTER OF POLY(ETHYLENETEREPHTHALATE-CO-P-HYDROXIBENZOATE)

K. Friedrich, M. Heß, R. Kosfeld
FB 6/Physikalische Chemie, Universität -GH- Duisburg

ABSTRACT

The development of polymer blends prepared from rigid-rod or semiflexible polmyers with coil-shaped polymers is a way which might lead to self-reinforcing polymer systems. The mixing behaviour, thermic properties and studies on behaviour of ternary systems with a solvent were studied with the thermotropic copolyester from p-hydroxibenzoic acid and poly(ethyleneterephthalate) and with poly(bisphenol A-carbonate)

INTRODUCTION

The development of thermotropic polymers leads to the concept of reinforcing plastics by introduction of rodlike molecules ("coil- rod-blends"). The system used for our measurements consisted of a thermotropic copolyester (COP) from poly(ethyleneterephthalate) (PET) and p-hydroxibenzoate (PHB) and a coil-shaped component poly(bisphenol-A-carbonate) (PC).

To achieve maximum reinforcement the dispersion of the stiff molecules in the coil matrix is an essential feature [1-3]. Since the investigated blends were all prepared from a ternary solution, the knowledge of the phase diagram is very important, too.

RESULTS AND DISCUSSION

The copolyesters were prepared by melt transesterification of PET and p-acetoxybenzoic acid (PAB):

Ten different copolyesters with PHB-contents between 18 and 66% were prepared. The lower limit for liquid crystalline behaviour is a content of 30% oxibenzoate. ^{13}C-NMR-spectroscopy showed that the distribution of PHB units is almost statistical [5]. Two phase diagrams were measured with two different copolyesters, namely COP 8 with 35% PHB and COP 9 with 27% PHB content (fig. 1, temp. 295 K). As coil-like component and solvent in each case PC and chloroform were used respectively. The starting point of the measurements were isotropic solutions with concentrations slowly increased by evaporation of the solvent. The turbidity was determined by measuring the extinction by UV spectroscopy.

The phase behaviour of ternary systems of stiff rod-like molecules, flexible coil polymers and solvent were calculated by Flory by means of statistical thermodynamics (fig. 2) [3,6]. The miscibility gap depends on the axis ratio of the rod and the contour length of the coil. The measured phase diagrams do not agree with Flory's calculations, since the copolyesters, because of their flexible ethyleneglycol-spacers, probably could not be treated as totally stiff rods. To determinate the compatibility in the binary polymer system co-polyester/polycarbonate two different kinds of foils were used which were prepared from isotropic solutions of COP 8 and PC in chloroform by two different procedures:
- slow evaporation of solvent
- quick quenching
Quenching was achieved by shifting water over the chloroform phase. The polymer film could be drawn from the phase boundary. Microscopic investigations showed pronounced phase separation in the case of the foils obtained by evaporation. In contrast to this observation the quenched foils optically indicated no phase separation, they were clear and transparent. Basing on these results above it cannot be concluded that the system shows polymer-polymer-compatibility. If phase separation occurs in domains smaller in size than the wavelength of visible light, no turbidity of the foils is to

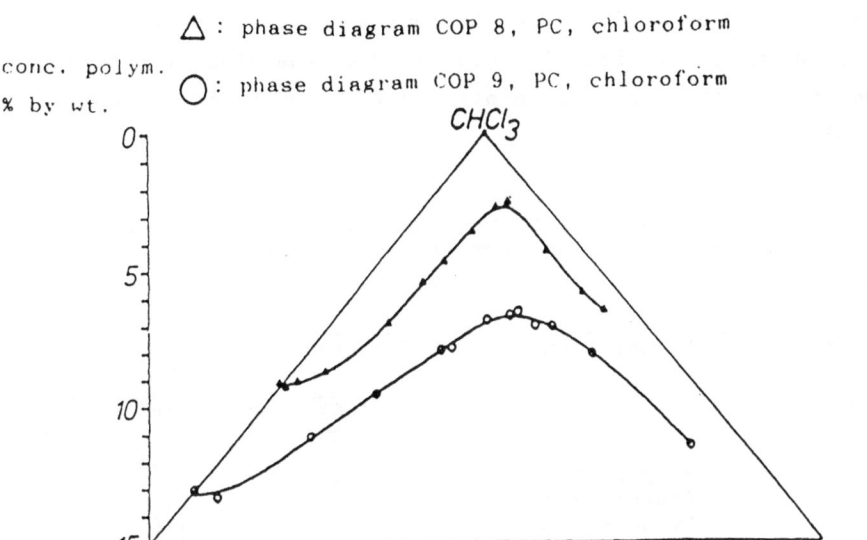

Fig. 1: Turbidity diagrams of the systems COP 8/PC/CHCl₃ and
COP 9/PC/CHCl₃ at 295 K

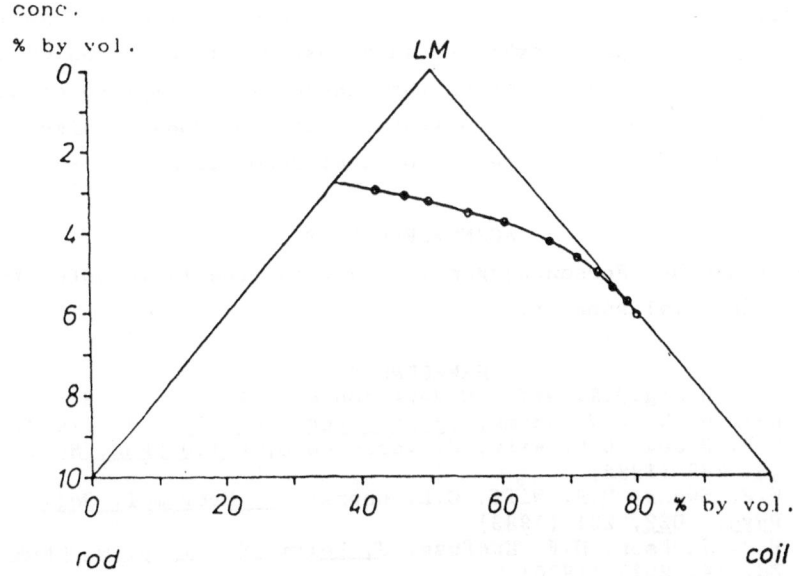

Fig. 2: Phase diagram according to Flory (axis ratio = contour
length = 300) [3, 6].

be expected as well. So the thermal behaviour of the foils
was examined by DSC. The behaviour of the glass transition
regions of the pure components and the blends is shown in
table 1:

TABLE 1

Glass transition temperature in Kelvin (K) for the copolyester component (T_{g1}) and for the polycarbonate component
(T_{g2})

Sample	Preparation	COP content	T_{g1}	T_{g2}
COP 8	-	100%	345	-
PC	-	0%	-	422
F1	Evaporation	25%	-[a)]	416
F2	Evaporation	50%	346	415
F3	Evaporation	75%	346	413
F4	Quenching	25%	-[a)]	416
F5	Quenching	50%	348	412
F6	Quenching	75%	348	408

[a)] not detectable

The shift of the T_{g2} values shows that there is an appreciable miscibility of COP in the PC-phase while it seems
that the COP-phase exhibits an almost complete exclusion of
PC. This analysis indicates that there is no complete compatibility between the two polymers, but the domain size of
the quenched foils is beyond optical resolution.

ACKNOWLEDGEMENT
The Deutsche Forschungsgemeinschaft is greatly indepted for
her financial support.

REFERENCES
1. W.F. Hwang,D.R. Wiff, C.Verschoore, G.E. Rice, T.E. Helminiak, W.W. Adams, Polym. Eng. Sci. 23, 784 (1983)
2. W.F. Hwang, D.R. Wiff, C. Verschoore, Polym. Eng. Sci. 23, 789 (1983)
3. W.F. Hwang, D.R. Wiff, C.L. Benner, J. Macromol. Sci. Phys. B22, 231 (1983)
4. W.J. Jackson, H.F. Kuhfuss, J. Polym. Sci., Polym. Chem. Ed. 14, 2043 (1976)
5. R.W. Lenz, J.-I. Jin, K.A. Feichtinger, Polym. 24, 327 (1983)
6. P.J. Flory, Macromol. 11, 1138 (1978)

STRUCTURED POLYMER BLENDS-1

P.H.M. Elemans, J.G.M. van Gisbergen, H.E.H. Meijer*
Laboratory of Polymer Technology
Eindhoven University of Technology
P.O. Box 513
5600 MB Eindhoven, The Netherlands
* DSM Research BV, Geleen

ABSTRACT

The influence of viscosity ratio and volume fraction on the morphology
of a blend of two incompatible polymers is shown for the model system
polyethylene/polystyrene (PE/PS). An attempt has been made to fixate
the morphology by crosslinking the dispersed (PE) phase using electron
beam or gamma irradiation. Due to shielding of the PS matrix, the
crosslinking in the PE phase is less effective up to now, to achieve
complete fixation of the morphology.

INTRODUCTION

In the search for new materials blending of polymers is important. Not
only compatible, but also incompatible systems can provide interesting
properties. In the latter case the morphology, the relative
distribution of the constituents, plays a dominant role. When a blend
of two incompatible polymers is subjected to shearing forces e.g. in a
corotating twin screw extruder, droplets of the dispersed phase will
deform into long, threadlike particles (1) which can break up again
into smaller droplets. The ultimate morphology depends on volume or
weight fraction (2), viscosity ratio of both polymers and type of flow
(3). These aspects are shown in blends of the model system PE/PS.
In actual practice one could envisage the production of polymer blends
possessing an optimum morphology by using dedicated mixing equipment to
produce a particular blend in for example pelletized form. However,
subsequent processing, for example injection-moulding, the original
morphology is subject to change due to the inherent non-equilibrium
situation with respect to morphology. In the model system PE/PS we have
tried to fixate the morphology using electron-beam irradiation. Since
the dispersed phase polyethylene will crosslink and the matrix
polystyrene is but hardly affected by electrons, one might fixate a
particular morphology (4,5). In fact, the usage of high-energy
radiation could be applied for any blend system provided that the
matrix is not affected (polystyrene, SAN, polyester), or degrades in a
controlled manner (polypropylene), whereas the dispersed phase
selectively crosslinks at comparable radiation doses. In this paper the
system PE/PS is investigated.

Dispersion process

Once a certain length scale is reached by shearing forces, the dispersion process mainly consists of the formation of long threadlike particles (1) which finally break up into droplets under influence of interfacial tension (capillary instabilities (6,7)).

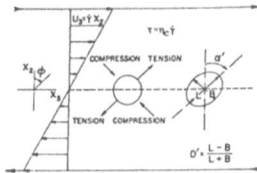

Figure 1. Deformation of a droplet undergoing a plane shear flow

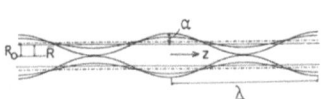

Figure 1a. Capillary instabilities on a liquid cylinder (Elmendorp 1986)

These droplets can be deformed as long as there is no equilibrium between shearing and interfacial forces (1,8,9). The equilibrium droplet size, given by ratio of shear stress (τ) and interfacial tension (σ/r) depends on the viscosity ratio p and the type of flow (3).

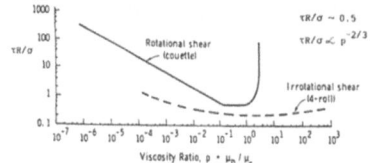

Figure 2. Comparison of effect of viscosity ratio on critical shear in rotational and irrotational shear fields

The morphology thus formed is unstable. Not only because of changes in shear stresses during post-processing but mainly since with increasing volume fraction of the dispersed phase the droplets will coalesce (1,10). This can give rise to co-continuous structures and phase inversion (2), see Figure 3.

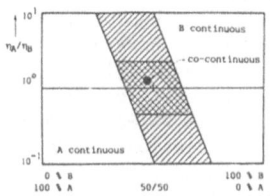

Figure 3. Phase continuity for an "A-B" blend

EXPERIMENTAL

Blends of PS and PE were made on a corotating twin-screw extruder. Irradiation was performed with the Co^{60} gamma facility of the "Inter-Universitair Reactor Institute" (IRI) in Delft (NL) and with a 3 MV electron beam generator of GBS in Whiel (FRG). Doses up to 25 MRad (= 250 KGray) were used. Both types of irradiation showed about the same results and minor differences will not be discussed.

RESULTS

Phase Inversion Diagram

Figures 4a and b show this phase inversion diagram for the system HDPE/PS for one processing condition on a corotating twin screw extruder. As can be inferred from these micrographs the same areas as in Figure 3 can be seen. However, the co-continuous part is larger and the imaginary lines indicating the co-continuous area are not parallel.

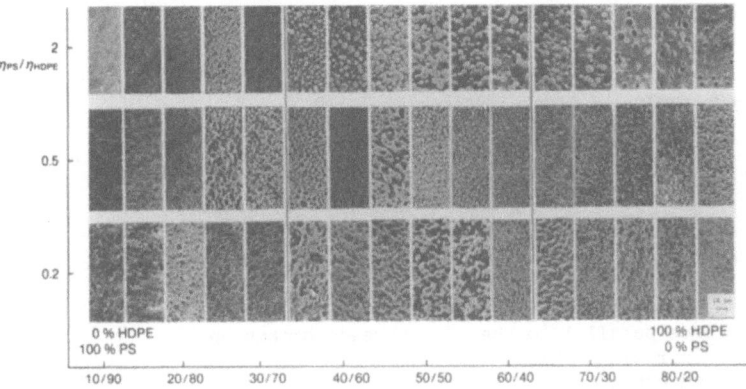

Figure 4a. Scanning electron micrographs of the microtomed extrudate surface perpendicular to the direction of extrusion (magnification:1000x)

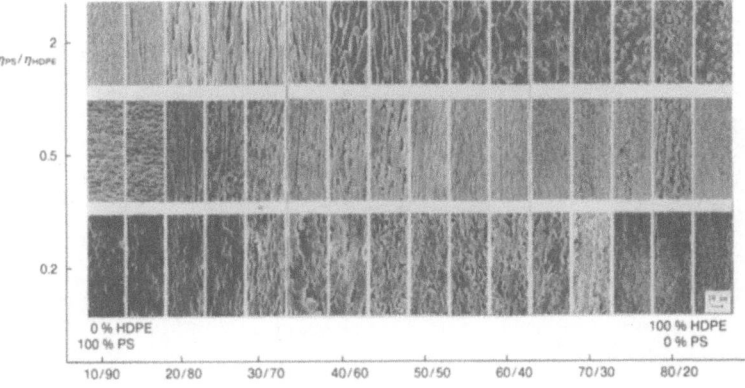

Figure 4b. As Figure 4a, parallel to the direction of extrusion

Some details of Figures 4a and b, which show the different stages of
thread formation and break up, are shown in Figures 5a, b, c and d.

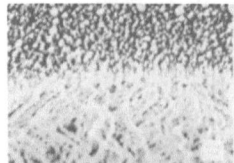

Figure 5a. Scanning electron micrograph of the edge of a microtomed
surface of a 55/45 HDPE/PS blend
(p ~1). Capricious (PS) threads
are visible in the direction of
extrusion (4000x)

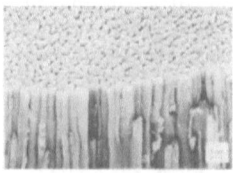

Figure 5b. As 5a of a 27/75 HDPE/PS
blend (p ~ 2). HDPE ("Black")
still forms the continuous phase

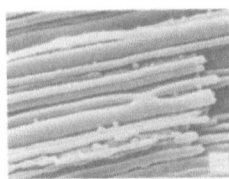

Figure 5c. As 5a, but now of a
fracture surface parallel to the
direction of extrusion. Thread-like
particles of PS are shown in
different stages of break-up and
coalescence (10000x)

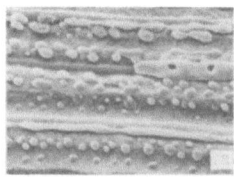

Figure 5d. As 5c but with more
threads broken up

Fixation of Blend Morphology

It has been investigated to fixate a thread-like structure with
electron beam irradiation (no crosslink accelerators were used yet).

Figure 6. Scanning electron micrograph of a
85/15 PS/LDPE blend, parallel to the direction of extrusion (1000x)

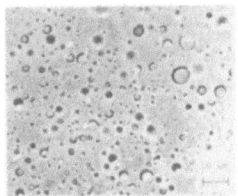 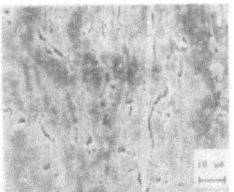

Figure 7a. As Figure 6: annealed   Figure 7b. As Figure 6: annealed
(170°C) for 30 minutes, unirradiated   (170°C) for 30 minutes, irradiated
                                          with 25 MRad

If no flow is being introduced, the unirradiated blend morphology
develops into a droplet-matrix structure. The irradiated sample,
although slightly changed, globally shows the same morphology.

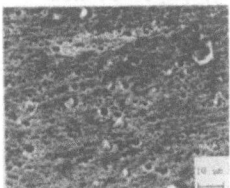 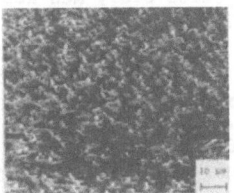

Figure 8a. As 6: compression moulded   Figure 8b. As 6: compression
for 3 minutes, unirradiated   moulded for 3 minutes, irradiated
                                          with 25 MRads

When the blend is subjected to small deformations in compression
moulding (Figure 8) or to high shearing and elongational forces
occurring in the surface layer during injection moulding the morphology
is not or insufficiently fixed. This insufficient fixation is probably
due to the low obtained crosslink efficiency of the PE phase (less than
75% is crosslinked). Shielding of the PS matrix makes the PE phase less
sensitive to the irradiation (12).

CONCLUSION

As a result of inducing crosslinks in the dispersed phase the morphology
can be preserved to some extent. In the case of PE/PS blends, the PE
phase can be crosslinked by high energy radiation such as EB or gamma
rays whereas the PS (matrix) is hardly affected. Due to shielding
effects of the PS matrix, the crosslinking of the dispersed PE particles
is less effective.

However the concept of selectively crosslinking a dispersed phase without changing the matrix is generally applicable to, in principle, any combination of low T_g (rubbery) dispersed phase in a high T_g (glassy) matrix. The ultimate goal is to induce initially the optimum morphology via radiation crosslinking to preserve the desired structure during subsequent processing.

ACKNOWLEDGEMENT

The authors wish to thank S. Nadorp of DSM-Research BV (dept. FA-OM) and H. Ladan of Eindhoven University of Technology (dept. TCK) for making the scanning electron micrographs.

REFERENCES

1. J.J. Elmendorp, Dissertation Delft University of Technology (1986).
2. G.N. Avgeropoulos e.a., ACS Rubber Division Meeting, paper 3, New Orleans (1975).
3. H.P. Grace, Chem. Eng. Comm., 14 (1983) 225-277.
4. A. Chapiro, "Radiation Chemistry of Polymers", John Wiley (1962).
5. A. Charlesby in "Integration of Fundamental Polymer Science and Technology", ed. by L.A. Kleintjens and P.J. Lemstra, Elsevier (1986) 648-662.
6. Lord Rayleigh, "On the Capillary Phenomena of Jets", Proc. Roy. Soc., 29 (1879) 71-97.
7. S. Tomotika, Ibid., A150, (1935) 322-337.
8. G.I. Taylor, Ibid., A146, (1934) 501-523.
9. R.G. Cox. J. of Fluid Mech., 37 (1969) 601-623.
10. N. Tokita, ACS Rubber Division Meeting, paper 26, San Francisco (1976).
11. Acrivos and Lo, J. Fluid. Mech., 86 (1973) 18.
12. J.G.M. van Gisbergen and P.H.M. Elemans, poster presented at European Symposium on Polymer Blends, Strasbourg, 25-27 May 1987.

Part 4
NETWORKS/GELS

Part 4

NETWORKERS

THERMODYNAMICS OF SWELLING OF MODEL NETWORKS

Zoran S. Petrović (Faculty of Technology, University of Novi Sad, Yugoslavia),
William J. MacKnight, Ronald Koningsveld (Polymer Sci.& Eng. Dep., Univ. of
Massachusetts, Amherst, MA 01003, USA),
Karel Dušek (Inst. Macromol.Chem., Czechoslovak Acad. Sci., 16206 Prague 6)

Synopsis

An approach is proposed for the treatment of equilibrium swelling of networks
obtained by crosslinking polypropylene glycol (PPG) chains with a triisocyanate
crosslinker, 'Desmodur RF' (DRF). Three models are used for the analysis of
swelling of these polyurethane networks in toluene and methanol. It is shown that the
χ parameter of the copolymer network depends on the interactions between the
solvent and the individual components of the network and between these components
themselves. Swelling data in methanol and toluene appear to yield the same value for
the interaction parameter of the network components provided a) the differences in
molecular surface area between the various building blocks of the systems and b) the
molecular association of methanol are taken into account. Two networks, identical in
number average chain length between crosslinks but differing in chain length
distribution, have identical enthalpic contributions to the effective interaction
parameter but different entropic terms. The conclusions are affected neither by the
choice of the model used for the front factors nor by possibly necessary corrections
for the number of elastically active network chains. A simple solubility parameter
treatment provides a fair prediction of swelling in mixtures of methanol and toluene.

Introduction

Under certain simplifying conditions the elastic free energy change (ΔG) of a
network during deformation can be represented by the expression [1]:

$$\Delta G_{el}/RT = (A\nu_e/2)(\lambda_x^2 + \lambda_y^2 + \lambda_z^2 - 3) - B\nu_e \ln\lambda_x\lambda_y\lambda_z \tag{1}$$

where R = the gas constant, T = the absolute temperature, ν_e = the number of elastically active network chains in moles per unit volume of dry rubber, and λ_i = the
deformation ratio in direction i. Revision of the classic theory of rubber elasticity
[2-4] has shown that the factors A and B, previously considered as constants, may
depend on ν_e as well as on the degree of swelling. In the junction-fluctuation model
of Flory the limiting values of A and B are $(f_e - 2)/f_e$ and zero, respectively, for
the phantom network limit (free fluctuations of crosslinks), and 1 and $2/f_e$ for the
limit of fully suppressed fluctuations of crosslinks (original Flory-Wall model [5]).

Here, f_e is the number-average functionality of an elastically active crosslink. Similar conclusions can be drawn from the tube models, although there the effect of interchain constraints is not limited by $A = 1$ (A may be > 1). Both models predict that $A \to (f_e - 2)/f_e$ and $B \to 0$ with increasing swelling and increasing tensile deformation.

There may also be an uncertainty in the value of ν_e if the network structure is not well defined. For **perfect** networks with ν_e^0 elastically active network chains and a density ρ, prepared from telechelic polymers, the molar mass between crosslinks, M_c, given by

$$M_c = \rho/\nu_e^0 \qquad (2)$$

can be identified with the molar mass of the telechelic polymer. However, a state of network perfection is probably impossible to achieve and a very low level of imperfection may have a considerable effect on elastic properties [6]. Also, there may be a contribution to ν_e by permanent topological constraints - trapped entanglements - which may not disappear even at high degrees of swelling.

These points are raised to show that it is difficult to evaluate the polymer-solvent interaction parameter χ from swelling measurements. Another complication is represented by the usual assumption that the network contains a single type of interacting polymer segment. This assumption is not often fulfilled, the crosslinks or bridging molecules and the elastomeric chains may have a different nature. As a consequence, the solvent-polymer interactions may be more complex than is usually assumed. Swelling of such 'copolymer' networks will be governed by the interaction between the solvent molecules and the two types of polymer segments, and at least three χ parameters are necessary to describe the system. We studied the swelling of such networks in order to find out whether, in spite of the complications described above, it might still be possible to separate the various χ parameters contributing to the effective interaction parameter χ_{ef}, characterizing the various interactions between the components in the swollen network. The treatment we developed was applied to polyurethane networks based on polypropylene glycols (PPG) and a triisocyanate, 'Desmodur RF' (DRF), having the structural formula:

$$(OCN\langle\ O\ \rangle O)_3P=S$$

Theory

The swelling equation derived for networks composed of a single type of segment, on the assumption of ΔG_{el} being given by eq.(1) and ΔG_{mix} by the expression of Flory [7], Huggins [8] and Staverman and Van Santen [9], reads

$$\ln(1 - \phi) + \phi + \chi\phi^2 + \nu_e \tilde{V}_1(A\phi^{\frac{1}{3}} - B\phi) = 0 \qquad (3)$$

where ϕ is the volume fraction of the polymer in the swollen network and $\tilde{V}_1$ the molar volume of the solvent. The total free energy change upon swelling is assumed to be the sum of ΔG_{el} and ΔG_{mix}, the Gibbs free energy of mixing.

Swelling of the copolymer networks in this study involves three different interactions: between solvent and PPG units (χ_{12}), solvent and DRF (χ_{13}), and PPG units and DRF (χ_{23}). A classical analysis of such swelling data with eq.(3) will produce an overall χ value, here designated by χ_{ef}. If the network is treated as a copolymer then χ_{ef} might be considered as an average of two contributions [10,11]

$$\chi_{ef} = \phi_2\chi_{12} + \phi_3\chi_{13} \tag{4}$$

where ϕ_2 and ϕ_3 are the volume fractions of repeat units 2 and 3 in the dry network. This **two-parameter** model ignores a possible interaction between the repeat units 2 and 3 and has been found to be too simple an approximation to explain experimental data on solvent-copolymer solutions [12,13] and blends of homopolymers with copolymers [14]. A better treatment of the situation has been developed for copolymer solutions [15,16] and for blends containing copolymers [14,17-19], and this leads to a **three-parameter** expression for χ_{ef}:

$$\chi_{ef} = \phi_2\chi_{12} + \phi_3\chi_{13} - \phi_2\phi_3\chi_{23} \tag{5}$$

The three-parameter model suggests that χ_{ef}, and thus the degree of swelling, will depend not only on interactions between solvent and the two repeat units, but also on the attraction or repulsion between the repeat units themselves. Eq.(5) is better than Eq.(4), but it has been found to be inadequate in some cases [13].

Several possibilities exist for the modification of eq.(5) [19-22]. Improvement is possible i.a. by introduction of terms accounting for the differences between the coordination numbers (numbers of nearest neighbors) of the various repeat units in the system. We use a first approximation proposed many years ago by Staverman [23] who suggested setting the numbers of nearest neighbor contacts proportional to the attainable molecular surface areas. Application to the present system yields [19,24]

$$g_{ef} = \phi_2 g_{12} + \phi_3 g_{13} - \phi_2\phi_3 g_{23}/\alpha \tag{6}$$

where $\alpha = \phi_2 + s_{32}\phi_3$,

$$g_{ef} = \chi_{ef}\{1 + (\beta - 1)\phi\}^2/\beta = \chi_{ef}Q^2/\beta, \tag{7}$$

$\beta = s_{21}\alpha$

The introduction of disparity in contact numbers involves an overall concentration dependence of the effective interaction parameter χ_{ef} determined from swelling equilibrium data with Eq.(3). As a consequence, χ_{ef} must be corrected (Eq.(7)) and we denote the interaction parameters by g_{ij} to distinguish between this **modified three-parameter** model and the three-parameter model defined by Eq.(5). The quantities s_{ij} stand for σ_i/σ_j, the ratios of molecular surface areas σ_i and σ_j.

The aim of this work is to examine the applicability of the three models expressed by Eqs (4)-(6) to real copolymer networks of different crosslink density and different composition ratios, *i.e.* different volume ratios of chains and crosslinker. With the two-parameter model we do not need to compare χ values between the two solvents, Eq.(4) does not contain the 2-3 interaction and linearity of $\chi_{ef}(\phi_2)$ is the criterion for validity of the model. With the two three-parameter models, however, the criterion is much more demanding, *viz.* agreement within experimental accuracy of the value of the 2-3 interaction parameter, obtained from swelling equilibrium in two different solvents.

In the present study we have used two solvents with a large disparity in molecular size (methanol and toluene). If the comparison of the χ_{23} and g_{23} values between the two swelling solvents is to be meaningful, the data analysis with Eq.(3) must be carried out with the same value of the basic volume unit (BVU), within the rigid lattice model identical to the size of the lattice sites. It would seem obvious to use the molar volume of methanol for the purpose, since it is the smallest of the various species. Methanol, however, is known to associate and form quite stable tri-mers in the crystalline and liquid states [25]. Its molar volume may therefore be expected to be a little larger than that of toluene which would seem to be a better choice for the BVU.

The preceding equations must be amended so that a direct comparison of χ or g values is possible. We find [24] that the following expression accounts for the various aspects discussed here; it is a modified Flory-Rehner equation [26]

$$ln(1 - \phi) + \phi + m_1\chi_{ef}\phi^2 + \nu_e m_1 V_b(A\phi^{\frac{1}{3}} - B\phi) = 0 \qquad (8)$$

where V_b = the molar volume of the BVU, $m_1 = \hat{V}_1/V_b$ and $\chi_{ef} = g_{ef}\beta/Q^2$.

Evaluation of interaction parameters

Figure 1 shows the equilibrium swelling data for five polyurethane networks in toluene and in methanol. M_c was taken to be the molecular weight of the polyol plus two thirds of the molecular weight of DRF. The functionality f_e was set equal to 3.

If the networks were perfect, Eqs (3) or (8) could be used. However, the sol fractions were found not to be neglible, except with network 1. Using a universal relationship between the number of network chains and the gel fraction, derived ear-lier on the basis of branching theory [27] and applied to polyetherurethane networks [28], one can estimate that ν_e should be corrected by a decrease of 25% and 40% for networks 2 and 3, respectively. We incorporate such corrections in the A and B values (then denoted by an asterisk) and use the perfect ν_e value calculated with Eq.(2). Further, using $A = \frac{1}{3}$ and $B = 0$ (see below), we have adapted A* values of 1/3 (= $\frac{1}{3}$*1), 1/4 (= $\frac{1}{3}$*0.75) and 1/5 (= $\frac{1}{3}$*0.6) for the three networks, respectively.

There may also be a contribution to ν_e from trapped entanglements which,

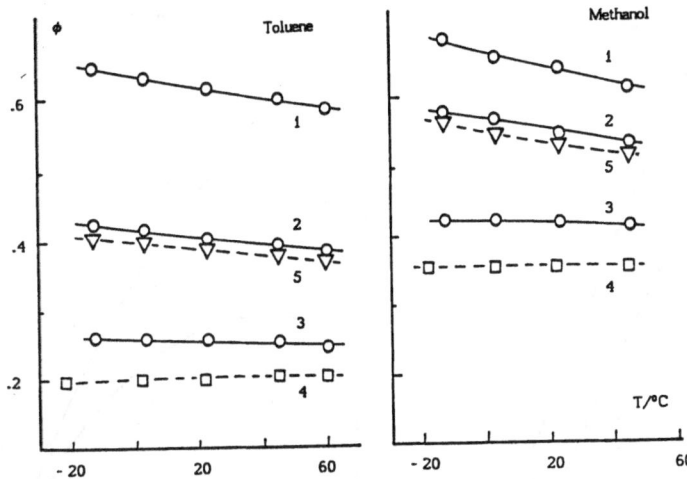

Figure 1.   Volume fraction of polymer in swollen PPG/DRF networks 1 - 5
in toluene (left) and methanol (right) as a function of temperature.
M_c values: 1: 781, 2: 1334, 3: 2395, 4: 4006, 5: 1334 (bimodal).

according to an estimate based on data in ref. [28], may amount to *ca.* 30% in the
case of PPG 2000/DRF. To see how this feature affects the interaction parameters
we include a calculation on the basis of estimated values for A^* ($= 1*1.3$) and B^* (=
$0.9 = 1.3*\frac{2}{3}$).

A possible dependence of A and B on the swelling ratio [2-4,29] may be
approximated roughly by linear functions of ϕ [24], a manoeuvre that does not
introduce new parameters since we have values for A and B at $\phi = 0$ (1/3 and 0),
and at $\phi = 1$ (1 and 2/3, resp.):

$$A = A_1 + A_2\phi; \quad B = B_1 + B_2\phi \tag{9}$$

The equation for swelling equilibrium now reads

$$ln(1 - \phi) + \phi + m_1\chi_{ef}\phi^2$$
$$+ \nu_e m_1 V_b\{A_1\phi^{\frac{1}{3}} - B_1\phi + A_2\phi(3\phi - \phi^{\frac{1}{3}})/2 - B_2\phi^2(1 + ln\phi)\} = 0 \tag{10}$$

Sets of χ_{23} and g_{23} values have been calculated for the two solvents on the basis
of various A/B combinations.

Two- and three-parameter models

In the two- and three-parameter models the coordination numbers of all species in
the system are assumed to be identical and χ_{ef}, calculated with Eq.(8), can be used
to test Eqs.(4) and (5). We base the evaluation on toluene as the BVU and assume the
trimer of methanol to be stable within the full range of temperatures used. Then the

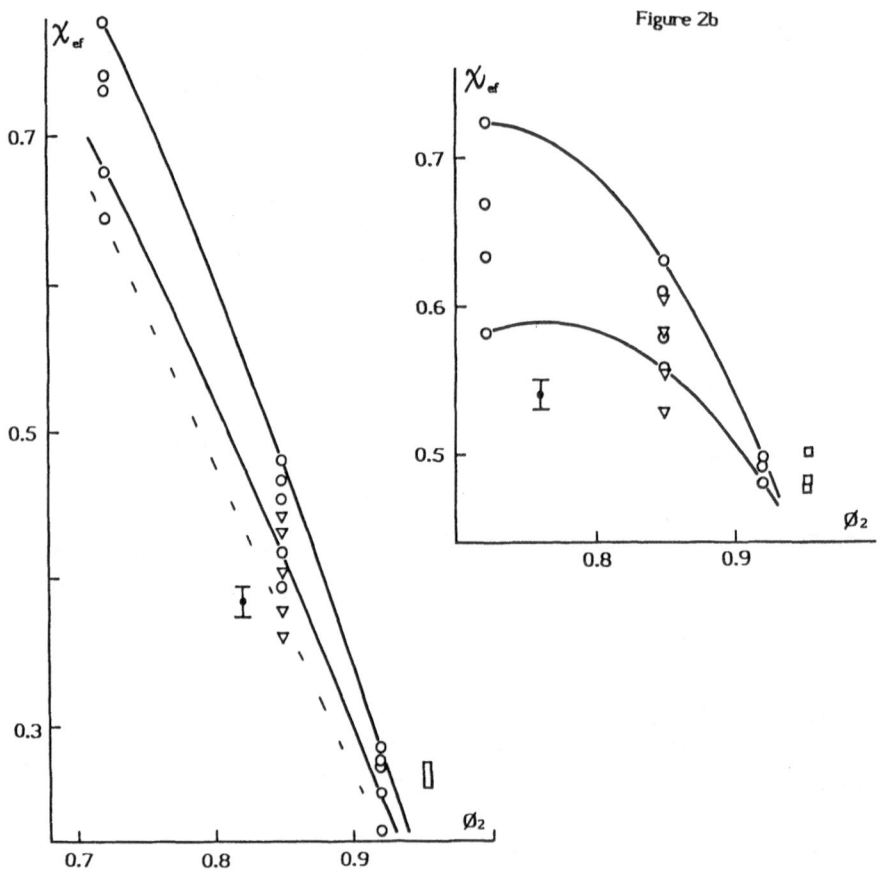

Figure 2. χ_{ef} as a function of the PG content ϕ_2 of the dry network. Networks 1, 2 and 3: o; network 4: □, network 5 (bimodal): ∇. The lower measuring temperatures go with the higher χ_{ef} values. Effect of probable experimental errors indicated by ⍊. a: toluene, b: methanol - - - - - -: straight-line relation. Calculation based on A = 1, B = ⅔, BVU: toluene; trimeric methanol.

number of occupied BVUs per trimer is $m_1 = 3V_1$(methanol)$/V_1$(toluene). Figure 2 demonstrates the inadequacy of the two-parameter model, which requires χ_{ef} to be linear in ϕ_2. The effect of estimated experimental errors is indicated.

Using the data on systems 1-3 we can solve Eq. (5) for χ_{12}, χ_{13} and χ_{23}. Thus we find at - 13°C and 45°C, χ_{23} = - 5.7 and -4.5 for methanol and χ_{23} = - 2.1 and - 1.3 for toluene (A = 1; B = ⅔). It is clearly not possible to describe both the toluene and methanol data with the same value of χ_{23}. Hence, the three parameter model cannot be valid.

Table 1
Three-parameter model (Eq.(5))

BVU	$A^{(*)}$	$B^{(*)}$	χ_{23}^{**}	
			methanol	toluene
toluene (meth.3)	1	2/3	-5.7;-5.7;-4.1;-4.5	-2.1;-2.5;+0.2;-1.3;-2.0
toluene (meth.1)			-14;-14;-10;-10	-2.1;-2.5;+0.2;-1.3;-2.0
methanol (3)			-6.5;-6.5;-4.7;-5.1	-2.4;-2.9;+0.2;-1.5;-2.3
methanol (1)			-5.3;-5.3;-3.8;-4.1	-0.8;-1.0;+0.1;-0.5;-0.8
toluene (meth.3)	1/3	0	-4.5;-4.6;-3.3;-3.7	+0.6;+0.3;+2.0;+0.8;+0.4
toluene (meth.1)			-13;-13;-9;-9	+0.6;+0.3;+2.0;+0.8;+0.4
methanol (3)			-5.4;-5.5;-3.9;-4.3	+0.7;+0.3;+2.2;+0.9;+0.4
methanol (1)			-4.9;-4.9;-3.6;-3.9	+0.2;+0.1;+0.7;+0.3;+0.1
toluene (meth.3)	1.3	0.9	-6.0;-6.0;-4.3;-4.8	-3.4;-3.8;-0.6;-2.2;-3.0
toluene (meth.3)	var.***	0	-4.7;-4.8;-3.7;-4.1	+1.4;+1.0;+2.5;+1.4;+1.1
toluene (meth.3)	$A(\phi)$	$B(\phi)$	-4.7;-4.8;-3.4;-3.7	+1.2;+0.8;+2.6;+1.3;+0.9

** χ_{23} values from left to right for lower to higher measuring temperatures
*** var. means different A* values for networks 1, 2 and 3

Neither the particular choice of the A and B values in figure 2 (A = 1, B = 2/f = ⅔) nor that of the BVU are significant for the rejection of the three-parameter model for the systems in hand. Table 1 lists some examples for other combinations of A, B and m_1, and the conclusion remains unaltered. Detailed listing of values for χ_{12} and χ_{13} was therefore not considered meaningful.

We shall see that the present data allow conclusions to be drawn about the ϕ_2 and T dependence of χ_{ef} and/or g_{ef}, but the values of the individual interaction parameters are not significant enough to justify detailed specification. We make an exception for χ_{23} and g_{23} but are interested in their values only to find out which model makes them roughly identical for the two solvents.

Modified three-parameter model

Use of Eq.(6) presupposes knowledge of the molecular surface areas. We draw on Bondi's estimations [30] and have, expressed in 10^9 cm²/mole: toluene, $\sigma_{1T} = 7.45$; methanol, $\sigma_{1M} = 3.58$; PG, $\sigma_2 = 4.65$ and $\sigma_3 = 7.2*10^9$ cm²/mole for each of the three groups in DRF.

Regarding the surface area of trimeric methanol we use σ_1 as a parameter and find that a reasonable decrease of $3*\sigma_{1M}$ by about 40% to $6*10^9$ cm²/mole brings the g_{23} values for methanol and toluene close enough together to consider them identical in view of the experimental uncertainties involved. Table 2 shows a few combinat-

Table 2

Modified three-parameter model (Eq.(6))

BVU: toluene; methanol: trimer; A = 1; B = 2/3

σ_{1M} [10^9 cm^2/mole]	g_{23}	
	methanol	toluene
10.74	-0.2;-0.3;+0.1;-0.2	-3.5;-3.7;-1.6;-2.5;-3.3
8	-1.2;-1.3;-0.6;-0.9	
6‡	-3.2;-3.3;-2.2;-2.6	

‡ Value of σ_{1M} for which agreement for g_{23} between methanol and toluene is obtained

Table 3

Modified three-parameter model

A = 1; B = 2/3

BVU	σ_{1M}	g_{23}	
		methanol	toluene
methanol (3)	10.74	-0.2;-0.3;+0.2;-0.2	-4.0;-4.2;-1.9;-3.0;-3.9
	8	-2.0;-2.2;-1.3;-1.6	
	6‡	-3.7;-3.8;-2.5;-3.0	
methanol (1)	3.58	-5.7;-5.9;-4.1;-4.6	-1.3;-1.4;-0.6;-1.0;-1.3

‡ Value of σ_{1M} for which agreement for g_{23} between methanol and toluene is obtained

ions. In Table 3 we demonstrate that a) setting the BVU equal to $3*V_1$(methanol) does not change this result and b) use of monomeric methanol as BVU does not lead to agreement in g_{23} at all. These results indicate that the modified three-parameter model provides a meaningful treatment of the present data provided the association of methanol is accounted for.

Selecting the g_{ef} values corresponding to $\sigma_1 = 6*10^9$ cm^2/mole, we can construct $g_{ef}(\phi_2)$ curves to compare with $\chi_{ef}(\phi_2)$ in figure 2. We see in figure 3 that a similar result is obtained.

We believe it to be a significant finding that the introduction of the physical fact of disparity in size and numbers of nearest neighbor contacts between the various molecular entities in the system brings such an obvious improvement in the description. This conclusion is consistent with recent findings with polymer solutions [31] and polymer mixtures [32].

Swelling in solvent mixtures

If the mixing behavior of the system methanol-toluene could be modelled by the present treatment one could try to predict the equilibrium swelling of the network in the mixed solvent. While this approach is being studied we draw attention here to a

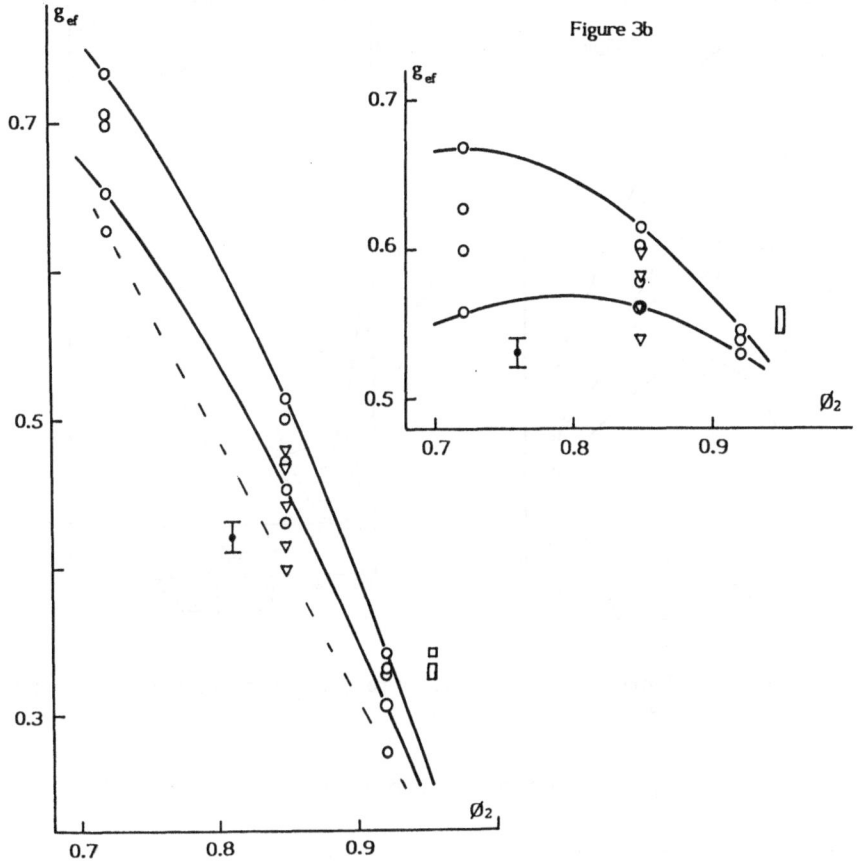

Figure 3. g_{ef} as a function of ϕ_2. See caption figure 2.

possibly less accurate but useful approximation that does not depend on knowledge of the behavior of methanol-toluene and might for that reason offer a procedure of practical interest.

If the interaction parameters of a system are not known the solubility parameter approach may be used for their estimation. Solubility parameters, δ, can be calculated from cohesive energy densities, and the volume contribution of atomic groups [33,34].

It is an interesting point that the occurrence of swelling maxima can be predicted if a solvent mixture is looked upon as a single solvent with an average solubility parameter: $\delta_{ef} = \phi_{tol}\delta_{tol} + \phi_{me}\delta_{me}$, where ϕ_{tol} and ϕ_{me} are the volume fractions of toluene and methanol, and δ_{tol} and δ_{me} the corresponding solubility parameters. Figure 5 shows calculated and experimental swelling curves in terms of χ_{ef} vs solvent composition. In view of the simplicity of the calculation the agreement can

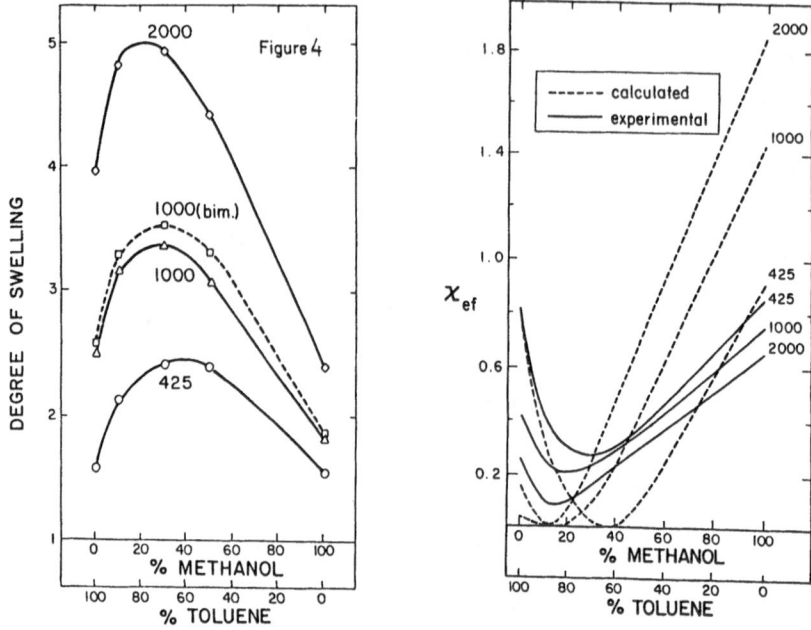

Figure 4.   Degree of swelling vs solvent composition in mixtures of methanol and toluene (experimental).

Figure 5. χ_{ef} vs solvent composition; predicted: – – – –, experimental: ———.

be considered quite satisfactory.

Discussion

The theory of rubber elasticity, which supplies one part of the equilibrium swelling equation, contains two front factors A and B, the values of which and their dependence on the degree of swelling are still very much under discussion. The present analysis does not discriminate between the various suggestions, but this is partly due to the remaining uncertainty with respect to the concentration of elastically active network chains and the attainable surface area of the associated alcohol molecules. It is conceivable that swelling measurements on well-defined networks in a number of non-associating solvents, differing widely in molecular surface area, might offer possibilities here, but they would have to be very accurate (see Table 4), and the evaluation should be supported by independent measurements of the enthalpic part of g_{23}. Table 5 gives examples for the influence of network imperfection and a swelling-ratio dependence of A and B. The better approximations seem to remove the need to treat the surface area of trimeric methanol as an adjustable parameter.

Table 4

Modified three-parameter model

Effect of A and B

BVU: toluene; methanol: trimer

σ_{1M}	$A^{(*)}$	$B^{(*)}$	g_{23}	
			methanol	toluene
10.74	1	2/3	-0.2;-0.3;+0.1;-0.2	-3.5;-3.7;-1.2;-2.6;-3. 4
7			-2.5;-2.6;-1.6;-2.0	
6‡			-3.2;-3.3;-2.2;-2.6	
10.74	1/3	0	+1.6;+1.3;+1.3;+1.0	+0.5;+0.3;+1.3;+0.6;+0.3
9‡			+0.4;+0.1;+0.4;+0.1	
8			-0.4;-0.6;-0.2;-0.5	
10.74	1.3	0.9	-0.8;-0.9;-0.3;-0.8	5.3;-5.5;-2.9;-4.1;-5.0
7			-2.9;-3.0;-2.0;-2.4	
4.5‡			-4.8;-5.0;-3.5;-4.0	

‡ Value of σ_{1M} for which agreement for g_{23} between ethanol and toluene is obtained

Table 5

Modified three-parameter model

ν and ϕ dependence of A and B

BVU: toluene; methanol: trimer

σ_{1M}	$A^{(*)}$	$B^{(*)}$	g_{23}	
			methanol	toluene
10.74‡	var.**	0	+1.9;+1.5;+1.4;+1.1	+1.9;+1.6;+2.4;+1.8;+1.5
6			-2.1;-2.3;-1.6;-2.0	
10.74	$A(\phi)$	$B(\phi)$	+1.4;+1.1;+1.3;+1.0	+1.0;+0.8;+1.9;+1.1;+0.8
10‡			+0.9;+0.6;+0.9;+0.6	
6			-2.2;-2.4;-1.5;-1.8	

** var. means different A^* values for networks 1, 2 and 3

‡ Value of σ_{1M} for which agreement for g_{23} between methanol and toluene is obtained

So far the emphasis has been on the three networks 1, 2 and 3, prepared from propylene glycols with narrow molar mass distributions [24]. The $g_{ef}(T)$ relations for these three and the other two swollen networks show an interesting feature (figure 6). It is seen that the usual relationship [35]

$$g_{ef} = g_S + g_H/T \qquad (11)$$

appears to be reasonably well obeyed. The enthalpic term g_H varies between samples 1 - 3 and 5, which probably reflects a systematic dependence on ϕ_2, the latter being somewhat stronger with methanol. It is of particular interest to note the

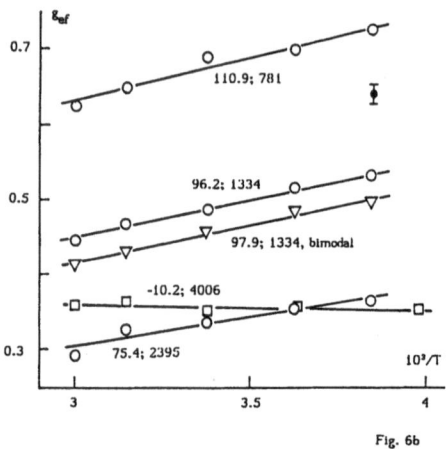

Fig. 6b

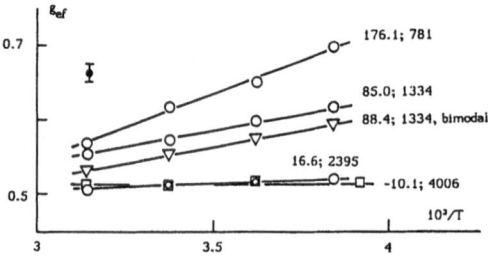

Figure 6.   Swelling in single solvents, g_{ef} vs $1/T$. Values of g_H (Eq.(10)), determined by least squares analysis, and M_c are indicated. Effect of
probable experimental errors indicated by $\bar{\bar{\Phi}}$. a: toluene, b: methanol
Calculation based on $A = 1$, $B = \frac{2}{3}$, BVU: toluene; trimeric
methanol.

correspondence for **both** solvents between the unimodal and bimodal networks with
identical M_c, the bimodal network having been prepared from a mixture of two
narrow-distribution polypropylene glycols. The enthalpic terms (g_H) are essentially
identical but they differ in the entropic contribution g_s to g_{ef}. It is not obvious why
this should be so. The enthalpic term may be assumed independent of the chain
geometry (linear, long-chain branched, crosslinked) but depends - within the strictly
regular approximation - on ϕ_2 which has the same value in networks 2 and 5. The
same situation is encountered with other A/B combinations. Little is known about
the entropic consequences of the chain length distribution between crosslinks which,
if we use the modified Flory-Rehner equation, might show up in g_s. The little
information available at present deals with stress-strain behavior only [36-38] and
predicts no influence at all from elastic sources [38]. The mixing term in the
swelling equation can hardly be considered responsible for the significant differences

in g_S and we have to leave the problem open for the moment. The same holds for network 4, which shows a differing $g_{ef}(T)$ behavior in both solvents for which we cannot offer an explanation though we suspect that a relatively large number of dangling chains are present that might partially invalidate the analysis.

Conclusion

The objective of the study, separation of the χ parameters in χ_{ef}, or the g terms in g_{ef}, has been achieved up to a point only. The experimental accuracy limits the discussion to rough values of χ_{23} or g_{23} only, but they are good enough to reject two of the three interaction models considered. The results stress a) the importance of accounting for disparities in numbers of nearest neighbors in the analysis of swelling data, b) the necessity of comparing such data in different solvents, c) the need for more decisive information on the front factors A and B and d) the importance of attempting to prepare perfect networks, particularly when the emphasis is on the evaluation of interaction parameters.

References

1. Dušek, K.; Prins, W. Adv.Polym.Sci., **1969**, *6*, 1.
2. Flory, P.J. Proc.Roy.Soc.(London), **1976**, *351*, 351
3. Gottlieb, M.; Gaylord, R.J. Macromol., **1984**, *17*, 2024
4. Heinrich, S.; Straube, E.; Helmis, G. Adv.Polym.Sci., in press
5. Flory, P.J.; Wall, F.T. J.Chem.Phys., **1950**, *18*, 108.
6. Dušek, K. Rubber Chem.Techn., **1982**, *55*, 1
7. Flory, P.J. J.Chem.Phys., **1941**, *9*, 660
8. Huggins, M.L. J.Chem.Phys., **1941**, *9*, 440
9. Staverman, A.J.; Van Santen, J.H. Rec.Trav.Chim., **1941**, *60*, 76
10. Topchiev, A.V.; Litmanovich, A.D.; Shtern, V.Ya. Dokl.Akad.Nauk. SSSR, **1962** *147*, 1389.
11. Lautout, M.; Magat, M. Z.Physik.Chem., **1958**, *16*, 292.
12. Glöckner, G; Lohmann, D, Faserforsch.Textiltechn., **1973**, *24*, 365.
13. Glöckner, G. Faserforsch.Textiltechn., **1974**, *25*, 476.
14. Ten Brinke, G.; Karasz, F.E.; MacKnight, W.J. Macromol., **1983**, *16*, 1827.
15. Simha, R.; Branson, H. J.Chem.Phys., **1944**, *12*, 253.
16. Stockmayer, W.H., Moore Jr., L.D., Fixman, M; Epstein, B.N. J.Polym.Sci., **1955**, *16*, 517.
17. Kambour, R.P.; Bendler, J.T.; Bopp, R.C. Macromol., **1983**, *16*, 753.
18. Paul, D.R.; Barlow, J.W. Polymer, **1984**, *25*, 487.
19. Koningsveld, R.; Kleintjens, L.A. Macromol. **1985**, *18*, 243.
20. Balazs, A.C.; Sanchez, I.C.; Epstein, I.R.; Karasz, F.E.; MacKnight, W.J. Macromol., **1985**, *18*, 2188.
21. Balazs, A.C.; Karasz, F.E.; MacKnight, W.J. Croat.Chim.Acta, in press.

22. Howe, S.E.; Coleman, M.M. Macromol. **1986**, *19*, 72.

23. Staverman, A.J. Rec.Trav.Chim., **1937**, *56*, 885.

24. Petrović, Z.S.; MacKnight, W.J.; Koningsveld, R; Dušek, K. Macromol., in press

25. Glasstone, S.; Lewis, D. *Elements of Physical Chemistry*, MacMillan, London, **1946**, p. 193.

26. Flory, P.J.; Rehner, J.J. J.Chem.Phys., **1943**, *11*, 521.

27. Dušek, K.; Hadhoud, M.; Ilavský, M. Brit.Polym.J., **1977**, 9, 172

28. Ilavský, M.; Dušek, K. Polymer, **1983**, *24*, 981

29. Queslel, J.; Mark, J.E. *Encycl.Polym.Sci.Technol.*, Wiley, New York, **1986**, 2nd Ed., Vol. 5, 3625.

30. Bondi, A. J.Phys.Chem., **1964**, 68, 441.

31. Nies, E.; Koningsveld, R.; Kleintjens, L.A. Progr.Coll.Polym.Sci., **1985**, *71*, 2

32. Voigt-Martin, I.G.; Leister, K.-H.; Rosenau, R.; Koningsveld, R. J.Polym.Sci. Part B:Polym.Phys., **1986**, *24*, 723.

33. Van Krevelen, D.W.; Hoftijzer, P.J. *Properties of Polymers*, Elsevier, Amsterdam, **1976**, p. 137.

34. Brandrup, J.; Immergut, E.A. *Polymer Handbook*, Wiley-Interscience, New York, **1975**, p.IV, 343.

35. Rehage, G. Kunststoffe, **1963**, *53*, 605.

36. Case, L.C. Makromol.Chem., **1960**, 37, 243; **1960**, 39, 119.

37. Curro, J.G.; Mark, J.E. J.Chem.Phys., **1984**, 80, 4521.

38. Kilian, H.-G. unpublished results.

THEORETICAL TREATMENT OF NETWORK FORMATION IN A THREE-STAGE PROCESS

B.J.R. Scholtens, G.P.J.M. Tiemersma-Thoone
DSM Research, PO Box 18, 6160 MD GELEEN, Netherlands

and

K. Dušek
Institute of Macromolecular Chemistry, Czechoslovak Academy of
Sciences, 16206 PRAGUE, Czechoslovakia

ABSTRACT

A stochastic description of network formation in a three-stage process has recently been derived with the theory of branching processes with
cascade substitution [1,2]. The distributions of units in the different
reaction states are calculated with differential equations, so that
substitution effects may be incorporated. Some preliminary results based
on this theory are presented in this contribution.

INTRODUCTION

Crosslinking processes are very important from a technological point
of view: they are used in a wide variety of fields in polymer technology,
e.g. adhesives, coatings, construction materials and elastomers. In these
processes a monomeric or polymeric liquid is transformed into a permanent
network by an interlinking reaction. Actually, most of these processes
must be regarded as multistage processes in which prepolymers are formed
in one or several stages, upon which a network is formed by crosslinking
the polymers in a final stage.

Recently, a theoretical scheme has been proposed for network formation in a three-stage process [1,2]. It was derived with the theory of
branching processes with cascade substitution [3-6]. As far as we know it
is the first theoretical attempt to describe the molecular mass distribution of polymers and the network characteristics of thermosetting systems

prepared in more than two consecutive stages.

In this contribution some preliminary results are presented based on this theory.

THE THREE-STAGE PROCESS

A schematical description of the three-stage process of network formation is given in Figure 1 and explained further below.

Stage 1: Difunctional monomers A, with functional end groups called c, react exclusively with an excess mixture of difunctional D and trifunctional T monomers, which have the same functional end groups, called h (and thus are equally reactive), to (mainly) h-terminated prepolymers P1.

Stage 2: Prepolymers P1 are modified with an excess of difunctional C monomers, also with functionality c, into (mainly) c-terminated prepolymers P2. Unreacted functional groups of the A monomers are assumed not to react in this stage.

Stage 3: The unreacted functional c groups in P2 react in this last stage with a mixture of difunctional E and trifunctional F hardeners, which have the same functional end groups called e (and thus are equally reactive). The h end groups are assumed not to react in this stage.

Both in the A and C monomers substitution effects can be taken into account, as explained elsewhere [2]. The substitution effect factor K_{IJ} indicates with which factor the reaction rate between I and L is multiplied if I did already react once before with J (first shell substitution effect, see refs. 2 and 6). In all three stages cyclisation is postulated not to occur in the sol fraction.

PROGRAMS

Two computer programs were written in Fortran for a mainframe IBM 4381: KINREL, in which the various bond probabilities or reaction states are calculated as a function of the conversion by solving differential

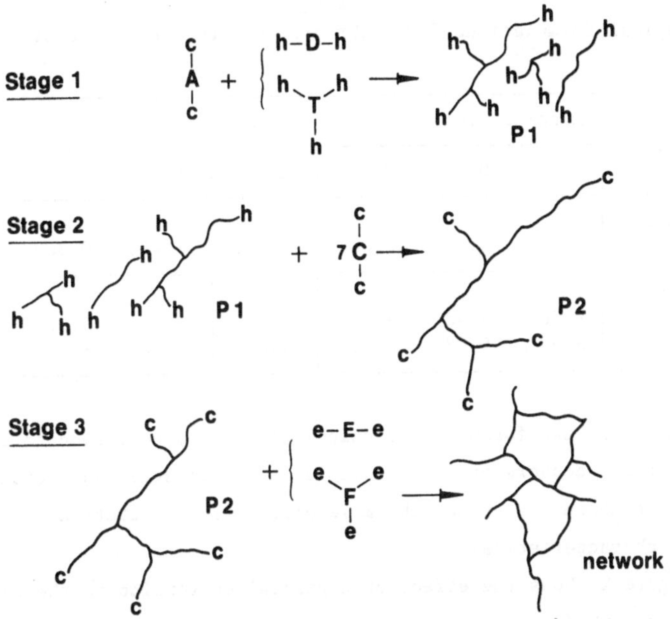

Figure 1. Schematic representation of the three-stage process of network formation.

equations with Gear-stiff's method, and POLYM, which uses the output data of KINREL to calculate the various average molecular masses and functionalities in the pre-gel region and the network characteristics in the post-gel regime. These programs were tested extensively with examples for which the results were known from the literature. More details on the underlying principles of the programs were given elsewhere [2].

RESULTS AND DISCUSSION

Table 1 summarizes two typical formulations with which the present calculations were performed. Formulation F1 results in a linear prepolymer in stages 1 and 2, which is crosslinked subsequently with a trifunctional monomer. Formulation F2, on the other hand, results in a branched prepolymer in stages 1 and 2, which is crosslinked subsequently with a difunctional monomer. More details of the prepolymers are collected in Table 2.

TABLE 1

Two typical formulations for a three-stage process of network formation

	monomers (moles)					
	A	D	T	C	E	F
molar mass (kg)	0.166	0.068	0.080	0.166	1.450	0.300
formulation F1	8.98	9.98	–	2.00	–	0.67
F2	8.68	9.28	0.40	2.40	1.20	–

For these two formulations calculations were performed to find out how a partial conversion in the second and third stage and substitution effects in monomer C affect the molecular mass distribution, gel point and network characteristics.

Figure 2 shows the effect of a partial conversion in the second stage on the network formation and network properties in the third stage for the linear prepolymer P2 crosslinked with the trifunctional monomer F. Similar results are given in Figure 3 for the branched prepolymer P2 which forms a network with the difunctional monomer E.

Figure 4 shows the results of a negative, no and a positive substitution effect in the C monomers in stages 2 and 3 for formulation F1. As shown in Figure 5 a positive substitution effect in the C monomers in formulation F2 may even cause gelation in the second stage, which is usually highly undesirable.

TABLE 2

Prepolymer characteristics after full conversion in stages 1 and 2

formulation	P1				P2			
	$\bar{M}_n$ $\bar{M}_m$ (kg/mol)		$\bar{\phi}_n$[1]	$\bar{\phi}_e$[2]	$\bar{M}_n$ $\bar{M}_m$ (kg/mol)		$\bar{\phi}_n$[1]	$\bar{\phi}_e$[2]
F1	2.2	4.4	2.0	2.0	2.5	8.8	2.0	2.0
F2	2.1	6.6	2.4	2.9	2.5	121.0	2.4	10.2

[1] number average functionality
[2] end-group average functionality

The results given above indicate that the viscosity, gel point and network properties of thermosetting systems may be adjusted between certain limits by variations in the conversion in the second or third stage or by variations in the substitution effect parameters. The range which can be covered is apparently much wider for the branched than for the linear prepolymer systems.

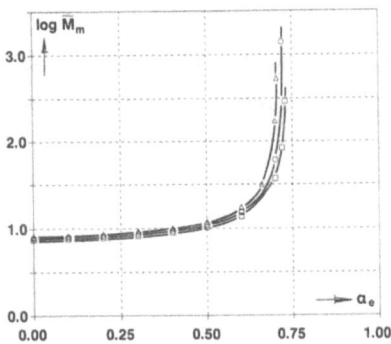

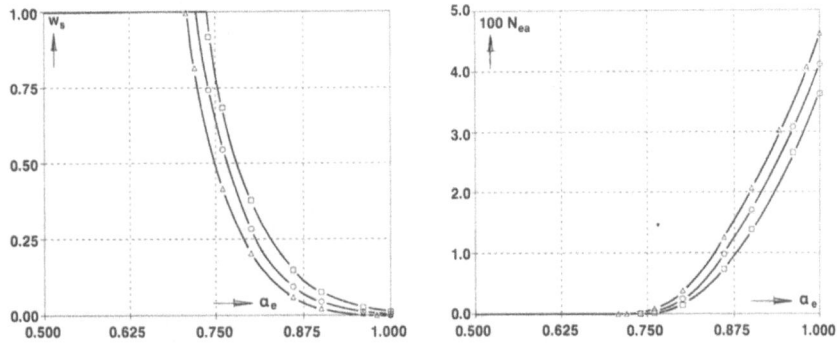

Figure 2. Variation of the mass average molecular mass (top), sol fraction (left) and number of elastically active network chains per monomer (right) as a function of the conversion in the third stage for formulation F1. The conversion in the second stage is changed systematically: □ : 0.96, ○ : 0.98 and ▲: 1.00.

CONCLUSIONS

1. In principle, a three-stage process of network formation can be described theoretically in the pre- and post-gel region with the theory of branching processes with cascade substitution.
2. Substitution effects can be incorporated in this theory.
3. Incomplete conversion in the second stage has a marked effect on the network properties in the third stage, in particular for branched pre-polymers.
4. Positive substitution effects have a more pronounced effect on the network formation and properties than negative substitution effects.

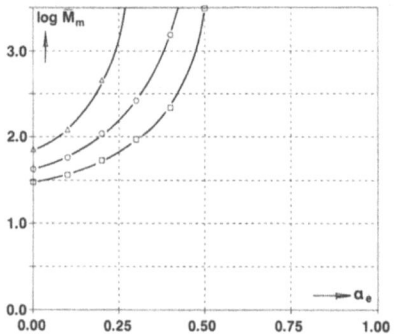

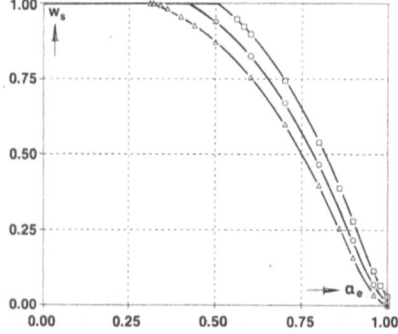 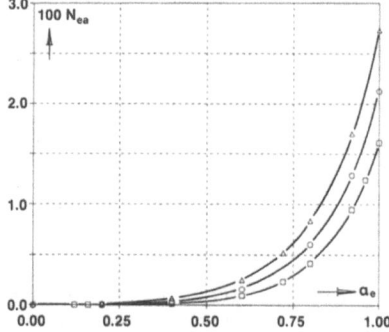

Figure 3. See legend of Figure 2, but now for F2.

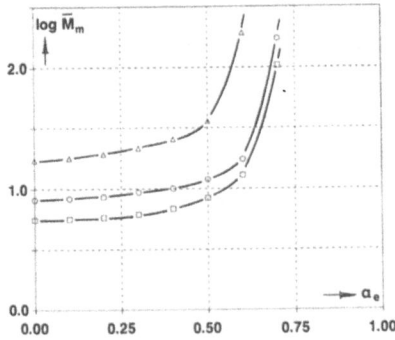

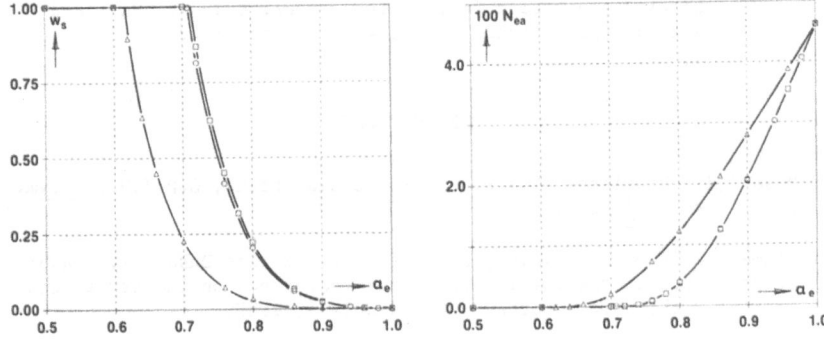

Figure 4. Variation of the mass average molecular mass (top), sol fraction
(left) and number of elastically active network chains per
monomer (right) as a function of the conversion in the third
stage for formulation F1. The substitution effect factors for
monomer C in stages 2 and 3 are changed systematically:
□: 1/4, ○: 1 and △: 4.

ACKNOWLEDGEMENT

The authors are grateful to the management of DSM Resins for their
permission to publish this work.

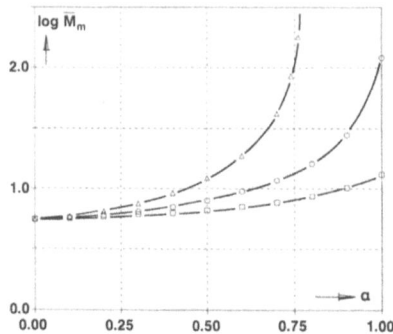

Figure 5. Variation of the mass average molecular mass as a function of the conversion in the second stage for formulation F2. The substitution effect factors for monomer C in stages 2 and 3 are changed systematically: ◇ : 1/4, ○: 1 and △ : 4.

REFERENCES

1. Dušek, K., Scholtens, B.J.R. and Tiemersma-Thoone, G.P.J.M., Polym. Bull., 1987, 17, 239.

2. Tiemersma-Thoone, G.P.J.M., Scholtens, B.J.R. and Dušek, K., paper to be presented at the 1st International Conference on Industrial and Applied Mathematics, Paris, June 29–July 3, 1987.

3. Gordon, M., Proc. Roy. Soc. London, 1962, A268, 240.

4. Gordon, M. and Malcolm, G.N., Proc. Roy. Soc. London, 1966, A295, 29.

5. Dušek, K., Adv. Polym. Sci., 1986, 78, 1.

6. Gordon, M., and Scantlebury, G.R., Proc. Roy. Soc. London, 1966, A292, 380.

AZO-GROUP CONTAINING STYRENE-BUTADIENE-
COPOLYMER SYSTEM FOR SELF-CROSSLINKING

Oskar Nuyken, Richard Weidner, Arthur Grunow
Lehrstuhl für Makromolekulare Chemie I
Universität Bayreuth

ABSTRACT

Azomonomers of the general structure

Nr.	R_1	R_2
1	CH_3	CH_3
2	CH_3	Ph
3	Ph	Ph

have been terpolymerized with styrene and butadiene. Rate of polymerization, polymer composition and molar masses were determined. The resulting azopolymers were crosslinked by heating or by UV-irradiation. The network characterization was carried out by swelling experiments and vulcanometric studies.

1. Azomonomers

1-(3-aminophenyl)ethanol was converted into 3-aminostyrene by distillation

over Al_2O_3 (300 °C). Then the amine was reacted with $NaNO_2$ yielding

the corresponding diazonium salt, which was reduced to the 3-hydrazino-

styrene. This step was followed by the reaction of the hydrazine

derivative with either acetone, acetophenone or benzophenone. Finally,

the hydrazones were oxidized by $Pb(OAc)_4$.

2. Azopolymers

Terpolymerization of the azomonomers 1-3 with styrene and butadiene

led to polymers containing both azo groups (which can be used subsequently

as radical sources) and C=C-double bonds:

Using a similar procedure as is used for "cold rubber" various mixtures
of 1-3 with styrene and butadiene (25 weight % : 75 weight %, whereby
various amounts of styrene were substituted by azo compounds) have
been polymerized in emulsion at 5 °C using a redox initiating system.
The incorporation of the azo monomers 1-3 into the terpolymer was
"ideal", i.e., equal to their proportions in the initial monomer
mixture - and independent of conversion. The retarding effect of
monomers 1 and 2 on the overall polymerization rate can be explained
by the formation of inreactive hydrazyl radicals.

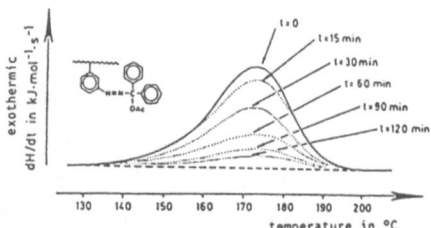

For the characterization of the terpolymers ^{1}H-NMR, IR, UV, DSC,
elementary analysis and GPC were used. In particular, DSC measurements
gave interesting information about the content of azo function in
the polymer.

DSC measurements of azopolymers after different heating time

3. Crosslinking

Thermolysis:

Depending on their structure the azo monomers decompose at different
temperatures. Thermolysis of polymers, containing 1-3, yield polymeric
and low molar mass radicals. Both types of radicals can initiate
polymerization reactions through the double bonds of the polymer

backbone, resulting insoluble polymers.

For the characterization of the network, swelling experiments were carried out.

$$1/Q = \frac{b}{a-b}$$

a: weight of the swollen polymer

b: weight of the polymer dried after swelling

a-b: weight of the occupied solvent

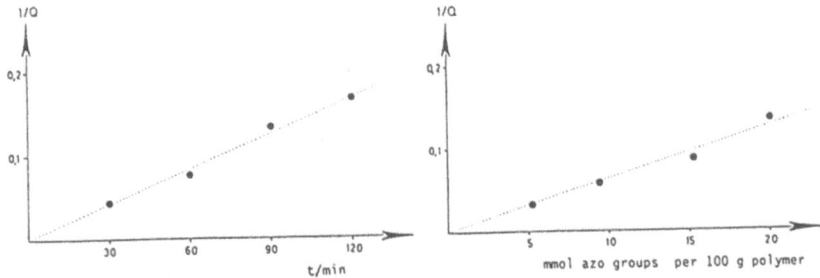

1/Q as function of reaction time (constant temperature)

1/Q as function of the azo concentration

A linear relation was found between the 1/Q value and the crosslinking time at a certain temperature.

When the azo concentration was changed but the reaction time and temperature were kept constant a linear increase of the 1/Q-value with the azo concentration was observed.

Incorporation of azo compounds of different stability into a polymer, allows selective, stepwise crosslinking.

Photolysis

UV-irradiation of a polymer film made from butadiene, styrene and 3 at room temperature yields crosslinked material.

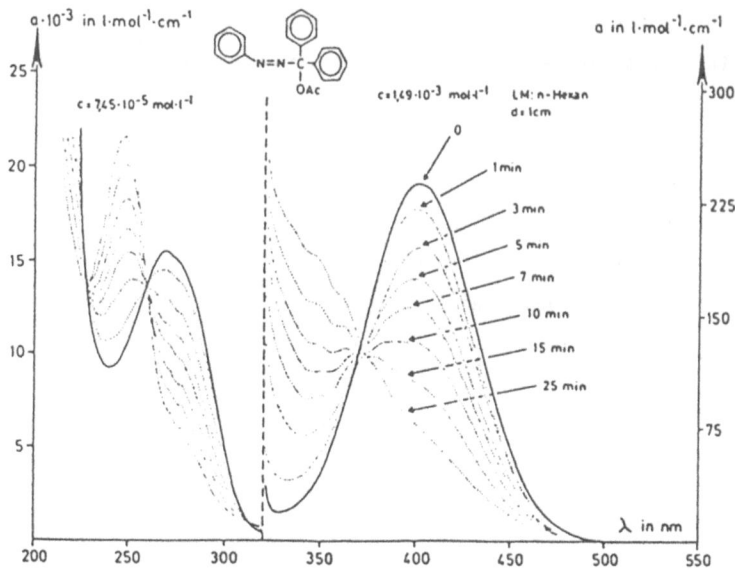

UV-irradiation of polymers containing 1 or 2 results in trans-cis-isomerization at room temperature. At higher temperature they also can be converted into networks.

4. Vulcanometry

Typical stress-strain curves of this type of measurements are given below.

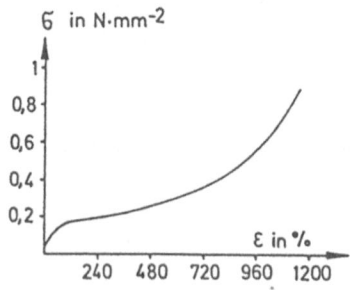

Stress-strain diagram of Poly-(Bu(St/3); 5.2 mmol groups per 100 g polymer

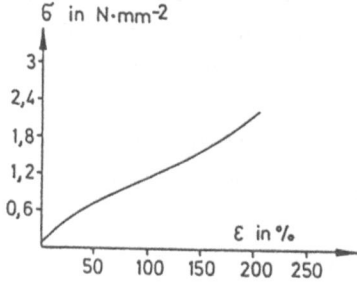

Stress-strain diagram of Poly(Bu/St/3) 13.3 mmol groups per 100 g polymer

REFERENCES

R. Weidner, PhD-Thesis, TU München, 1986

O. Nuyken, R. Weidner, Adv. Polym. Sci. 73/4, <u>145</u> (1985)

A. Grunow, Diploma Thesis, TU Müchen, 1986.

THERMOREVERSIBLE GELATION OF SYNTHETIC POLYMERS

H. Berghmans, Laboratory for Polymer Research, University of Leuven, Belgium

ABSTRACT

The thermoreversible gelation of polymer solutions is discussed. Four different gelation mechanisms can be distinguished. They originate from a crystallization, a combination of liquid-liquid phase separation with a crystallization or a glass transition and a molecular conformational change.

INTRODUCTION

The thermoreversible gelation of polymer solutions is a well known physicochemical phenomenon. It is widely used by nature as well as by man to prepare "structurated" polymer solutions[1]. Some of these gels also show interesting mechanical properties and find an important application in industry. The most illustrative example is polyethylene[2]. By passing through the gel state, solutions of this polymer in e.g. decaline can be transformed in ultra strong fibres. Gels of biopolymers like carrageenan, are extensively used in food, pharmaceutical and cosmetics industry[3].

For synthetic polymers, the phenomenon was already reported in the early days of their production. Although it was not termed at that time as thermoreversible gelation, the description of swollen PVC, with small crystals as physical cross-links between the molecules, fits very well the concept of physical gelation[4]. Many homo- and copolymers have been reported since that time to form thermoreversible gels in different solvents (e.g. ref.5,6). It concerns mostly crystallizable systems, although atactic polymers like polystyrene have been reported to form gels in many solvents[7].

DEFINITION OF A GEL

The first problem that arises when discussing this phenomenon, is the exact definition of a thermoreversible gel.

The concept "gel" originates from Polymer Chemistry. It refers to a polymer network, swollen by low molecular weight solvents. The cross-links can be introduced during or after the polymerization, in presence or absence of the solvent. The way by which these cross-links are introduced, together with their concentration and the interaction polymer-solvent will determine the degree of swelling. Networks obtained in this way will show a molecular connectivity throughout the polymer sample. This can easily be revealed by mechanical tests that are very sensitive to the presence of loose chain ends, molecular entanglements and non cross-linked molecules.

Thermoreversible gels are often defined in the same way. The most important, apparent difference is the reversibility of the cross-links. Cooling the polymer solution will result in gel formation while heating will turn it back into solution. The interconnection of the molecules is the result of specific interactions that very often lead to the formation of crystallites. Consequently, the problem is generally approached from this side. Experimental data are obtained from X-ray scattering analysis, spectroscopic investigations and calorimetric measurements. They are often combined with the results of optical observations and of simple mechanical tests. An important argument sometimes used to differentiate between suspensions of lamellar crystals and crystalline networks or real gels is their deformatibility to prepare highly oriented systems [8]. Although these experiments clearly bring about differences in these intermolecular connections, they do not reveal the exact nature of the network. This can only be done by a detailed mechanical analysis[9]. Dynamic mechanical tests will reveal the contribution of the real (G') and imaginary (G'') part of the complex modulus and their dependence on temperature, frequency, amplitude, gelation time etc. This type of investigation can lead to surprising results as to the presence of a network and its relation to some of the apparent differences in properties of the gels [10].

In most of the literature reports however, these types of experiments are not carried out. The formation of a gel is generally based on simple experimental observations like the absence of flow, changes in penetration characteristics, etc.

Because of the complex character of most of the thermoreversible gels, it seems to us that this "network" definition, derived from the "chemical" definition, is much too limited. It would be more convenient to start with a very general concept in which different types of thermoreversible systems can be fitted. A good starting point could be the statement of Flory in his introductory lecture of the Faraday Discussions in 1974 [11]. An elementary characteristic of a gel is "its solid like behaviour". Gels have an elastic response as long as the stress remains under a certain limit. Above this limit yield stress, plastic flow should occur. Generally, the modulus is quite low. They should possess some degree of continuity in their structure with a certain degree of permanency.

Starting from this very broad definition, a large number of systems can be included. They all result from a thermal transition that can occur on cooling polymer solutions.

THERMAL TRANSITION IN POLYMER SOLUTIONS AND THEIR RELATION TO THERMOREVERSIBLE GELATION.

The most discussed transition in relation to thermoreversible gelation is crystallization[8]. This liquid-solid phase separation is a first order, nucleation controlled, transition. It occurs at a certain degree of undercooling with the formation of a lamellar morphology. Consequently, crystallization and melting occur at different temperatures (resp. T_c and T_m). The onset of crystallization during a cooling experiment will depend on many factors such as cooling rate, concentration, heterogeneities left in the solution. It can only be considered as a reproductible parameter if these experimental conditions are kept rigorously the same. Because of the metastable character of the structure formed, T_m will depend on the crystallization conditions such as T_c , scanning rate during melting, and any further treatment given after the initial crystallization. It will always show up at temperatures well above T_c. The presence of a solvent will decrease T_m as observed with low molecular weight substances. If the solvent has its melting point far below that of the polymer only one part- of the entectic diagram can be seen. The onset of crystallization will decrease in a parallel way.

Because one is working in solution, a liquid-liquid phase separation can also occur [12]. If it occurs in the vicinity of the crystallization and melting, it will affect both phenomena [13]. We will only consider the binodal floculation curve for an idealized two component system. It is well known that real polymeric systems represent a multi-component system with a much more complex demixing behaviour. However, the conclusions that are obtained from the discussion of an idealized two component system can be very useful in the understanding of the behaviour of real systems.

A possible combination is represented in figure.1. When the demixing domain is reached, the solution splits into two solutions with strongly different concentration. When the temperature of intersection of the crystallization - concentration line with the binodal is reached, the concentrated phase will crystallize. As long as this demixing precedes the crystallization, the melting point will remain constant. This crystallization will result in a stiffening of the polymer solutions and a thermoreversible gel is obtained. Its structure will depend on the initial concentration and the cooling rate and the corresponding properties.

This liquid-liquid phase separation can interfere with still another transition that can occur in solution. Amorphous, non crystallizable polymers like atactic polystyrene, are characterized by glass transition temperature. It is very sensitive to the presence of a low molecular weight substance like a solvent. It has been demonstrated both experimentally and theoretically that T_g is a continuous decreasing function of the solvent concentration [14]. Interesting phenomena can be expected when a T_g -concentration curve intersects with a floculation curve (fig.1). A freezing of the concentrated domains, analogous to the crystallization will be observed and rigid gels are expected to be formed.

A last important transition occuring in polymer solution is often encountered with biopolymer solutions[3]. Many of these systems show a coil to helix transition on cooling their solutions. Stabilization of this regular conformation can be obtained by specific interactions with the solvent or e.g. ions in solution. These regular conformations can then aggregate to form an intermolecular network or gel. This type of gelation is very different from the previously mentioned crystallization gelation. During crystallization, a regular conformation is only obtained on incorporating the chain in the lattice. Here, the regular conformation is formed first, followed by an intermolecular association.

When considering these different transitions that can occur on cooling polymer solutions, we can differentiate between at least four different mechanisms of gelation.

EXAMPLES OF THERMOREVERSIBLE GELATION

1. THERMOREVERSIBLE GELATION BY CRYSTALLIZATION

Gelation by this mechanism has been frequently reported in literature[6]. It occurs with highly crystallizable polymers like polyethene (PE) as well as with irregular atactic chains like polyvinylchloride[15]. Important industrial applications have already been found in the production of ultra strong fibres.

The morphology and properties of these gels are strongly influenced by the experimental conditions. A typical example is PE. Cooling its solutions in xylene results in the formation of paste like dispersions of single crystals. They do not provide a continuous elastic network. On stirring solutions at much higher temperatures, transparent, elastic gels are obtained. They result in the formation of the well known shis-kebab morphology.

Besides the folded chain morphology, fringed micellar crystallites can also be obtained. Their relative contribution depends strongly on the chain composition and microstructure [5]. They were also reputed to be much better physical cross-links in the formation of a network [8].

Some of the well crystallizable polymers like isotactic polystyrene (i-PS) show an even more complex behaviour[16]. Besides the gelation by formation of a lamellar suspension of lamellar crystals with a 3_1 helical conformation, transparent elastic gels are formed on quenching to room temperature. They possess a close to the all-trans helical conformation.

2. THERMOREVERSIBLE GELATION BY LIQUID-LIQUID PHASE SEPARATION AND CRYSTALLIZATION

When liquid-liquid phase separation occurs in the same temperature domain as the liquid-solid transition, the gelation mechanism becomes much more complex.

A typical example is polyvinyl alcohol (PVAL) in ethylene glycol (EG)[1]. Concentrations between 5 and 10% are well suited to prepare gels. Their properties depend strongly on the experimental conditions. When a 8% solution is cooled below 90°C, a paste like opaque gel is very quickly formed. It has very poor mechanical characteristics reflected in its brittleness. Gelation above this temperature takes about three days. They are much less opaque and more elastic. Their response to orientation is also quite different[1]. Gels prepared at low temperature show a very poor molecular orientation on stretching. Only a very small fraction of the molecules have their molecular axis parallel to the stretching direction. When prepared at high temperature, orientation is much more pronounced but in this case with the molecules perpendicular to the stretching direction.

In order to fully understand this gelation, the concentration dependence of the behaviour of PVAL was investigated.

In figure 2 the different transitions that can be observed are plotted as a function of the polymer concentration .

The melting point decreases practically in a linear way with increasing solvent content. In the lower concentration domain it has a tendency to remain constant. Only in this concentration domain, the solution becomes opaque and this floculation occurs at temperatures above the onset of crystallization. This onset follows very well the floculation curve. At higher concentrations, only crystallization occurs while the samples remain practically transparent. While the onset of whitening is practically cooling rate independent, the onset of crystallization decreases in an appreciable way with increasing cooling rate.

When liquid-liquid phase separation occurs, the initially formed concentrated domains of very reduced dimensions, crystallize and any further increase of their size is stopped. The presence of these particles is responsable for its stretching behaviour[1]. When the gelation is carried out above this demixing domain, gels are formed very slowly by crystallization from solution. Its stretching behaviour is that of a highly crystalline, lamellar system. The presence of this demixing is further supported by the influence of the molecular weight and the quality of the solvent.

3. THERMOREVERSIBLE GELATION BY LIQUID-LIQUID PHASE SEPARATION AND VITRIFICATION

When a polymer is mixed with a solvent, a decrease of the glass transition temperature (T_g) is observed. A typical example is given in figure 6 for atactic polystyrene in a mixture of cis and trans decalin[17]. Because decalin is a poor solvent for polystyrene, a floculation curve is located in the same temperature domain. This floculation curve was obtained by a combination of optical and calorimetric observations. In a first approximation, no attention has to be paid to the difference in floculation and coexistence curve. This will be very small in the concentration domain of importance to our work. The floculation curve and the curve representing the decrease of Tg with concentration intersects at $W_{a-ps} = 0.60$. When a solution of lower concentration is cooled, it will separate into two liquid phases of different concentration. At $T < -20°C$, the concentrated phases will turn into a glassy phase, "freezing" the mobility of the solution. A thermoreversible gel will be obtained. The T_g of the solution will remain constant, independently of the starting polymer concentration. This was shown by calorimetric as well as by volume-temperature measurements. Because of T_g is at the origin of this gelation it will show the characteristics of an equilibrium phenomenon.

4. CONFORMATIONAL GELATION

The possibility of gelation by a conformation change will be illustrated by the behaviour of predominantly syndiotactic polymethyl methacrylate [18].

The two steps mechanism proposed was deduced from different experimental observations. The rapid intramolecular conformational change was deduced from infrared observation as a function of the temperature. The gelation and melting temperatures and the corresponding hysteresis were deduced from calorimetric obsevations. The change of the modulus with time at constant temperature informs about the second step in the gelation mechanism. The increase of G' is characteristic for the formation of a network. The value of G' for a 10% solution is of the order of 10^4 Pa. Its behaviour is very thixotropic. It was further shown that the gelation behaviour is very sensitive to differences in chain microstructure. Polymers analogous tacticities, as revealed by NMR, can show a very different gelation behaviour, depending on their method of preparation.

REFERENCES

(1) Berghmans, H. and Stoks, W. in "Integration of Fundamental Polymer Science and Technology", Eds. L.A. Kleintjens and P.J. Lemstra, Elsevier Appl. Science, 1986, p. 218
(2) Lemstra, P.J. and Kirsbaum R., Polymer, 1985, 26, 1372
(3) Rees, D.A. in "Polysaccharide Shapes", 1977, John Wiley and Sons
(4) Alfrey, T., Wierhorn, N., Stein R. and Tobolsky, A., Ind. Eng. Chem., 1949, 41, 701
(5) Berghmans, H., Overbergh, N. and Govaerts, F., J. Polymer Sci., Polymer Phys. Ed., 1979, 17, 1251
(6) Rehage, G., Prog. Colloid Polymer Sci., 1975, 57, 7
(7) Tan, H.M., Hiltner, H. and Baes, E., Macromolecules, 1983, 16, 28
(8) Keller, A., in "Structure-Property Relationships of Polymeric Solids", Ed. A. Hiltner, Plenum Press, 1983
(9) Ferry, J.D., in "Viscoelastic Properties of Polymers", Ch. 17, John Wiley and Sons, 1980
(10) Stoks, W., Berghmans, H., Moldenaers, P. and Mewis, J., to be published
(11) Flory, P.J., Faraday Discussions, 1975, 57, 7
(12) Koningsveld, R., Kleintjens, L.A. and Schoffeleers, H.M., Pure Applied Chem., 1979, 39, 1
(13) Richards, R.B., Trans. Faraday Soc., 1946, 42, 10
(14) Kanig, G., Koll. Z. z. Polymere, 1963, 190, 1
(15) Lemstra, P.J., Keller, A. and Cudby, M., J. Polymer Sci., Phys. Ed., 1978, 16, 1507
(16) Girolamo, M., Keller, A., Migasaka, K. and Overbergh, N., J. Polymer Sci., Phys. Ed., 1977, 15, 211
(17) Arnauts, J. and Berghmans, H., Polymer Communications, 1987, 28, 66
(18) Berghmans, H., Donkers, A., Frenay, L., Stoks, W., De Schryver, F.C., Moldenaers, P. and Mewis, J., Polymer, 1987, 28, 97

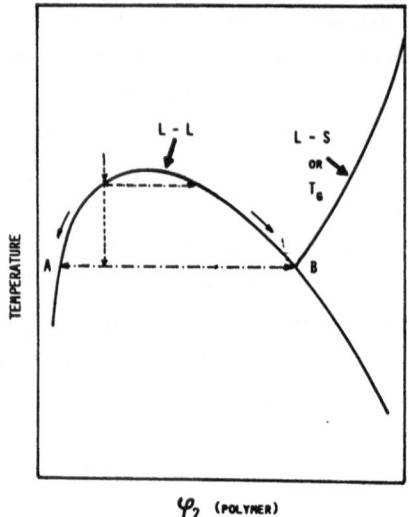

Fig. 1. Schematic representation of the combination of a liquid-liquid phase separation and a crystallization or glass transition temperature-concentration line.

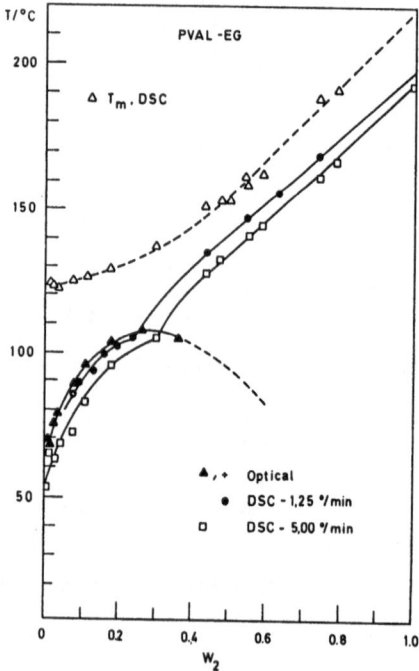

Fig. 2. Melting point (Δ), floculation curve (▲) and onset of crystallization on cooling (□,●) of the system PVAL – EG.

305

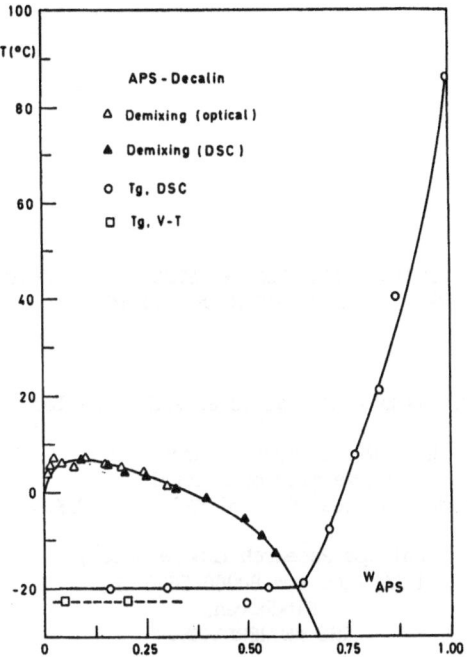

Fig. 3. Liquid–liquid demixing (▲,Δ) and glass transition–concentration
line (O,□) for the system a–PS – decalin.

THERMOREVERSIBLE GELATION OF ATACTIC POLYSTYRENE:
PHASE TRANSFORMATION AND MORPHOLOGY

R.M. Hikmet+, S. Callister and A. Keller

H.H. Wills Physics Laboratory,
University of Bristol,
Tyndall Avenue, Bristol BS8 1TL   U.K.

+Philips Research Laboratories,
B.V. P.O. Box 80000 5600 JA,
Eindhoven,
THE NETHERLANDS.

ABSTRACT

It has been convincingly demonstrated, that thermoreversible gelation of atactic polystyrene (a-PS) is the consequence of liquid-liquid phase segregation arrested by vitrification in accordance with the scheme of Arnauts and Berghmans [1], in this case for cyclohexanol as solvent. Matrix inversion from a solvated rubbery to a glassy matrix with increasing polymer concentration has been demonstrated with bicontinuous network structures in the intermediate stages, both by macroscopic constituency and by SEM study of the morphology. The onset of bicontinuous network structures is associated with passing from the metastable to the spinodal region of the phase diagram.

INTRODUCTION

At one stage it appeared to us [2] that localised interchain crystallization as a junction forming agency between different chains could provide a general scheme for physical gelation. However, since totally uncrystallizable atactic polystyrene (a-PS) was found to gel [3] this could not be upheld as a sole and universal basic source for physical gelation any longer. Attention was, therefore, focussed on a-PS in the belief that it could provide a universal gelation scheme in which special junction forming features in their numerous varieties [4] could then possibly represent additional features overlying the basic trend.

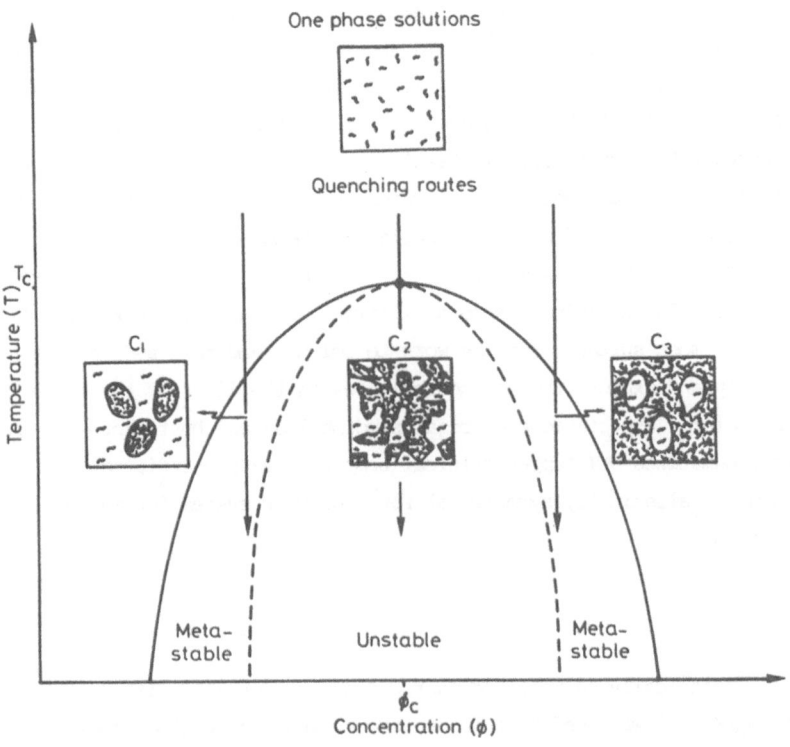

Figure 1. Schematic UCT phase diagram illustrating the expected variation in gel morphology with concentration c. A dispersed polymer rich glassy phase in a solvated matrix at low c (inset c_1), a dispersed solvated phase in a glassy matrix at high c (inset c_3) and bi-continuous inter penetrating networks at intermediate c (inset c_2).

In the seminal studies by Arnauts and Berghmans [1] it was proposed
that gelation resulted from localised liquid-liquid phase segregation
arrested by vitrification at a stage where the microphases are connected.
For this, in our words, three requirements need satisfying. 1) On
cooling, the system has to traverse the biphasic coexistence line in the
T(temperature)-c(concentration) phase diagram. 2) The polymer rich phase
in the segregated structure should attain the glass transition (T_g) of the
corresponding solvent containing system, when further phase segregation
will cease. 3) If at this stage there is mechanical connectedness
throughout the system it will 'set' as a gel. It follows from the above
that $T_{gel} \equiv T_g$, a temperature at which, cooling rate determined kinetic
effects apart, should be concentration independent.

Our own objective was first to confirm and establish the above
conditions for ourselves using a different solvent (cyclohexanol c.f.
Decalin) most suitable for the work to follow, and next and chiefly, to
identify and explore the gel morphologies in the light of the above new
view point. Expectations as regards morphology are best expressed through
the schematized UCT (upper critical temperature) phase diagram of Fig. 1.,
drawn unrealistically symmetrical for a polymer system for easier
illustration.

MATERIALS AND METHODS

Anionically polymerized narrow distribution a-PS with $M_w = 2.75 \times 10^6$
and $M_w/M_n = 1.1$ was used throughout, the solvent was cyclohexanol.

The phase diagram was determined by observing the onset of turbidity
on cooling in simple light scattering apparatus. DSC was used to
determine T_g as a function of c. The gelation temperature was taken to be
the onset of fluidity on heating a tilted test tube. Morphological
investigation was undertaken using Scanning Electron Microscopy (SEM) on
samples that had been freeze-dried prior to metal coating in order to
remove solvent without damaging the internal structure.

RESULTS AND DISCUSSION

The biphasic coexistence curve is shown in Fig 2a with enlarged
detail in Fig. 2b. As expected, it is of UCT type and highly asymmetrical

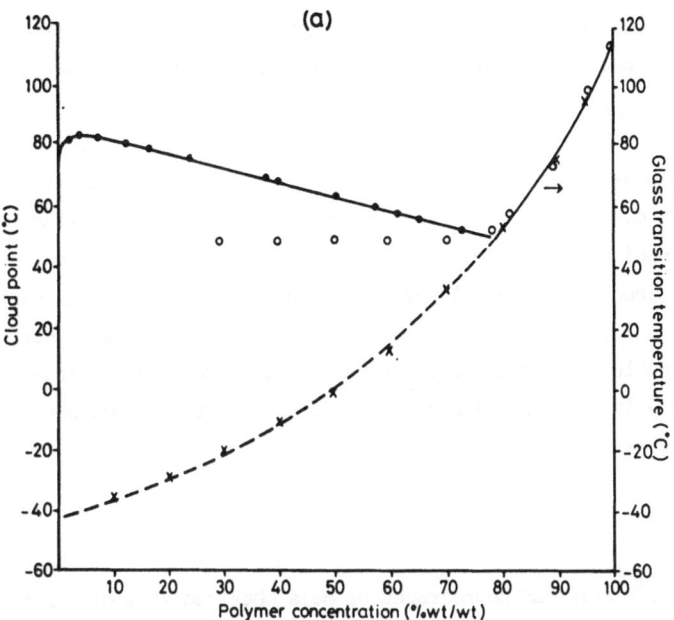

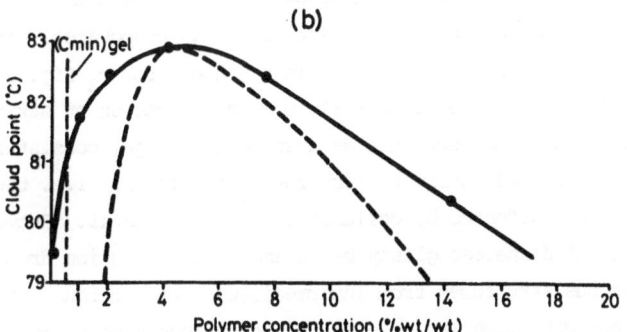

Figure 2a) The phase diagram for the a-PS/cyclohexanol system $M_w = 2.75\times10^6$ depicted in relation to its glass transition curve (the dashed portion is a theoretical extrapolation, see text).

2b) Enlarged detail of Fig. 2a about the commensurate point. The boundary (c_{min})gel together with the solid dots along the coexistence curve relate to experiments described in the text. The dashed line represents the expected (see text) spinodal boundary.

with a commensurate point at $83^{O}C$. T_g drops with decreasing polymer concentration becoming invariant below 78%. This behaviour is depicted in relation to the phase diagram in Fig. 2a.

Gel setting occurred over the whole c range from a lower limit $(c_{gel})_{min} = 0.6\%$ up to 78%. The gelation temperature was found to be closely $50^{O}C$ over the whole of this range. Beyond 78% the setting effect was indistinguishable from conventional vitrification. All gels from 4% upwards retained their overall macroscopic dimensions on solvent exchange and drying, while the 1% gel shrunk. The latter, even if very weak, was more stretchable and elastic than the rest. Thus both the shrinkage and handling behaviour suggest a rubbery matrix below 4% and a glassy matrix at and above 4%.

SEM Morphology

Characteristic SEM photographs of gels obtained from different solutions on rapid cooling directly from the solution state, are shown by Figs 3,4,5,6. Taking extremes first, Fig. 6, from the most concentrated solution of the series (57%) is a clear case of the dilute phase forming droplets (seen as holes left by the solvent as it has evaporated) within a glassy matrix, in accordance with expectations (insert for c_3 Fig. 1). Fig. 3 from the most dilute solution (1%) is seen to consist of strings of small beads (1-2μ).We interpret this as arising from a gel consisting of glassy beads in a solvated matrix (insert for c_1 in Fig. 1), i.e. the inverse of Fig. 6 but affected by collapse on solvent removal. Between the two extremes, of dispersed glassy beads and dilute solution droplets, are the bicontinuous structures from intermediate concentrations. Fig. 5 is like Fig. 6 but with much larger holes, it corresponds to a rigid foam. The continuity of the glassy phase is obvious but that of the dilute phases is not, and depends on whether we have an open or closed cell foam.

The gel structures from the 2% and 4% concentrations are different from the above. The glassy phase, which gives it overall, (even if fragile) rigidity appears like a 3-dimensional, rather spiky wire network with the intervening empty spaces (the original dilute phase) also displaying continuity, Fig. 4. Thus, here the combination of dilute and concentrated phases are bicontinuous networks as expected (insert c_2 Fig. 1).

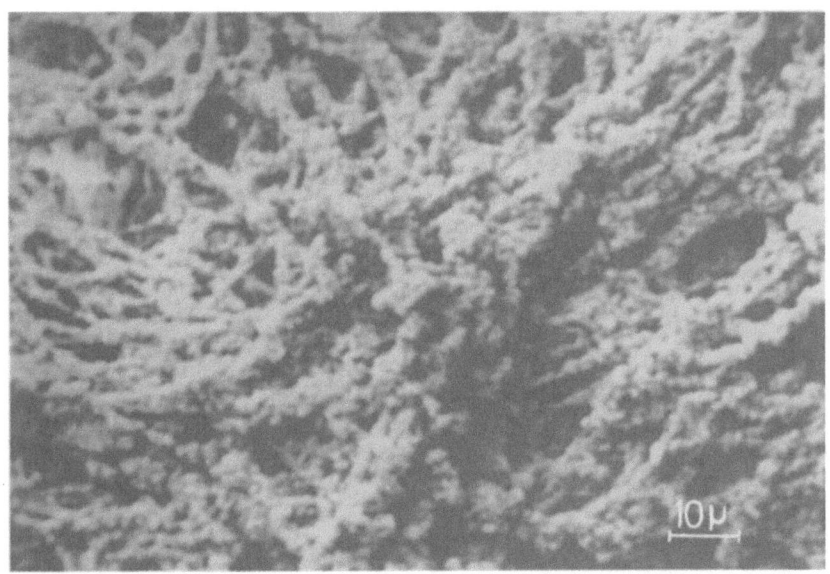

Figure 3.   Scanning electron micrograph of a 1% gel sample directly
quenched prior to freeze-drying.

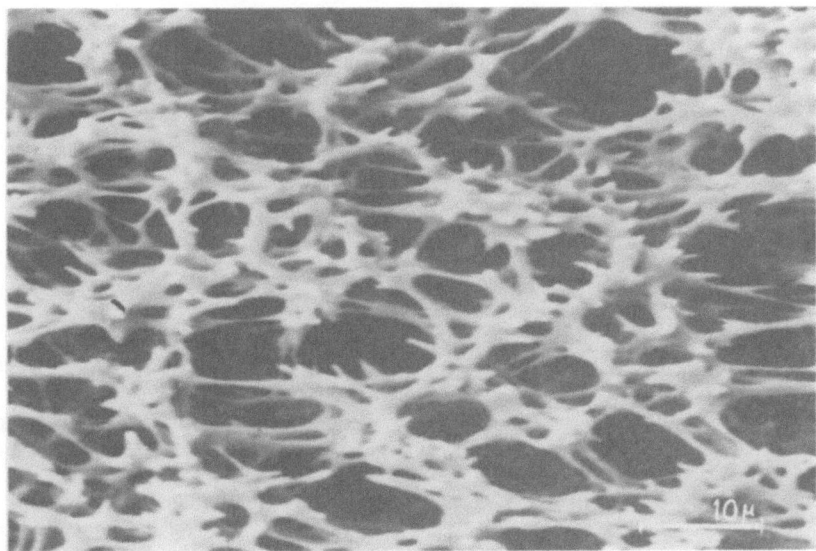

Figure 4.   Scanning electron micrograph of a 4% gel sample slowly cooled
prior to freeze-drying.

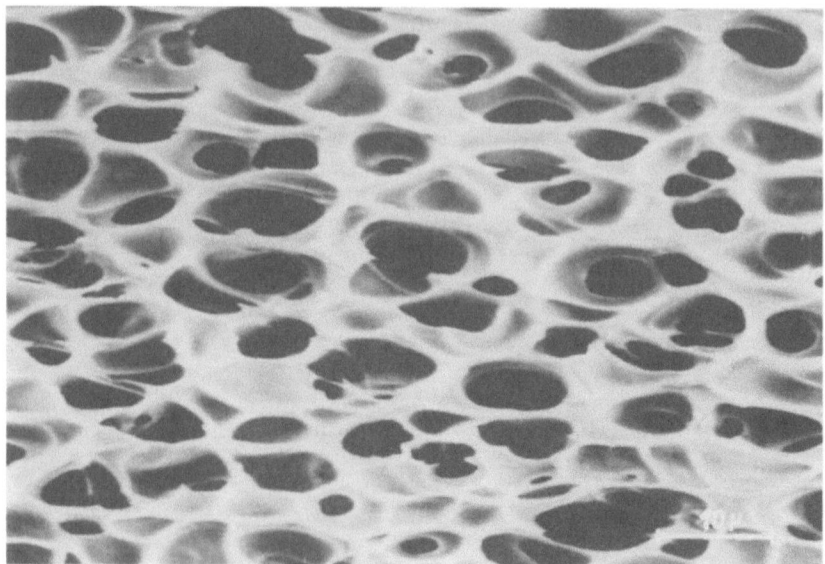

Figure 5.   Scanning electron micrograph of a 16% gel sample directly
             quenched prior to freeze-drying.

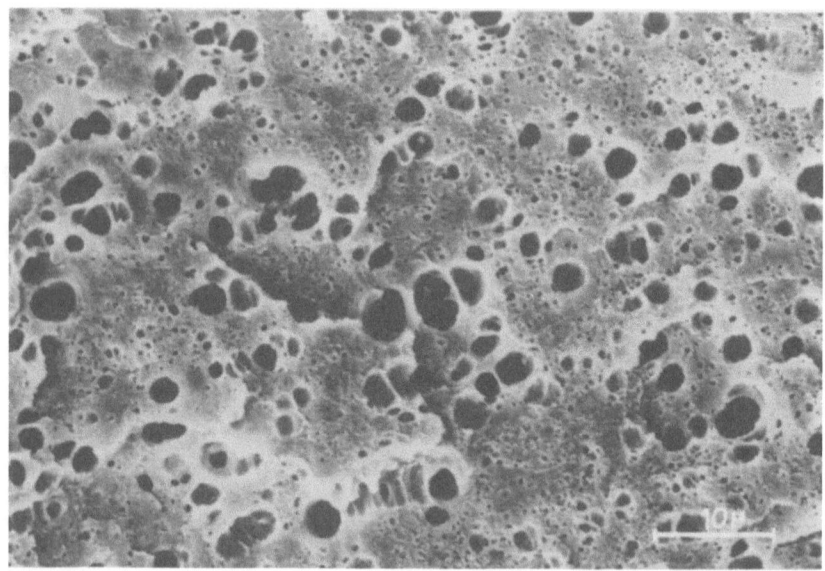

Figure 6.   Scanning electron micrograph of a 57% gel sample directly
             quenched prior to freeze-drying.

Referring to the tentative metastable – spinodal boundary line in Fig. 2b, its actual position is not known but it is expected with certainty that: i) at the commensurate point which is at 4% polymer, we are within the spinodal region at all temperatures below UCT. ii) At the extremes of concentration and at points lying closely within the cloud point line we should be in the metastable region where phase segregation is via nucleation. Figs 3,5 and 6 from the dilute and concentrated solutions are consistent with ii). The different character of the phase segregated structure in the 4% case of Fig. 4, where we know that the system at least must have passed through the spinodal region, induces us to suggest that the image seen could be at least a reflection of spinodal decomposition, if not the spinodal structure itself.

The above assignment of observed structures to nucleation and spinodal mechanisms respectively gains further strong support through some two step quench experiments. Here solutions are first held at temperatures below the cloud point curve but above the gelation temperature (i.e. T_g) so as to induce phase segregation without the arresting effect of vitrification. Fig. 7 shows a 2% sample held at 81°C for 5 minutes and reveals a two component morphology. The first consists of 10μm ellipsoids which we maintain formed during storage at 81°C via a nucleation and growth mechanism. The second component is a network of fine strings of beads ($\sim/\mu$). As seen it clearly resembles the gel structure from the 1% solution obtained by direct quenching (Fig. 3). We maintain that this structure, corresponding to the gel forming component, arises on quenching and originates from the dilute phase product of the phase segregation at 81°C. The large ellipsoids adhere to this network thus being supported by it. Both components are consistent with having arisen through nucleation and are distinctly different from the spiky network structure given by the same initial concentration on single stage quenching, Fig. 8. The implication is clear: at 81°C the 2% solution is within the metastable and at 50°C (T_g) it is within the spinodal region, i.e. in passing from 81°C to 50°C we have crossed the spinodal boundary.

Storage of a 4% solution produced no such drastic changes in morphology only a slight coarsening of the spiky wire structure. To sum up, on the basis of visual images we conclude that: for c = 1%, 16%, 57% we are always in the metastable region, for c = 2% we may be either in the metastable or spinodal region according to temperature, while for 4% we

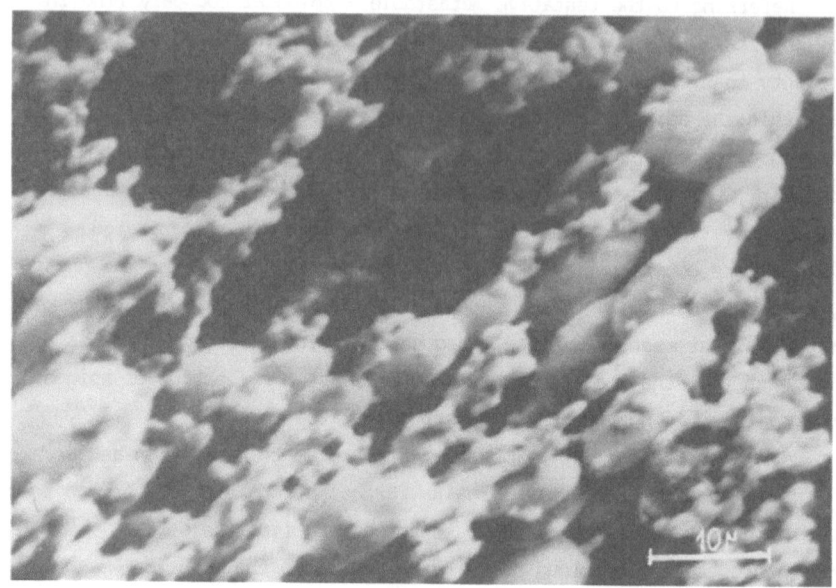

Figure 7. Scanning electron micrograph of a 2% gel sample stored at 81°C for 5 mins prior to freeze-drying.

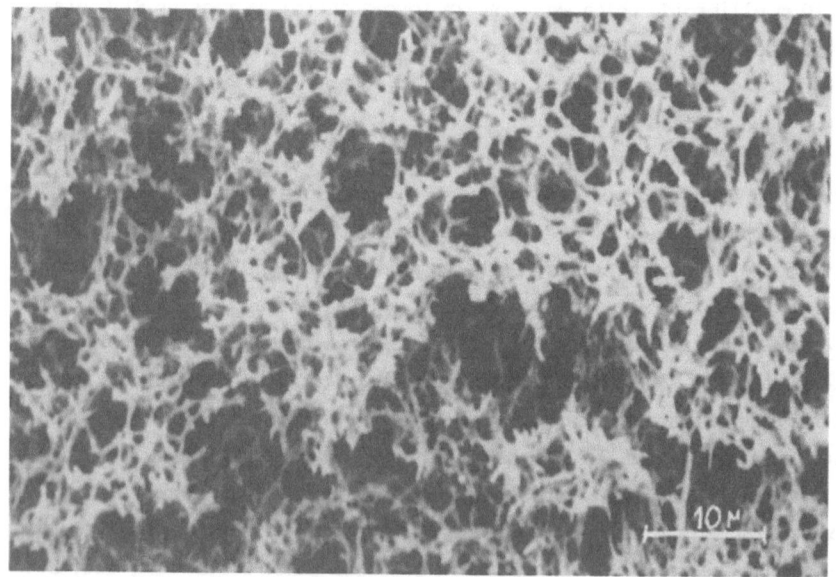

Figure 8. Scanning electron micrograph of a 2% gel sample directly quenched prior to freeze-drying.

metastable or spinodal region according to temperature, while for 4% we are in the spinodal region throughout. This is by no means a complete description of the range of work undertaken, for a full account see ref. 5, but it serves adequately to outline our proposed gelation scheme for a-PS.

WIDER ISSUES AND CONCLUSIONS

The concept of a spinodal mechanism has been invoked previously in connection with gelation of the more rod like molecules of gelatin and poly(benzyl glutamate) [6]. It was also mooted in broad general terms for a-PS in CS_2 [7] but not followed up when the suggestion of a liquid-liquid phase segregation had been abandoned altogether [3], we believe due to the difficult experimental nature of their system. The issue however, has recently come up in favour of the latter for a-PS in other solvents e.g. decalin [1] and cyclohexane [8]. In the latter work the phenomenon was being used to produce microporous low density foams, clearly this approach invites further exploration.

That a-PS can gel, and this via a liquid-liquid phase segregation, raises wider issues in the field of thermoreversible gelation of crystallizable polymers, e.g. i-PS [9], PVC [10], PVA [11] and PE [12]. Clearly, liquid-liquid phase segregation should at least be possible in these systems also; if so crystallization could then follow, possibly in the polymer rich phase. Indeed, such behaviour has recently been reported for PE in various solvents [13].

Within the framework of the present study the definition of a physical gel has widened considerably. Entropy elastic network elements, imparting rubbery characteristics, are now seen to be a limiting case restricted to the low concentration side of the phase diagram. At the high concentration side the connectedness is provided by a glassy matrix, imparting a glass-like property to the system. The same overall glassy character is displayed also at intermediate concentrations, but nevertheless with the strong suggestion of being generated by spinodal decomposition. It is hoped that by looking at it in this way the diversity of structures characterized by three dimensional connectivity, in the broadened sense of either molecular or phase continuity, can be brought together under the widened definition of physical gels.

ACKNOWLEDGEMENTS

We wish to thank Professor A. Berghmans, Leuven, for interim
information communicated to us before publication which substantially
benefited the present work. Two of us (R.M.H. and S.C.) wish to
acknowledge support from the Science and Engineering Research Council, and
(in the case of S.C.) also from Edeco Co. Ltd.

REFERENCES

1. Arnauts J., Berghmans H., Polymer Communications, 1987, 28, 66.

2. Keller A., in 'Structure-Property Relationships of Polymer Solids',
   Ed. A. Hiltner, Plenum Press, 1983, Amsterdam.

3. Tan H.,Moet A., Hiltner A. and Baer, E., Macromolecules, 1983, 16,
   28.

4. Miles M.J., in 'Developments in Crystalline Polymers', Ed. Bassett
   D.C., Vol. 2, Elsevier Science Publishers B.V., Amsterdam, 1987.

5. Hikmet R.M., Callister S., and Keller A., submitted to Polymer.

6. Miller W.G., Kon L., Tohyama K., and Voltaggio V., J. Polymer Sci.,
   Polymer Symposium, 1978, 65, 91.

7. Wellinghof S.T., Shaw J., and Baer E., Macromolecules, 1979, 12, 932.

8. Aubert J.H., Clough R.L., Polymer, 1985, 26, 2047.

9. Atkins E.D.T., Hill M.J., Jarvis D.A., Keller A., Sarhene E., and
   Shapiro J.S., Colloid and Polymer Sci., 1984, 262, 22.

10. Guerrero S.J., Keller A., Soni P.L., and Geil P.H., J. Polymer Sci.,
    Phys. Ed., 1980, 18, 1533.

11. Berghmans H., and Stoks W., in 'Interaction of Fundamental Polymer
    Science and Technology' Edts. Kleintjens L.A., and Lemstra P.J.,
    Elsevier Applied Sci. Publ., 1986, p.218.

12. Smith P. and Lemstra P.J., J. Mat. Sci., 1980, 15, 505.

13. Schaaf P., Lotz B., and Wittmann J.C., Polymer, 1987, 28, 193.

POLYELECTROLYTE-SURFACTANT COMPLEXES

H. Hoffmann, O.El Seoud, G. Huber and R. Bächer
Lehrstuhl für Physikalische Chemie I der Universität Bayreuth
FRG

ABSTRACT

The chloride ions in aqueous solutions of the polyelectrolyte Poly-diallyldimethylammoniumchloride (DADMAC) were substituted by different sur-factant and other small anions. Both perfluoro- and normal hydrocarbon surfactants were used for the substitution. The resulting complexes were water soluble as long as the alkylcarboxylate-surfactants had less than 7 CH_2-groups or 5 CF_2-groups, respectively. The compounds with larger alkylgroups precipated out. These complexes were however soluble in ethanol or other small chain alcohols or in alcohol-water mixtures. Clear isotropic solutions were obtained for low concentrations of the polyelectrolyte, while liquid crystalline phases were formed at high concentrations. A few phase diagrams were established. The aqueous polyelectrolyte and the soluble complexes were characterized by electric birefingence, light scattering and rheological measurements. The liquid crystalline phases were characterized by polarization microscopy.

INTRODUCTION

Polyelectrolytes with hydrophilic counterions are soluble in water. Their dissociation degree is dependent on the charge density along the chain and on the concentration [1]. When ionic surfactants are used as counterions the surfactants bind in a cooperative way on the polyelectrolyte and the charge of the polyelectrolytes is therebye completely compensated [2]. As a result the polyelectrolyte-surfactant complex usually precipitates. Such measurements have been reported by Dubin et al [3]. In such investigations typically surfactants with a chain lengths between 10 and 18 CH_2-groups are used.

We have started systematic investigations on polyelectrolyte-surfactant systems in order to find out what happens when the chainlength of the surfactant ion is increased stepwise from very small to larger chains. The acetate ion can be expected to behave like a hydrophilic counterion and the complex to be soluble while for typical surfactants precipitation occurs. Somewhere along the increasing chain length the system has to switch from the normal polyelectrolyte behavior to the behavior which is dominated by charge neutralisation. The increased neutralisation is likely to influence the conformation of the polyelectrolyte in solution. As a consequence of their high charge density normal polyelectrolytes in dilute solutions are assumed to be completely stretched and they are assumed to coil when their rotational volums begin to overlap. It is therefore conceivable that polyelectrolytes with strongly binding counterions could already be coiled in the dilute concentration region.

EXPERIMENTAL RESULTS

Our investigations were carried out on the Poly-DADMAC system which was a gift from the Hoechst AG/Werk Gendorf. The polyelectrolyte was received as a 35 % solution and used as received. When higher concentrations were needed the solution was freeze dried. The polyelectrolyte solutions were charactericed by light scattering, viscosity and electric birefringence measurements. Some parameters which were evaluated from the different techniques are given in Table 1.

TABLE 1

Characteristic parameters of Poly-DADMAC

light scattering:	molecular weight M_w = 35.000 g·mol^{-1} degree of polymerization P = 216 length of all-trans chain L_s = 1.090 A
viscosity:	intrinsic viscosity $[\eta]$ = 4.000 ml·g^{-1}
electric birefringence:	relaxation time τ = 15 μs corresponding length L_b = 1.100 A

The contour length which was calculated from the molecular weight as obtained from the light scattering value agree remarkably well with the length of the polyelectrolyte which was determined from the time constant of the decay of the electric birefringence.

For a given voltage the birefringence passes over a maximum with increasing concentration (Fig. 1). The maximum corresponds to the crossover from the dilute to the semidilute concentration region. Similar results have been observed before on an other polyelectrolyte by W. Oppermann [4].

The decrease of the birefringence with increasing concentration is usually interpreted as the onset of the coiling of the polyelectrolyte. This can also be seen from the specific Kerr constant and the relaxation time of Poly-DADMAC (Fig. 2). For low concentrations the specific Kerr constant and the relaxation time are nearly independent of the polymer concentration, which indicates, that the length of the molecule does not change in this dilute concentration range. Here the length of the molecule corresponds to

the fully stretched polyelectrolyte chain. For higher concentrations the
specific Kerr constant and the relaxation time decreases indicating a
steady coiling of the chain. The onset of this decrease corresponds to the
concentration range, in which the rotational volumes of the single
polyelectrolyte molecules begin to overlap. In the semidilute concentration
range a much greater decrease of the relaxation time would be expected from
the calculated mean distances between the single molecules. This indicates
that the relaxation process in this concentration range is strongly
influenced by the increasing electrostatic repulsion between the polyelectrolyte molecules.

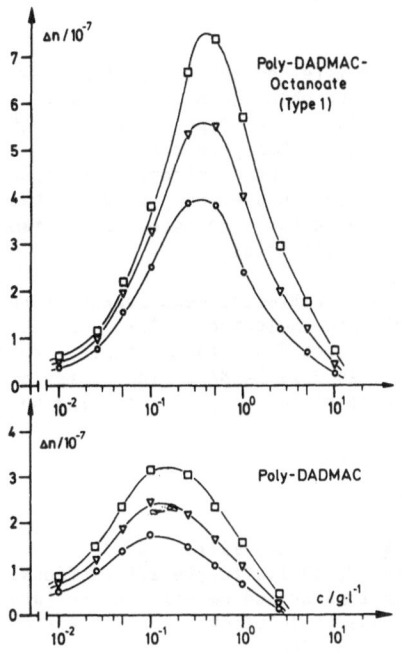

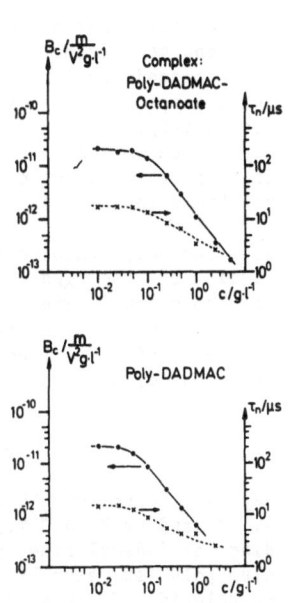

Figure 1.
Electric birefringence Δn of
Poly-DADMAC and Poly-DADMAC-
Octanoate (Type 1) as a function of the concentration c
for various values of the
electric field strength E.

O : E = 0.71·10⁶ Vm⁻¹
▽ : E = 0.95·10⁶ Vm⁻¹
□ : E = 1.19·10⁶ Vm⁻¹

Figure 2.
Specific Kerr constant B_c and
relaxation time τ_n of Poly-
DADMAC and Poly-DADMAC-Octano-
ate (Type 1) as a function of
the concentration c.

When anionic surfactant solutions are added to the polyelectrolyte there is immediate precipitation when the chainlength is above a certain value. For hydrocarbon surfactants the octanoate-system was still soluble while the decanoate system precipitated from the aqueous phase as an amorphous solid. For the perfluoro-surfactant even the perfluoroctanoate precipitated out. These solids can be dissolved again in organic solvents, or in particular in intermediate chain alcohol/H_2O mixtures. For a high complex concentration the solutions are birefringent. The texture under the polarisation microscope of the liquid crystalline phase was fine coarse but did not show well developed characteristic patterns as normal lyotropic liquid crystalline phases do. In the system with precipitation the surfactant begins to bind to the polyelectrolyte at very small concentrations. The binding was studied with ion selective electrodes. For the sodium dodecylsulfate the e.m.f. - concentration plot gives a clear break at a concentration of $1 \cdot 10^{-5} M$ SDS what is almost two orders of magnitude below the CMC for SDS in the presence of 0.1 M NaCl.

We studied in detail solutions of polyelectrolyte octanoate system. Two types of solutions were prepared. In Type 1 the chloride of the polyelectrolyte was replaced by ion-exchange procedure. In Type 2 equivalent amounts of Na-octanoate were added to Poly-DADMAC. The Type 2 solutions are single phase isotropic up to 40% of the complex. The isotropic solution borders on a two phase system which shows a l.c. phase. Surprisingly, the solution of Type 1 did not give a l.c. phase but separated into two isotropic solutions for smaller surfactant concentrations.

Electric birefringence, rheological and light scattering measurements were performed again on the single phase isotropic solutions. Some of the results are also given in Fig. 1 and 2. As seen from the figures, the octanoate system behaves very much similar as Poly-DADMAC. There is no cooperative binding of the octanoate on the polyelectrolyte and coiling was not observed in the dilute range.

REFERENCES

1. Katchalsky, A., Polyelectrolytes. Pure Appl. Chem., 1971, 26, 32
2. Hayakawa, K. and Kwak, J.C.T., Surfactant-polyelectrolyte interactions. 1. Binding of dodecyltrimethylammonium ions by sodium dextran sulfate and sodium poly (styrenesulfonate) in aqueous solution in the presence of sodium chloride. J. Phys. Chem., 1982, 86, 3866
3. Dubin, P.L. and Davis, D., Stoichiometry and coacervation of complexes formed between polyelectrolytes and mixed micelles. Colloids and Surfaces, 1985, 13, 113
4. Oppermann, W., Untersuchungen zur molekularen Gestalt gelöster Polyelektrolyte mit der Methode der elektrischen Doppelbrechung. Habilitationsschrift, Technische Universität Clausthal, 1986

THE PHOTOACOUSTIC EFFECT FOR THE POLYMERIC NETWORK GENERAL THEORY.

Szorek Ryszard

Silesian Center of Cardiology, 41-800 Zabrze, Poland.

ABSTRACT

In the papers recently published, elsewhere, some theoretical aspects of the Photo-acoustic Effect only for the linear polymer chain has been presented. Now, we present its general form obtained for general type of a polymeric network. Starting from the diffusion equation obtained in the form respective for intended research goal, the relaxational bands λ_1 corresponding to the excitation process of an arbitrary taken linear polymer chain and the like-phantom polymeric network to 1st level were found. The calculations of the λ_1 were carried out under assumption that the excitation process is generated by the action of the acoustical periodic field with the relatively high amplitude, A and the frequency, Q coherent to the wave vector lying in the region: $k < k_0$. Futhermore, the boundary conditions and some structural defects of the network were taken into account. Hence, the λ_1 is obtained by

$$\lambda_1 = QA/2KT\left[\langle\Psi|e^{i\varrho h}h|\varrho_0\rangle\langle\Psi|e^{i\varrho h}\nabla_h|\varrho_0\rangle + \langle\Psi|e^{i\varrho d}d|\varrho_0\rangle\langle\Psi|e^{i\varrho d}\nabla_d|\varrho_0\rangle\right]$$

where: h, d - coordinates.
Due to the fact that these bands may appear only after fullfiling the following selection rule
$$k = Q \pm q$$
where: Q, q - fonons and fotons wave vectors, respectively, consequences of the fonon-foton (acoustical) transformation process were considered.
Assuming that the fotons created are completly absorbed by the polymer material and that they produce an excited states of the polymer valence electrons, the expression for the dielectric constant dispersion curve was possible to derived.

INTRODUCTION

As , it was earlier anounced in the papers published, elsewhere [1-3], discussion of a polymer network dynamics at θ condition and, for the wave vector lying in the region: $k < k_0$ must necessary involve consideration of the valence electrons dynamics and, eventually, the electrons-network interactions. Thus, the self-states of this system are described by the eigensolutions of the following equation

$$\frac{\partial \Psi}{\partial t} = H\Psi \qquad (1)$$

where: $H = H_{NET} + H_{EL} + H_{INT}$ $\qquad (2)$
H_{NET} - the diffusional operator obtained in the form expressed by eq.(6)

H_{EL} - the Hamiltonian characteristic for the polymer valence electrons
H_{INT} - operator of the electron-polymeric network interactons.
The· eigenproblem of this equation was solved separately for two following cases: firstly, for the linear polymer chain, and, secondly, for the two- dimensional polymeric network.

THE LINEAR POLYMER CHAIN DYNAMICS
Model
The hopping bead-spring model consists with the N+1 beads (monomeric units) immersed in the θ solvent and fixed on the infinite rods parallel to the Y direction, along which they are slided. The rods simulate a stiffnes against strain of the polymer backbone. The distribution of the beads as well as the rods are governed by the random walk statistics. However, the beads distribution is only changed during the local motion of polymer chain. Unfortunately, the polymer chain motion is produced by the action of the diffusional force on the several beads and, hence, the polymer dynamics is described by the Langevin-Diffusion equations system. At last, it was assumed that the interbead interactions are described by the following potential

$$U = \sum_n U(h_n) \qquad (3)$$

where: $h_n = 2 b_n \cdot c_n$; $b_n = y_n - y_{n-1}$, $c_n = y_n - y_{n-2}$,
y_n -nth beads coordinate,
and that the polymer valence electrons are weackly connected with the polymer backbone having both the bonds and the angles non-equivalence. On the basis of this model it was shown that the acoustic periodical field, with the relatively high amplitude, A and the frequency, Q coherent to the wave vector of the region: $k < k_0$, acting on the linear polymer chain may excites it to the higher levels than the equivalence levels obtained from the non-perturbational form of diffusion equation. Then, by putting into the ordinary form of diffusion equation obtained by

$$\frac{\partial P(y_0, \ldots, y_N, t)}{\partial t} = \frac{1}{f} \sum_{n=0}^{N} [\frac{\partial P}{\partial y_n} \frac{\partial U}{\partial y_n} + P \frac{\partial^2 U}{\partial y_n^2} + KT \frac{\partial^2 P}{\partial y_n^2}] \qquad (4)$$

where: $P = \varrho \cdot \exp(-U/2kT)$ - the beads distribution function,

the following perturbation

$$U' = U - f A c^{iQh} \cdot h \qquad (5)$$

we hawe

$$\frac{f}{KT}\frac{\partial \varrho}{\partial t} = \sum_{n=0}^{N}\left\{\left[\frac{\partial^2 \varrho}{\partial y_n^2} + \frac{1}{2KT}\left[\frac{\partial^2 U}{\partial y_n^2} - \frac{1}{2KT}\left(\frac{\partial U}{\partial y_n}\right)^2\right]\varrho + \frac{fA}{KT}a_{n,n+2}(1+iQh)e^{iQh}\left(\frac{\partial}{\partial y_n} - \frac{1}{2KT}\frac{\partial U}{\partial y_n}\right)\varrho + \frac{fAQ}{KT}a_{n,n+2}^2(2i-Qh)e^{iQh}\cdot\varrho\right\} \quad (6)$$

Hence, after introducing new variables of the type: h_n, into eq.(6) we find

$$\frac{f}{KT}\frac{\partial \varrho}{\partial t} = \sum_{n=1}^{N}\left\{\left[\left(\frac{\partial^2}{\partial h_n^2} + \frac{1}{2KT}\left[\frac{\partial^2 U}{\partial h_n^2} - \frac{1}{2KT}\left(\frac{\partial U}{\partial h_n}\right)^2\right] - \frac{\partial^2}{\partial h_n \partial h_{n+2}} + \frac{1}{(2KT)^2}\frac{\partial U}{\partial h_n}\cdot\frac{\partial U}{\partial h_{n+2}}\right]\varrho + \left[\frac{fA}{KT}(1+iQh)e^{iQh}\cdot a_n\frac{\partial U}{\partial h_n} + \frac{fAQ}{KT}(2i-Qh)e^{iQh}a_{n,n+2}^2\right]\varrho\right\} \quad (7)$$

where: a_n, a_{n+2} -a proportionality factors
A -the complex amplitude of the acoustic field applied.

Covering both sides of eq.(7) with the ϱ function obtained in the following form

$$\varrho_P = \varrho_0 + \sum_{\ell=1}^{\infty}a_\ell \psi(L,K) \quad (8)$$

where: $\psi(L,K) = \frac{1}{\sqrt{N}}\sum_{n=1}^{N}|\xi_\ell(h_n)\rangle e^{iKn}$

ξ_ℓ -the eigenfunctions of the stationary part of eq.(7)

$\varrho_0 = \prod_{n=1}^{N}|\xi_\ell(h_n)\rangle$ -the ground state function

f - the Stokes' friction coefficient
a_ℓ -a proportionality factor,

and integrating it, we find

$$\lambda_1 = \frac{4\varrho}{f}(1-\cos 2K) + \frac{QA}{2KT}\left[\langle\psi|e^{iQh}h|\varrho_0\rangle\langle\psi|e^{iQh}\cdot\varrho|\varrho_0\rangle\right] \quad (9)$$

where:a - oscillators constant.
Futhermore, taking into account the earlier assumption about the dominant role of the mentioned above perturbation, the last expression keeps the second component, only. As, it may be seen, the λ_1 band obtained by eq.(9) expression may be expressed only after fullfiling the following selection rule

$$k = Q \pm q \quad (10)$$

Due to this fact the fonon-foton transformation process should be occurred. According to that, it was assumed that the fotons created in this process are completly absorbed by a polymer chain, proceeding to the valence electron excitation from the ground state to higher levels. Because, these electrons are taken to be weackly connected with the polymer chain backbone , their self-stationary states are obtained by the solutions of the following equation

$$i\hbar\frac{\partial\phi}{\partial t} = H_{EL}\phi \quad (11)$$

where: $H_{EL} = -\hbar^2/m_e\left\{\sum_{n=1}^{N}\left[\frac{\partial^2}{\partial r_n^2} + W(r_n)\right] + \sum_{n\neq m}\sum W(r_{nm})\right\} \quad (12)$

$r_n = y_n - y_n'$; y_n' -nth electron coordinate of the equilibrium position
m_e -electrons mass

$\phi = \phi_0 + \sum_\ell a_\ell \psi(L,q)$-the eigenfunctions of H_{EL} operator.

Unfortunately, in the case when the following perturbational field acts on the system

$$E = E_0 c^{i\omega t} \tag{13}$$

where: E -the intensity of the electric field induced by the acoustic periodical field ($E_0 \sim P_A$) [5]

the equation (11) takes the form

$$i\hbar \frac{\partial \phi}{\partial t} = (H_{EL} - eyE)\phi \tag{14}$$

where: e -the electrons elementary charge.
Hence, using the ϕ function expressed in the perturbational form we were found the formulae for the $\mathcal{E}(\omega)$ and the H_{INT} operator contribution to the self-values of H operator. Finally, it is worth to note, that all calculations above presented have been carried out for the polymer chain closed in a loop. It was appeared, that the same solution are also available for the opened relatively long polymer chain with both ends fixed at the equilibrium position. [4].

THE LIKE-PHANTOM POLYMER NETWORK DYNAMICS.

Model
The polymer chain model extends on two dimensional network having the (N+1)× (N+1) beads hung on the infinite rods parallel to the Y direction along which they slide. Then, the network is constructed in the "crossward" form in which crosses correspond to junctions (beads) governed by the random walk statistics. However, the beads distribution function can be changed during the local motion of network in the Y direction, only. Due to this fact and to assumption that the network motion is produced by the fluctuation forces, this motion , at the Θ conditions, is described by the diffusion equation obtained in the form characteristic for the potential

$$U = \sum_{n=1}^{N} U(h_n, d_n) \tag{15}$$

where:
$d_n = 2b'_n \cdot c'_n$; $b'_n = y'_n - y'_{n+1}$; $c'_n = y'_n - y'_{n-2}$; y'_n -nth beads coordinate,
hence, we found the following expression for the λ_1

$$\lambda_1 = QA/2KT \left[\langle \Psi | e^{i\varrho h} \cdot h | \varrho_0 \rangle \langle \Psi | e^{i\varrho h} \nabla_h | \varrho_0 \rangle + \langle \Psi | e^{i\varrho d} d | \varrho_0 \rangle \langle \Psi | e^{i\varrho d} \nabla_d | \varrho_0 \rangle \right] \tag{16}$$

However, this solution is characteristic only for the regular polymer network. While, in the case when the defected network is taken into consideration the eigenproblem of diffusion equation is solved by using Greens function , researching within its particular points which simultaneously are taken to be the point defects in the network considered. On the other hand, the self-states of the polymer valence electrons were similarly obtained, as those obtained for the polymeric network. Also, in this case the expression $\mathcal{E}(\omega)$ (expression) was derived in close analogy to the previous case.

CONCLUSION

On the basis the calculations presented above we may
conclude that the action of the ultrasonic field with the
relatively high amplitude, A and the frequency, Q coherent
to the wave vector from the region: $k < k_o$ on the polymeric
network has to involve the electric dispersion phenomena.

REFERENCES

1.Szorek,R.,IUPAC MACRO'85, Hague, 1986
2.Szorek,R., SPECIALITY POLYMERS'86, Baltimore, 1986
3.Szorek,R., J.Chem.Phys., in press
4.Zimm,B.H., J.Chem.Phys., 24 (2) ,1956, p.269
5.Kittel,C., Introduction to Solid State Physycs,
   John Wiley & Sons, Inc., N.Y., 1966

Part 5

CRYSTALLIZATION

ANALYSIS OF CONDIS CRYSTALS

Bernhard Wunderlich

Department of Chemistry

Rensselaer Polyechnic Institute

Troy, NY 12180-3590, USA

ABSTRACT

Conformationally disordered crystals (condis crystals) may present the most important mesophase of linear macromolecules. To distinguish condis crystals from the other two mesophases, liquid crystals and plastic crystals, one notes that there are no rigid rod or disc-like mesogens as one finds in liquid crystals, and there are no close-to-spherical molecules as one finds in plastic crystals. The liquid-like translational motion and positional disorder, characteristic of liquid crystals, is absent in condis crystals. Similarly, the orientational disorder and rotational motion of the molecules in plastic crystals is not present in condis crystals. The analysis of condis crystals is based on thermal analysis to find the degree of conformational disorder, on X-ray diffraction to elucidate the structure and limits of intramolecular motion, on Raman and infrared spectroscopy to evaluate vibrational motion, and nuclear magnetic resonance for detailed analysis of the motion in the condensed state. Discussed examples will be: cycloalkanes of 12 to 26 ring atoms, polyethylene, paraffins of 9 to 44 chain atoms, polytetrafluoroethylene, poly-trans-1,4-butadiene, polyparaxylylene, and oligomeric polyphenylenes.

INTRODUCTION

In a recent review of thermotropic mesophases and their transitions [1] it was suggestes that a crystal with dynamic conformational disorder does not fit into the standard classification of mesophases as either plastic crystals (with orientational disorder and long-range positional order) or liquid crystals (with positional disorder and some long-range orientational order). At that time it was proposed that a third mesophase exists, that of the condis crystals (with conformational disorder and long-range positional and orientational order). Figure 1 illustrates schematically the various condensed phases

Figure 1. Schematic diagram of the relationship between the three limiting phases in double-outline and the six mesophases.

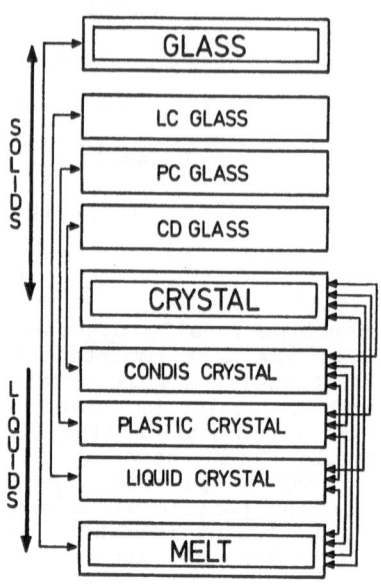

and their interrelations. Condis crystal, plastic crystal, and liquid crystal are increasingly less solid (more liquid). The arrows on the right-hand side indicate the possible transitions between the phases that are often first order transitions, i.e. transitions that involve a discontinuity in the first derivative of the Gibbs energy function, the entropy. The entropy incrreases going to successively lower states drawn in Fig. 1 when starting with the crystal. The overall entropy of fusion of a crystal can be separated approximately into contributions arising from positional disordering, orientational disordering, and conformational disordering so that

$$\Delta S_f = \Delta S_{pos} + \Delta S_{orient} + \Delta S_{conf}. \tag{1}$$

It is possible to make a first judgment of a phase classification based on the thermal analysis of the first order transitions. For a more detailed description of structure and motion methods such as X-ray, electron and neutron diffraction; IR-, Raman- and NMR-spectroscopy are needed.

In this discussion an attempt will be made to describe in greater detail the structure and motion for several condis crystals. A special effort will be made to point-out the differences between condis crystals on the one hand, and liquid and plastic crystals on the other. It seems reasonable, and has been illustrated on several examples [2], that molecules with dynamic conformational disorder in the liquid state show such conformational disorder also in the liquid crystalline and plastic crystalline states. The major goal in distinguishing condis crystals from other mesophases is thus the identification of positional mobility and disorder of the molecular center of gravity in the case of liquid crystals, and of molecular rotation in the case of plastic crystals. To distinguish condis crystals from classical crystals will prove more difficult than initially thought. It has been shown in the last 20 years through NMR analysis of crystals that rotational and even translational jumps of motifs are possible far below the melting or disordering transition temperatures. Several examples will be discussed in detail. If such jumps lead to identical symmetry positions, there is little change detectable through structure analysis. Without entropy change there is no latent heat to be found by thermal analysis. Direct information on the state of motion through solid-state NMR must thus always be coupled with structural analysis and thermal data to safely identify the nine condensed phases listed in Fig. 1.

The thermal analysis data on cycloalkanes are summarized in Figure 2. It is quite evident that cycloalkanes that show plastic crystallinity have a more or less constant ΔS_d per mole, i.e. ΔS per CH_2 as plotted in Fig. 2 decreases rapidly with ring size. The condis crystals, on the other hand, approach a constant ΔS_d per CH_2. Conformational mobility is proven by second moment and T_1 data of ^{1}H NMR and by line-shape analysis of ^{13}C MAS NMR and ^{2}H NMR. In connection with X-ray structure data it can be shown that the molecules may jump below the disordering transition T_d between symmetry related postions (C_3H_6, C_6H_{12} - $C_{12}H_{24}$). The larger ring-sizes show conformational mobility during these jumps (rotoreptation). In the plastic crystalline state (molecules C_3H_6 - C_9H_{18}) the conformational disorder is similar to that in the liquid. In the condis phase ($C_{12}H_{24}$ - $C_{34}H_{68}$) the rings are increasingly

Figure 2.

Plot of the entropy-change per CH_2 for the cycloalkanes. Filled circles: total entropy change (of disordering and of fusion). Open circles: the entropy-change to the plastic or the condis crystal states, ΔS_d.

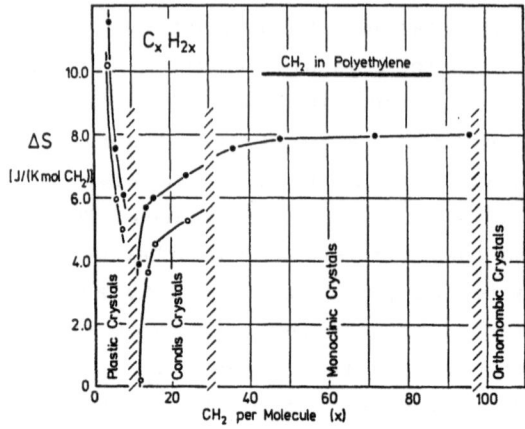

hindered by the long trans sequences that pack in a monoclinic crystal. For $C_{38}H_{76}$ to about $C_{96}H_{192}$ the equilibrium monoclinic phase shows no condis crystallinity. The higher ring-sizes have an orthorhombic polyethylene-like crystal structure and condis crystallinity is coupled with the appearance of a hexagonal phase. This hexagonal phase is known at high temperature and pressure for polyethylene and for paraffins of 9 to 43 chain atoms. Irradiation or copolymerization can also stabilize the hexagonal condis phase to much lower temperatures. Finally it was observed in superheated fibrillar crystals in the stretched condition. Analysis of the dynamic kink concentration seems to indicate a continuous increase with temperatue and cross-section of the CH_2-stems from much less than 1 % in paraffins (excluding surface gauche conformations) to about 30 % in the high temperature and pressure condis crystals of polyethylene.

The more gradual development of condis crystallinity with only a small or no thermodynamic transition is illustrated in Figure 3 for the homologous series of the polyparapheny-lenes. Most of the evidence for the conformation disorder comes from X-ray and neutron scattering analysis and NMR data that could identify the large-angle libration between the conforma-tion isomers within the crystals. The connection to the liquid crystalline states is obvious. Similarly, it can be found for many typical low molecular weight liquid crystals that their

333

Figure 3. Transition temperatures of seven polyparaphenylenes. Open circles: Low entropy transitions. In parentheses: weak or questionable transitions in a thermal analysis. Shaded areas are change in motion indicated by NMR.

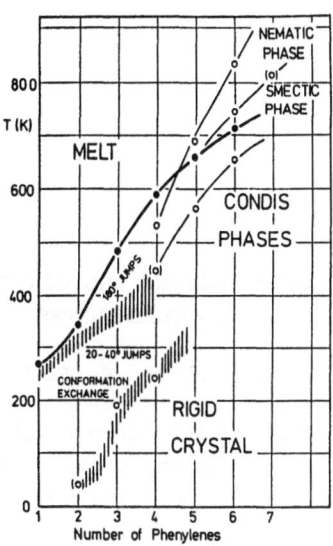

low temperature crystal forms permit still dynamic conformational disorder.

The high temperature crystal of polytetrafluoroethylene, poly-trans-1,4-butadiene and polyparaxylylene can be related easily to the condis crystals of the paraffins and the polyparaphenylenes.

Acknowledgments: Much of this work was supported at RPI by the Polymers Program of the US National Science Foundation (Grant Number DMR 83-17097) and at the U. Freiburg by the Deutsche Forschungsgemeinschaft within the SFB 60. The cooperation, discussion and ultimate writing of this paper was generously helped by the Alexander von Humboldt Stiftung through a Senior US Scientist Award (B.W., 1987).

REFERENCES

1. Wunderlich, B. and Grebowicz, J., Adv. Polymer Sci., 1984, 60/61, 1.

2. A detailed listing and correlation of the hundreds of references needed to document these conclusions are prepared for publication in the Adv. Polymer Sci., 1988.

CHAIN FOLDING AND CONFORMATIONAL DISORDER IN CYCLOALKANE CRYSTALLITES - A MODEL FOR POLYETHYLENE

M. Möller*, R. F. Waldron[+], H. Drotloff, G. Kögler

Institut für Makromolekulare Chemie, Universität Freiburg
Hermann-Staudinger-Haus, Stefan-Meier-Str.31
D-7800 Freiburg, Federal Republic of Germany

SYNOPSIS

Chain folding and conformational disorder in crystallites of cyclodoheptacontane and cyclohexanonacontane have been studied by high resolution solid state NMR and differential scanning calorimetry. The melt and solution crystallized samples differ with respect to the packing of the all-anti zig-zag stems. For the solution crystallized modification, a well defined ggagg fold was found. In the orthorhombic melt crystallized modification the adjacent reentry fold is strained and conformationally disordered. As a consequence, the arrangement of the methylene groups adjacent to the lamellar surface is disturbed. The fraction of the ordered inner segment changes as function of temperature. The strained adjacent reentry fold and the temperature dependent dynamic disordering of the chain segments near the lamella surface may explain controversial observations of both regular adjacent reentry and disordering in the fold layer of polyethylene crystals.

INTRODUCTION

The discussion of polyethylene`s crystallization, melting, and solid state structure has been intensive and controversial. Part of the discussion dealt with the question of whether the folding can be predominantly a regular array of adjacent folds, or whether the folds are loose and disordered[1-3]. Electron microscopy investigations of solution crystallized polyethylene specimen and epitaxial crystallization on polyethylene crystals are indicative of a regular orientation of the folds [4,5]. However, conclusive evidence of an adjacent reentry has been re-

---

[+]present adress: Yale University, Department of Chemistry, New Haven, Connecticut, USA

ported only recently. Neutron scattering and IR-experiments on solution crystallized polyethylene samples are in agreement with a structure with approximately 75% adjacent reentry folds[6,7]. Well defined models for the structure of an adjacent reentry fold have been developed but were never verified[8,9]. Moreover there is increasing evidence that the lamellar surface of polyethylene single crystals with adjacent reentry is partially disordered[10].

n-Alkanes and cycloalkanes are well defined models for investigating the crystallization of polyethylene. Because of the cyclic structure, cycloalkanes invariably form lamellar crystals in which the adjacent stems are connected by tight folds. Consequently, they can provide detailed information about the formation, structure, and thermodynamics of crystallites in which tight folds form the lamellar interface. Experimental studies comparing cycloalkanes with polyethylene have been reported[11,12]. However, the analogy is not perfect and must be discussed in detail. Cycloalkanes with less than 100 carbon atoms preferentially form crystals in which the methylene groups are packed in a monoclinic subcell[13,14], while the most important and most discussed modification of polyethylene has an orthorhombic subcell[1]. In the case of the monoclinic cycloalkanes, a tight fold with a well defined ggagg conformation connects two parallel all-anti arranged chains by a distance of 4.2 Å (200 fold)[14,15]. There is strong evidence in the case of polyethylene that the fold connects two antiparallel zig-zag stems in the 110 plane[4,5] which are separated by 4.5 Å[1]. This implies that the fold configuration in the two cases is not the same.

RESULTS AND DISCUSSION

Recently we reported NMR and calorimetric data on a series of cycloalkanes, $(CH_2)_n$, with up to 96 carbon atoms[16]. In agreement with X-ray studies by Trzebiatowski and Strobl[17], the solid structure of rings with 72 and 96 methylene groups was shown to be different in solution versus melt crystallized samples. The methylene groups in the solution crystallized modification are packed monoclinically, while there is evidence of an orthorhombic subcell in samples crystallized from the melt. Melting occurs in one step in the solution crystallized modification. The entropy of fusion per methylene group is about 20% less than the ΔS value reported for polyethylene[1]. In contrast, a solid-solid transition is observed in the melt crystallized modifications, which corresponds to a change to a hexagonal structure[17]. Most remarkable is that the enthalpies of fusion decrease by more than 40% from the values of polyethylene or the solution crystallized modifications.. The samples are well crystallized and do not contain an amorphous fraction which could in principle explain the differences. Neither careful annealing below nor above the solid-solid transition leads to different results. The decrease in ΔH_i is not balanced by the solid-solid transition s ΔH_d, and is therefore not simply a result of splitting-up the one-step melting of the solution crystallized modification into a two step process in the melt crystallized samples.

| | $C_{72}H_{144}$ | | $C_{96}H_{192}$ | |
	melt crys.	solut. crys.	melt crys.	solut. crys.
T_d[a]	353.1	-.-	363.9	-.-
ΔH_d[b]	21.9	-.-	36.7	-.-
ΔS_d[c]	0.86	-.-	1.05	-.-
T_i	378.1	380.2	387.1	389.1
ΔS_i	4.60	8.05	4.46	8.03
ΔH_i	117.0	220.3	168.0	302.0

[a]temperatures in K, [b]ΔH per molecule in kJ mol^{-1}
[c]ΔS per methylene unit in J K^{-1}mol^{-1},

Following the fact that there is certainly no difference in the entropy after reaching the melt, the entropy of the "orthorhombic" modification must be considerably larger than the entropy of the solution crystallized sample. Because entropy is correlated with disorder, the "orthorhombic" modification is thus, at least at high temperatures, less ordered than the monoclinic modification.

The question then arises as to whetherthe solution crystallized and melt crystallized samples have different specific heats. Figure 1 shows DSC data on the specific heat of the melt and the solution crystallized modification of cyclohexanonacontane. As expected the heat capacity becomes identical above the transition to the isotropic melt. However, throughout a considerable temperature range below the melt transition, a higher heat capacity of the melt crystallized sample is evident. The two curves converge below 250 K. Integration of the heat capacities between 248 K and 423 K results in a total enthalpy difference of 34.8 kJ mol^{-1} between the melt and the solution crystallized samples. Thus, with respect to the states existing at 248 K, the difference in the transition entropies upon melting is considerably smaller than calculations from the transition enthalpies alone suggest. For the melt crystallized sample, the experimental data indicate that here is a continous ordering of a partially disordered structure as the temperature is decreased further. However, the perfection of the "orthorhombic" crystals is not equivalent to the monoclinic modification even at very low temperatures.

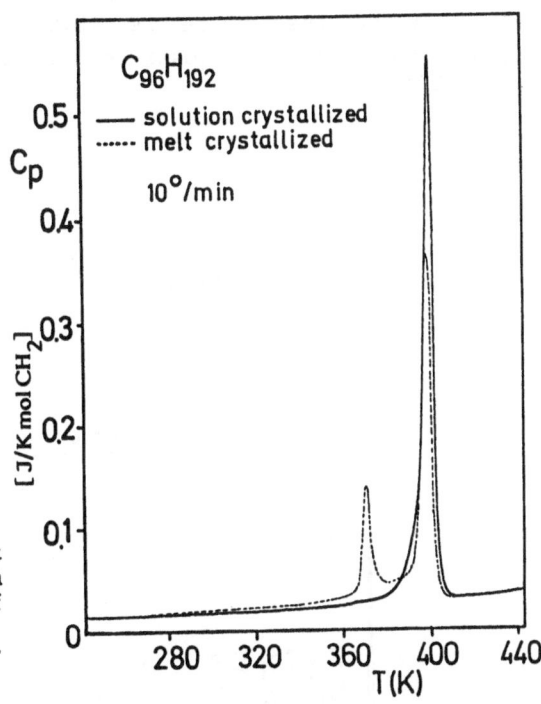

Fig. 1:
Heat capacities of a melt crystallized and solution crystallized sample of cyclohexanonacontane. (For the experimental details see ref. 17).

The different arrangement of methylene units in an orthorhombic and a monoclinic subcell does not explain the variations discussed above. Hence, it is tempting to attribute the thermo-analytical observations to partially disorderd adjacent reentry folds in the melt crystallized modification.

High resolution ^{13}C-NMR experiments can yield detailed information on the molecular structure and packing of hydrocarbon molecules in the solid state including information on the chain conformations in the fold. Figure 2 shows high resolution ^{13}C-NMR spectra of the different modifications of cyclohexanonacontane. The spectra were obtained by magic angle sample spinning and high power decoupling techniques as described previously[16]. The room temperature spectrum of a solution cry-

stallized sample is shown on the left side, with. the main
signal at 34.7 ppm belonging to the all-anti arranged stems.
The low intensity signals can be assigned to the well defined
ggagg fold described for monoclinically packed cycloalkanes
(see ref. 16,18 for details of the assignment). A different
picture emerges rrom the spectrum of the melt crystallized
modification. At high temperatures when the sample is in the
hexagonal modification, one sharp signal is observed. The
spectrum does not discriminate between stems and folds. ^{2}H-NMR
data indicate that there is significant and fast molecular
motion similar to that observed in the rotator phase of normal
alkanes[19].

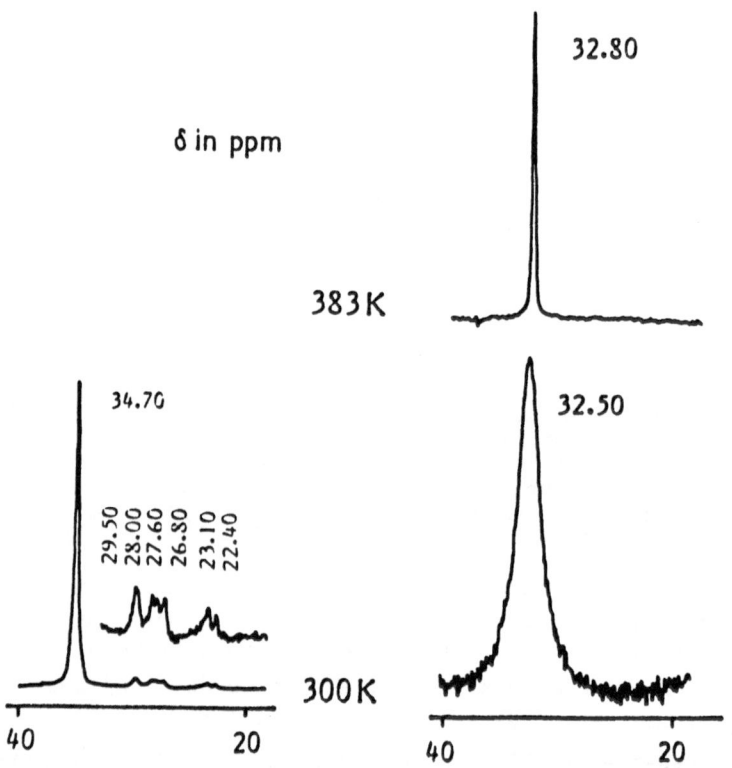

Fig. 2: MAS ^{13}C NMR spectra of cyclohexanonacontane. Left:
solution-crystallized sample at 300 K. Right: melt crystallized sample at 300 K and 383 K. (For experimental
details see ref 17.)

The spectrum represents the average signal of carbons undergoing fast exchange between different conformational sites. At
room temperature, the signal is remarkably broadened. This is

indicative of a state in which the nuclei occupy different ill-defined sites, for which the exchange is slow with respect to the NMR time scale. A sharp signal due to orthorhombically packed all-anti chains may be hidden within the broad resonance. However, spin-lattice relaxation experiments did not result in a corresponding discrimination. Lowering the temperature further results in the spectrum shown in Figure 3. Similar to the spectrum of the solution crystallized sample, two types of carbons are now discriminated. A sharp resonance of high intensity at 33.8 ppm is assigned to the orthornombically packed all-trans stems. The broad low intensity signal at higher field belongs to the carbons in the tight adjacent reentry fold. Clearly, the broad resonance indicates a fold conformation with non-staggered, undefined rotational angles which do not fit in a diamond lattice.

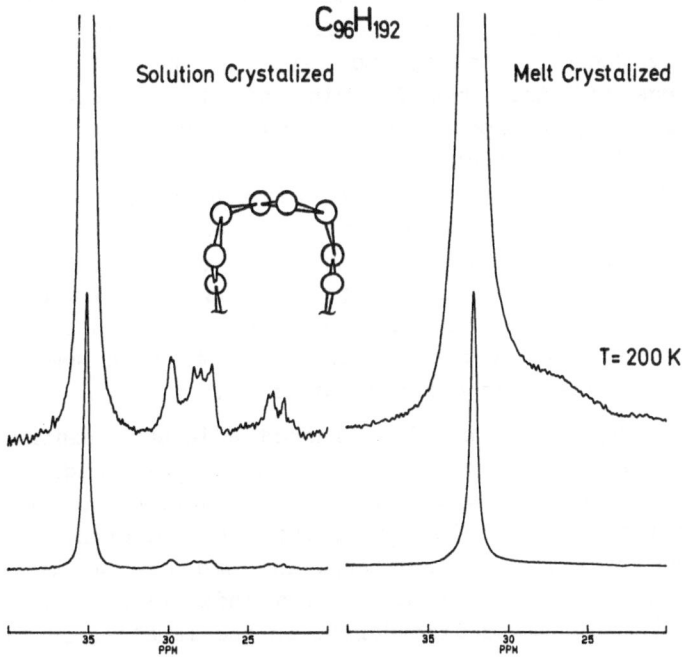

Fig. 3: Low temperature spectra of the monoclinic solution crystallized (left) and the orthorhombic melt crystallized modification of cyclohexanonacontane. The drawing on the left shows the well defined ggagg structure for the 200 fold, which is assigned to the the low intensity signals of the spectrum on the left according to ref 17 and 18. Only a broad undefined resonance is shown for the fold in the orthorhombic modification.

These NMR spectra confirm the results obtained by DSC experiments. In summary, the following picture emerges. The different melting behavior of the cyclic alkanes is a result of the incompatibility of the optimal packing of the stems and the minimum energy arrangement of the folds. The orthorhombic subcell is the most stable arrangement of all-anti CH_2-stems. Folds with a minimum number of closure atoms can be arranged in a diamond lattice only if the subcell of the stems is monoclinic. The length of the cycloparaffines discussed in this paper is obviously such that neither the fold nor the stem arrangement dominates the crystallization. Different crystal preparations yield either the orthorhombic or the monoclinic modification. In the orthorhombic modification, the folds are highly strained and disordered. As a consequence, the packing of the methylene groups adjacent to the lamellar surface is severely disturbed. Orthorhombic diffraction patterns observed for the melt crystallized modification[20] represent the inner part of the lamellae where the methylene units are perfectly packed. The thickness and thus the fraction of this ordered inner part changes as a function of temperature. Only at very low temperatures is the disordering restricted to the few carbons which form the actual fold. At high temperatures, the orthorhombic form changes completely to a conformationally disordered hexagonal phase. Thus, ordering occurs over a broad temperature range and will never be as perfect as observed in the monoclinic modification. The 32.8 kJ mol^{-1} difference in the total enthalpy of the two modifications at 248 K is some indication of the strain in the orthorhombic fold.

The picture of a highly strained adjacent reentry fold and the corresponding temperature dependent dynamic disordering of the chain segments near the lamella surface in orthorhombic crystals should be relevant to chain folded orthorhombic polyethylene crystals. It may explain some of the different and controversial experimental results indicative of both regular adjacent folding and disordering in the lamellar surface.

Acknowledgement
Financial support was provided by the Deutsche Forschungsgemeinschaft within the SFB 60

LITERATURE

1. B. Wunderlich, "Macromolecular Physics", Vols. 1 and 3, Academic Press: New York 1973 and 1980

2. P. J. Flory, J. Am. Chem. Soc. $\underline{84}$, 2857 (1962)

3. A. Keller, Rep. Progr. Phys. $\underline{31}$, 623 (1968)

4. D. M. Sadler, in "Structure of Crystalline Polymers", Ed. I. Hall, Applied Science: London 1984

5. J. C. Wittmann and B. Lotz, Makromol Chem. Rapid. Comm. $\underline{3}$, 733 (1982)

6. S. J. Spells and D. M. Sadler, Polymer $\underline{25}$, 739 (1984)

7. S. J. Spells, A. Keller and D. M. Sadler, Polymer $\underline{25}$, 749 (1984)

8. T. K. Oyama, K. Shiokawa, T. Ishimary, J. Macroml. Phys. $\underline{B8}$, 229 (1973)

9. P. E. MacMahon, R. L. McCullough, A. A. Schlegel, J. Appl. Physics $\underline{38}$, 4123 (1967)

10. G. N. Patel, A. Keller, J. Polym. Sci. Phys. Ed. $\underline{13}$, 2259 (1975)

11. I. Ando, T. Yamanobe, T. Komoto, H. Sato, K. Deguchi, and M. Imanari, Polymer $\underline{26}$, 1864 (1985)

12. H. P. Grossmann, R. Arnold, and K. R. Bürkle, Polym. Bull $\underline{3}$, 135 (1980); H. P. Grossmann, Polym. Bull. $\underline{5}$, 137 (1981); H. P. Grossmann, Polym. Bull. $\underline{7}$, 413 (1982)

13. H. F. Kay and B. A. Newmann, Acta Cryst.ogr. B $\underline{24}$, 615 (1968)

14. T. Trzebiatowski, M. Dräger, and G. R. Strobl, Makromol. Chem. $\underline{183}$, 731 (1982)

15. K. S. Lee, Dissertation, Freiburg 1984

16. H. Drotloff, D. Emeis, R. F. Waldron, and Martin Möller, Polymer 1987 in press

17. T.Trzebiatowski, Dissertation Mainz 1980

18. M. Möller, W. Gronski, H.-J. Cantow, and H. Höcker, J. Am. Chem. Soc. $\underline{106}$, 5093 (1984)

19. M. G. Taylor, E. C. Kelusky, I. C. P. Smith, H. L. Casal, and D. G. Cameron, J. Chem. Phys. $\underline{78}$, 5108 (1983); G. Kögler, Diploma Thesis, Freiburg 1987

POLYMER CRYSTALLIZATION :

A Monte Carlo Simulation of Lamellar Growth

Frank van Dieren

Max-Plack-Institut für Polymerforschung
Postfach 3148, D-6500 Mainz, Federal Republic of Germany

Artur Baumgärtner

Institut für Festkörperforschung
der Kernforschungsanlage Jülich
Postfach 1930, D-5170 Jülich,
Federal Republic of Germany

ABSTRACT

Semicrystalline polymers formed from the melt generally consist of lamellar crystals seperated by amorphous regions. Typically, the lengths of the polymer chains are much larger than the lamellar thickness. Hence each molecule must pass through the same or different lamellae many times.

The probability for a polymer emerging from the face of the lamella and returning into the same lamella at some distance from the crystalline "stem" is oftenly assumed to obey <u>equilibrium</u> statistics. This basic assumption is in apparent disagreement with the fact of crystallization as a <u>non-equilibrium</u> phenomena.

Therefore the question whether the re-entrance distribution is affected significantly by kinetic effects is of great interest. In the present we investigate this by studying a 2-D kinetic crystallization model using a novel Grand Canonical Monte Carlo technique.

Estimates of the re-entrance distribution, prallel and perpendicular to the front of growth, exhibit quite different behavior. This anisotropy becomes stronger the higher the dilution of the system is. From estimates of the velocity of growth, we found at a particular temperature a transition from slow to rapid crystallization.

INTRODUCTION

The morphology of semicrystalline polymers can be described to a good approximation as consisting of lamellar crystallites separated by amorphous regions. The way in which a single macromolecule traverses the crystalline and amorphous

phases can be investigated for instance by neutron scattering
[1]. However, some quantities such as the number of "re-en-
tries" of a chain molecule into the same lamella, are derived
from experiment using some model. Here we present an idealized
model, which will allow us to do (computer) calculations from
first principles.

It has been noticed experimentally that mainly one
lamella at the time is formed. We therefore will treat the
problem like many others [2-4] and model the growth of one
lamella independent of the existence of others. As a consequence we are not able to study the properties of the amorphous layer between two crystallite ones, which we assume to
be locally in equilibrium, i.e. a piece of polymer in the
amorphous region has Gaussian (melt like) behavior.

MODEL AND SIMULATION TECHNIQUE.

The crystal region may be described by two dimensional
"stem" co-ordinates, which we will assume to lie on a quadratic lattice.

We introduce two system parameters, μ_{st} and ϵ_{st}. The stem
chemical potential μ_{st}, is the energy balance when one stem is
created. It contains e.g. the stretching - and crystal-melt
interfacial energy, but not entropy. The stem-stem interaction
energy ϵ_{st}, is the gain of energy for two stems being nearest
neighbours.

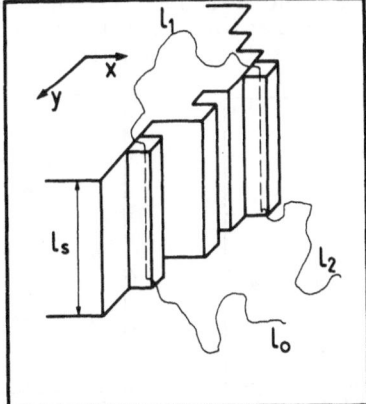

Fig. 1

In fig. 1 we show two states in which a flexible piece of
polymer (non stem) can be, namely as a tail (ℓ_0 and ℓ_2) with
only one end fixed, and as a loop (ℓ_1) with both ends fixed.
For a loop of length ℓ we have taken the entropy to be the one
of a Non Reversal Random Walk (NRRW) in 3-d on a cubic lattice
with both ends r_1 and r_2 fixed at an impermeable wall. The
number of states $N_1(\ell, r_{12})$, depending on ℓ and $r_1 - r_2$ only, is

calculated iteratively. Equivalently the number of states for a tail of length ℓ is defined to be $N_t(\ell)$. Hence, the free energy is given by $F_s = E_{st} - Tk_B(\Sigma \ln N_1(\ell_1, r_{12}) + \Sigma \ln N_t(1))$.

The dynamics of the system is simulated by using a novel Grand Canonical Monte Carlo technique. A new state is generated as follows: a randomly selected stem is moved to an other position, or is removed from a chain, or is added at the end of a chain with equal probabilities. The transition rate is given by $W(\text{old} \rightarrow \text{new}) = \min (1, \exp(-\Delta F_s/k_B T))$ where $\Delta F_s = F_{s,new} - F_{s,old}$. A new state is accepted if $\exp(-\Delta F_s/k_B T) > \xi$, where ξ is a random number $0 < \xi < 1$.

Simulations have been performed using a finite L x L box. In x-direction we have fixed -, in y-direction periodic boundary conditions. Simulations start with a plane crystallization front at x=1. More computational details will be published elsewhere [5].

RESULTS

We define $c(r,\ell)$ to be the fraction of re-entrances at a distance r in the same lamella, connected by a piece of polymer in the amorphous phase having length $\ell = \ell_1$ (fig. 1).

The random switchboard model [6] predicts that re-entrances are distributed isotropically and r has Gaussian statistics, i.e. $\langle r \cdot r \rangle \sim \ell$. The average length $\langle \ell \rangle$ depends on the width of the amorphous region. Models starting from kinetics [2,3] yield a strong anisotropy in r for $c(r,\ell)$.

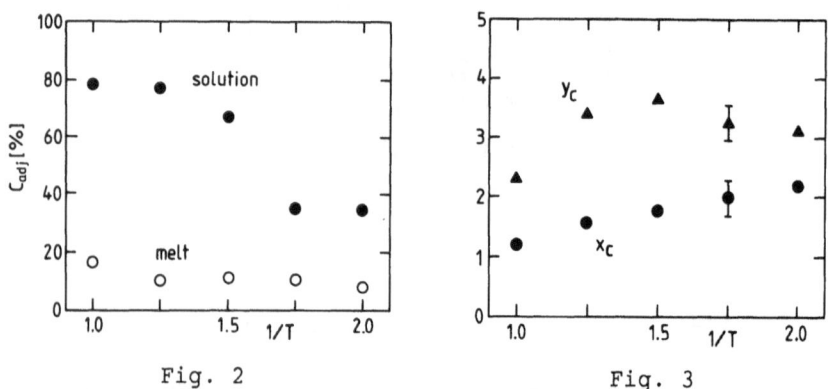

Fig. 2   Fig. 3

Here we present simulation for $\ell_{st} = 10$ and $\ell_{max} = 10$ while L=40 and the polymers have 200 monomers. Other values of these parameters gave no qualitative different results. In fig. 2 we present for $\mu_{st}/\ell_{st} = -0.5$ as a function of $1/T \equiv -\epsilon_{st}/\ell_{st}$ the fraction of adjacent re-entry c_{adj} in the cases of crystallization from dilute solution and from the melt. One finds for the dilute case a very anisotropic $c(r,\ell)$, and c_{adj} varies from 100% to 30% for T just below the melting point down to low temperatures. For the melt however we find a 20% to 8% decrease in c_{adj} for $T > T_m$. We estimated the melting temperature to $T_m = 1.25 \pm 0.03$

To measure the anisotropy in the melt we define:
$$x_c = \sqrt{<x_i^2>}_c = \sqrt{\int_1^\infty d\ell \int dr x_i^2 c(r,\ell)}$$ and y_c accordingly.

In fig. 3 we plotted x_c and y_c for $\mu_{st}/\ell_{st} = -0.5$ as a function of $1/T$. An interesting feature is that there is a maximum for some T in y_c. This may be seen as a result of 'dynamical roughening' [7] of the growth-front for larger undercooling.

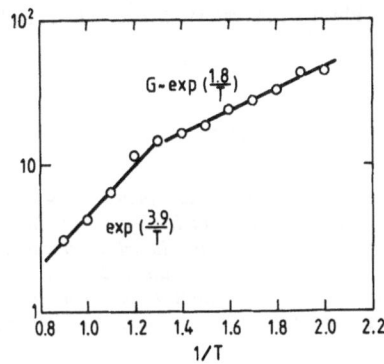

Fig. 4

In fig 4 we show a log-plot of the growth-velocity G (number of stems per time) as a function of $1/T$, again for $\mu_{st}/\ell_{st} = -0.5$. One clearly distinguishes Regime I growth, where one stem causes completion of the entire substrate, and Regime II growth, where a multiple of stems crystallize [2,3]. The transition occurs at $T_b \simeq 0.77$. We found a change of slope in $\log(G)$ vs. $1/T$ of a factor 2, as predicted by nucleation theory [3].

REFERENCES

[1] E.W. Fischer, Polymer Journal, Vol. 17, No. 1 (1985), 307.
[2] J.I. Lauritzen Jr and J.D. Hoffman, J. Res. Nat. Bur. Stand., 64A (1960), 73.
[3] J.D. Hoffman, C.M. Guttman and E.A. Dimarzio, Faraday Discuss. 68 (1979), 177.
[4] D.M. Sadler and G.H. Gilmer, Phys. Rev. Lett. 56, No. 25 (1986).
[5] F. v. Dieren, A. Baumgärtner (to be published)
[6] P.J. Flory and D.Y. Yoon, Nature 272 (1978), 226.
[7] W. van Saarloos and G.H. Gilmer, Phys. Rev. B 33, No. 7 (1986).

WHAT DO LONG ALKANES REVEAL ABOUT POLYMERS ?

G. Ungar [+]

H.H.Wills Physics Laboratory, Bristol BS8 1TL, U.K.

ABSTRACT

The recently synthesised very long n-alkanes are shown to be ideal model systems for crystallizeable polymers. They provide qualitatively new insights into the nature of chain folding, fold-surface order and disorder, lamellar thickening and thinning, etc. Some interesting new observations (crystallization rate minima, critical nucleus size) provide crucial data favouring the fundamentally new concept that activation entropy determinines the morphology and crystallization kinetics of polymers.

INTRODUCTION

While polymers crystallize by chain folding, nonpolymeric chain-like molecules do not. Thus even the longest linear paraffins available until recently (with up to ca. 100 C-atoms) invariably maintain their extended conformation in the crystalline state; the chains pack in layers with end groups at the layer surface. It is of considerable interest to establish at what chain length the molecules start to fold and, in this respect, behave like polymers. Furthermore, if such sufficiently long chains could be synthesised through a series of closely controlled steps to yield strictly uniform molecular weights, they would provide highly desirable model systems with numerous advantages over the intrinsically polydisperse materials produced through polymerization. Polydispersity appears to have obscured important details of polymer crystallization and morphology in the past and precluded unambiguous information being obtained, even with the most advanced experimental techniques. Recently, pure n-alkanes up to 400 carbon atoms in length have indeed been prepared [1,2] and found to fulfil the above requirements for model crystalline polymers. This paper reviews salient results of physical studies on alkanes prepared by Bidd and Whiting [1], placing particular emphasis on new information

---

[+]Present address: Rudjer Boskovic Institute, P.O.B. 1016, Zagreb, Yugoslavia

that enables a clearer understanding of macromolecular crystallization and morphology. The importance, in turn, of these latter issues for processing and properties of so many polymeric materials needs no commentary.

NATURE OF CHAIN FOLDING

Chain-folded crystallization in the paraffins is already found to occur with C150 (n-hectapentacosane), and it is increasingly more favoured with increasing chain length [3]. The fold length, or lamellar thickness, l, follows approximately the same pattern as in ordinary polymers: it decreases with decreasing crystallization temperature. However in paraffins l takes up only discrete values identical or very close to $l=L/n$, where L is the full extended chain length and n is an integer. It is clear that such "integer folding" is energetically favoured in monodisperse systems as it allows the chain ends to conglomerate along smooth layer surfaces (see below). Up to four folds per molecule (n=5) can be obtained by solution crystallization of the longest available paraffin C390 [3]. All the integer chain-folded forms observed so far in our alkanes are listed below (E = extended, F = folded):

chain conform.	E	·F2	F3	F4	F5
paraffin					
C 102	+				
C 150	+	+			
C 198	+	+	+		
C 246	+	+	+	+	
C 294	+	+	+	+	
C 390	+	+	+	+	+

The fold lengths are determined by two independent methods: from the Raman frequency of the longitudinal acoustic mode, l(LAM), and from small-angle X-ray scattering, l(SAXS). While the LAM method measures the length of the extended portion of a chain traversing the lamellae, SAXS measures the average layer periodicity and thus includes the folds as well as the thickness of any amorphous or void layer that might be present. The values of l(LAM) and l(SAXS) in our paraffins were found to be very close to each other and to the integer fractions L/n, indicating the absence of an amorphous layer and of any appreciable amount of loose folds or loose ends (cilia).

Just how well the measured LAM frequencies of chain-folded paraffins agree with those for the corresponding fully extended stems of length L/n is demonstrated in Fig. 1 [4]. Here the straight line corresponds to the best fit of the chain-extended paraffin frequencies (marked by crosses) vs. reciprocal carbon number i(total); the data points for chain-folded paraffins are denoted by the appropriate graphic symbols indicating the number of folds, the abscissas corresponding to i=i(total)/n. It can be seen that only the thinnest folded crystals containing the largest number of folds per molecule (3 or 4) depart somewhat from the reference line; the bulk of the chain-folded data matches it within the limits of experimental error. Thus e.g. for a once folded paraffin the LAM frequency is exactly the same as for an extended paraffin of half the length. It follows that the fold cannot contain more than just a few (e.g. 5 or 6) carbon atoms in most samples, hence the folds are adjacent and reasonably tight.

RAMAN LAM FREQUENCY VS. LAMELLAR THICKNESS: THE ULTIMATE ELASTIC MODULUS OF POLYETHYLENE

A digression concerning the important issue of LAM frequency vs. stem length and of the longitudinal elastic modulus of polyethylene may be appropriate at this point. The early LAM studies on shorter paraffins [5] came up with the simple relationship between the first-order LAM frequency shift ν and the straight stem length l:

$$\nu \simeq \sqrt{E/\rho} \ (1/(2cl))$$

with E being the elastic modulus of the chain, ρ the density, and c the speed of light. However, it has been argued subsequently [6] that corrections are needed to account for surface effects. No consensus on this issue has yet been reached. The two approaches yield significantly different fold lengths when applied to polymers in practice and they assume different values of E. The extended chain crystals of the present long paraffins provide the ideal means for testing the two relationships, since (a) l is known exactly, and (b) the increased chain length renders the end effects less important. The full straight line in Fig. 1 evidently fits the experimental data for extended chain paraffins very well. Its numerical form is

$$\nu(cm^{-1}) = 2470/i$$

with i being the carbon number ($l = 0.1275$ i); it corresponds to the simple original relationship [5] with the elastic modulus E = 350 GPa. The alternative formula, which introduces corrections due to the lamellar surface perturbation, assumes a lower polyethylene modulus of 290 GPa in order to fit the experimental data on the previously available lower alkanes. However, as the broken line in Fig. 1 shows, the latter relationship fails for long alkanes. Thus we conclude that the originally proposed expression [5] is adequate, at least as far as the first order LA mode is concerned, and that the end-effect correction is unnecessary. The higher E value means that a somewhat higher ultimate elastic modulus is achievable in polyethylene, in close agreement with calculation [7].

INTEGER FOLDING

We return now to the main theme of chain folding in paraffins. As already mentioned, the SAXS measurements are in full agreement with the presence of integer folding in mature crystals as observed at room temperature. Later in the course of our studies [8], it was found that this does not apply to the crystals during their primary formation, as will be discussed further below. Presently, the particularly close match between measured l(SAXS) and calculated L/n values for sedimented mats of solution-grown crystals is pointed out. In such cases, provided that crystallization took place at moderate rates, the match was within 0.1 to 0.2 nm. For melt crystallized material such close correspondence could, on the whole, not be claimed with certainty as the occurrence of chain tilt complicates the issue to some extent.

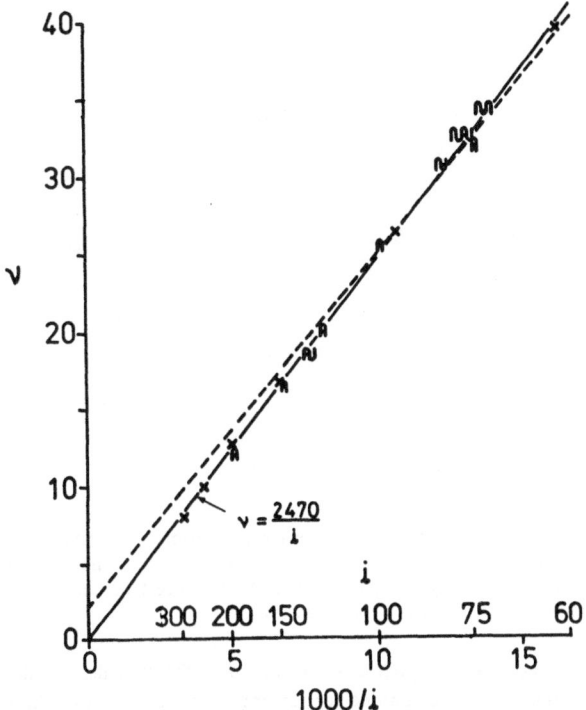

Figure 1 - Measured frequencies of the first LAM mode vs. carbon number i(total) (chain-extended crystals, denoted x) or integer fraction thereof i=i(total)/n (chain-folded crystals, denoted by appropriate symbols) [4]. Dashed line represents the alternative relationship $\nu(cm^{-1}) = [2236/(i-1.6)] + 2.2$ [6].

As a typical example of solution grown crystals we mention the once-folded crystals of C198, where 7 small-angle diffraction orders were observed with the measured periodicity l(SAXS) = 12.5 nm, as compared with the calculated L/2 value of 12.7 nm [9]. This clearly excludes the possibility of folds or cilia of any appreciable looseness. Nevertheless, as discussed in the next section, truly conformationally uniform folds are found only at low temperatures. Electron microscopy provides additional evidence of fold regularity, as shown in Fig. 2 [10,11]. Here the once-folded chain crystal of C198 had been surface decorated by the method of Wittman and Lotz [12]. The decoration indicates a strong preference for the folds in each of the four crystal sectors to lie parallel to the particular {110} growth plane of the sector, and further, that no significant amorphous overlayer is present as this would prevent the transmission of the orientational influence of folds to the decorating particles.

Figure 2 - Once-folded-chain crystal of C198 grown from solution and decorated with polyethylene, showing preferential {110} folding [11]. Bar = 1 μm. (The small central area, apparently undecorated, probably arose through a temperature fluctuation during growth and is not a typical feature).

It should be evident from the foregoing that chain-folded paraffin crystals, particularly those grown from solution, offer a unique opportunity for examining chain folds themselves, since the latter are not obscured by an amorphous overlayer as exists in even the best single crystals of polyethylene. What we have here consists of (a) straight crystalline chain stems and (b) chain ends and (c) folds at the surface. Since the same paraffin could be crystallized in the extended form, a subtractive spectroscopic technique such as FTIR could, for the first time, isolate the features of pure chain folds. The result of such a study [11] is presented in Fig. 3 which shows a selected range of the subtracted (folded minus extended-chain, denoted "F-E") IR spectrum of C198 crystals recorded at 110 K. For comparison the spectra of polyethylene single crystals (PESC) and of bulk linear polyethylene, recorded at the same temperature, are also shown. Particular attention is drawn to the band at 1345 cm^{-1} which is characteristic of tight regular folds [11], whereas the neighbouring 1352 cm^{-1} band is associated with unconstrained gauche-gauche sequences in the more disordered or liquid-like state. As can be seen in Fig. 3 the fraction of tight folds in the non-crystalline phase decreases in the order: paraffin $\rightarrow$ PESC $\rightarrow$ bulk polyethylene. The newly established reference spectrum of tight folds has already been applied in a more detailed study of chain folding in polyethylene crystals as a function of crystallization temperature and annealing [13].

351

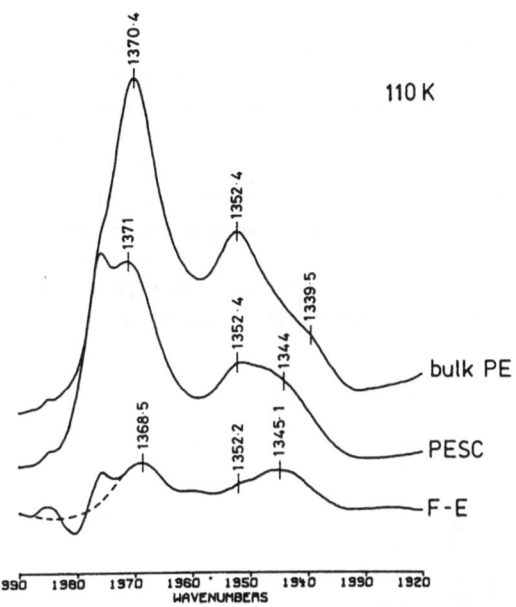

Figure 3 - Characteristic FTIR spectra of {110} chain folds in paraffin C198 (folded-chain minus extended-chain crystals, F-E), of polyethylene single crystals (PESC) and of bulk linear polyethylene [11].

ADJACENT VS. TIGHT FOLDS: EQUILIBRIUM ORDER / DISORDER

Observing the temperature induced changes in the IR spectrum of the folds is rather interesting. In Fig. 4 low and room temperature F-E spectra are compared [11]. The main difference between the two, i.e. the increase of the 1352 cm^{-1} band at the expense of that at 1345 cm^{-1} on heating, may be attributed to a reversible disordering of the fold surface with increasing temperature: the appearance at room temperature of the 1352 cm^{-1} component, arising from unstrained gg groupings, is probably due to a limited loosening of a proportion of folds. As the 1345 cm^{-1} band intensity at 110 K is far higher than that at 1352 cm^{-1}, it is concluded that the great majority of folds are of the tight type at low temperature.

It is important to note that the attainment of such high predominance of tight folds on cooling, as is indicated by the 1345 cm^{-1} band in the present 110K F-E spectrum, can only occur if the folds span adjacent stems in the first place. Nonadjacent "switchboard" folds could not regularize on cooling and their low-temperature spectrum would resemble that of the amorphous phase in polyethylene (cf. Fig. 3, top).

We touched here upon an important distinction that is not usually made when discussing the issue of chain folding in polymers: i.e. that between *adjacency* and *tightness* of folds.

While the former is a precondition for the latter, the latter does not necessarily follow from the former. The degree of adjacency (as well as fold orientation, e.g. (110) or (100)) is established during crystallization and determined by the kinetics of the process. The tightness of adjacent folds, on the other hand, ought to be an equilibrium property (of course, chain-folded crystals can only attain a metastable equilibrium). Tight folds are those with the minimum energy conformation, hence they are uniform. To satisfy the increased entropy requirement at higher temperatures the minimum energy condition may be sacrificed for somewhat looser folds with greater conformational freedom but with a decreased overall crystallinity. In our particular example the loosening at room temperature appears to be sufficient to be detected spectroscopically, but not pronounced enough to affect the measured l(SAXS) or l(LAM) values.

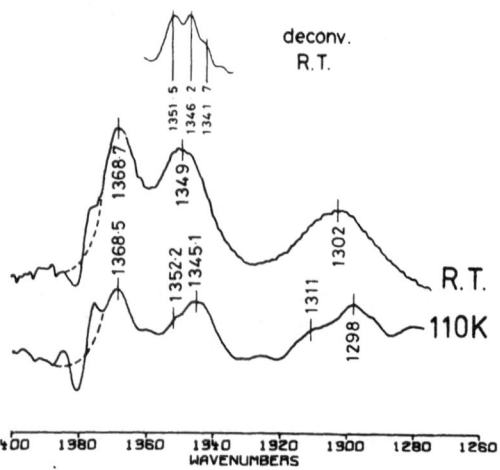

Figure 4 - FTIR difference spectra (F-E) of folds in C198 recorded at room temperature (R.T.) and at 110 K. Inset: Fourier deconvoluted 1330-1360 cm^{-1} band of the R.T. spectrum.

The problem of smooth vs. rough lamellar surface in a crystal of once-folded monodisperse chains has been treated quantitatively [14]. We based our calculation on the model schematically depicted in Fig. 5. In the figure on the left l is exactly equal L/2 (strictly integer folding). Only two configurations per molecule are allowed: hairpin up and hairpin down (only one orientation is shown in Fig. 5). However, if the layer surface is allowed to be rough many more configurations become available, as sketched on the right. A self-consistent mean field approach is adopted where the interaction potential profile ψ is iteratively matched against the calculated density profile ρ. ψ takes the form of the error function whose width is defined by the standard deviation δ (step function with $\delta = 0$ in the extreme case on the left).

Protruding ends are allowed rotational isomeric freedom, while the folds are kept tight in this simplified model. Usual values for the lattice and the gauche-trans energy difference for polyethylene are used and the resulting free energies vs. δ are plotted in Fig. 6a. Strict folding in two with smooth surfaces ($\delta=0$) is favoured only below ca. 200 K, the minimum in free energy moving to larger deltas (increased roughness) with increasing temperature. The corresponding energies are shown in Fig. 6b, the dashed curve delineating the temperature dependence of the equilibrium state energy. Although the model is relatively simple and takes no account of the different fold conformations or intralamellar defects (prominent near the melting point), the surface melting or the so called roughening transition is evident.

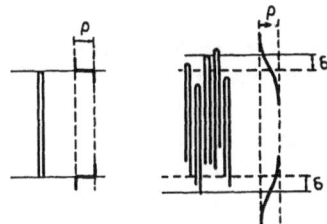

Figure 5 - Schematic representation of the model used to calculate equilibrium surface disorder in chain-folded paraffin crystals.

The experiments do indeed show premelting effects. Fig. 7a shows the DSC heating thermogram (10 K/min) of once-folded paraffin C246. The chain-folded crystals first melt, then recrystallize as chain-extended crystals (exotherm), which finally melt producing the second endotherm. Especial attention should be paid here to the extended low-temperature

end of the first endotherm: as the material is of high purity and the crystals are all of the same thickness to start with, only surface disordering could account for the broad melting range. Until the crystal cores have melted fully (ca. 395 K) recrystallization of the extended form cannot take place; hence the exotherm starts only above 395 K.

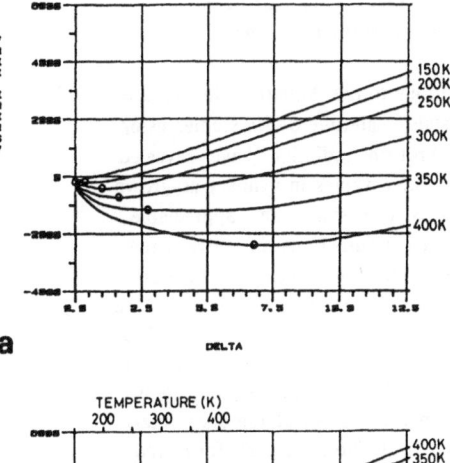

a

The above interpretation of the C246 thermogram has been vindicated by parallel real-time SAXS experiments using synchrotron radiation. The same once-folded alkane was heated at the same rate as in the DSC run (10 K/min); the series of scattering curves thus obtained are shown in Fig. 7b. Exactly in the temperature range where the heat capacity curves upwards (Fig. 7a) the SAXS peak markedly intensifies and shifts to larger l values, both phenomena indicating an increasingly disordered surface layer.

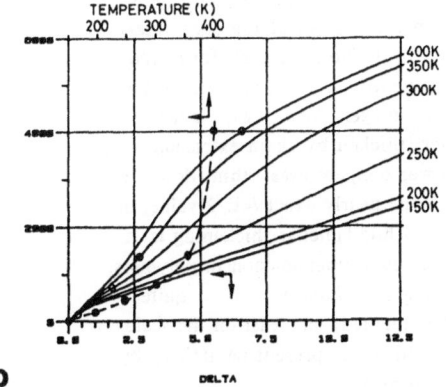

b

Figure 6 - Molecular free energies (top) and energies (bottom) vs. surface roughness parameter δ for once-folded paraffin crystals (model in Fig. 5) at different temperatures. Circles mark the free energy minima. Dashed line shows the temperature dependence of equilibrium energy.

KINETIC VS. EQUILIBRIUM DISORDER: NON-INTEGER FOLDING

So far we have been looking at mature crystals which were all either in the extended chain form or in one of the integer folded forms. The term "integer form" is taken here in a broader sense, tolerating l values approximately in the range $L/n-0.2nm < l < L/n+1nm$. As a rule strict integer folding ($l = L/n \pm 0.2nm$) was found in solution-grown crystals. Larger upward deviations from integer l-values were found only a few degrees below the melting temperature of the folded form, the effect being attributed to reversible surface melting.

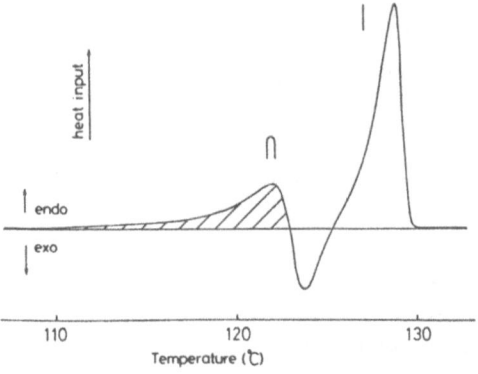

Contrary to the situation in mature crystals, real-time SAXS studies using synchrotron radiation have revealed that the l-values in primary as-grown crystals can span an appreciable (albeit not full) range in between the integer fraction values [8]. In our studies of paraffins C198 and C246 that meant primary fold periods of up to 50 A in excess of the L/2 value if crystallization was carried out below the melting point of the once-folded form. This "non-integer" form (NIF) subsequently transforms into a more stable integer form: the lamellae either thicken by chain extension or, at lower temperatures, thin down to reach the nearly exact $l=L/2$ value, or both. While lamellar thickening is a well-known phenomenon in polymers, isothermal *thinning* is quite unprecedented, at least on the scale observed in the present paraffins (by up to 4 nm).

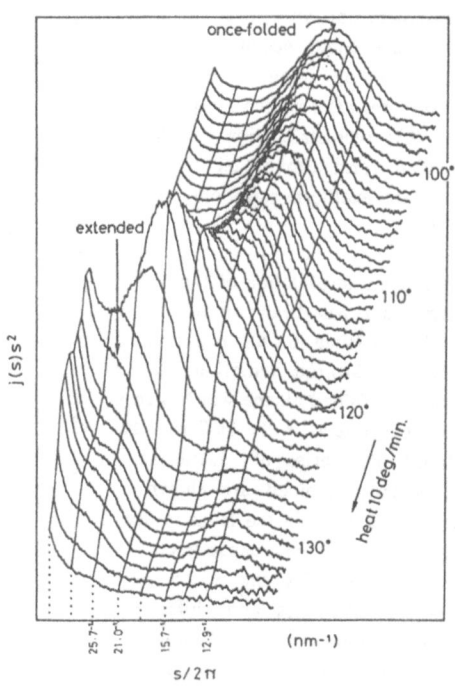

Figure 7 - Top: DSC heating thermogram of once-folded melt-crystallized paraffin C246; shaded area marks surface pre-melting. Bottom: Corresponding synchrotron SAXS traces. Heating rate 10 deg/min.

Our interpretation of these effects is presented schematically in Fig. 8. The key point is that the number of ways in which a once-folded molecule can deposit on the growth face increases very steeply with an increase in l beyond L/2, as already discussed in relation to the equilibrium surface disorder. However, the present case of kinetically induced disorder is distinct from the previous case of equilibrium disorder in that the more highly disordered layers will grow fastest even if they are not the most stable ones, as long as their free energy of crystallization is sufficiently negative [14]. Thus,

although the driving force for crystallization of the non-integer form is less than that for the integer form, the *activation entropy*, ΔS^+, for NIF is a smaller negative quantity; this is equivalent to saying that the frequency factor in the crystallization constant, which is proportional to $\exp(\Delta S^+/R)$, is higher for NIF than it is for integer folding. Thus, kinetic disorder will be higher than equilibrium disorder and primary lamellae will have a larger ("non-integer") l value than that corresponding to maximum stability (cf. Fig. 6a). The importance of activation entropy in crystallization of chain molecules will become even more apparent in subsequent sections.

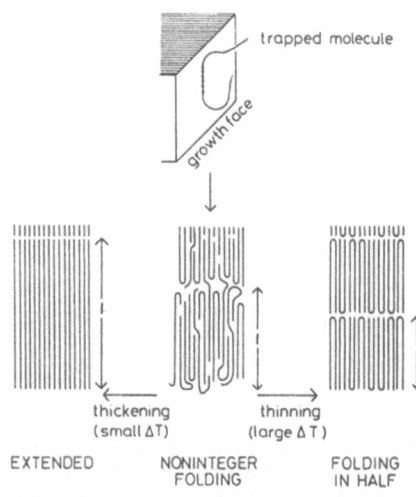

Figure 8 - Graphic description of primary crystallization with non-integer folding (kinetic surface disorder) and of subsequent transformation to integer forms by layer thickening or thinning.

The true equilibrium state in crystals of even the longest paraffins is, of course, the chain-extended form. Starting from NIF, this can indeed be achieved at sufficiently high annealing temperatures [8] through a complex solid-state diffusion process which must involve nucleation. At lower temperatures only the simpler thinning process can take place [8], the system thus approaching the local free energy minimum (approach from the right in Fig. 6a). A SAXS example of secondary ordering by thinning of primary chain-folded lamellae is shown in Fig. 9b.

As opposed to chain-folded crystals, in chain-extended lamellae little configurational freedom is gained with l increasing beyond the integer value L, while much lattice interaction energy is lost. Experiments do in fact show that chain-extended crystals always have l's very close to the calculated chain length value L. Primary crystallization produces such an integer form directly and there is little subsequent change (see Fig. 9a).

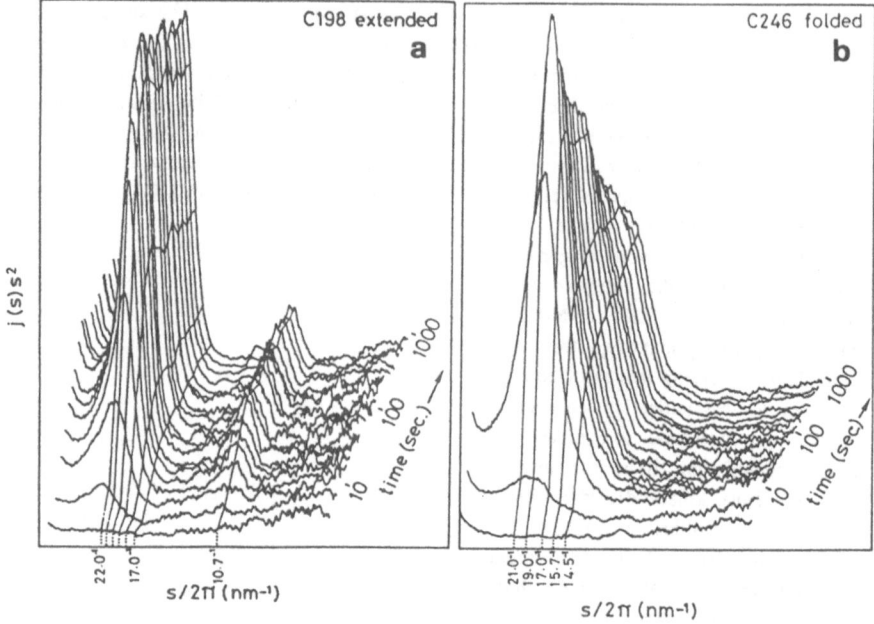

Figure 9 - SAXS curve evolution during isothermal crystallization (a) of the chain-extended form (C198, 115°C) and (b) chain-folded form (C246, 104°C). In (b) both diffraction intensity and layer spacing decrease after primary crystallization due to ordering of fold layers; l(SAXS) decreases from 19 to 15.7 nm (=L/2). Note log time scale.

CRYSTALLIZATION RATE MINIMA: "MISLED" MOLECULES

Rather remarkable effects are observed with crystallization kinetics of long alkanes. Some of them have already been reported previously on low molecular weight fractions of polyethylene oxide [15], notably the discontinuities in the temperature gradient of the growth rate at changeover points between extended and once folded, once folded and twice folded etc. crystallization. These discontinuities can be explained, at least qualitatively, within the framework of current crystalization theories. However, what has not been observed previously and presents quite a challenge to the theory is the fact that in paraffins the crystallization rate not only undergoes discontinuous changes but actually *passes through minima* at the crossover points [16]. For illustration Fig. 10 shows a parameter roughly proportional to the primary nucleation rate for melt crystallization of C246. As the crystallization temperature is lowered from the extended-chain melting point T_m^E the rate starts increasing exponentially, as expected. However, it soon levels off and on approaching the melting point of the once-folded form (ca. 122°C) it starts *decreasing* to reach a deep minimum at 121°C. Below this temperature it rapidly picks up again; parallel synchrotron experiments reveal that the minimum corresponds to the changeover between extended and folded chain (NIF) nucleation. Similar minima have been observed also in solution crystallization [17], and they occur in both nucleation and growth rates [16].

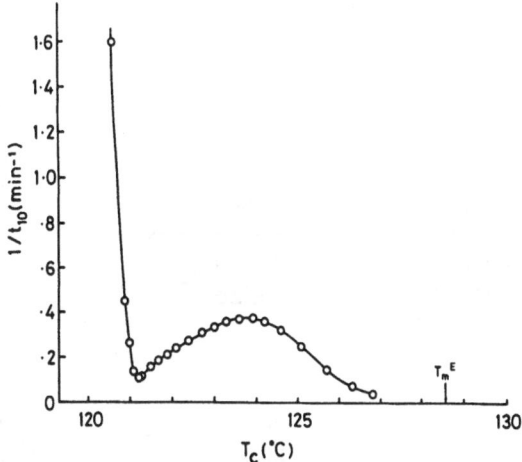

Figure 10 - Inverse induction period vs. temperature of crystallization of C246 from the melt, as determined by DSC [16].

Apart from the effects near the glass transition temperature, a decrease in crystallization rate with increasing supercooling has, to our knowledge, not been observed previously in any crystallizeable system, polymeric or non-polymeric. As the driving force for crystallization increases monotonously with increasing supercooling, the slow-down effect must be attributed to a second non-productive process which interferes with productive crystallization. In the case of primary extended-chain nucleation shown in Fig. 10 this second process (II) ought to be chain-folded growth of an embryonic nucleus. Above the folded-chain melting point it cannot lead to stable chain-folded crystals, but this process must have a significantly lower free energy barrier than the productive extended chain nucleation (process I), if it was to interfere with it effectively. These ideas are sketched in Fig. 11. The embryonic nucleus is perceived here as not yet having the chains fully extended (Fig. 11b). It "does not know" at the outset which way will be the ultimately productive one; its most probable fluctuations are those with the least steep free energy ascent. It is thus being "misled" into non-productive chain-folded lateral growth which reduces the chance for extension of the molecules trapped in the interior. The nearer the folded-chain melting point, the larger the chain-folded fluctuations and the longer their lifetime; hence the more effective their interference with the chain extension process.

The minimum in the rate of crystal growth can be interpreted along similar lines. Here the chain-extended substrate already exists but further growth is hindered by the unstable but frequent chain-folded depositions which block the growth sites. Depending whether the "rough surface" growth [18] or the secondary nucleation mechanism [19] is adopted, respectively, the folded molecules would either form a belt along the whole growth face (shown in Fig. 12a in side view) or it would suffice that they obstruct only the niches

358

on the growth face to retard effectively the completion of a new depositing lateral layer (Fig. 12b). Recent rate equation treatment of "rough surface" crystal growth has shown that minima similar to those observed here can indeed be expected if the free energy of crystallization has discrete minima at specific values of crystal thickness [20].

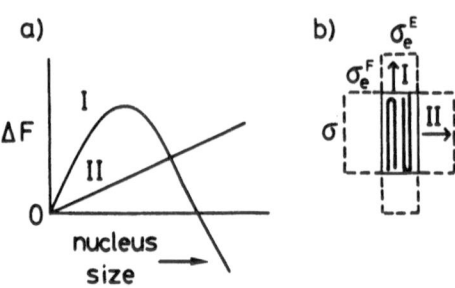

Figure 11 - Schematic illustration of the competing processes in the formation of a chain-extended nucleus: I - chain extension, II - chain-folded lateral growth. a) free energies of crystallization, b) visualization of the nucleus.

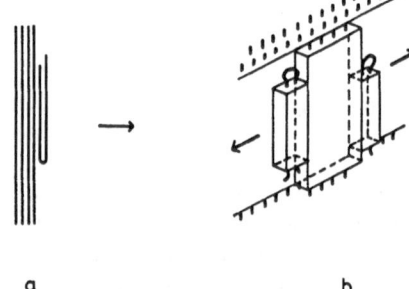

Figure 12 - Unstable chain - folded deposition obstructing the chain-extended crystal growth. a) "rough-surface" growth mechanism, side view of the growth face; b) secondary nucleation mechanism, obstruction of niches. Arrows mark the direction of new chain deposition.

WHAT GOVERNS POLYMER MORPHOLOGY AND CRYSTALLIZATION KINETICS ? THE ROLE OF ACTIVATION ENTROPY

We now turn to the more specific problem of primary nucleation as it happens to be very revealing about fundamental principles of polymer crystallization. In particular we are interested in the critical thickness $l^{\dagger}$ of a primary nucleus, i.e. that which corresponds to the lowest free-energy path towards stability. The established theory of primary nucleation [21], based on the same principle of surface energy barriers as the secondary nucleation theory of polymer crystal growth, predicts that $l^{\dagger}$ for primary nucleation should always be larger than $l^{\dagger}$ for further crystal growth. This should hold in all but rather special instances of epitaxial nuclei that are more stable than the fully grown crystals themselves. However, the foregoing prediction has never been tested experimentally, since the thickness of the primary nucleus is not perpetuated by further growth of the crystal; the l values as normally measured are determined by crystal growth kinetics. Due to the lack of experimentally verifiable parameters primary nucleation has not received much attention lately.

New light on the problem of primary nucleation is shed by the behaviour of our long paraffins because of the large stepwise changes that occur at crossovers between different fold modes. Thus, in the case of C246 we have evidence that the primary nuclei which appear below 121°C (Fig. 10), i.e. only a couple of degrees below the folded-chain melting point T_m^F, are chain-folded [16]. Moreover, the DSC crystallization exotherms in combination with real time SAXS indicate that, at least in a limited temperature range below T_m^F, primary nucleation favours chain-folded crystals while growth favours chain-extended ones; thus l for nucleation is *smaller* than l for crystal growth [16]. These observations disagree with the predictions of the theory.

To visualize the disparity between theory and experiment, the calculated free energy surface ΔF for a primary nucleus of paraffin C246 is shown in Fig. 13 as a function of the fold length l and of the lateral nucleus dimension a for two temperatures. These plots are the equivalents of the "saddle" surfaces in polymer nucleation theory [21], except that for monodisperse paraffins, depending on temperature, there may be several

"saddle points" (marked A, B, C) due to the stepwise change in the end surface free energy σ with l. According to the classical theory, the stable nucleus is achieved by lateral growth (increasing a), the most likely l being that which corresponds to the lowest saddle. Fig. 13, top, refers. to a temperature already below T_m^F, where the experiment shows that chain-folded nuclei are preferred to extended-chain ones. However, as can be seen, the calculated free energy barrier for once folded chains (point B) is several times *higher* than that for extended chains (point A). In fact the calculated free energies of the two saddles A and B only become equal below the melting point of twice folded chain crystals (Fig. 13, bottom). Incidentally, the shape of the ΔF surface is unaffected by whether homogeneous or heterogeneous nucleation is considered as long as the lowest free energy shape is maintained during the nucleus growth; only the absolute scales of ΔF and a are affected.

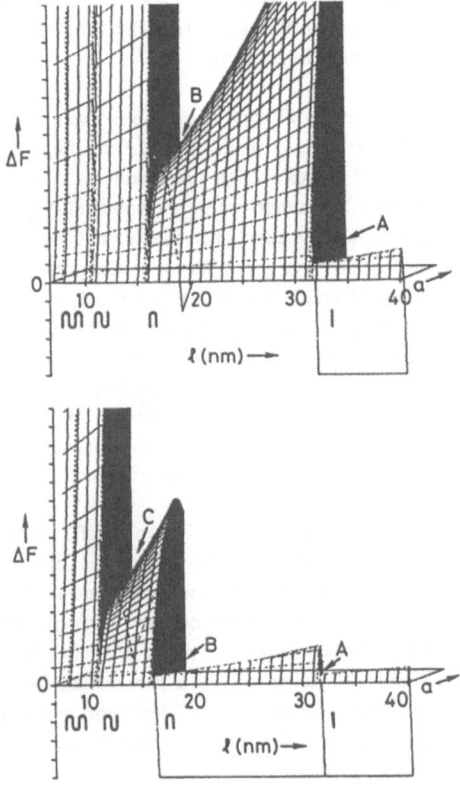

Figure 13 - Free energy of crystallization, ΔF, for a primary nucleus of paraffin C246 as a function of its dimensions along (l) and normal (a) to the chain direction. Top: at T = T_m^E- 8.0 = 120.5°C, Bottom: at T = T_m^E-15 = 103.5°C. Simple step function for σ_e was assumed [22]: $\sigma_e = \sigma + \sigma_\infty(n-1)/n$; n-1 is the number of folds per molecule, σ=10 erg/cm² and σ is the free energy due to folds for a surface containing no chain ends, only folds; σ_∞ =62 erg/cm², as determined from the melting point depression in folded paraffins.

The above stated disparity in l's between theory and experiment would appear to suggest that the surface energy of the nucleus might not be the primary factor determining the activation free energy for primary nucleation.

As space prohibits a detailed elaboration of this point, only the final conclusion and some implications will be stated. It is believed that both the major "anomalies" described in this paper, i.e. the crystallization rate minima and the unexpectedly small l of the primary nucleus, indicate the important effect of activation entropy in the crystallization of chain molecules. Thus, for instance, the main restriction to lamellar thickness might come from the low probability of simultaneous extension of all segments in a portion of a flexible polymer chain of length l that is required for successful incorporation of the stem into the crystal; the larger the l the lower this probability and the higher the "entropic barrier" $-T\Delta S$ [24]. At least under certain conditions this probability consideration may be more important in limiting the l value than the energy barrier due to the creation of additional side surface on deposition of a new stem, a quantity proportional to $l\sigma$.

The issue of chain alignment probability as a factor influencing l was first invoked by Point [23] in his extension of the nucleation based theory of crystal growth to account for limiting l values at high supercoolings. Lately it is introduced as one of the main ingredients of an altogether new theoretical departure, the "rough surface" growth theory of Sadler [24]. The concept has so far not been applied to primary nucleation and it remains to be seen whether it would predict a relationship between l (nucleation) and l (growth) different than that of the purely surface energy based theory (cf. Fig. 13). It may be speculated that the entropic barrier in nucleation would increase very steeply with l, since chain extension may be particularly improbable for the first isolated stems which are not supported and stabilized by lattice interaction with the pre-existing crystal substrate as in the case of further crystal growth. If calculation proves this to be the case, the activation entropy concept will indeed provide an explanation for the observed lower l in primary nucleation compared to that in growth.

As the crystallization rate minima in paraffins indicate, the depositing stem not only has to achieve the proper extended conformation through a series of random attachment / detachment steps, but each such step must also be coordinated with a simultaneous detachment of the unstable overlay. Thus the activation entropy for stem deposition ΔS would actually consist of two parts: (1) that which relates to the extension of the stem itself, and (2) the co-operative part arising from the requirement for detachment of the overlay before rearrangement of the stem can proceed. Accordingly, it would be the increased co-operative part (2) near T that produces the minima in the crystallization rates of long paraffins.

In view of the current developments in polymer crystallization theory, the emergence of the important role of an activation entropy on an experimental basis with the newly available paraffins is very topical. It is being invoked here to account for three qualitatively novel effects which otherwise would remain inexplicable or would require more than one ad hoc assumption. These effects are: (i) non-integer folding as the initial crystal form, (ii) minima in crystallization rate vs. ΔT and (iii) - at least within a limited ΔT range - smaller l pertaining to primary nucleation than to crystal growth. The cumulative experimental support provides an explicit starting point for new theoretical developments and concrete material in the evaluation of existing theories. Also, and chiefly, it provides explicit pointers for further experiments.

CONCLUSION

Although of no direct commercial importance, the new long alkanes have already proved to be invaluable models providing a number of revealing insights into the nature of chain-folding, principles of polymer crystallization and, not least, spectroscopic standards for structure characterization. Just as noble gases were indispensable for the understanding of fundamental principles relevant to all matter, the present simplest and purest long chain compounds may well play a similar role in the domain of crystalline polymers. It was the purpose of this review to stress this potential.

ACKNOWLEDGEMENT

I am indebted to Professor M.C. Whiting for placing these unique alkanes at our disposal, and to Professor A. Keller who initiated the whole project and whose continuous involvement and backing made this work possible. Financial support from Science and Engineering Research Council is appreciated.

REFERENCES

1. Bidd, I. and Whiting, M.C., **J. Chem. Soc., Chem. Commun., 1985**, 543.

2. Lee, K.-S. and Wegner, G., **Makromol. Chem., Rapid Commun.**, 1985, **6**, 203.

3. Ungar, G., Stejny, J., Keller, A., Bidd, I. and Whiting, M.C., **Science**, 1985, **229**, 386.

4. Ungar, G. and Keller, A., to be published.

5. Mizushima, S. and Shimanouchi, T., **J. Am. Chem. Soc.**, 1949, **71**, 1320.

6. Strobl, G.R. and Eckel, R., **J. Polym. Sci., Polym. Phys. Ed.**, 1976, **14**, 913.

7. McCullough, R.L., Eisenstein, A.J. and Weikart, D.F., **ibid.**, 1977, **15**, 1837.

8. Ungar, G. and Keller, A., **Polymer**, 1986, **27**, 1835.

9. Ungar, G., Organ, S.J. and Keller, A., to be published.

10. Organ, S.J. and Keller, A., **J. Polym. Sci., Polym. Phys. Ed.**, in press.

11. Ungar, G. and Organ, S.J., **Polym. Commun.**, in press.

12. Wittmann, J.C. and Lotz, B., **J. Polym. Sci., Polym. Phys. Ed.**, 1985, **23**, 205.

13. Ungar, G. and Organ, S.J., in preparation.

14. Ungar, G., to be published.

15. Buckley, C.P. and Kovacs, A.J., in **Structure of Crystalline Polymers**, ed. I.H. Hall, Elsevier Applied Science Publishers, London, 1984, p. 261.

16. Ungar, G. and Keller, A., **Polymer**, in press.

17. Organ, S.J., Ungar, G. and Keller, A., in preparation.

18. Sadler, D.M. and Gilmer, G.H., **Polymer**, 1984, **25**, 1446.

19. Hoffman, J.D., Davis, G.T. and Lauritzen, J.I.,Jr., in **Treatise in Solid State Chemistry**, ed. N.B. Hannay, Plenum, New York, 1976, p. 497.

20. Sadler, D.M. and Gilmer, G.H., **Polymer Commun.**, in press.

21. For review see: Wunderlich, B., **Macromolecular Physics**, Vol. 2, Academic Press, New York, 1976.

22. Hoffman, J.D., **Macromolecules**, 1986, **19**, 1124.

23. Point, J.J., **ibid.**, 1979, **12**, 770.

24. Sadler, D.M., **Nature**, 1987, **326**, 174.

THE EFFECT OF SIDE GROUPS ON THE CRYSTALLIZATION OF POLYMERS; AN X-RAY DIFFRACTION APPROACH

C.G. Vonk
formerly DSM Central Laboratory
P.O. Box 18, 6160 MD Geleen
the Netherlands

ABSTRACT

As an extension to a previous publication by Vonk and Pijpers [1], a revised scheme for the crystallization of polymers containing side groups is presented. The crystallization is considered to proceed according to the process leading to the highest rate of crystallization. Furthermore, lattice constants of a number of linear as well as of branched polyethylene samples are compared with the corresponding values of paraffins, as determined by Davis et. al [8]. It is concluded that the greater part of the expansion with respect to the ideal, infinite chain lattice is caused by the limited thickness of the crystalline lamellae.

INTRODUCTION

In a recent study by Vonk and Pijpers [1] of a number of samples of High-Density Polyethylene (HDPE), Low-Density Polyethylene (LDPE) and Ethylene-Vinylacetate copolymers (EVA), which were cooled at different rates from the melt, the following data were obtained.

a) By Wide Angle X-ray Scattering (WAXS): the crystallinity ϕ_w and the lattice constants a, b, c of the crystalline part at room temperature.

b) By Small Angle X-ray Scattering (SAXS): the invariant, the width of the transition layer between the crystalline and amorphous phases, and the thicknesses C and A of the crystalline and amorphous layers respectively.

c) By flotation: the overall density.

For experimental details and a description of the various methods used in this investigation, the reader is referred to the orginical publication.

Here, in view of a further discussion of some of the issues, the pertaining results and conclusions are summarized.

Firstly, from the values of the invariant, the overall density and the crystallinity, properly corrected for the width of the transition layer, the crystalline and amorphous densities ρ_{cS} and ρ_{aS} were obtained. Comparison of ρ_{cS} with the value ρ_{cW} of the crystalline density, calculated from the lattice constants, led the authors to the conclusion that the concentration of the side groups in the crystalline regions of the EVA samples was 20-40 % of the overall concentration. This conclusion is in agreement with the results obtained by nitric acid etching techniques [2], from which this concentration in LDPE samples was estimated to be in the same range.

Secondly, a plot of the crystalline lamellar thickness C versus the number average length L of the linear chain segments between side groups (LDPE and EVA samples) or endgroups (HDPE samples) was presented, which is reprinted here as fig. 1. It should be mentioned that the values of C in this plot are of higher quality than those presented generally, which are based on the relation $C = \phi D$, where D is the identity period, and ϕ the crystallinity by volume. The C-values in the present study are derived from the slope of the correlation function near the origin, which implies that these values are based on the specific surface of the phase boundary, and as such are number-averaged values. The C-values obtained from $C = \phi D$ on the other hand do not correspond to some discrete type of average, and furthermore suffer from the circumstance that determination of D from the very broad peaks often met in SAXS curves, also after proper correction, is a rather ambiguous procedure.

In the discussion of fig. 1 the strong dependence on cooling rate found for the HPDE sample (L > 500A) was brought in connection with the thickening of the crystals occuring on annealing. This is thought to be governed by kink motions, involving both rotation and longitudinal displacement of the chains in the crystalline lattice. Here, it may be added that such movements in the LDPE and EVA samples would be impeded, as the rotation of a chain carrying a bulky side group in a crystalline environment is virtually impossible. Also, reptation-like displacements of a chain carrying a side group in the amorphous phase would stop as soon as these displacements would bring a side group to a crystalline surface. Accordingly, samples carrying side groups can be expected to exhibit greatly reduced sensitivity to cooling rate. This is actually shown by the convergence of the lines 'slowly cooled' and 'quenched' at low values of L in fig. 1.

The slope of these curves shows that for high concentrations of side groups the thickness C of the crystalline lamellae is completely determined by the length of the linear chain segments between the side groups. In discussing this dependence, also a limited degree of chain folding, as dictated by the overcrowding effect [3], was taken into account. Furthermore, on the basis of the evidence given before, a limited inclusion of side groups in the crystalline phase was assumed. On the basis of these conditions a scheme for the crystallization of the polymers under consideration was proposed, which was derived from the 'solidification model' (Erstarrungsmodel, [4]). According to this scheme the chains, offered at a growing crystal face, would solidify regardless of the presence of a side group in the crystallizing sequence. The side groups thus dragged into the growing crystal would have to be encapsulated in the crystalline regions. However, after folding only linear segments were assumed to crystallize.

In this communication firstly a somewhat different scheme of the crystallization is put forward, which at present is thought to be better in agreement with various observations on the crystallization of polymers in general. Secondly, as a seperate issue, the effect of crystalline thickness on lattice constants is discussed.

AN ALTERNATIVE TO THE PREVIOUSLY PROPOSED SCHEME OF CRYSTALLIZATION

A shortcoming of the previously proposed scheme is, that it leaves no room for any selection of the chains prior to crystallization. That such a selection takes place can however be inferred from several observations, such as the molecular weight segregation which may take place during crystallization [5], and the crystallization of one component from mixtures of two components which in the molten state are completely miscible [6]. Here, crystallization would be almost completely suppressed if it occurred according to the solidification model. Also convincing in this respect is the separate crystallization of at least one of the components of block copolymers in which there is no phase separation prior to crystallization [7].

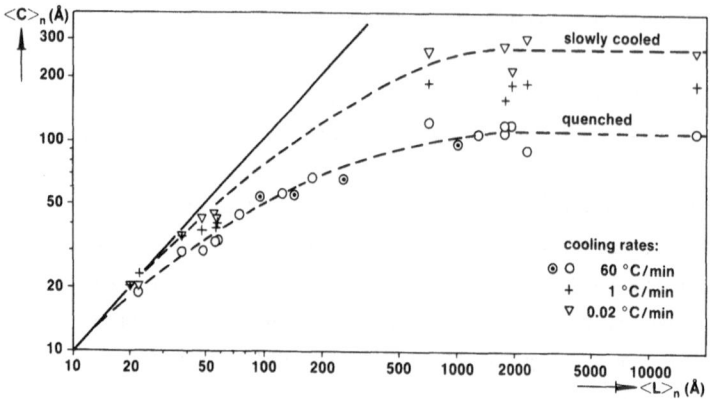

Figure 1. Plot of the number average lamellar thickness versus number average of linear segements between side groups (L < 500 A) or end groups (L > 500 A)

According to the scheme to be proposed next, during the crystallization of the first stem of a newly added chain, sidegroups are excluded from
the crystalline lattice. Thus, if a chain segment, which is in the right
position for crystallization, carries a side group, crystallization will
not proceed until this group has diffused away from the region of
crystallization by reptation of the chain. However, once a first stem has
crystallized, reptation is supposed to be greatly reduced, and any side
group which is connected to the stems to be laid down after folding has to
be incorporated in the lattice. As about half of the stems in a crystalline
lamella can be assumed to have crystallized after folding [3], and as these
stems are assumed to carry the average number of side groups, the concentration of the side groups in the crystalline regions can be expected to
be about 50 % of the overall concentration. (Here, the effect of a possible
enrichment of side groups in the second crystallizing stem, due to removal
from the first stem, has not been taken into account. Such side groups may
be assumed to be accomodated in the fold surface, and thus to be positioned
in the transition layer.) Though this concentration is at the upper side of
the range of concentrations deduced from the experiments described above,
one may in view of the large uncertainties in these experiments consider it
to be in agreement with observations.

The figure of 50 % derived above is independent of the thickness of
the crystalline layers, and thus is not related to the final crystallinity.

For a given thickness of the amorphous layers, one may consider the crystallinity to be determined by the thickness of the crystalline lamellae, and thus by the length of the first stems which are added by newly crystallizing chains. In the following we will consider the consequences of the assumption that the final morphology is determined by the rate of crystallization. For a lamella of thickness C this will be proportional to the thickness C, and furthermore to the rate of addition of new chain segments carrying no side groups. All other circumstances being equal, one may assume as a simple approximation that this rate is proportional to the probability P that a newly offered segment of length C carries no side groups. As can easily be shown, for a random distribution of the sidegroups, P is proportional to $\exp(-C/\langle L \rangle)$. The growth rate would thus be proportional to $C\exp(-C/\langle L \rangle)$, for which the maximum occurs at $C = \langle L \rangle$. This might explain the observation that the data points in fig. 1 approach the tangent $C = \langle L \rangle$ at decreasing values of $\langle L \rangle$.

The present scheme, like the previous one, is not meant to describe the crystallization process accurately as to details. Furthermore, the effect of temperature variations during crystallization has been disregarded, and needs further consideration. However, it may possibly serve as a starting point for further discussions.

THE EFFECT OF CRYSTAL THICKNESS ON LATTICE CONSTANTS

In many studies the expansion of the crystalline lattice of polymers has been regarded as an indication for the incorporation of chain defects in the crystalline regions. However, Davis et. al [8] in 1974 suggested that in this expansion also the thickness of the crystalline lamellae might play an important role. In 1948 Vand and de Boer [9] had pointed to the contraction of the lattice of n-paraffins occurring at increasing chain length, and had ascribed this to the net increase of attractive forces between units of neighbouring chains in the lattice. Davis et al. very accurately measured the lattice constants of a number of n-paraffins in the orthorhombic state, and indeed observed the expected effect.

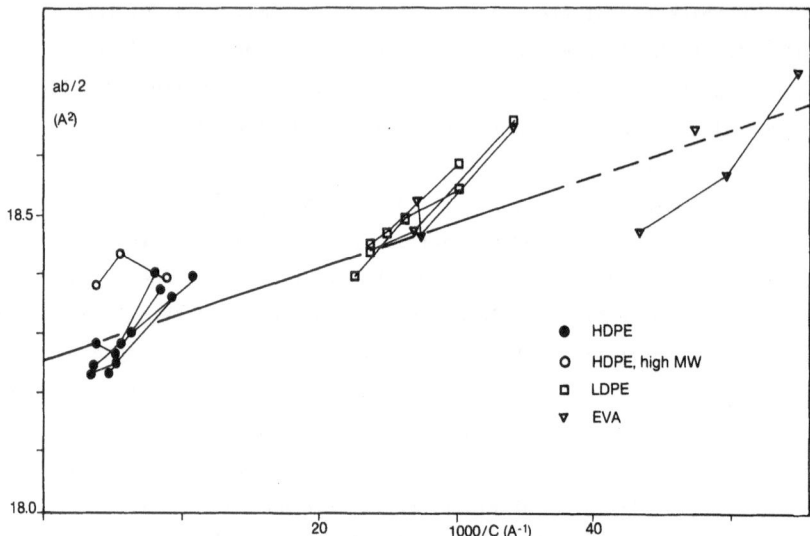

Figure 2. Plot of chain cross section area versus inverse of lamellar thickness. The heavy line represents the equation for n-paraffins, determined by Davis et al. [8].
Samples cooled at different rates are connected by lines; the highest cooling rate in all cases corresponds to the most right data point.

The results obtained in the investigation of Vonk and Pijpers on the HDPE, LDPE and EVA samples allow a comparison with those of Davis et al. In fig. 2 values of ab/2, which is the surface of the chain in projection along the c axis, is plotted versus 1000/C. Here, data points referring to one sample, which has been cooled at different rates from the melt, are connected by straight lines. Also plotted in fig. 2 is the line representing the results of Davis et al. on n-paraffins of length C.

The general agreement between the data points and the line shows that probably the thickness of the crystalline layers is the major effect determining the differences of the lattice constants from the ideal values. However, some systematic deviations can be observed. Thus, the majority of the linear samples gives rise to a slope of the individual curves which is larger than the slope of the paraffin curve. A similar effect was also observed by Davis et al. Only the data points for the high molecular weight sample do not obey this pattern, for reasons which are so far unknown.

The data point for the LDPE samples and the EVA sample with the lowest vinylacetate content are mostly just above the paraffin line. This may be considered as an indication that the lattices of these samples are expanded by the presence of side groups as well. It is to be noted that only a small part of the expansion with respect to the ideal cell ($ab/2 = 18.26$ A^2) would be caused by this effect. If however the side groups are incorporated by substitutional solution, as was suggested in the paper of Vonk and Pijpers, they do not necessarily cause a large expansion of the lattice.

Finally the four data points for the EVA copolymers of higher vinylacetate content are in the correct region with respect to the paraffin line. The number of points however is too small, and the respective accuracies are too low to justify any conclusion with regard to these samples.

In conclusion, it may be stated that according to these results side groups such as occurring in LDPE and EVA polymers play only a minor role in the expansion of the lattice, the major effect being caused by the reduced thickness of the crystalline lamellae. The presence of such defects in the crystalline phase has to be concluded from studies by other methods, of which the nitric acid etching technique probably is the most promising one.

REFERENCES

1. Vonk, C.G. and Pijpers, A.P., An X-ray diffraction study of non-linear polyethylene. I. Room-temperature observations. J. Polym. Sci. Polym. Phys. Ed., 1985, 23, 2517-37

2. Vile, J., Hendra, P.J., Willis, H.A., Cudby, M.E.A. and Bunn, A., Chain branching in high pressure polymerized polyethylene: 2., Polymer, 1984, 25, 1173-7

3. Vonk, C.G., On the overcrowding in the amorphous phase of semicrystalline polymers, J. Polym. Sci. Polym. Lett., 1986, 24, 305-9

4. Stamm, M., Fischer, E.W. and Dettenmaier, M., Chain conformation in the crystalline state by means of neutron scattering methods, Disc. Far. Soc., 1979, 68, 263-78

5. Wunderlich, B., Macromolecular Physics, vol. 2, Acad. Press, New York, 1973, section 5.3

6. Russell, T.P. and Stein, R.S., An investigation of the compatibility and morphology of semicrystalline poly(ϵ-caprolactone)-poly(vinylchloride) blends, J. Polym. Sci. Polym. Phys. Ed., 1983, 21, 999-1010

7. Wunderlich, B., *Marcomolecular Physics, vol. 3*, Acad. Press, New York, 1973, section 10.3.3

8. Davis, G.T., Weeks, J.J., Martin, G.M. and Eby, R.K., Cell dimensions of hydrocarbon crystals: surface effects, *J. Appl. Phys.*, 1974, 45, 4175-81

9. Vand, V. and de Boer, J.H., Intermolecular forces in soap crystals, *Proc. Kon. Ned. Ak. Wet.*, 1947, 50, 991-1002

KINETICS AND MECHANISMS OF FLOW-INDUCED CRYSTALLIZATION

A. J. McHugh
Department of Chemical Engineering
University of Illinois
Urbana, Illinois 61801
USA

ABSTRACT

This paper presents a discussion of flow-induced crystallization emphasizing studies of fiber growth in tubular flow and the highly related rheological phenomenon of shear-thickening. Studies of seeded growth have demonstrated that crystallization proceeds within a concentrated, liquid precursor phase which forms as a first step in the process. The growth kinetics are monitored in-situ using video recording and digital image processing to convert polarized light images to kinetic isotherms. Viscosity studies have been concerned with shear-thickening in fiber forming systems and its analysis in terms of kinetic network models to describe the effects of the chain entanglement process.

INTRODUCTION

The realization that fibrous crystallization in flowing solutions always proceeds within a concentrated liquid or highly entangled gel-like precursor phase has had a profound impact on our understanding of the character of the transformation as well as our ability to monitor the process in-situ [1-3]. Details of precursor formation depend on the particulars of the growth method, however, its importance to the crystallization cannot be overemphasized since it establishes both the existence of the fiber-forming phase as well as its molecular weight and entanglement density-factors known to play a critical role in the final properties. The development of oriented crystallinity and the role of stress can be directly studied using flow visualization methods [4], thus allowing the generation of much-needed quantitative information on this important aspect of the process. Likewise, study of the thermodynamic and rheological aspects of precursor formation provides an important link between the fields of rheology and fiber crystallization. Entanglement formation and precursor crystallization, and their role in fiber property development are in fact central to a number of the techniques for producing high strength and modulus fibers from flexible chain polymers such as gel spinning and hot drawing.

In this paper we present an overview discussion of our recent studies concerning these topics.

MATERIALS AND METHODS

Crystallization Studies

A general schematic of the fiber growth experiment and associated video imaging and display hardware is shown in Figure 1, ref. [4]. The system consists of a 2 mm i.d. capillary tube sandwiched between 2 polarizers whose optical axes are crossed 90° relative to each other and are oriented at 45° relative to the tube axis. A high intensity source is used for illumination. Fiber formation initiates by streaming solution past a stationary seed near the tube entrance. Details including solution preparation techniques can be found elsewhere [2,3]. Table 1 lists the polymer systems and range of conditions we have studied to date. Unless otherwise indicated, concentrations were 0.01 wt%.

TABLE 1

Polymer systems and crystallization conditions

Polymer	Solvent	$[\eta]_0 dl/g$	Temperatures T_c (°C)	Wall Shear Rates $\dot{\gamma}_R$ (s^{-1})
Polyethylene: UHMWPE	xylene	21	101–108	1300–6500
Polypropylene: UHMWPP	tetralin	19.4	84–100	500–2700
Poly(ethylene oxide): (0.1–0.2%)	ethanol	(MW – $4x10^6$)	31–40	500–2700

The analysis hardware consists of a video camera and recorder interfaced through a digitizing board to an IBM computer. The digitizing board decomposes each video frame into a number of picture elements (pixels) approximately 3x5 μm and assigns a value in the range 0 to 255, proportional to the light intensity, for storage and access by the computer. A semi-automated software searches for the pixels associated with a fixed axial position in the fiber and converts the intensity, $I(t)$, relative to the initial (precursor) state, I_o (see Figure 3, ref. [2]), to a fiber retardation, δ_f (t), according to

$$\delta_f (t) = \cos^{-1} \left[1 - \frac{I(t)}{I_o} (1 - \cos 2 \delta_p) \right] - 2 \delta_p \qquad (1)$$

Here, δ_p represents the retardation due to all elements in the optical path other than the fiber and can be obtained by Senarmont compensation.

Viscometry Studies

Steady state viscosity behavior of the fiber forming solutions has been investigated using capillary (Ubbelohde-type, 0.24 mm i.d., L/D = 375) rheometry. Solutions were stabilized with antioxidant and experiments were carried out under a nitrogen atmosphere. Special care was taken to eliminate memory effects (i.e. entanglements) by allowing settling times ($\sim 1/2$ hr) at the respective dissolution temperatures. Appropriate corrections were employed to obtain the wall shear rate and relative viscosity [5]. All drainage time measurements were repeated at least three times and fresh solutions prepared at different times were tested for repeatability. Derived quantities were computed based on data with deviations less than 0.7%.

RESULTS

Transformation Character and Fiber Properties

Figure 3 of ref. [2] shows a typical example of the transformation sequence one sees with the UHMWPE at a fixed axial position in the flow tube (3 cm downstream from the seed tip). One finds the transformation initiates through the formation of a concentrated, optically isotropic precursor which exhibits zero birefringence under all orientations of the crossed polars and is therefore both amorphous and unoriented. Oriented crystallization proceeds within the anchored precursor fiber, induced by stresses transmitted to it from the streaming solution. This precise pattern occurs under all conditions where crystalline fibers form with both the UHMWPE and UHMWPP solutions [2,3]. Time constants for precursor formation, the initiation of birefringence and completion of the axial development of birefringence display a marked dependence with flow rate

at a fixed growth temperature [2,3]. Of particular importance is the
flow plateau region where above some value of average flow rate, $\langle v \rangle$,
time intervals become essentially constant. Likewise at very low flows
(or higher temperatures) time constants rapidly climb to values on thè
orders of minutes and eventually one finds no precursor formation. On
the other hand fiber melting points (obtained by hot stage microscopy)
and room temperature birefringence do show increases with $\langle v \rangle$ over the
full range of flow rates [2,3]. One also finds a strong dependence with
temperature at fixed flow which similarly shows in the crystal
orientation from wide angle x-ray diffraction [2]. The fractionation
which occurs during precursor formation leads to an increase in the
intrinsic viscosity of the fiber phase [6]. The data display trends with
flow rate and temperature which are consistent with expectations from
thermodynamic calculations for the effect of the applied stress on the
"shifted" liquid binodal curves [7].

Crystallization Kinetics

Conversion of transmitted intensity to fiber retardation, δ, at
a fixed axial position is carried out as discussed in the previous
section. To convert such data to a crystallization isotherm one uses
the expression for the birefringence, Δ, of a uniaxially oriented
semicrystalline system in the absence of form effects [8]. Combination
with the definition of δ ($= 2\ \pi\Delta d/\lambda$), along with the fact that
fiber diameter, d, remains nearly constant, gives the following

$$\frac{x}{x_\infty} = \frac{\delta_f(t) - \delta_f(0)}{\delta_f(\infty) - \delta_f(0)} \qquad (2)$$

where x is the volume fraction crystallinity at time t relative to the
initial (0) and final (∞) times. Figure 1 shows examples of such data
for the UHMWPE system plotted according to the Avrami equation [9]

$$-\ln\left(1 - \frac{x}{x_\infty}\right) = \frac{k}{x_\infty}\ t^n \qquad (3)$$

where the coefficients k and n relate to temperature effects, and growth
geometry and nucleation effects respectively. The behavior patterns
shown here are representative of the linear fits obtained for the data in
the high flow rate plateau region. In all cases, Avrami plots were well
fit by slopes corresponding to time exponents n = 2 or 2.5 similar to
earlier observed patterns of strain-induced crystallization of chemically
crosslinked systems [10,11].

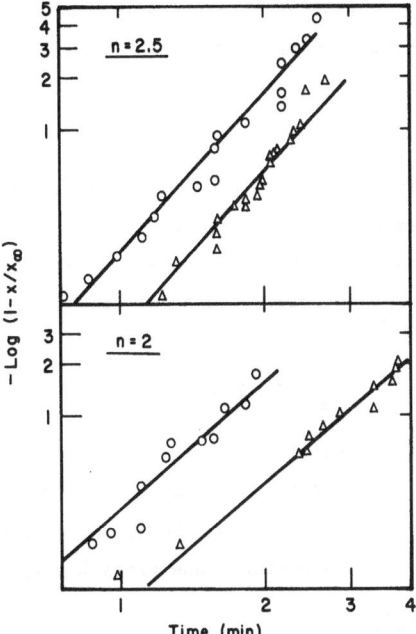

Figure 1. Transformation plots, for n = 2.5 0 - 103.1°C, Δ = 106.2°C,
for n = 2, 0 = 102.5°C, Δ = 103.9°C.

To analyze the temperature coefficient one needs the equilibrium melting temperature, T_m, for an infinite crystal in the precursor phase, which should depend on the applied stress and solvent concentration as well as the precursor entanglement density. The latter two quantities should depend on the initial conditions and flow history leading to precursor structure formation and phase separation [3]. The plateau behavior with ⟨v⟩ suggests that these effects reach a constant value which precludes the need for their precise analysis. Likewise, since stress on the precursor is low (~10^{-3} GPa), the effect on the melting point elevation is completely negligible [2]. Thus, to a good approximation, T_m is constant in the plateau region. Following standard arguments [9,11] the rate constant k can be plotted versus the undercooling parameter for heterogeneous nucleation and growth, $T_m/T_c \, \Delta T$, where T_c is the crystallization temperature and $\Delta T = T_m - T_c$, resulting in a good linear fit based on T_m = 134°C [4].

For the n = 2 data a molecular interpretation can be given to the slope in terms of a one-dimensional growth with sporadic nucleation [9]. Thus

$$k \sim \nu G \qquad (4)$$

where G is the linear (1-D) growth rate and ν is the heterogeneous nucleation rate which may be taken to have the same temperature dependence as G. According to the model suggested earlier by Pennings [12], the longitudinal growth rate, G, can be written as

$$G \sim \exp \left(\frac{-4L\sigma_s^2}{k_B \Delta H} \frac{T_m}{T_c \Delta T} \right) \qquad (5)$$

where L is the length of the axial growth nucleus, σ_s is the associated side surface energy, k_B is the Boltzman constant, and ΔH is the equilibrium crystal heat of fusion. From the slope, together with typical values for the other parameters, one finds $L \cong 12 \ A^o$.

Solution Viscosity Behavior

An indication of the structure formation leading to the precursor state can be seen in the high shear rate viscosity behavior of the fiber-forming systems. Figure 2 shows an example of the reduced viscosity $(\eta_{red} = \eta_{sp}/c = (\eta_r - 1)/c)$ behavior of the UMWPE solution. At lower

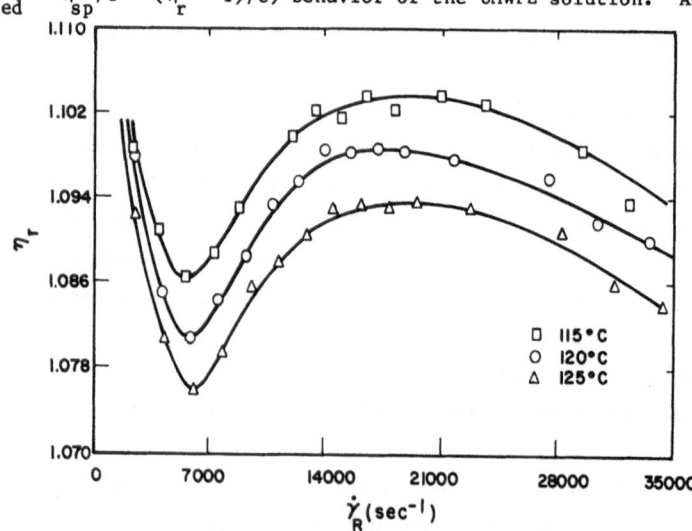

Figure 2. UHMWPE-xylene solution (0.005%) viscosity behavior for various indicated temperatures.

shear rates one sees shear thinning behavior, while above some critical shear rate, $\dot{\gamma}_c$, the solution viscosity begins increasing, remaining nearly constant over an intermediate range of shear rates, before finally decreasing again. This shear thickening pattern, as well as the value of $\dot{\gamma}_c$, varies with solution concentration and temperature. Molecular weight dependence was investigated using polyethylene fractions having zero shear intrinsic viscosities of $7 \frac{d\ell}{g}$ (denoted F50-06) and $2.5 \frac{d\ell}{g}$ (denoted F60-600), respectively. As expected, the magnitude of the thickening decreases with decreasing molecular weight and the critical shear rate increases. Results are summarized in Table 2. Data are also shown for the two high molecular weight poly(ethylene oxide)/ethanol solutions whose fiber forming characteristics were studied [3]. In this case, shear thickening, as such, was not observed, however the viscosity did reach a constant plateau value at around $3000s^{-1}$.

TABLE 2

Summary comparison of critical shear rates for onset of shear thickening

Polymer	Solvent	wt%	T(°C)	$\dot{\gamma}_c(s^{-1})$
Polyethylene:	xylene		110	
UHMWPE		0.01		3300
		0.005		5200
F50-06		0.01		6200
F60-600		0.01		1400
Polypropylene	tetralin		105	
UHMWPP		0.10		2900
Poly(ethylene oxide)	ethanol		33	
PEO ($M_W = 9x10^5$)		0.3		No
PEO ($M_W = 4x10^6$)		0.1	Plateau ~ 3000	

DISCUSSION

Shear Thickening

The data summarized in Tables 1 and 2 indicate that a connection exists between shear thickening and liquid phase precursor formation. We have given careful consideration to possible artifacts which could give rise to anomalies in the observed behavior [5], and believe that the viscosity increase is due to molecular associations and entanglements

which form above the critical shear rate $\dot\gamma_c$ (as opposed to an intramolecular effect due to hydrodynamic resistence and nonuniform coil
expansion [13]). Flow-induced entanglements of this type are also the
likely cause of precursor formation as suggested by the overlap in shear
rates where the two phenomena occur. This picture of intermolecular
interactions is also supported by the relaxation time that is needed for
the solutions to return to their original state. Supporting evidence for
structure formation in similar flow fields is available in a number of
works (see [5] or [14]).

Several authors have suggested that shear-thickening can be
attributed to a conformational transition from the randomly coiled
state to a deformed coil state. Above the critical Weissenberg number
associated with this transition [14], the coils should remain deformed,
thus interpenetration and entanglement formation should be enhanced.
One can estimate the coil-stretch transition condition from the
relaxation time, θ for the coil given by

$$\theta = \frac{6}{\pi^2} \frac{M}{cRT} \; \eta_s \cdot (\eta_{r_0} - 1) \tag{6}$$

where η_s is the solvent viscosity, η_{r_0} is the zero-shear relative
viscosity, M is molecular weight, c is concentration and R and T have
their usual meaning. The deGennes criterion [15] suggests that $\theta \cdot \dot\varepsilon \sim 1$
at the coil-stretch transition, where $\dot\varepsilon$ is the fluid <u>stretch</u> rate whose
maximum value for capillary entrance flow may be estimated as
$\sim 0.5 \; \dot\gamma_R$ [5]   where $\dot\gamma_R$ is the tube wall shear rate. The associated
$\dot\gamma_c$ values would be 6000 s^{-1} for the 0.01% UHMWPE system, 84000 s^{-1}
for the F50-06 system and approximately 2800 s^{-1} for the UHMWPP and
PEO systems. While the value for F50-06 is clearly unrealistic
and, as noted, the PEO solutions did not exhibit shear thickening,
the range over which thickening in the UHMWPE and UHMWPP systems
occurs would be consistent with this argument.

Shear thickening can be shown to arise as a natural consequence of
the non-Gaussian nature of short chain segments in
the transient network model due to Yamamoto [16]. In such a case
one starts with the continuity equation for the segment distribution
function $f(\underset{\sim}{x}, N, t)$

$$\frac{\partial f}{\partial t} + \underset{\sim}{\nabla} \cdot (\dot{\underset{\sim}{x}} \, f) = G(\underset{\sim}{x}, N) - \beta\,(\underset{\sim}{x}, N)\,f \tag{7}$$

where f($\underset{\sim}{x}$, N, t) is the number of segments per unit volume consisting of N subunits with normalized end-to-end vector, $\underset{\sim}{x}$ (= $\underset{\sim}{R}/N\ell$, ℓ = subunit length), $\dot{\underset{\sim}{x}}$ is the rate of change of the segment end-to-end vector under flow, and the functions G, β are respectively the rates of entanglement formation and destruction.

Estimating the segment free energy by a two term non-Gaussian expansion leads to the following expression for the network stress tensor, $\underset{\sim}{\tau}_p$ [16]

$$\underset{\sim}{\tau}_p = 3kT \sum_N N\langle\underset{\sim}{xx}\rangle + \frac{9}{5} kT \sum_N N \int\!\!\int\!\!\int_{-\infty}^{\infty} \underset{\sim}{x}\,\underset{\sim}{x}\,(x^2 + y^2 + z^2)\, f(\underset{\sim}{x},N,t)\, dxdydz \qquad (8)$$

from which the shear viscosity and normal stress coefficients can be derived.

Figure 3 shows an example of the calculated viscosity - shear rate behavior as a function of the non-affine slip coefficient, ξ , and segment length N. Since β_o is constant it takes on the meaning of an inverse relaxation time, L is the entanglement rate coefficient, and $\alpha = \dot{\gamma}/\beta_o$. One sees that, depending on N, shear-thickening can be predicted. Also, at fixed N, the effect decreases with ξ[16].

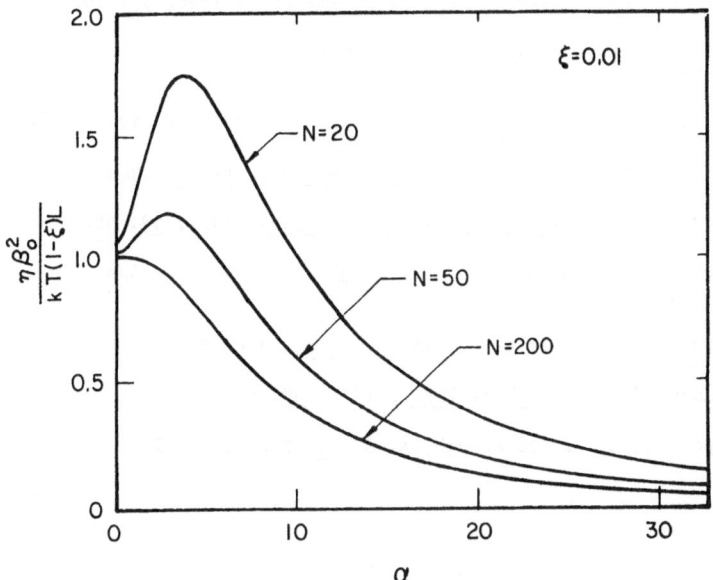

Figure 3. Normalized shear viscosity versus dimensionless shear rate (see discussion).

380

These studies have been supported by the National Science Foundation (DMR 84-04968) and the 3M Company.

REFERENCES

1. McHugh, A. J. and Rietveld, J., J. Polym. Sci., Polym. Lett. Ed., 1985, 23, 2359.

2. Rietveld, J. and McHugh, A. J., J. Polym. Sci., Polym. Phys. Ed.,, 1985, 23, 2339 and 2359.

3. McHugh, A. J. and Blunk, R. H. Macromolecules, 1986, 19, 1249.

4. McHugh, A. J. and Spevacek, J. A., J. Polym. Sci.: PtC: Polym. Lett., 1987, 25, 105.

5. Vrahopoulou, E. P. and McHugh, A. J., J. Non-Newtonian Fluid Mech., 1987, in press.

6. McHugh, A. J., Vrahopoulou, E. P., and Edwards, B. J., J. Polym. Sci., Polym. Phys. Ed., in press.

7. Vrahopoulou-Gilbert, E. and McHugh, A. J., Macromolecules, 1984, 17, 2657 and J. Appl. Polym. Sci., 1986, 31, 399.

8. Stein, R. S and Norris, F. H., J. Polym. Sci., 1958, 27, 567.

9. Mandelkern, L., Crystallization of Polymers, McGraw-Hill, New York, 1964, Chap. 8.

10. Gent, A. N., J. Polym. Sci. PtA-2, 1965, 3, 3787 and, 1966, 4, 447.

11. Kim, H. G. and Mandelkern, L., J. Polym. Sci. PtA-2, 1968, 6, 181.

12. Pennings, A. J., J. Polym. Sci. Polym. Symp., 1977, 59, 55.

13. Peterlin, A., J. Chem. Phys., 1960, 33, 1799 and Pure Appl. Chem., 1966, 12, 563.

14. Choplin, L. and Sabatie, J., Rheol. Acta, 1986, 25, 570.

15. de Gennes, P. G., J. Chem. Phys., 1974, 60, 5030.

16. Vrahopoulou, E. P. and McHugh, A. J., J. Rheol., 1987, in press.

STRUCTURE AND CRYSTALLIZATION OF HOMOGENEOUS AND HETEROGENEOUS ETHYLENE COPOLYMERS INCLUDING VERY LOW DENSITY POLYETHYLENES

V.B.F. Mathot, M.F.J. Pijpers
DSM Research
P.O. Box 18
6160 MD Geleen
Netherlands

ABSTRACT

Homogeneous and heterogeneous ethylene copolymers with very low densities are described. The heterogeneous Very Low Density Polyethylenes (VLDPE's) differ from the homogeneous ethylene-propylene and ethylene-octene copolymers in that their crystallization and melting ranges are extremely wide, as found by quantitative DSC heat capacity measurements. Enthalpy curves show that the melting process is already in progress from -60 °C onwards and is completed at the remarkably high temperature of about 130 °C. Further, the crystallization and melting curves of these materials show several peaks.

INTRODUCTION

Since the commercial introduction of LLDPE around 1980, this type of polymer has received increasing scientific attention. Especially its molecular structure and crystallization behaviour were found to be complex [1]. It has been found that comonomer incorporation in LLDPE is intermolecularly heterogeneous, and the polymer is in fact a blend of molecules ranging from highly branched to virtually linear. This information, in combination with DSC measurements, yields clues to morphology and properties of the polymer.

Recently, homogeneous and heterogeneous ethylene copolymers of even lower density were introduced on the market. They are not only industrially important but are also interesting from a theoretical point of view, on account of intriguing relationships between their molecular structure and their crystallization behaviour.

MATERIALS AND RESULTS

Recently, it became possible to polymerize Very Low Density Polyethylenes (VLDPE's) with densities to well below 900 kg/m^3. Figure 1 shows such a VLDPE; the most remarkable features of this polymer are its wide crystallization and melting ranges, especially compared with an ethylene-propylene copolymer with virtually the same density, whose heat capacity curve has been included in the figure.

Such ethylene-propylene copolymers can be classified as homogeneous, since the distribution of the propylene units is intra- and intermolecularly homogeneous. Since propylene can be incorporated as normal or inverted propylene, these EP-copolymers can be regarded as terpolymers. With the aid of C-13 NMR and a specially developed model, reactivity values have been determined for these polymers [2]. These values enable accurate calculation of chain structure and methylene and propylene sequence length distributions, which are important for crystallization.

Meanwhile, similar homogeneous ethylene-propylene copolymers and ethylene-butene copolymers of very low density have become commercially available.

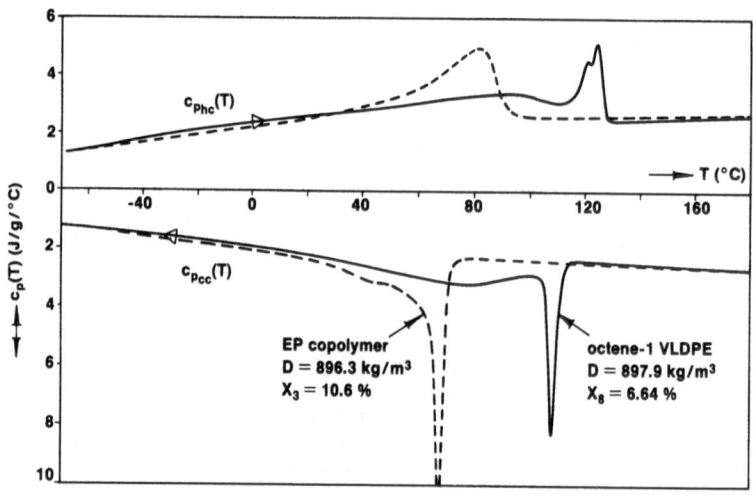

Figure 1. DSC-2 continuous heat capacity curves for cooling ($c_{p_{cc}}$) at 10 °C/min from 180 to -70 °C and subsequent heating ($c_{p_{hc}}$) at 10 °C/min. Isothermal stays of 5 minutes.

Figure 2 shows heat capacity curves for a homogeneous and a heterogeneous ethylene-octene copolymer with virtually the same octene content, which is higher than that of the VLDPE of Figure 1.

The relative width of the temperature ranges for the VLDPE in comparison with the homogeneous copolymer is also illustrated in Figure 3, which was obtained via a procedure previously applied to LPE [3]. For the VLDPE of Figure 2, the $c_{p_{hc}}(T)$ curve (hc stands for heating curve) was integrated, yielding a specific enthalpy curve $h_{hc}(T)$, which was compared with reference curves for 100 % amorphous and 100 % crystalline linear polyethylene [$h_a(T)$ and $h_c(T)$, respectively]. The quantity $w^c(T) = [h_a(T) - h(T)]/[h_a(T) - h_c(T)]$ is an accurate measure of the degree to which these extreme conditions are realized and is very sensitive to experimental problems.

The $w^c_{hc}(T)$ curve shows a continuous decrease from −60 °C. Its value at 23 °C, which is 25 %, is considerably lower than its value at −60 °C. This illustrates that the melting process is already in progress from −60 °C onwards, while for this specific sample it is completed at the remarkably high temperature of about 130 °C.

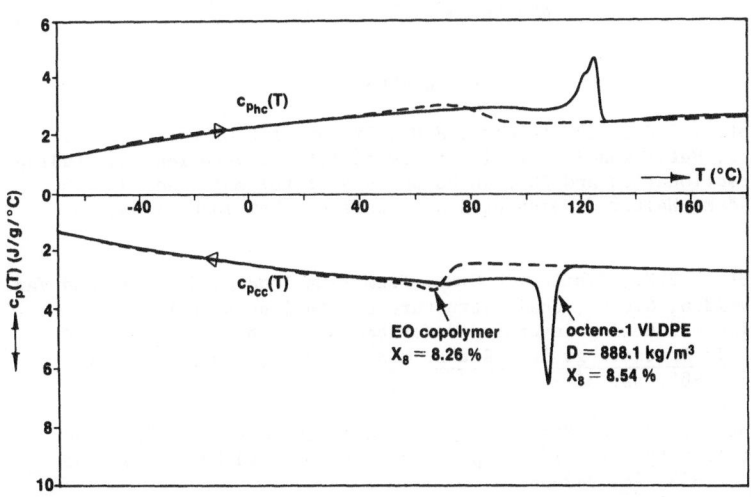

Figure 2. DSC-2 continuous heat capacity curves for cooling ($c_{p_{cc}}$) at 10 °C/min from 180 to −70 °C and subsequent heating ($c_{p_{hc}}$) at 10 °C/min. Isothermal stays of 5 minutes.

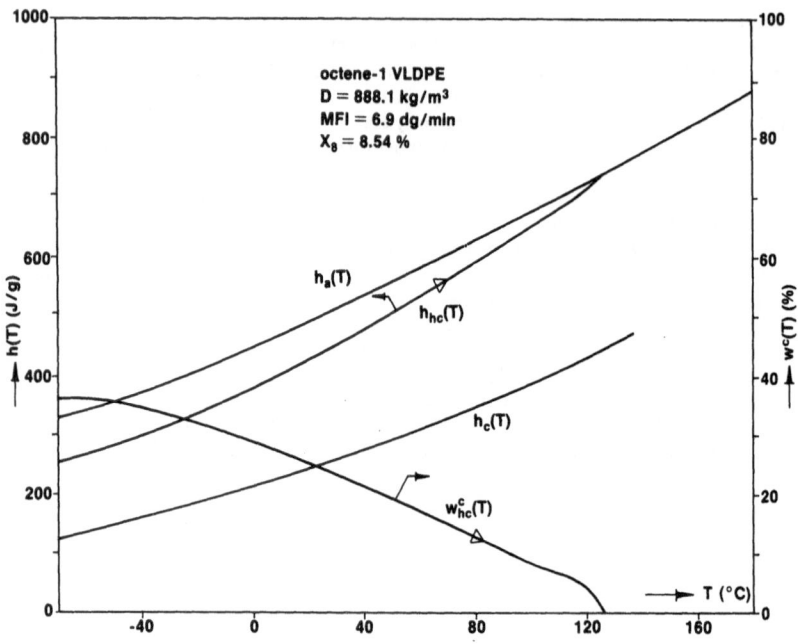

Figure 3. Specific enthalpy curve, reference specific enthalpy curves and the $w^c{}_{hc}(T)$ curve.

REFERENCES

1. Mathot, V.B.F., Schoffeleers, H.M., Brands, A.M.G. and Pijpers, M.F.J., Heterogeneity of linear low density polyethylene as studied by fractionation and DSC. In Morphology of Polymers, ed. B. Sedláček, Walter de Gruyter & Co., Berlin – New York, 1986, pp. 363-70.

2. Mathot, V.B.F., Fabrie, Ch.C.M., Tiemersma-Thoone, G.P.J.M. and Van der Velden, G.P.M., Chain structure of ethylene-propylene copolymers showing inversion described on the basis of C-13 NMR data. In Proceedings Int. Rubber Conf., Kyoto, Japan, October 15-18, 1985, pp. 334-40.

3. Mathot, V.B.F. and Pijpers, M.F.J., Heat capacity, enthalpy and crystallinity for a linear polyethylene obtained by DSC. J. Thermal Anal., 1983, 28, 349-58.

SOLUTION INDUCED CRYSTALLIZATION OF POLYCARBONATE BY THF

H. Schorn, M. Heß, R. Kosfeld
Universität -GH- Duisburg
FB 6/Physikalische Chemie

ABSTRACT

Solution induced crystallization of Poly(bisphenol-A-carbonate) with THF was investigated using combined chromatographic and light-scattering procedures.
The distribution functions of the molar masses of crystalline and amorphous material from solvent induced processes and of spherulitic material, which - on the contrary was obtained by evaporation - are discussed.

INTRODUCTION

Crystallization of polymeric material is not only known due to the thermic history of the sample but also to the contact with solvents of special quality. This type of "solvent induced crystallization" has been investigated in the case of Polyethyleneterephthalate with different solvents [1,2] and in the system Polycarbonate/Aceton [3].

Analysing Polycarbonate based on Bisphenol A (PC) in Tetrahydrofuran (THF) by size-exclusion chromatography (SEC) equiped with low-angle laser light-scattering (LALLS) photometer [4,5] polymodale distribution of PC was observed. Fractions of higher molar masses were observed depending on the procedure of preparing the solution. As the PC under analysis was known to be of unimodal mass distribution origionally, a fractionation process had occured during the process of solution-forming. Our investigations started from the assumption that solvent induced crystallization took place producing unsoluble microcrystallites and thus pretending high molecular material.

In contrast to the work done by Mercier et al. [3] with the non-solvent Aceton the THF used in our experiments is known as a low quality solvent which, nevertheless, is often used in combination with PC.

RESULTS AND DISCUSSION

Depending on dissolution ratio SEC-LALLS analysis of PC in
THF resulted in chromatograms of different shape. Quickly
forming the solution at elevated temperatures under stirring
resulted in solutions showing the expected unimodal polymer
distribution.

Slow solution processes left small unsolved polymer particles
which could not be separated by a 0,5 μm filter. Chromatographic analysis of these suspensions showed a quickly eluted
prefraction pretending a bimodal polymer distribution with a
high molecular component [6].

The mass fraction of the microcrystalline material varied
depending on preparation the solution and decreased slowly
with increasing time.

It is well known that crystalline polymer phases are of considerable lower solubility than amorphous phases [2,3,7]. The
residues, we obtained in our experiments, were almost
unsoluble in THF unless they were heated up to about 60 $^{\circ}$C.
At this temperature clear solutions were obtained, which
remained in this state even at subsequent cooling. SEC-LALLS
analysis in this case did not show higher fractions,
indicating that the assumption of solvent induced growth of
microcrystalline material is of some sense.

Up to 40 % (w/w) of unsoluble crystalline material could be
obtained from granules by simply pouring THF over PC for one
day.

Calorimetric analysis of the residue revealed an unusual high
melting range localized at 240-250 $^{\circ}$C, indicating a high
degree of crystallinity. Common values for the melting region
from literature are located between 220 and 230 $^{\circ}$C [5].

Microscopic investigations in a raster electron microscope
clearly show the brittleness of the cystalline material
(figure).

Fractionation processes during crystallization have been
published by Sadler [8] for different systems. Similar processes are obviously occuring in the system PC/THF, too.

Calculation of the distribution function of the molar masses from SEC-LALLS data of unsoluble crystalline residue, the soluble fraction, and by 60 ^{o}C obtained complete solutions resulted in significantly different distribution functions.

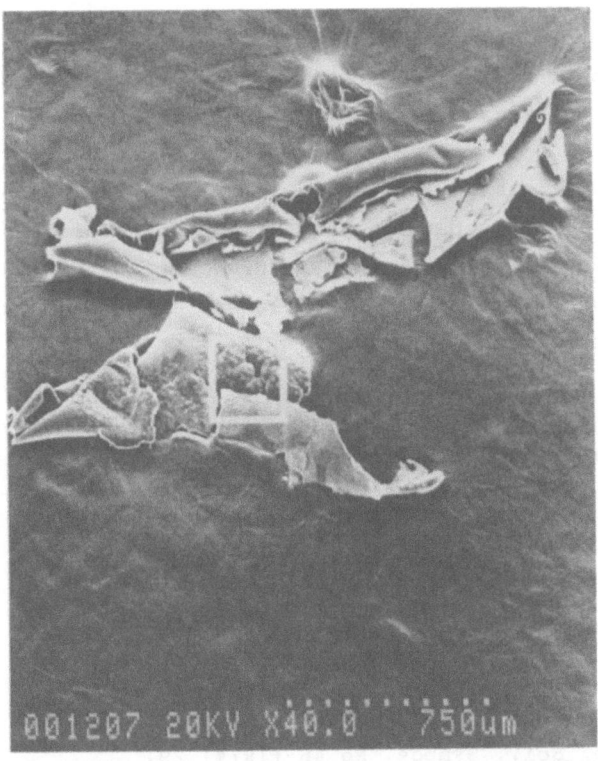

electron micrograph of the crystalline residue on support

It was found that long chain molecules preferably build up crystalline phases. The distribution function is more narrow in the crystalline material and the average values of the molar masses are higher than in the heat treated solutions of the whole sample. The PC dissolved at room temperature in contrary exhibits lower values of average molar masses but shows a more narrow distribution function as well.

Crystalline PC prepared by solvent evaporation forms the well known spherulitic structures visible in polarization microscopy. Spherulitic PC only shows a partial crystallinity and was not found in samples prepared by solvent induced crystallizaton.

Spherulitic PC is not easily dissolved in THF. This is quite in contrast to what one finds if totally amorphous PC in thin films is investigated which quickly dissolves in THF, generally.

This spherulitic materials are quite stable in THF so that it is possible to separate them from the amorphous matrix. Electron microscopy shows a structured surface of the spheric particles where amorphous PC obviously has been extracted from the crystalline lamellas.

Spherulits contacting each other during growing process build up spherulite aggregates with planar contact interfaces. During dissolution in THF these aggregates divide into the spherulitic parts, which then are of great stability against THF as well.

REFERENCES

1. W.R. Moore, R. P. Sheldon
   Polymer, 2 (1961) 315
2. A.B. Desaï, G.L. Wilkes
   J. Polym. Sci., Sympos. No 46 (1974) 291
3. J.P. Mercier, G. Groeninckx, M. Lesne
   J. Polym. Sci., C 16 (1967) 2059
4. A.C. Duano, W. Kay, J. Polym. Sci. Polym. Chem. Ed., 12 (1974) 1151
5. H. Schorn, R. Kosfeld, M. Heß, J. Chromatogr. 282 (1983) 579
6. H. Schorn, R. Kosfeld, M. Heß, J. Chromatogr. 353 (1986) 273
7. W.F. Christopher, D.W. Fox
   Polycarbonates, Reinhold Publishing Corporation, New York (1962)
8. D.M. Sadler, H.H. Wills
   J. Polym. Sci., Part A-2, 9 (1971) 779

Part 6

STRUCTURE/MORPHOLOGY

RIGID ROD POLYMERS WITH FLEXIBLE SIDE CHAINS:
A NEW CLASS OF LIQUID CRYSTALLINE POLYMERS

M. Ballauff
Max-Planck-Institut für Polymerforschung
Postfach 3148, 6500 Mainz, FRG

ABSTRACT

A series of fully aromatic polyesters, polyamides, and polyimides bearing n-alkoxy side chains (n=2-18) have been investigated by optical microscopy, X-ray analysis and DSC. All members of these series have a rigid backbone and exhibit a decreasing melting range with increasing length of the side chains. The polyester with short side chains (n=2-6) form nematic melts. Longer side chains lead to the observation of a novel layered mesophase for all systems under consideration here. There is a linear relationship between the length of the side chains and the layer spacing observed in the mesophase. All findings can be explained by assuming a packing model with interdigitated side chains.

INTRODUCTION

Rigid chain polymers usually exhibit melting points far above the temperature of decomposition which makes these materials difficult to process and to study. Recently [1-4], it has been shown that appending flexible side chains to rigid backbones will lower the melting point considerably leading to the observation of thermotropic mesophases. In this paper we wish to present a survey of recent structural studies on the stiff-chain

polyesters 1, polyamides 2, 3, and the polyimides 4. Details of the synthesis and characterisation of these systems may be found in references [4-7].

RESULTS AND DISCUSSION

In accord with the theoretical considerations [8] short alkyl side
chains disturb crystallization and lower the transition temperature [4,6] .
Thus the polyesters 1 based on hydroquinone and 2,5-dialkoxy terephthalic
acid will exhibit a nematic mesophase if the number n of carbon atoms in
the alkoxy side chains is located between 2 and 6. In case of n=2 the
length of the substituent is still too small to allow the observation of
the nematic-isotropic transition below the temperature of thermal decomposition ($\geq 350^\circ$C).Appending of propoxy side chains already is followed by a
depression of the ordering transition to temperatures around 300°C [6],
longer alkyl chains lead to even lower values of the transition point.

If long side chains (n 12) are appended the formation of layered
structures in the solid state as well as in the mesophase appears as a
novel feature [4] . Figures (1a-c) displays the diffractograms resulting
from the mesophases of polyesters 1 and the polyamides 2 and 3   5 . The

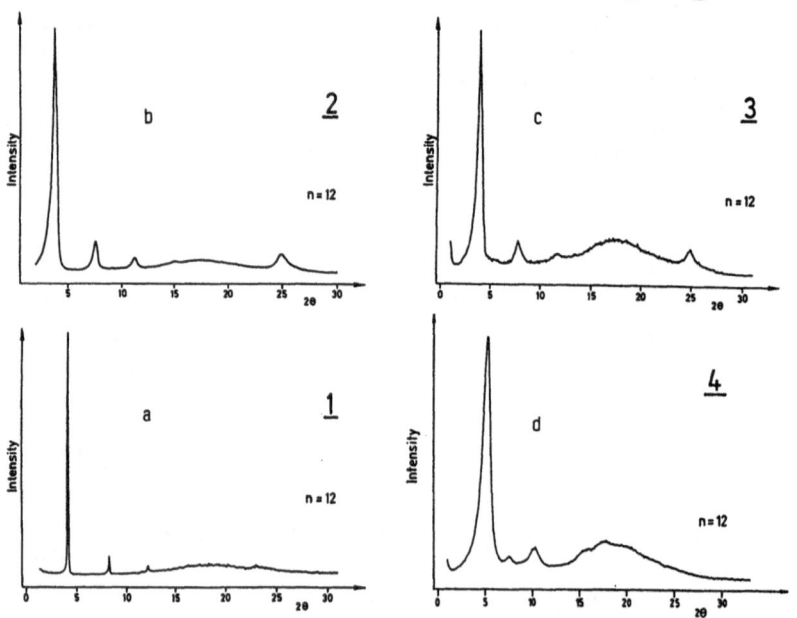

Fig. (1a-d): X-ray diffractograms for polyesters 1 (2a), Polyamides 2,3
(2b,c) and polyimides 4 (2d) bearing dodecyloxy side chains
(n=12).

strong reflections together with their higher orders indicate a layered structure to be present in the mesophase. A broad halo typical of a liquid-like arrangement of the molecules is seen at higher scattering angles. For the polyamides 2 and 3 an additional reflection not being observed in the diffractograms of the corresponding polyesters 1 appears in the region of wide angles. This reflection may be related to prefered distances between the main chains due to hydrogen bonding. In addition to these observations it is interesting to note that even polyamides can be transfered into the molten state through appending of flexible side chains [3,5].

If the layer spacing d obtained from the mesophase is plotted versus n, the number of carbon atoms in the side chains, a straight line results with a slope of 1.25Å per CH_2 unit (see Fig.(2)).

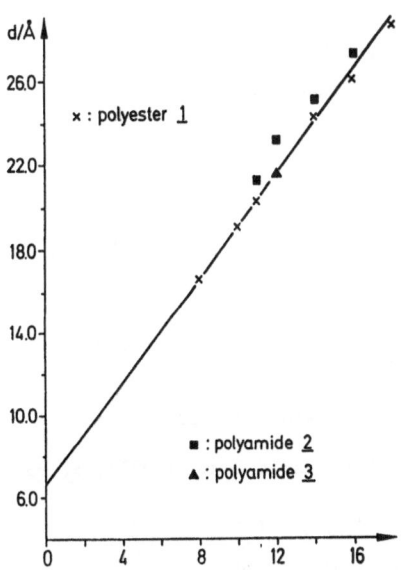

Fig. (2): Layer spacing d calculated according to Bragg's law from the low-angle X-ray reflection observed in the mesophases of the polyesters and polyamides 1 - 3 [8].

Since this value corresponds approximately to the projection of the C-C distance onto the hydrocarbon axis, the alkyl side chains must protrude perpendicularly to the main chain axis. The intercept of ca.6.5Å agrees very well with the diameter of the aromatic main chain and a certain increment for the oxygen atom. Figure (2) furthermore shows that approximately the same layer spacing is found for polymers 1 - 3 despite the different natur of the main chain. The residual differences may be traced back to the different temperatures. A similar investigation of the polyimides 4 5 again lead to the conclusion that these materials can be transfered into a mesophase with a layered structure (see Fig. (1d)). However, the spacing d as well as the increment per CH_2 unit is smaller than the results displayed in Fig. (2)). A comparison with results obtained from polyesters 1 suggests that the mesophase of the polyimides 4 is more akin to the crystalline structure B [6]. Hence, it may be regarded as high-temperature phase having features both of the liquid and the solid state.

From all these observations the following model of molecular packing in the mesophase may be set up (see ref. [5, 6]):

The liquid crystalline melt of the polymers 1 - 4 consists of well-defined layers. Both the confinement of the side chains by the layered structure as well as by the main chains leads to the observed linear relationship of layer spacing d and the length of the hydrocarbon chains (see Fig.(2)). A comparison of the overall breadth of the molecule with d suggests that the side chains interdigitate. The results of the polyamides 2 and 3 as well as of the polyimides 4 indicate that stronger interactions between the main chains is followed by the development of a higher correlation and order within the mesophases.

ACKNOWLEDGEMENT

Financial support by the Bundesministerium für Forschung und Technologie, Projekt "Steife Makromoleküle" is gratefully acknowledged.

REFERENCES

[1] J. Majnusz, J.M. Catala, R.W. Lenz, Europ. Pol. J. (1983), 19, 1043

[2] W.R. Krigbaum, H. Hakemi, R. Kotek, Macromolecules (1985), 18, 965

[3] O. Hermann-Schönherr, J.H. Wendorff, H. Ringsdorf, P. Tschirner, Makromol. Chem. Rapid Comm. (1986), 7, 791

[4] M. Ballauff, Makromol. Chem. Rapid Comm. (1986), 7, 407

[5] M. Ballauff, G.F. Schmidt, Makromol. Chem. Rapid Comm. (1987), 8, 93

[6] M. Ballauff, G.F. Schmidt, Mol. Cryst. Liq. Cryst., in press

[7] M. Wenzel, M. Ballauff, G. Wegner, to be published

[8] M. Ballauff, Macromolecules (1986), 19, 1366

SOLID STATE NMR STUDIES OF THE RELATIONSHIP BETWEEN POLYMER STRUCTURE/DYNAMICS AT THE SEGMENTAL LEVEL AND MACROSCOPIC PHYSICAL PROPERTIES

Alan D. English
Central Research and Development Department
Experimental Station
E. I. Du Pont de Nemours and Company
Wilmington, Delaware 19898

ABSTRACT

Solid State NMR has been used to study polymer structure, morphology, and dynamics for nearly 40 years. In the past 15 years with the advent of various methods to manipulate nuclear spin Hamiltonians and advances in theoretical approaches, solid state NMR spectroscopy has advanced to the point that it is the most detailed probe of polymer structure and dynamics at the segmental level. These methods are also of use in examining macromolecular dynamics on length scales comparable to an entanglement spacing. Additionally, various methods have been developed to discriminate between spin populations of varying mobility that are identified with crystalline, truly amorphous, and phases of intermediate mobility. Recent results on the relationship between polymer segmental dynamics (from solid state NMR, dielectric relaxation, and mechanical characterization) and excess free volume illustrate that our understanding of the relationship between microscopic properties and macroscopic equilibrium properties is improving.

INTRODUCTION

Solid state NMR methods have been used to investigate polymer structure and dynamics for nearly forty years. Conventional methods[1,2] have been used to examine not only the chemical structure of macromolecules, but also to explore the relationship between nuclear spin relaxation and mobility of chain segments. These methods have mainly been confined to studies of abundant spin systems (^{1}H and ^{19}F) where the nuclear dipole interaction is dominant. Due to the many body nature of the nuclear spin dipolar Hamiltonian, this approach is usually not amenable to detailed investigation of the types of motion present. With the advent of coherent averaging methods[3], various techniques have been developed which allow for the manipulation of the nuclear spin Hamiltonian to permit detailed investigation of the structure and dynamics of polymer chains on different length scales.

Various types of coherent averaging methods have been employed to investigate polymer structure and dynamics[4]. There are now more than 100 different polymer systems that have been investigated with these methods and only a few illustrative examples are given here. Multiple pulse methods (WAHUHA, MREV-8, etc.) have been used to suppress homonuclear dipolar interactions and thus allow the observation of the chemical shift anisotropy in both ^{1}H (polyethylene and polyethyleneterephthalate) and ^{19}F (polytetrafluoroethylene homopolymers and copolymers) NMR experiments; in conjunction with magic angle spinning (CRAMPS) to give isotropic chemical shift spectra [poly(ethyleneterephthalate)]. Continuous wave heteronuclear decoupling methods (Dipolar Decoupling) have been used to observe chemical shift anisotropy in ^{13}C NMR experiments (polyethylene, polytetrafluoro-ethylene, polyethyleneterephthalate, polycarbonate, etc.); and combined with magic angle spinning (CP/MAS--cross polarization/dipolar decoupling and magic angle spinning) to yield spectra dominated by the isotropic chemical shift values (virtually every synthetic polymer). Additionally, one of the most successful methods has been the use of ^{2}H NMR when used with specifically labelled polymer.

These methods have been used to investigate chain microstructure, solid state packing effects on chain conformations, orientation at the segmental level, influence of morphology upon chain structure and chain dynamics, the origin of mechanical and dielectric relaxations in both crystalline and noncrystalline domains, and terminal relaxation times in polymer melts. The objective of these investigations is to obtain structural/dynamical information on a segmental length scale that can assist in developing a more fundamental understanding of the relationship between these microscopic properties and macroscopic properties observed in mechanical, rheological, thermal, or dielectric measurements. Recently we have begun to investigate the relationship between chain dynamics in polymer melts as inferred from ^{1}H NMR data and the excess free volume as obtained from the Simha-Somcynsky[5,6] equation of state.

MATERIALS AND METHODS

^{1}H NMR spectra were acquired at 200 MHz with a Bruker MSL 200 NMR spectrometer with a $\pi/2$ radiofrequency pulse length of 2 usec, a 4 usec blanking time before data acquisition, and repetition times larger than three times the spin-lattice relaxation times at all temperatures. Data on pressure, volume and temperature relationships were acquired with an apparatus previously described[7]. The polymers were as-polymerized polyoxymethylene (M_n ~4×10^4), melt-recrystallized poly(ethylene terephthalate) (M_n ~3×10^4), as-polymerized polyethylene (M_w ~3×10^6), and melt-recrystallized nylon-66 (M_n ~1.7×10^4). All polymers were dried for 40-60 hours in a vacuum oven with a nitrogen bleed at 100°C; residual water content was less than 0.2% by weight as determined by thermogravimetric analysis.

RESULTS

Definition of the fraction of a semicrystalline polymer that is "mobile" in a conventional NMR experiment depends substantially upon temperature. The less mobile spins have a near Gaussian lineshape and the more mobile spins have a near Lorentzian lineshape. The relative amounts and the temporal character of the two components in the lineshape may be determined in the time domain and thus the relative amount of mobile material (δ) determined. The temperature dependence of the relative amount of mobile material present in the ^{1}H NMR spectrum of as-polymerized polyoxymethylene is given in Table 1 and illustrates a monotonic increase with temperature; the highest temperature for which δ is measured is determined by the temperature at which the smallest crystallites begin to melt (T_{max}) as observed by differential scanning calorimetry.

TABLE 1
Temperature Dependence of the Mobile Fraction (δ)
As Oberved by ^{1}H NMR in Polyoxymethylene

Temperature(°C)	δ
60	.20
77	.23
95	.26
113	.28
130	.30
148	.33

Table 2 illustrates that the fraction of immobile material $(1-\delta)$ in POM as determined from ^{1}H NMR agrees with values of crystallinity determined by other methods if T_{max} is chosen as the reference temperature. This agreement indicates that all of the noncrystalline component can undergo sufficiently large amplitude motions to be distinguished from the crystalline component before any of the crystallites melt. Data in Table 2 also shows similarly good agreement between $(1-\delta)$ at T_{max} and other measures of crystallinity for PE. The fortuitous nature of this agreement between $(1-\delta)$ at T_{max} and the crystalline content as observed by other methods for POM and PE is illustrated by the data for PET and nylon 66. In these cases, $(1-\delta)$ = 82%, 55% crystalline) and nylon 66 [$(1-\delta)$ = 67%, 45% crystalline]. This disagreement is not uncommon and is the motivation for examining other methods of analyzing the temperature dependence of δ to extract a meaningful value for the crystalline content of the polymer.

TABLE 2
Crystallinity Values Determined by Various Methods

Polymer	$(1-\delta)$[a]	T_{max}	$(1-\delta_o)$	% Crystallinity
POM	67%	148	62%	56%-67%[b]
PET	82%	238	55%	53%-58%[d,e]
PE	75%	103	68%	76%[f]
Nylon 66	67%	220	49%	45%[g]

a) At T_{max}
b) DSC (ΔH_f = 78 cal/g or 93 cal/g for 100% crystallinity)
c) ^{13}C NMR (unpublished results)
d) WAXS
e) DSC (ΔH_f = 32.5 cal/g for 100% crystallinity)
f) DSC (ΔH_f = 70 cal/g for 100% crystallinity)
g) DSC (ΔH_f = 45.5 cal/g for 100% crystallinity)

DISCUSSION

Previous efforts[8] to use thermophysical data to aid in the inter-pretation of conventional NMR spectra have centered on the idea that above some temperature (T_o) there exists sufficient free volume in the lattice to permit some fraction of the material (δ) to become more mobile; at high enough temperatures all noncrystalline material (δ_o) becomes sufficiently mobile to be differentiated from the crystalline material. We have modified this treatment and have used more precise data on the temperature dependence of the excess free volume $[(1-Y)\equiv Y]$ as obtained from the Simha-Somcynsky equation-of-state analysis of pressure-volume-temperature data[9]. Analyses of PVT data for PET[10], POM[11], PE[12] and nylon

66[11] yielded values of the reducing parameters (T*, V*)(°K, cm^3/g) of (11710, 0.7408), (11024, 0.7686), (9772, 1.1417), and (11470, 0.8935) respectively; the reducing parameters and then used to calculate the excess free volume (Ψ) as a function of temperature at P = 0 [5,6]. We propose that there exists some value of Ψ [e.g., $\Psi(T_o)$] below which δ is zero and above which δ will increase to a plateau value of δ_o; such a relationship may be expressed as:

$$\delta(T) = \delta_o \left\{ 1 - e^{-[(\Psi(T)=\Psi(T_o))/\Psi(T_o)]} \right\}$$

The results given in Table 2 for the determination of $(1-\delta_o)$ illustrate that a value for the crystallinity derived from the NMR data using the method previously described, gives values which are in reasonable agreement with other methods for all four polymers. This agreement indicates that above the glass transition temperature an empirical correlation exists between the temperature dependence of the fraction of material that is characterized by NMR as being mobile and the temperature dependence of the excess free volume in a polymer melt.

REFERENCES

1. Slichter, W. P. and Davis, D. D., J. Appl. Phys., 1963, 34, 98.

2. McBrierty, V. J. and Douglas, D. C., Macromol. Rev., 1981, 16, 295.

3. Haeberlen, U. and Waugh, J. S., Phys. Rev., 1969, 185, 185.

4. See for example: Fyfe, C. A., Solid State NMR for Chemists, C.F.C. Press, Guelph, 1983.

5. Simha, R., and Somcynsky, T., Macromolecules 1969, 2, 342.

6. Simha, R., Macromolecules 1977, 10, 1025.

7. Zoller, P., Bolli, P., Pahud, V., and Ackermann, H., Rev. Sci. Instr. 1976, 47, 948.

8. Eichhoff, U., and Zachmann, H. G., Ber. Bunsenges, Phys. Chem. 1970, 74, 919.

9. Zoller, P., PVT Relationships and Equations of State of Polymers, "Polymer Handbook", 3rd Ed., Wiley, New York, in press.

10. Zoller, P., and Bolli, P., J. Macromol. Sci. - Phys. 1980, B18(3), 555.

11. Zoller, P., unpublished results.

12. Olabisi, O., and Simha, R., Macromolecules, 1975, 8, 211.

Acknowledgements

The pressure/volume/temperature results and analysis are the work of Professor P. Zoller, University of Colorado.

ELEMENT SPECIFIC MICROSCOPIC IMAGING
OF POLYMERIC MATERIALS

Hans - Joachim Cantow

Institut für Makromolekulare Chemie der Universität Freiburg
Hermann - Staudinger - Haus, Stefan - Meier - Strasse 31
D - 7800 Freiburg i. Br., Federal Republic of Germany

ABSTRACT

For the first time the morphology of synthetic multiphase polymer systems has been visulalized by the element net distribution in an electron microscope, the ZEISS EM-902 in the electron
spectroscopic imaging mode (ESI). The net distribution of the
elements has been performed by electronic subtraction of the
images obtained before and after the respective absorption edges. Studies of poly(styrene-b-2-vinylpyridine) block copolymers, of polyethylenes and of multiphase membranes are reported.
C, N, O, S, Cl, and Ru have been visualized. In addition, imaging based on elastic electrons is discussed.

INTRODUCTION

Elastically and inelastically scattered, as well as unscattered electrons are engaged when imaging with conventional transmission electron microscopes (TEM). Brightfield images may be
discussed briefly: The contrast is caused by excluding the electrons scattered at wide angles by use of a diaphragm. Wide angles are favoured in sample positions with elements of high order
number or of high density. Consequently, the intensity of the
electron irradiation passing through such positions is decreased
in comparison with the effect due to light atoms. The contrast
is reduced by the bright background caused by electrons which
are inelastically scattered at small angles and by unscattered
electrons. Given below is the interrelation between the chromatic
aberration disk, Δ_C, the chromatic aberration constant of the
objective lens, c_C, the objective aperture angle, α, the electron energy loss, ΔE, and the acceleration energy, E_O:

$$\Delta_C = c_C \alpha \, \Delta E / E_O$$

The use of high acceleration voltage electron microscopes - with up to 1 MV - is the classical method of reducing the chromatic aberration. Castaing, Henry [1] and Ottensmeyer [2] have introduced a combination prism - mirror - prism into the projector system of a TEM. The scattered electrons are first deflected in the magnetic prism, then reflected by the electrostatic mirror and reflected again by the second magnetic prism. They are consequently dispersed in the energy dispersive plane, according to their energy. Introducing a diaphragm into this plane, the inelastically scattered electrons as a source of chromatic aberrations are removed. Thus, elastic brightfield yields marked improvement of images for thick samples, at moderate acceleration.

Additionally, in the ZEISS EM-902 inelastically scattered electrons may be depicted in the optical axis, independent of their energy loss ΔE. This is done by increasing the acceleration voltage from E_O to $E_O + \Delta E$, applying a microprocessor system. Because of their properly adjusted additional acceleration the selected inelastically scattered electrons are sharply focused. Thus, element specific imaging (ESI) is possible, with a resolution which corresponds to that of conventional TEM, and which is far superior to conventional electron energy loss spectrometers.

The net distribution of elements within the sample may be observed by electronic subtraction of images obtained before and after an absorption edge of the respective element. Due to the ESI technique unique possibilities of picture generation are created since staining with heavy elements, which often generates artefacts, may no longer be necessary. Examples of element net distribution analyses on multiphase polymer systems as well as of elastic bright field images are presented in the following.

RESULTS AND DISCUSSION

POLY(STYRENE - b - 2 - VINYLPYRIDINE) BLOCK COPOLYMERS

The net element distribution on those block copolymers which contain two glassy phases has been visualized for nitrogen, and - after its quaternization, also with iodine - and for carbon. In films obtained from dimethylformamide solution, at 50° C, hexagonally arranged cylinders have been observed for samples with 80 % styrene content. Unequivocal and quantitative assignment of the phases is possible in different ways. The morphology of this system was strongly dependent on the solvent and temperature used for film preparation.

In specimens with identical amounts of styrene and vinylpyridine (DMF, 50° C) lamellae are obtained. Nitrogen, carbon and iodine net distributions have been obtained. The value of such studies is evident because of the interdependence of morphology and mechanical properties [3].

POLYETHYLENES

Outstanding well resolved elastic bright field images have been realized with RuCl$_3$/NaOCl staining. The rigid-amorphous regions [4] are stained, whereas the mobile-amorphous phase remains almost unreacted. Element specific analysis with ruthenium, chlorine and oxygen confirms the diagnosis. An interrelation with the molecular mass is evident. Furthermore, the influence of chain orientation has been studied on cold - extruded polyethylenes.

After treatment with chlorosulfonic acid well resolved elastic bright field images and element net distributions also have been obtained - with and without staining with uranyl acetate.

Consequently, the rigid - amorphous phase, which has been discussed by Wunderlich [4] on the basis of calorimetric measurements, and which has been confirmed by NMR studies [5], has been visualized directly by ESI electron microscopy. Thus, one may replace the unsatisfying two-phase model for semicrystalline polymers - crystalline, amorphous - by a three-phase model - crystalline, rigid-amorphous, mobile-amorphous. This may enhance the potential for a quantitative interpretation of morphology - property - interrelations for semicrystalline polymers.

MULTIPHASE MEMBRANES

With elastic brightfield imaging, hydrophilic channel structures with a diameter of about 3 nm have been visualized in hydrophilic-hydrophobic membranes, which contain a small amount of oligo(ethyleneoxide) in a polybutadiene matrix [6].

The chlorine net distribution has been analyzed in asymmetrical anion exchange membranes which have been obtained by quater-

nization of 4-vinyl-co-styrene with poly(vinylbenzylchloride).
Again, the dimensions of the chlorine containing phase are in
the order of 3 nm.

CONCLUSIONS

The ESI technique offers new capabilities not only for the
morphological analysis of biological systems. For the first
time the element net distribution in synthetic polymeric materials has been analyzed quantitatively also. Contrasting with
heavy elements seems to be no longer required. Light elements
are accessible for electron spectroscopic imaging. Simple chemical reactions can be employed in order to create contrast.
Photometric analysis of element distribution in phase separated
systems is realizable, as well as element specific electron
diffraction.

In addition, the novel electron microscope yields excellent
results through imaging only with elastically scattered electrons. The resolution and brightness are greatly superior to
that obtained with the conventional global brightfield imaging,
which includes inelastically scattered electrons.

ACKNOWLEDGEMENTS

The electron microscope ZEISS EM-902 has been financed within
the Schwerpunkt 29 by the LAND BADEN-WÜRTTEMBERG. The work has
been carried out within the Sonderforschungsbereich 60 of the
DEUTSCHE FORSCHUNGSGEMEINSCHAFT.

I cordially thank my coworkers Dr. U.-R. Heinrich, M. Kunz and
Dr. M. Möller. To Mrs. C. Aehnelt I am indebted for her friendly
assistance.

REFERENCES

1. R. Castaing and L. Henry, C. R. Acad. Sci. Paris
   B 76, 255 (1962)
2. R. M. Henkelmann and F. P. Ottensmeyer, J. Microscopy
   102, 79 (1974); A. L. Arsenault and F. P. Ottensmeyer
   J. Microscopy 133, 69 (1984)
3. M. Kunz, M. Möller and H.-J. Cantow, Polymer Bull. in press
4. H. Suzuki, J. Grebowicz and B. Wunderlich, Makromol. Chemie
   189, 1109 (1985)
5. H.-J. Cantow, D. Emeis, W. Gronski, A. Hasenhindl, D. Lausberg
   M. Möller and Y. Shahab, Makromol. Chemie, Suppl. 7, 63 (1984)
6. M. Weber, Diplomarbeit Freiburg (1987)

OPTICAL ANISOTROPY OF POLYCARBONATES

G.H. Werumeus Buning, R. Wimberger-Friedl
Philips Research Laboratories
P.O.Box 80000, 5600 JA Eindhoven, The Netherlands

H. Janeschitz-Kriegl
J. Kepler-Universität
Altenberger Strasse 69, A-4040 Linz, Austria

T.M. Ford
E.I. du Pont de Nemours & Co.
PPD, Wilmington, DE 19898, US

ABSTRACT

The large optical anisotropy of bisphenol-A polycarbonate is mainly caused by the presence of the anisotropically polarising phenyl groups all located in the polymer backbone. By introducing an aromatic group in the polymer side-chain via the bisphenol-moiety we have tried to increase the perpendicular polarisability to obtain a smaller anisotropy of the polymer. With the aid of the flow-birefringence technique, combined with a rheological characterization, the stress-optical coefficients of these modified polycarbonates were determined. One of the modified polycarbonates shows an important, 50% reduction in stress-optical coefficient in comparison with bisphenol-A polycarbonate. To explain the flow-birefringence results in relation with molecular structure we have used molecular mechanics calculations, which give insight in the most favourable conformation of the polycarbonates. This combination of techniques is very useful for the design of polycarbonates with a low stress-optical coefficient.

INTRODUCTION

Since the development of optical information storage systems as LaserVision and Compact Disc the interest in polymers, suitable for optical applications, has grown enormously. This interest has resulted in various papers in which the special demands for this sort of polymers are described [1-5]. Bisphenol-A-polycarbonate is the most suitable disc substrate material at the moment because of its excellent thermal and mechanical properties and its low water absorption. The polycarbonate substrates are usually made by injection or compression moulding. Special moulding conditions are required to reduce the residual birefringence below the tolerance for the laser read-out system. This residual birefringence is created during processing by frozen-in thermal stresses and orientation [6,7]. Thorough investigation lead to the conclusion that orientation of the optically anisotropic polycarbonate chains is the most important source of birefringence [6]. By using low molecular weight polycarbonate the amount frozen-in orientation is lowered by fast relaxation processes thus leading to smaller birefringence values [3,8,9].

Bisphenol-A polycarbonate I, the structure of which is shown in Figure 1, has a large optical anisotropy, for the greater part caused by the phenyl-groups in the polymer backbone. This is reflected in a large, positive value of the stress-optical coefficient, determined in a broad temperature range, $C = 3.5\text{-}3.7 \cdot 10^{-9}\ Pa^{-1}$ in the molten state [6], which has been measured with the flow birefringence technique [10,11]. In this technique polymeric melt is sheared in the gap of a modified cone and plate apparatus. A linear relation exists between the applied shear stress $\Delta\sigma$ and the birefringence Δn caused by that shear stress, the stress optical rule:

$$\Delta n = C\ \Delta\sigma$$

where $\Delta n = n_I - n_{II}$ is the difference of principal refractive indices and C is the stress-optical coefficient. This coefficient is a material constant. An expression, relating C to the optical anisotropy of polymer molecules, can be derived from the theory of ideal rubbers [12]:

$$C = \frac{2\pi}{45kT}\ \frac{(n^2 + 2)^2}{n}\ (\alpha_{//} - \alpha_{\perp})$$

where n is the average refractive index of the material, k is Boltzmann's constant, T is the absolute temperature and $\alpha_{//} - \alpha_{\perp}$ is the difference between parallel and perpendicular polisaribility of a Kuhn's statistical segment. The latter factor determines the sign and to a great extent also the size of the stress-optical coefficient. We have tried to reduce the optical anisotropy of polycarbonate by introducing an aromatic group in the polymer side chain. This should result, compared with bisphenol-A polycarbonate, in an increase of $\alpha_{\perp}$, thus leading to a smaller stress-optical coefficient.

EXPERIMENTAL

Polycarbonates II and III were prepared via phosgenation of the corresponding bisphenol or a mixture of bisphenols, respectively, in homogeneous solution [13]. The "benzyl-substituted" bisphenol was synthesised from benzylmethylketone and excess phenol in toluene whilst gaseous hydrogen-chloride was passed through the solution [14]. Polycarbonate IV was kindly provided by Dr. F. Kleiner and Dr. U. Grigo (Bayer AG). The molecular weights were determined by GPC and were all of the order of Mw = 60.000 relative to polystyrene standards. The flow-birefringence measurements were carried out with a modified cone and plate apparatus [10,11]. Molecular mechanics energy minimisation was done utilizing Allinger's MM2 force field [15] modified with a Steffensen minimisation routine [16] for increased speed. No new parameters were required.

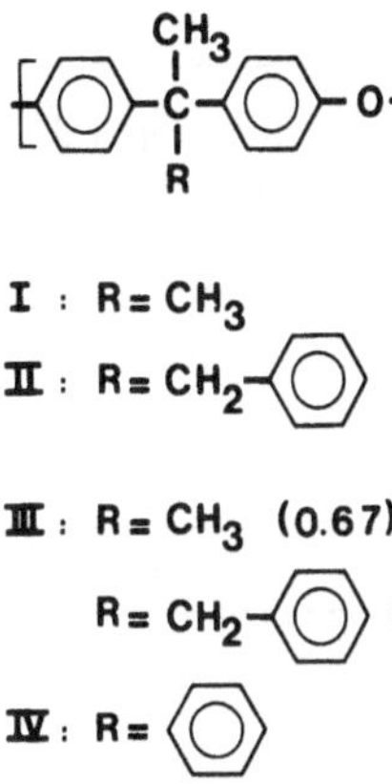

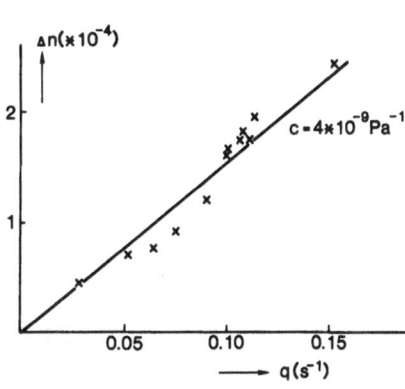

I : R = CH₃

II : R = CH₂—⬡

III : R = CH₃ (0.67)

R = CH₂—⬡ (0.33)

IV : R = ⬡

I : Tg = 145°C
I : Tg = 165°C
III : Tg = 156°C
IV : Tg = 188°C

Figure 1. Molecular structure of the modified polycarbonates.

Figure 2. Birefringence Δn as a function of shear rate q at 216°C for benzylmethylpoly-carbonate II.

RESULTS

The stress-optical coefficient for polymer melts can be determined via the following formula:

$$C = \frac{\Delta n \; \sin 2\chi}{2\eta q}$$

where Δn is the birefringence, χ is the extinction angle, η is the Cox-Merz viscosity and q is the shear rate. The birefringence and the extinction angle are measured as a function of the shear rate in an optical measurement; for determining the viscosity as a function of the shear rate, a separate rheological measurement is necessary. These viscosity measurements will not be discussed in this paper. For all the polycarbonates studied the extinction angle had, at the low shear rates used, a value close to 45° so this makes the $\sin 2\chi$ equal to 1.

The results of the optical measurements are plotted in Figure 2. The sudden increase of birefringence at a shear rate near 0.10 s^{-1} appeared to be reproducible and we found a similarly shaped curve, only with a smaller deviation from linearity, for copolymer III. We don't think that this non-linear behaviour described above is due to experimental scatter. For the polycarbonates I and IV such a behaviour has not been observed and therefore we think that this deviant behaviour is originating from the benzyl substituent. Further investigations will be necessary to explain this phenomenon. Combined with the viscosity at 216°C, $\eta = 1.8 \cdot 10^5$ Pas, the resulting stress-optical coefficient of benzylmethylpolycarbonate II had a value of $4 \cdot 10^{-9}$ Pa^{-1}, this means a small increase in comparison with bisphenol-A polycarbonate. For copolymer III and the phenyl-substituted polycarbonate IV the same optical and rheological characterisations were carried out. The resulting stress-optical constants are summarised in Table 1.

Table 1.

Polycarbonate no.	Temperature (°C)	Stress-optical constant ($\cdot 10^{-9}$ Pa^{-1})
I	170 - 230	3.5 - 3.7
II	216	4.0
III	220	3.4
III	240	3.6
IV	239	2.1
IV	253	1.8

DISCUSSION

From all the measured stress-optical coefficients, resumed in Table 1, we can conclude that the anisotropy of polycarbonates is largely influenced by introducing an aromatic group in the side chain. The introduction of a benzylgroup in polycarbonate II leads, at first sight suprisingly, to a small increase of the stress-optical coefficient relative to bisphenol-A polycarbonate. From this it can be concluded that the factor ($\alpha_{//} - \alpha_{\perp}$) is increased by the introduction of the benzylgroup. Figure 3 shows the most-favourable conformation of this polymer as calculated by molecular mechanics. From this one clearly sees that the phenyl group in the side chain is oriented more or less parallel with the polymer backbone thus leading to a larger $\alpha_{//}$, which is reflected in a larger value for the stress-optical coefficient. Copolymer III shows a value of the stress-optical coefficient as can be expected from the values of the two homopolymers (within experimental error) confirming roughly additivity of polarisability in this copolymer. Figure 4 shows the favourable conformation of polycarbonate IV. It is clear that in this case the phenyl group in the side chain is oriented perpendicularly to the polymer backbone leading to an increase in $\alpha_{\perp}$ thus to a smaller stress-optical coefficient as measured via the flow-birefringence technique.

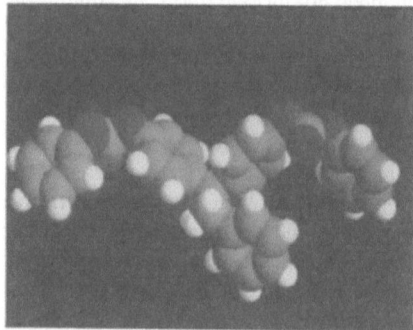

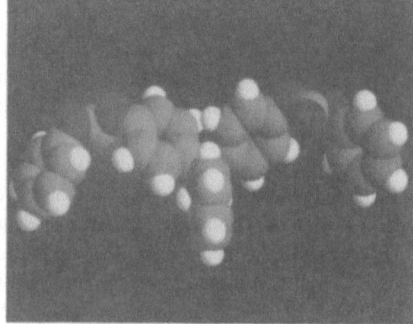

Figure 3. *Most favourable conformation of that part of the benzyl-substituted polycarbonate II-chain that is expected to be representative for the whole chain, as calculated by molecular mechanics.*

Figure 4. *Most favourable conformation of that part of the phenyl-substituted polycarbonate IV-chain that is expected to be representative for the whole chain, as calculated by molecular mechanics.*

CONCLUSION

The values of the stress-optical coefficient of some modified polycarbonates, determined via the flow-birefringence technique in the molten state, can be understood with the aid of molecular mechanics calculations. With these calculations insight in conformations is obtained thus allowing to explain qualitatively how the stress-optical coefficient is affected by the introduction of an aromatic group in the side chain. The conformation strongly affects the stress-optical coefficient via the the the important factor $(\alpha_{//} - \alpha_{\perp})$, the difference of parallel and perpendicular polarisability. Further it has been shown how strong the influence can be of one single CH_2 group, the only difference between polycarbonates II and IV. In the benzyl-substituted polycarbonate II this CH_2 group leads to a favourable conformation in which the phenyl group in the side chain is oriented parallel to the polymer backbone. In the phenyl-substituted polycarbonate IV the absence of the CH_2 group leads to a "wanted" conformation in which the side chain-phenyl group is directed perpendicularly to the polymer backbone thus leading to a smaller stress-optical coefficient. Optical discs with the same amount of frozen-in orientation, made of this polycarbonate, will give a 50% reduction in birefringence values in comparison with discs made of bisphenol-A polycarbonate.

ACKNOWLEDGEMENT

We thank Dr. F. Kleiner and Dr. U. Grigo from Bayer AG for providing polycarbonate IV.

REFERENCES

1. Hennig, J., Kunststoffe, 1985, **75**, 425.
2. Schrijver, J. and Werumeus Buning, G.H., Kunststoffen 1986, *Terugblik en Toekomst*, ed. H.M. Bruggemann, Wyt en Zonen, Rotterdam, 1986, 322.
3. Gossink, R.G., Angew. Makromol. Chem., 1986, **145/146**, 365.
4. Hennig, J., Angew. Makromol. Chem., 1986, **145/146**, 391.
5. Lippits, G.J.M. and Melis, G.P., *Integration of Fundamental Polymer Science and Technology*, ed. L.A. Kleintjens, P.J.Lemstra, Elsevier Applied Science Publishers, London, 1985, 663.
6. Wimberger-Friedl, R. and Janeschitz-Kriegl, H., to be published
7. Takeshima, M. and Funakoshi, N., J. Appl. Polym.Sci., 1986, **32**, 3457.
8. Riess, R. and Loewer, H., Proc. Antec 1985, p470.
9. Fujiwara, S., Japans Plastics Age, 1984, May-June, p21.
10. Janeschitz-Kriegl, H., *Polymer Melt Rheology and Flow Birefringence*, Polymers/Properties and Applications, vol.6, Springer-Verlag, Berlin, 1983.
11. Gortemaker, F.H., Hansen, M.G., de Cindio, B., Janeschitz-Kriegl, H., Rheol. Acta, 1976, **15**, 242.
12. Treloar, L.R.G., *The Physics of Rubber Elasticity*, 3 ed., Clarendon Press, Oxford, 1975.
13. Schnell, H., *Chemistry and Physics of Polycarbonates*, Interscience Publishers, New York, 1964.
14. Islam, A.M., Hassan, E.A., Rashad, M.E., Wassel, M.M., Egypt. J. Chem., 1977, **20**, 483.
15. Allinger, N.L., Chung, D.Y., J. Am. Chem. Soc., 1976, **98**, 6798.
16. Kalman, B.C., Wash. Univ. Tech. Memo 46, May 1982.

BIREFRINGENCE IN INJECTION-MOULDED OPTICAL COMPONENTS

R. Wimberger-Friedl

Philips Research Laboratories
P.O.Box 80000, 5600 JA Eindhoven, The Netherlands

H. Janeschitz-Kriegl

J. Kepler Universität Linz
Altenbergerstr. 69, 4045 Linz/D., Austria

ABSTRACT

The birefringence distribution in an injection-moulded Compact Disc is determined and discussed with respect to the retardation of a laser beam of normal incidence and the influence of the deformation history of the polymer on it.

INTRODUCTION

With the introduction of the Compact Disc (CD) and LaserVision systems on the consumer market a strong demand for transparent plastics with high optical quality has been created. Parallel with the realization of this technology the development of new optical recording systems such as write- once and erasable DRAW disks created even higher requirements for the quality of the polymer carriers [1,2]. Other challenging applications have also been introduced, e.g. injection-moulded lenses for the read-out system. For economic reasons thermoplastics are preferred to thermosets. Especially polycarbonate (PC) and polymethylmethacrylate (PMMA) are used at the moment. PMMA has the disadvantage of a high water absorbance and low heat deflection temperature, which makes it less suitable as a carrier for long-term data storage. PC shows superior mechanical properties but has one severe disadvantage. It is very sensitive to stress- birefringence. This is especially problematic in injection-moulded parts, where high stresses are built up in the polymer and frozen in. In this paper the distribution of bire-fringence in an CD, injection-moulded from PC, will be discussed.

BIREFRINGENCE

The refractive index at an infinitely small point can be represented geometrically by a three-dimensional ellipsoid. Such an ellipsoid is called an indicatrix. The axes of the indicatrix are parallel with the preferential directions of the medium. Their lengths are equal to the refractive indices in the respective directions of polarization. In order to calculate the birefringence the indicatrix has to be intersected with a plane through the centre of the indicatrix perpendicular to the direction of light propagation as sketched in figure 1. The axes of the resulting ellipse determine the direction of vibration and the speed of propagation of the ordinary and extraordinary beam.

If the direction of the beam coincides with an optical axis of the medium the cross-section becomes a circle and no birefringence occurs. When the sample is placed between crossed polarizers the transmitted intensity I is equal to [3]:

$$I = I_0 \sin^2 \left(2X \right) \sin^2 \left(\frac{\alpha}{2} \right) \qquad (1)$$

where X is the angle between a principal direction and the direction of the polarizer and α is the phase difference of the emerging light beams. By rotating the sample the principal direction can be found (zero intensity, isoclinic). The phase shift can be determined by introducing a birefringent medium with a variable phase shift (compensator) at 90 degrees with respect to the principal axis of the sample.

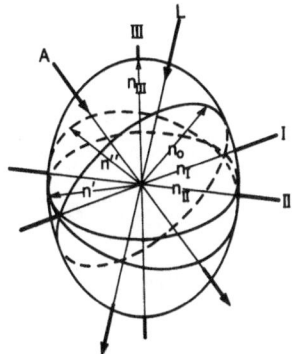

Figure 1. Indicatrix;
n_0 ... average refractive index
n_I, n_{II}, n_{III} ... principal refractive indices
n', n'' ... refractive indices for light propagating along the L-direction
A ... optical axis

THE STRESS OPTICAL RULE

Anisotropy can be introduced by applying a state of stress to an amorphous sample. It was shown that the principal axes of stress and the axes of the indicatrix coincide, i.e. stress and refractive index are proportional [3,4]. The constant of proportionality is called the stress-optical coefficient C:

$$\mathbf{n} = C \cdot \boldsymbol{\sigma} ; \quad n_{ij} = C\, \sigma_{ij} , \qquad (2)$$

with σ and $\mathbf{n}$ the stress- and refractive index tensors, respectively. The stress-optical coefficient of a polymer is a material constant and is temperature-dependent. In polymers the application of stress at temperatures above the glass transition leads to orientation of the chains of the macromolecules. There the linear stress-optical rule is valid. Upon cooling this orientation can be frozen in as well as the birefringence corresponding to it. In case the stress is applied or created below the glass transition temperature the intermolecular distances and the electron distributions are distorted and a different stress-optical coefficient is valid. At temperatures close to the glass transition the linearity of the relation is lost because of time-dependent orientation effects.

INJECTION MOULDING CYCLE

In injection moulding the hot polymer melt is injected at high speed into a cold cavity. The characteristics of this pressure-driven flow are shown in figure 2. The velocity is zero at the wall and highest in the centre. Since the flow front moves with average velocity a fountain-like flow is generated just behind the front [5]. The rate of elongation of a volume element in the z-direction, according to the simple picture sketched in fig. 2, is proportional to $1/r^2$ (r...radius of the disk). In the case of a central-gated disk shaped cavity there is also an elongational flow in the circumferential direction (θ-direction) with the rate of elongation also proportional to $1/r^2$. The biaxially stretched melt is then laid down, according to the fountain flow, at the cold wall and the orientation is frozen in. Behind the flow front the shear flow is built up as indicated in figure 2. A solid layer grows from the wall towards the centre at a rate determined by heat conduction to the wall, heat convection by the hot flowing core and dissipation of the mechanical energy as well as the heat conductivity of the polymer [4]. As soon as the cavity is filled the pressure approaches the one applied at the gate and the flow is determined by the thermal shrinkage of the cooling sample. After the gate is frozen off the pressure decreases and flow ceases. The thermal shrinkage of the still hot core is hindered by the rigid cold shell. This leads to a build-up of cooling stresses which are tensile in the core and compressive at the surface of the sample [6,7]. At the end of the cooling phase the sample is removed from the cavity and allowed to cool in ambient air. This complicated stress-temperature history determines the residual birefringence distribution in the specimen.

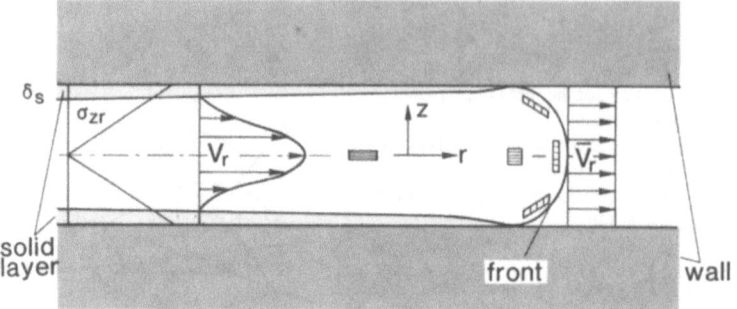

Figure 2. Mould filling; v ... velocity, σ_{rz} ... shear stress, δs ... solidified layer

BIREFRINGENCE DISTRIBUTION

The birefringence distribution of a CD, injection-moulded from PC Makrolon KL I-1189 (from Bayer), has been investigated. Since a rather complicated distribution through the thickness of the disk was expected, the oblique incidence technique seemed inapplicable. Assuming axial symmetry and a comparably small change of the indicatrix in the radial direction, the distribution was determined by cutting cross-sections out of the disk as indicated in figure 3. The radial cross-section, viewed in θ-direction, contains two principal axes and the tangential cross-section, viewed in r-direction, at least one. Since the refractive indices are expected to change continuously the following relation is valid:

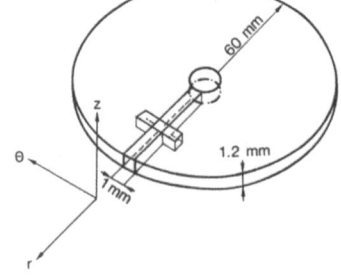

Figure 3. Radial and tangential cross-sections cut out of the CD

$$\Gamma_{r\theta} = \int_{-d/2}^{d/2} (n_{rr} - n_{\theta\theta})(z)dz =$$

$$\int_{-d/2}^{d/2} \left[(\Delta n_{rz} \cdot \cos 2X) - (n_{\theta\theta} - n_{zz}) \right](z)dz , \qquad (3)$$

with $\Gamma_{r\theta}$ the optical path difference of normal incident light and d the thickness of the sample. This gives an opportunity to check the results by measuring $\Gamma_{r\theta}$.

$$\Gamma = \frac{\alpha \cdot \lambda}{2\pi} \qquad \lambda. \,. \,.\text{wavelength of light} . \qquad (4)$$

A characteristic result from a CD is depicted in figures 4.A to 4.D for one radial position. They give the required information to calculate the terms of the integral in (3). Figure 4.A shows the birefringence profile in the r-z plane of a disk moulded under the conditions summarized in table 1. 4.B shows the distribution of $(n_{\theta\theta} - n_{zz})$ in the z-θ plane at the same radial position. Fig. 4.C shows the distribution of the angle between the direction of the highest refractive index and the r-direction (X), and finally 4.D shows the function $(n_{rr} - n_{\theta\theta})$ calculated according to (3). As can be seen, the birefringence distribution is very complicated. The retardation of a focussed beam can be calculated only numerically. Nevertheless it is found in practice that there is a good correlation between a collimated (fig. 4.D) and a focussed beam [8]. Another important consequence is, however, that by measuring the birefringence in the cross-sections the influence of processing can be made visible. This gives the opportunity to control and thus minimize birefringence.

INFLUENCE OF PROCESSING

From figure 4.D we see that the sample is uniaxial at the surface. Beyond the surface, determined by the filling phase, $n_{\theta\theta}$ is larger than n_{rr}. The frozen-in shear stresses contribute only partly to n_{rr} because the principal direction of stress is inclined at an angle X to the wall (r-direction). At the transition to the compression stage the elongational stress can relax more completely than the shear stress. During the compression stage a shear stress is built up directly at the solidifying layer, while elongational stresses are created first in the centre, where they can relax before solidification. In this way n_{rr} becomes larger than $n_{\theta\theta}$ over the remaining part towards the centre, where they become equal again. Elongational stresses are parallel to the mould wall, while shear stresses are directed at 45 degrees towards the wall (see fig.4.C). Cooling stresses are parallel to the wall and independent of direction (plane stress) because of the "one-dimensional" cooling and thus do not contribute to $n_{rr} - n_{\theta\theta}$. The art of processing lies therefore in finding the right balance between the filling and the compression stage. This is only true, however, as long as the cooling is "one-dimensional", that is where the thickness is very small compared to the width of the sample. Towards the rim of the disk the material feels the presence of an additional cooling wall

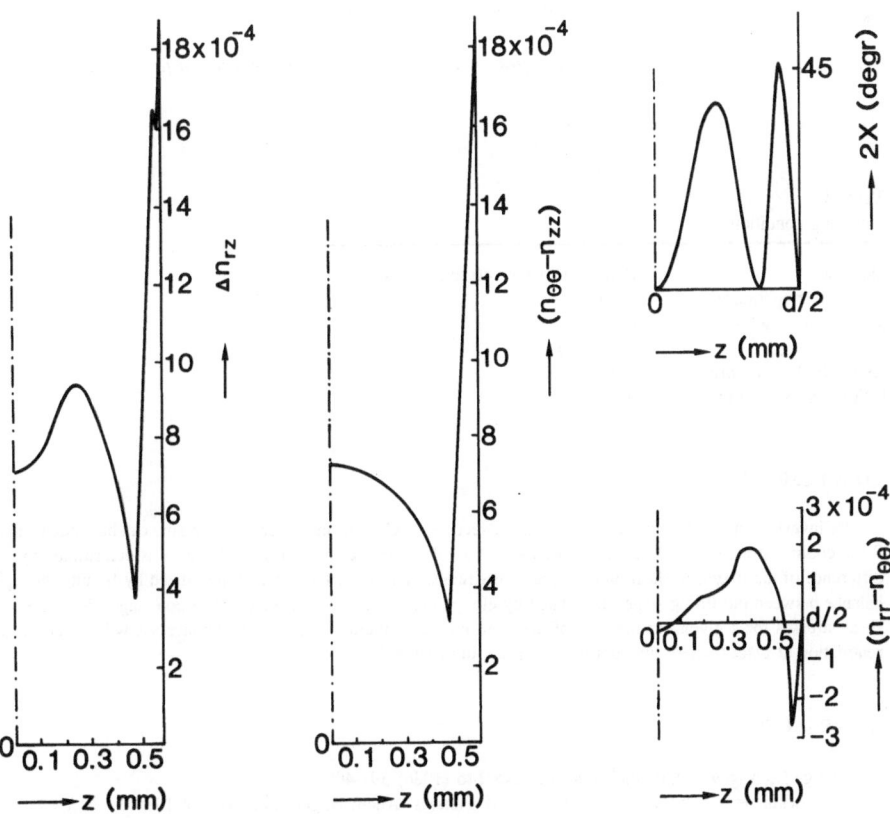

Figure 4.A. Birefringence Δn_{rz} in the radial cross-section at $r = 35mm$
Figure 4.B. $(n_{\theta\theta} - n_{zz})$ in the tangential cross section at $r = 35mm$
Figure 4.C. Angle $2X$ in the radial cross-section at $r = 35mm$
Figure 4.D. $(n_{rr} - n_{\theta\theta})$ calculated from figs. 4.A, B and C according to eq. (3)

parallel to the z-direction. The cooling stress field becomes very complicated there. Figure 5.A shows the isochromatics (lines of equal birefringence) in the r-z plane of a CD parallel with the direction of the polarizer and 5.B the direction of the isoclinics, indicated by dashed lines. The retardation of a beam in normal incidence can become very high locally. Fortunately this is restricted to the outer three millimetres of the disk, where no information is stored. For other optical components like lenses, which are "three-dimensional" as a whole, the minimization of cooling stresses becomes the most critical part of processing, however.

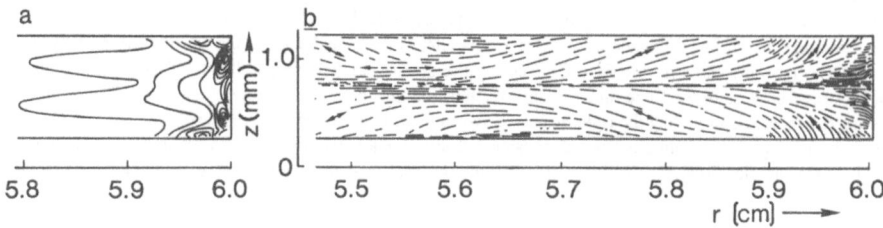

Figure 5.A. The distribution of isochromatics at the rim of a radial cross-section

Figure 5.B. The principal direction of the refractive index in the radial cross-section of figure 5.A

Table 1.
Moulding conditions

Machine	: Meiki dynamelter 40 mm screw diameter
Cylinder temperature	: 300°C
Mould temperature	: 65°C
Injection speed	: 125 ccm/s
Hydraulic holding pressure	: 55 bar
Holding pressure time	: 1s

CONCLUSIONS

By investigating the birefringence in the cross-sections of CDs the influence of processing can be made visible. Since cooling stresses are planar the retardation of a collimated beam of normal incidence is determined by the difference of the frozen-in shear and elongational stresses. This difference can be minimized by finding the right balance between the filling stage, dominated by elongational stresses, and the compression stage, dominated by shear stresses for a given combination of melt and mould temperatures. The cooling stresses will influence the retardation at places where the cooling is not one-dimensional.

REFERENCES

1. Hennig, J., Angew. Makromol. Chemie, **145/146** (1986) 391-409.
2. G. Bouwhuis et al.: *Principles of Optical Disk Systems*, Adam Hilger, Ltd., Bristol 1985.
3. Kuske, A., Robertson, G.: *Photoelastic Stress Analysis*, J. Wiley / Sons, Ltd., 1974.
4. Janeschitz-Kriegl, H.: *Polymer Melt Rheology and Flow Birefringence*, ed. Meissner, J., Springer, Berlin, 1983.
5. Tadmor, Z. J. Applied Polymer Science, **18**, (1974) 1753-1772.
6. Aggarwala, B.D., Saibel, E., Physics and Chemistry of Glasses, **2**, (1961) 137-140.
7. Struik, L.C.E., Polymer Eng. & Science, **18**, (1978) 799-811.
8. Grywatz, K., Physikalisches Labor PDO Hannover, private communications.

CURIE TRANSITION OF THE 70/30 POLYVINYLIDENE-TRIFLUOR ETHYLENE COPOLYMER AS REVEALED BY X-RAY SYNCHROTRON RADIATION

E. LOPEZ CABARCOS[*], A. GONZALEZ ARCHE[+], J. MARTINEZ SALAZAR[+], F.J. BALTA CALLEJA[+]

*Dpto. Química Física, Facultad de Farmacia, U.C.M. 28040 Madrid. Spain.
+E.U.I.T. Telecomunicaciones, 28031 Madrid. Spain.
+Instituto de Estructura de la Materia, C.S.I.C., Serrano 119. 28006 Madrid. Spain

SYNOPSIS

Real time WAXD and SAXS experiments from a 70/30 Polyvinylidene-Fluoride-Trifluor ethylene copolymer are reported as a function of temperature. The Curie transition (ferro-paraelectric) detected in the 80-110°C interval, is associated to a sudden decrease of the invariant. Comparison of the experimental and calculated invariant data assuming a simple model support the view of the two phases coexisting within the crystalline material, throughout the transition range studied.

1. INTRODUCTION

Since the early work of Lando and Doll[1] showing that vinylidene fluoride (VF_2)-trifluorethylene (F_3E) copolymer adopt the β-phase conformation of PVF_2, much attention has been recently directed towards these materials because of their relevant piezoelectric and pyroelectric properties. It is known that for concentrations of VF_2 between 50 and 80% these copolymers exhibits a Curie transition[2]. The ferroelectric paraelectric transition occurs at a temperature which depends on the copolymer concentration. Most of the work carried out on these materials has been devoted to their dielectric properties[3]. However, little effort has been done, as far as we know, on the actual mechanism of transformation. Lovinger et al[4-7] and Tashiro et al[8] have investigated the above transition in copolymers with 52/48, 65/35, 73/27 and 78/22 compositions using wide angle X-ray diffraction (WAXD). These authors conclude that the ferro-electric phase consists of chains in a polar trans conformation whereas in the paraelectric phase the chains are partly disordered exhibiting tg, tḡ and tt sequences. Legrand et al[9] investigated the 70/30 composition by NMR techniques suggesting that in the paraelectric phase each segment of the molecule has several possible equilibrium positions.

In the present study we report novel diffraction data concerning the microstructural behaviour through the Curie transition of the 70/30 copolymer. The present measurements have been carried out with help of a synchrotron radiation source having the advantage to yield real-time WAXD and SAXS patterns as a function of temperature.

2. EXPERIMENTAL

Commercial pellets of VF_2/F_3E (70/30) copolymer from Atochem were melted at 200°C during 8 minutes in a heating press. Films of about 100μm were subsequently crystallized by quenching the molten sample instantly into an ice-water bath. WAXD and SAXS measurements were performed using the synchrotron radiation which is produced by the positron beam of the DORIS storage ring at DESY. The machine was operated at 5.0GeV with beam currents between 10 and 30mA. The detector used was one-dimensional position sensitive proportional counter with a LC-delay line, filled with an Ar/CO-mixture at 3.5atm. pressure. Positional information from the counter was decoded using direct time digitalisation or a 500 MHz clock frequency. This information was fed into 256 position channels representing the 8cm length of the counter. The samples were heated at a rate of 10°C/min in a furnace, between two aluminium foils,to allow homogeneous heating from room temperature, through the transition temperature, beyond 110°C. DSC curves were recorded at 10°/min using a Mettler scanning calorimeter.

3. RESULTS AND DISCUSSION

Figure 1 illustrates the DSC transition from the ferro- to the paraelectric phase for the 70/30 copolymer in the 80-110°C range and the final endothermic melting peak at 153°C. Real time wide angle X-ray scattering (WAXS) patterns corresponding to the phase transition are shown in Figure 2 as a function of temperature. For the initial quenched sample, one observes the ferroelectric peak located at 2θ=19.9° coexisting with the amorphous halo together with an incipient shoulder which we attribute to the paraelectric phase. The shoulder gradually transforms into the paraelectric peak at 2θ=18.2° with increasing temperature. The transition is complete at about 110°C where the ferroelectric peak has totally vanished and only the paraelectric maximum remains.

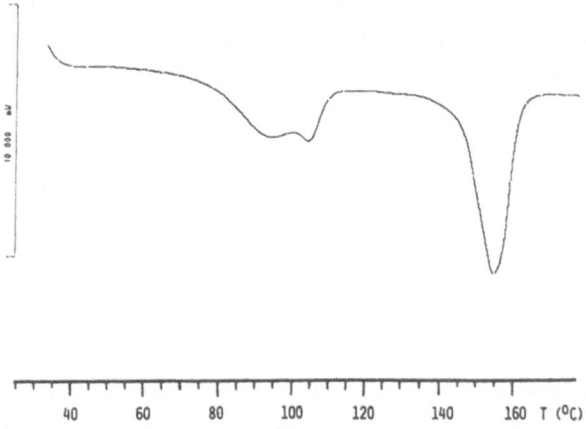

Fig. 1. DSC endotherm of 70/30 copolymer showing the ferro-paraelectric
transition (left) and the melting peak (right).

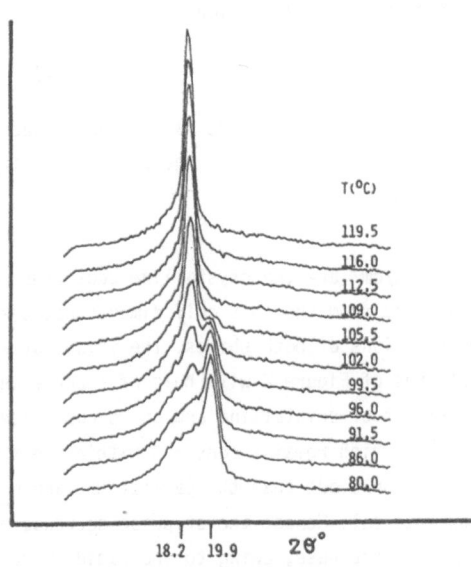

Fig. 2. WAXD patterns of the 70/30 copolymer as a function of
increasing temperature showing the ferro-paraelectric
transition.

The ferroelectric transition can also be conveniently detected following the small angle X-ray scattering (SAXS) maximum with increasing temperature. Figure 3 illustrates the evolution of the low angle scattering intensity with increasing heating temperatures. The intensity gradually rises from 50°C up to a maximum at about 96°C, shows a minimum at the end of the phase-transition and thereafter it increases with T up to the vicinity of the melting point, where it finally drops to zero. Figure 4 depicts the variation of the integrated intensity of the small angle peak in the temperature range studied. From these data it is concluded that the ferro-paraelectric transition is directly correlated with a sudden decrease of scattering intensity. The variation of the long period L, as a function of heating temperature is shown in Figure 5. The relevant result here is the nearly constancy of L over the transition region (90-110°C). Annealing beyond 110°C causes the long period to increase very rapidly as it occurs in common semicrystalline polymers[10].

The above results can be discussed in terms of the following model: it should suffice to assume that the low angle maxima provides a measure of the stacking of crystalline lamellae in which the coexistence of paraelectric and ferroelectric phases is postulated. According to this concept the invariant quantity Q can be calculated from the expression:

$$Q = (\rho_c - \rho_a)^2 \, w_c (1-w_c) \tag{1}$$

where ρ_a is the density of the amorphous layers sandwiched between the crystals, w_c is the crystallinity value and ρ_c is the average crystal density given by:

$$\bar{\rho}_c = \rho_c^f \, w_c^f + \rho_c^p \, w_c^p \tag{2}$$

where ρ_c^f=1.97 g/cm^3, ρ_c^p=1.83 g/cm^3 are the crystalline densities for para- and ferroelectric crystals, derived from the unit cell dimensions, ρ_a=1.68 g/cm^3 and w_c^f and w_c^p are the volume fractions of both phases. The values of w_c^f and w_c^p were derived from the WAXD patterns of Figure 2 with help of a curve analyzer using Gaussian profiles. The analysis of crystalline and amorphous phases yields a crystallinity value w_c=45-47% which remains constant within the transition range investigated. In summary, it turns out that the parameters L and w_c are constant throughout the transition interval. Hence, the stepwise decrease of Q in Fig. 4 is mainly the result of the $\bar{\rho}_c$ decrease, owing to the rapid increase of w_c^p at expenses of w_c^f in eq. 2; comparison of experimental results (Fig. 4) with the calculated Q-data from eq. 1 in the transition range shows an excellent

419

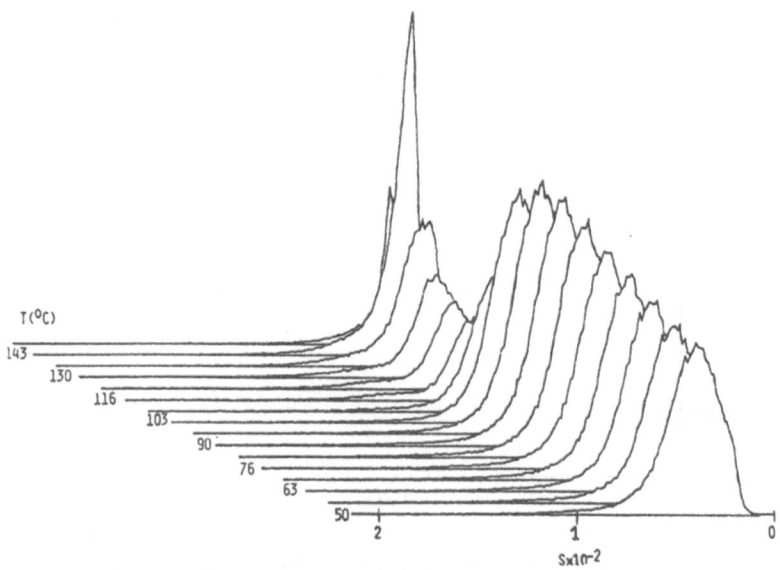

Fig. 3. Variation of SAXS maximum of the 70/30 copolymer with increasing temperature.

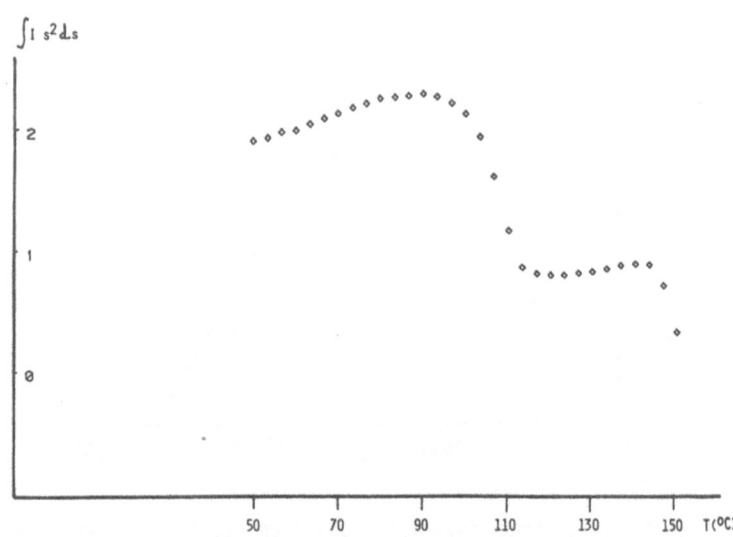

Fig. 4. SAXS integral scattering intensity in relative units vs. temperature (the integration limits are s=2x10^{-3} Å^{-1} and s=2x10^{-2} Å^{-1}).

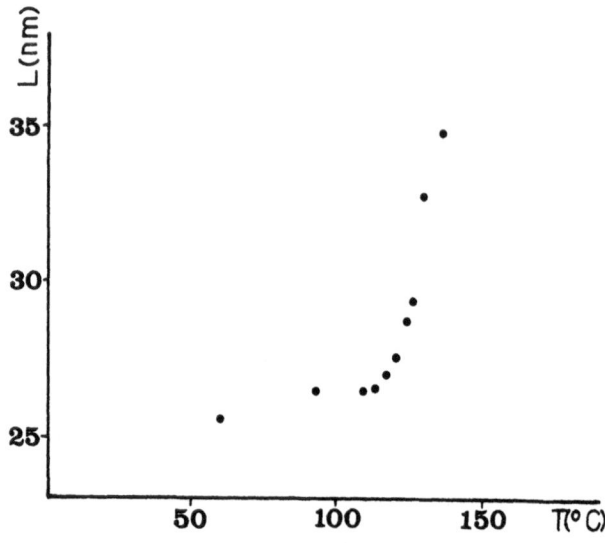

Fig. 5. Long period of the copolymer vs. heating temperature

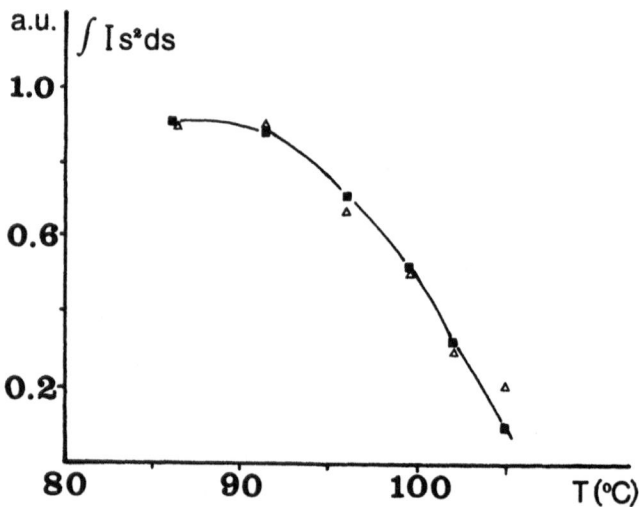

Fig. 6. Calculated invariant data from eq. 1 ($\triangle$) as compared with experimental results ($\blacksquare$) from Fig. 4 in the transition region.

agreement as evidenced in Fig. 6, thus, supporting the model of the two coexisting crystalline phases throughout the transition interval investigated. The only
question which still remains open is whether the ferro and paraelectric phases
may coexist within a crystal or rather contribute to the scattering as randomly
distributed stacks of independent lamellae.

4. CONCLUSIONS

A ferroelectric-paraelectric transition has been measured for the first
time in the copolymer with 70/30 ($CF_2/F_3E\%$ mol) composition in real time SAXS
experiments. Small angle intensity data have been quantitatively interpreted in
terms of a two-phase model assuming the coexistence of ferroelectric and paraelectric crystalline regions throughout the transition range of 80-110°C.

ACKNOWLEDGMENTS

Grateful acknowledgment is due to Prof. H.G. Zachmann for permitting the
authors to use the synchrotron facilities at DESY and Messrs. P. Bosecke and
S. Buchner for their valuable help during the experimental work. This work was
partly supported by CAICYT, Spain.

REFERENCES

1. J.B. Lando, W.W. Doll, J. Macromol. Sci. Phys. B, 2, 205 (1968)

2. T. Yagi, M. Tatemoto, J. Sako, Polymer J., 12, 209 (1980)

3. N. Koizumi, N. Haikawa, H. Habuba, Ferroelectrics, 57, 99 (1984)

4. A.J. Lovinger, G.T. Davies, T. Furukawa, M.G. Broadhurst,
   Macromolecules, 15, 323 (1982)

5. A.J. Davies, T. Furukawa, A.J. Lovinger, M.G. Broadhurst,
   Macromolecules, 15, 329 (1982)

6. A.J. Lovinger, T. Furukawa, G.T. Davis, M.G. Broadhurst,
   Polymer, 24, 1225 (1983)

7. A.J. Lovinger, T. Furukawa, G.T. Davis, M.G. Broadhurst,
   Polymer, 24, 1233 (1983)

8. K. Tashiro, M. Kobayashi, Polymer, 27, 667 (1986)

9. J.F. Legrand, P.J. Schuele, V.H. Schmidt, M. Miniuer, Polymer, 26, 1683 (1985)

10. D.R. Rueda, J. Martinez Salazar, F.J. Baltá Calleja,
    J. Mater. Sci., 20, 834 (1985)

A REAL-TIME X-RAY STUDY OF
THE DRAWING OF ULTRA-HIGH MOLECULAR WEIGHT POLYETHYLENE
USING SYNCHROTRON RADIATION

N.A.J.M. van Aerle
Department of Polymer Technology,
Eindhoven University of Technology,
P.O. Box 513, 5600 MB Eindhoven,
The Netherlands

and

A.W.M. Braam
DSM Research,
P.O. Box 18, 6160 MD Geleen,
The Netherlands

ABSTRACT

Orientational changes caused by the drawing of melt-crystallized ultra-high molecular weight polyethylene (UHMW-PE) at 100 °C were studied at low draw ratios via wide-angle X-ray scattering (WAXS). In order to avoid possible relaxation and/or re-crystallization effects, a real-time WAXS study was performed <u>during</u> the drawing process. To reduce the X-ray exposure times, highly intense X-radiation from a synchrotron source was combined with a two-dimensional X-ray detector. The observed real-time WAXS patterns will be discussed.

INTRODUCTION

One of the basic problems in the field of morphological and orientational studies of partly drawn polymers is the difficulty of studying samples at very low draw ratios and of avoiding the introduction of structural artefacts. Very often a polymer is drawn at elevated temperatures. Relieving the stretching force and cooling the sample to room temperature to study it can lead to artefacts caused by relaxation and/or re-crystallization effects. If, for example, a polymer like melt-crystallized UHMW-PE is drawn at 100 °C and the stretching force is subsequently relieved, the sample unavoidably shrinks to some extent. This shrinking can be regarded as a relaxation process. This means that orientational studies on such a sample can yield results which are not representative of the drawing process itself. Many relaxation studies on PE performed in the past indicate the possible introduction of strong orientational changes by relaxation [1-6].

Until now many drawing studies have been performed on PE samples which clearly exhibit neck-formation. In order to find out whether a non-necking PE system behaves differently with respect to orientational changes, we studied melt-crystallized UHMW-PE, which is known to exhibit homogeneous deformation without necking [7].

In order to eliminate possible relaxation and/or re-crystallization effects, a real-time wide angle X-ray scattering study was performed <u>during</u> the drawing. For this purpose a special heatable stretching device was used. To reduce the X-ray exposure time, we used a highly intense synchrotron radiation source in combination with a sensitive two-dimensional X-ray detector.

MATERIALS AND METHODS

The UHMW-PE, Hostalen Gur-412 ($M_w \cong 1700$ kg/mole), was obtained from Hoechst/Ruhrchemie. The as-received powder was compression moulded at 180 °C for 20 minutes to a thickness of about 0.3 mm, quenched to room temperature and subsequently cut into tapes of 100 x 2 mm^2.

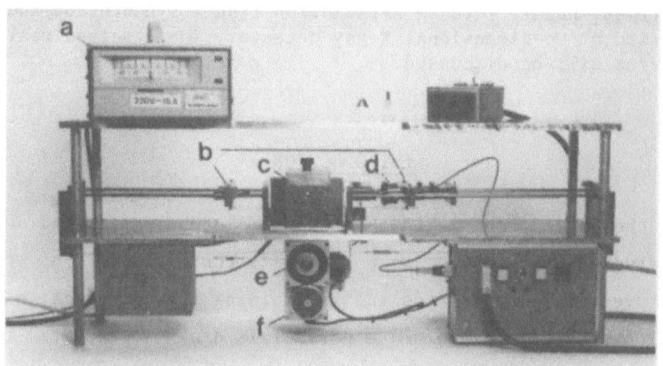

Fig. 1. Stretching device, used for real-time X-ray deformation
studies at elevated temperatures.
(a) temperature control unit; (b) sample clamps; (c) oven;
(d) stress transducer;   (e) gearbox;   (f) motor.

The drawing was performed in a heatable stretching device as shown in figure 1. The apparatus was designed to allow synchronous monitoring of SAXS or WAXS patterns, drawing temperature, clamp displacement and drawing force during the drawing process [8]. The special construction, with the sample clamps outside the oven, urged the use of a calibration curve to correlate the monitored clamp displacement to the actual draw ratio of the sample in front of the X-ray beam (figure 2). This calibration curve was determined empirically using ink-marked samples. The drawing temperature and clamp velocity were chosen to be 100 °C and 0.435 mm/sec, respectively.

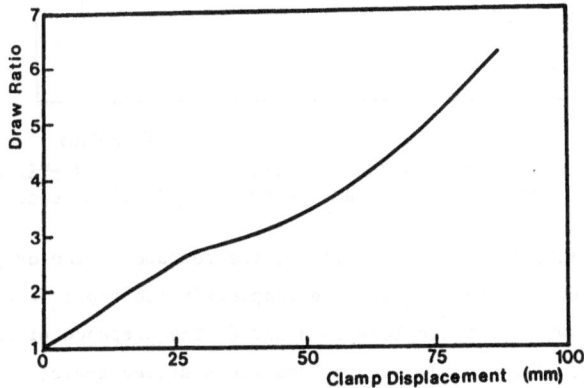

Fig. 2. Calibration curve to correlate the observed clamp displacement to the actual draw ratio for melt-crystallized Hostalen Gur-412 UHMW-PE, drawn at 100°C using a clamp velocity of 0.435 mm/sec.

The distance between the sample and the two-dimensional Westinghouse Vidicon X-ray detector was about 9 cm and the wavelength 1.61 Å. During the experiments the storage ring DORIS II at DESY-Hamburg was operated in single bunch mode at 5.260 GeV and 30 mA. The use of a highly intense synchrotron source in combination with the Vidicon detector enabled us to reduce the measuring cycle for a single WAXS pattern from about 20 minutes to 10 seconds.

RESULTS & DISCUSSION

In figure 3 a stress-strain curve is presented for melt-crystallized UHMW-PE drawn at 100 °C. The absence of a yield stress clearly indicates homogeneous deformation behaviour, i.e. absence of necking. The stress was monitored synchronously with the real-time WAXS patterns, which will be described below.

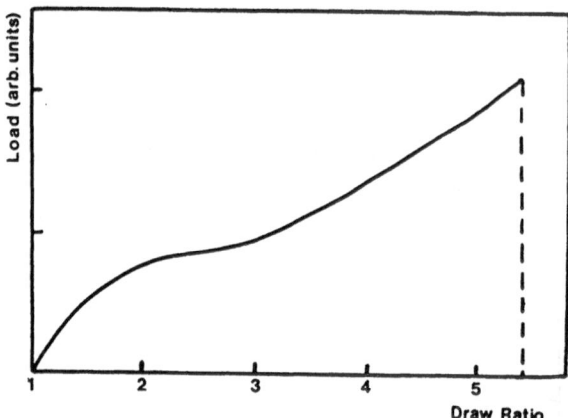

Fig. 3. Stress-strain curve of melt-crystallized UHMW-PE, drawn at 100°C, using a clamp velocity of 0.435 mm/sec.

Using conventional X-ray techniques, the texture of the original melt-crystallized samples was found to be completely isotropic. When such a sample was drawn to a ratio between 3 and 6, the presence of fibre symmetry could be observed. In view of the facts stated above, we assume that the sample will also exhibit fibre symmetry during the intermediate stages of the drawing process.

Figure 4 shows a series of real-time WAXS patterns obtained with the primary beam directed perpendicular to the tape surface. The patterns were monitored during deformation at 100 °C. Due to limited detector size, only one quadrant of the WAXS patterns was detected. As can be seen in figure 4, the first signs of anisotropy are observable at rather low draw ratios. At draw ratios up to about 2, both the orthorhombic (110) and (200) reflections are split into two arcs on both sides of the equator. In addition, an equatorial (200) reflection maximum can be discerned for all draw ratios, indicating that a fraction of all crystallites is preferentially oriented with the a-axis perpendicular to the drawing direction throughout the whole deformation process.

Furthermore, at the initial draw ratio of 1.05 weak equatorial reflection maxima for both the (110) and (200) can be observed. This might point to the presence of a small fraction of crystallites with the b-axis perpendicular to the drawing direction. Surprisingly however, as the

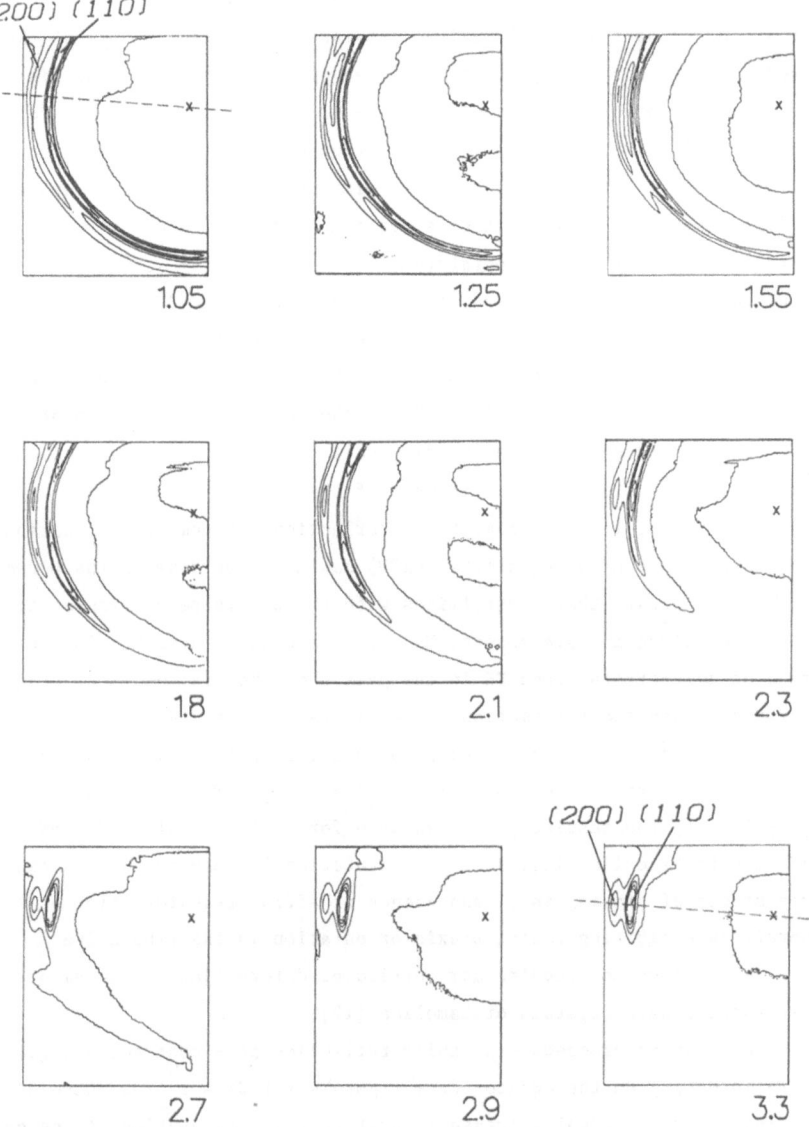

Fig. 4.   Series of isointensity contourplots of real-time WAXS patterns,
obtained by drawing melt-crystallized UHMW-PE at 100°C using
a clamp velocity of 0.435 mm/sec. The contourlines are drawn at
equal intensity intervals. Due to limited detector size, only
the lower left quadrant was detected. The drawing direction is
vertical. (---) and (x) denote the equator and the position of
the primary beam respectively. The corresponding draw ratios are
denoted in the lower right corner.

drawing proceeds, the weak equatorial (110) reflection maximum vanishes whereas the equatorial (200) maximum remains detectable. In view of the presence of fibre symmetry, this indicates an orientational change in which a small rotation of b-axes into the drawing direction occurs for the crystallites which were originally oriented with the b-axis perpendicular to the stretching direction. A similar finding has been reported for real-time FT-IR drawing studies performed on a low density PE (Vestolen A3512) at room temperature [13, 14].

Due to the oven construction, the maximum detectable 2θ-Bragg angle amounts about 30° [8]. Unfortunately, the orthorhombic (020) reflection of PE, which can give direct information about the b-axis orientation, exhibits a reflection angle of 36.5°, so the (020) reflection cannot be observed. Therefore, it is rather difficult to obtain independent information about the orientational changes of the b-axis.

The absence of an equatorial (110) reflection maximum in combination with the presence of an equatorial (200) maximum (for draw ratios between 1.2 and 2.1) proves that crystallites with the b-axis perpendicular to the drawing direction are absent. Many accurate WAXS investigations performed on melt-crystallized PE in the past show that in the first stages of drawing, when the deformation is still prevalently elastic, the a-axis is preferentially oriented perpendicular to the drawing direction and the b-axis more or less parallel [1,2,9-11,18]. These findings agree very well with the results presented here for melt-crystallized UHMW-PE. According to Peterlin [12], such an orientational behaviour during the early stages of drawing is a consequence of affine transformation of spherulites exhibiting radial b-axis orientation in the ribbonlike lamellae. It does not require any particular deformation mechanism of constituent single crystals or lamellae [12].

As the drawing proceeds, the split reflection arcs contract and gradually intensify on the equator (see figure 4 for draw ratios above 2). This indicates a gradual increase in preferential orientation of the c-axis, i.e., the molecular chain direction, parallel to the drawing direction.

Comparison of the orientational changes caused by the drawing of PE samples which exhibit neck formation and PE samples which do not (in our case UHMW-PE) shows that at corresponding draw ratios these changes are very similar (see e.g. [2,9,10]). This suggests that necking as such has no dramatic effect on the deformation process.

Finally, it is worth mentioning that during all stages of the deformation studied here no X-ray signals arising from a monoclinic PE lattice were observed. This implies that during drawing of melt-crystallized UHMW-PE at 100 °C no detectable amount of monoclinic PE is formed. On the other hand it is well known that deformation of PE at room temperature definitely leads to the formation of monoclinic PE (see e.g. [12] and the review of Bowden & Young [17]). The introduction of some monoclinic phase during the deformation of PE at room temperature has also been shown via real-time drawing studies at room temperature, using FT-IR and X-ray techniques [13,15,16]. Very probably, the drawing temperature strongly affects the formation of monoclinic phase. Confirmation of this statement requires drawing experiments at different temperatures.

CONCLUSIONS

- Real-time WAXS patterns obtained during the drawing of melt-crystallized UHMW-PE at 100 °C exhibit orientational changes which are rather similar to those reported for lower molecular weight PE-grades.
- The results presented here suggest that the orientational changes are related to the actual draw ratio rather than the presence or absence of necking in the PE sample.
- Drawing of melt-crystallized UHMW-PE at 100 °C does not lead to the formation of a detectable amount of monoclinic phase.

ACKNOWLEDGEMENTS

The authors wish to express their gratitude to Prof. H. Zachmann from the University of Hamburg for generously allowing them to use the synchrotron facilities at DESY-Hamburg. Furthermore, they wish to thank S. Buchner and P. Bösecke from the Institut für Technische und Makromolekulare Chemie Hamburg for their enthusiastic support during the experiments at DESY.

REFERENCES

1. Brown, A., J. Appl. Phys., 1949, 20, 552-8.

2. Horsley, R.A., & Nancarrow, H.A., Brit. J. Appl. Phys., 1951, 2, 345-51.

3. Keller, A., J. Polym. Sci., 1955, 15, 31-49.

4. Belbéoch, B., & Guinier, A., Makromol. Chem., 1959, 31, 1-26.

5. Rhodes, M.B., & Stein, R.S., J. Appl. Phys., 1961, 32, 2344-52.

6. Hay, I.L., & Keller, A., J. Mater. Sci., 1966, 1, 41-51.

7. Smith, P., Lemstra, P.J., Pijpers, J.P.L., & Kiel, A.M., Colloid & Polym. Sci., 1981, 259, 1070-80.

8. van Aerle, N.A.J.M., & Braam, A.W.M., submitted to J. Appl. Cryst.

9. Aggarwal, S.L., Tilley, G.P., & Sweeting, O.J., J. Polym. Sci., 1961, 51, 551-67.

10. Kasai, N., & Kakudo, M., J. Polym. Sci.: Part A, 1964, 2, 1955-76.

11. Sasaguri, K., Hoshino, S., & Stein, R.S., J. Appl. Phys., 1964, 35, 47-54.

12. Peterlin, A., J. Polym. Sci.: Part C, 1965, 9, 61-89.

13. Holland-Moritz, K., & van Werden, K., Makromol. Chem., 1981, 182, 651-5.

14. Holland-Moritz, K., Holland-Moritz, I., & van Werden, K., Colloid & Polym. Sci., 1981, 259, 156-62.

15. Heise, B., Riekel, C., & Stach, W., to be published.

16. Stach, W., private communication.

17. Bowden, P.B., & Young, R.J., J. Mater. Sci., 1974, 9, 2034-51.

18. Hoshino, S., Powers, J., Legrand, D.G., Kawai, H., & Stein, R.S., J. Polym. Sci., 1962, 58, 185-204.

^{13}C SOLID-STATE NMR AND DSC STUDIES OF THE EFFECTS OF γ-IRRADIATION ON LLDPE

J. H. O'Donnell and A. K. Whittaker[*],
Polymer and Radiation Group,
Department of Chemistry,
University of Queensland,
St. Lucia, Q, 4067,
Australia.

* current address:
GDPC-USTL,
34060 Montpellier,
France Cedex.

ABSTRACT

The effects of γ-radiation on the crystalline material in LLDPE was studied by solid-state ^{13}C NMR and DSC. Changes in the NMR spectra on irradiation indicated a decrease in the polymer crystallinity. The lamellae thickness was determined from the ^{1}H spin-lattice relaxation times in the rotating frame, and was found to decrease on irradiation. DSC meaurements showed a corresponding decrease in the overall crystallinity of the samples, and a mechanism of lattice disruption by crosslinking at the crystal surfaces was suggested.

INTRODUCTION

Poly(ethylene) is frequently irradiated in industrial processes to improve its physical properties for various applications. Although the radiation chemistry of PE has been extensively studied, precise details of the mechanism of crosslinking and the influence of sample morphology, have yet to be deduced, however the crystalline fraction of the polymer is recognised to influence the radiation chemistry.

Solid-state NMR has the potential to probe the dimensions

of phases in heterogeneous systems.   McBrierty and Douglass
(1) suggested that the ^1H NMR relaxation time in the rotating
frame, $T_{1\rho}$ (^1H), was determined by diffusion of the proton
spins, and therefore was a function of the distance of
diffusion, which for semicrystalline polymers such as PE
corresponds to the crystalline lamellar thickness (L).   Cudby
et al. (2) showed that $T_{1\rho}$ (^1H) was proportional to the square
of L for four PE samples.

The aim of this paper is to describe solid-state ^{13}C NMR
and DSC experiments on the effect of γ-irradiation on the
crystalline material in LLDPE.   The crystalline lamellar
thicknesses have been estimated from the NMR relaxation times,
while the crystalline melting behaviour was monitored by DSC.
A mechanism for the disruption of the crystalline material has
been proposed.

MATERIALS AND METHODS

A commercial LLDPE (UCAL), containing ca. 13 and 5 butyl
and methyl branches per 1000 carbons atoms respectively
(determined by ^{13}C solution NMR), was used as received.
Irradiation were performed in a ^{60}Co gamma source at ambient
temperature in vacuum, and the samples were allowed to stand
for several weeks before exposure to the atmosphere.   The
residual radical concentration was confirmed by ESR to be less
than 10^{15} spins/gram.

^{13}C solid-state NMR measurements were made with a Bruker
CXP300 operating at 75.46MHz.   The techniques of magic-angle
spinning, cross-polarization, and dipolar decoupling were used
to obtain the required resolution and signal-to-noise. $T_{1\rho}$
(^1H) was calculated by fitting curves of signal intensity
against cross-polarization contact time using the Simplex
procedure (3).   The melting behaviour of the polymers was
observed with a Perkin Elmer DSC-2 calorimeter, using a
heating rate of 20K/min.

RESULTS AND DISCUSSION

The ^{13}C NMR spectra of unirradiated LLDPE (Fig. 1A) showed
two unresolved peaks at 31.5 and 34.0ppm which were assigned
to backbone methylene carbons in disordered (amorphous +
intermediate regions) and crystalline regions respectively, in
agreement with earlier assignments for LDPE and HDPE (4).
On irradiation to 10MGy, the peak due to carbons in disordered
regions increased relative to the peak due to carbons in
crystalline regions (Fig. 1B).   The degree of crystallinity
is apparently reduced on irradiation.

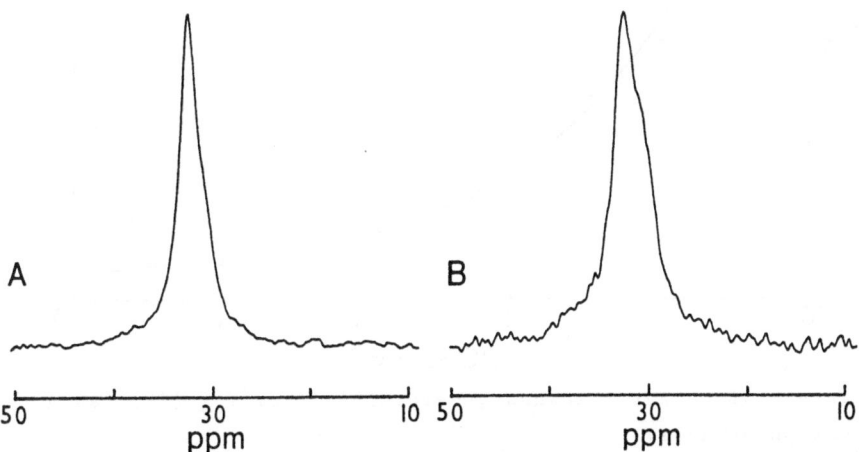

Figure 1. ^{13}C solid-state NMR spectra of A) unirradiated and
B) irradiated (10MGy) LLDPE.

^{1}H spin-lattice relaxation in the rotating frame ($T_{1\rho}$
(^{1}H)) for the unirradiated polymer, observed by varying the
cross-polarization contact time, showed evidence for two
distinct relaxation times.   The short-time relaxation was
assigned to relaxation of carbons in the disordered regions,
and the long-time relaxation to carbons in crystalline
regions, in agreement with previous workers (2).
Furthermore, relaxation in the crystalline regions is
dominated by spin-diffusion, and therefore related to the

crystalline lamellar thickness (L) (1).   Values of L,
calculated from $T_{1\rho}$ (long) and using the proportionality
constant obtained by Cudby et al. (2), were approximately
constant up to radiation doses of 2MGy, above which a
significant decrease was observed (Fig. 2).   This observed
decrease in the crystalline dimensions has been confirmed by
X-ray diffraction studies (5).

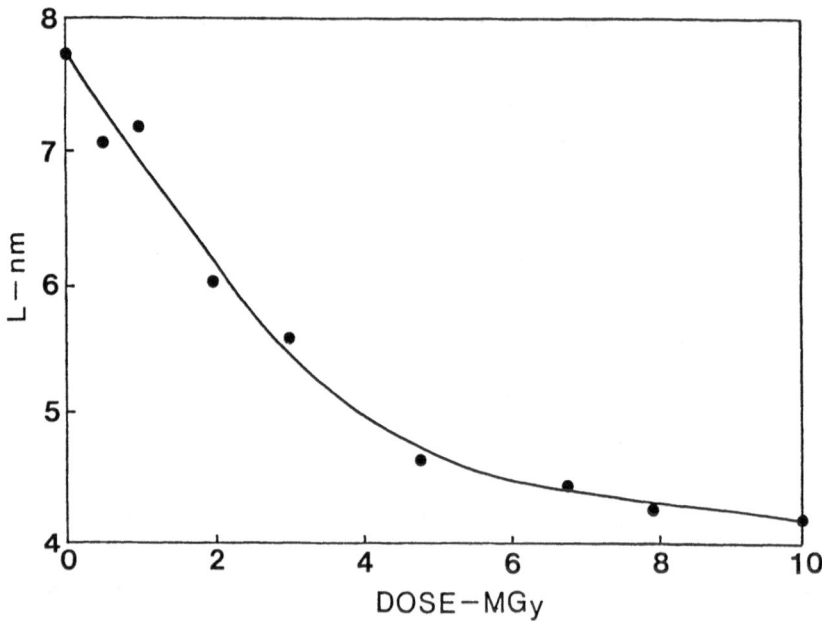

Figure 2. Lamellar thickness, determined from $T_{1\rho}$ (^{1}H), as a
function of dose.

The heat of fusion, ΔH_f, determined by DSC, decreased
initially on irradiation to 0.5MGy, and subsequently increased
to near the original value after 1MGy (Fig. 3).   Similar
behaviour has been observed for irradiated HDPE by Rijke and
Mandelkern (6).   The decrease in ΔH_f is ascribed to the
formation of imperfections in the crystal lattice, and the
increase to an increase in the melt entropy on crosslinking.
At higher radiation doses (>2MGy), ΔH_f decreased further to
61% of the initial value after 10MGy.   The large changes in
crystallinity and lamellar thickness suggest that crosslinking
in PE occurs in regions associated with the crystalline

material, most probably at the surfaces and at defects of crystals.

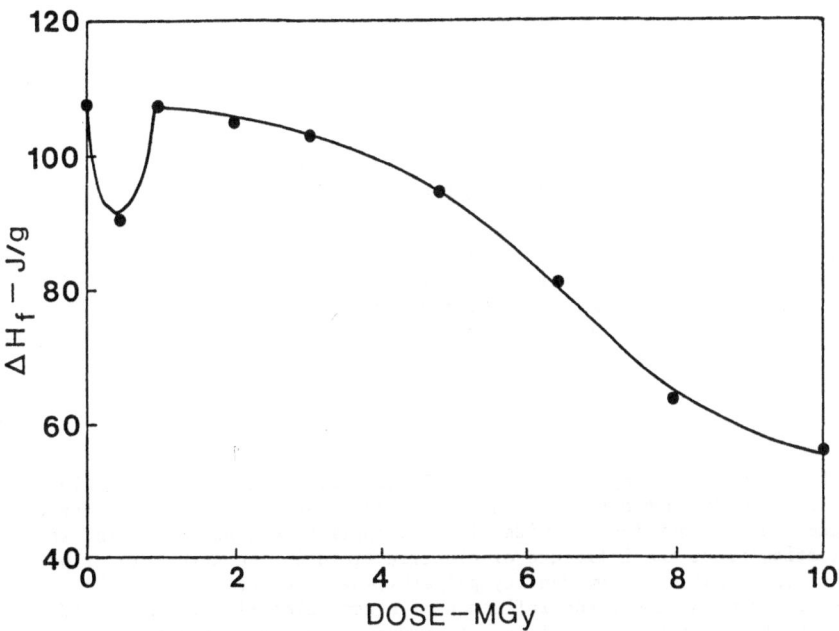

Figure 3. Heat of fusion of irradiated LLDPE, determined by DSC, as a function of dose.

REFERENCES

1. V.J McBrierty and D.C Douglass, J. Polym. Sci., Macromol. Rev., 16, 295 (1981).

2. M.E.A. Cudby, K.J. Packer, and P.J. Hensra, Polym. Commun., 25, 303 (1984).

3. J.A. Nedler and R. Mead, Comput. J., 7, 308 (1965).

4. C.A. Fyfe, J.R. Lyerla, W. Volksen, and C.S. Yannoni, Macromolecules, 19, 762 (1979).

5. O. Yoda and I. Kuriyama, Japan J. Appl. Phys., 16, 1447 (1977).

6. A.M. Rijke and L. Mandelkern, J. Polym. Sci., Polym. Lett. Ed., 7, 651 (1969).

ELECTRIC FIELD STIMULATED CHANGES
OF THE STRUCTURE OF LDPE

T.Sterzyński
Technical University of Poznań
Laboratory for Polymers
PL-61-752 Poznań, Poland

ABSTRACT

Polyethylene belongs to dielectric materials in which a strong electric field leads to the production of constant electric charges. Electrets
produced in this way found a wide field of application, but there is little
information about the influence of electric field on the structure of polymers. The samples of low density polyethylene were set in molten state
and during cooling under the influence of strong electric filed. The WAXS
measurements indicate a rebuilding of structure caused by electropolarization. As a result a decrease of unit cell parameters, a preferred orientation in the crystallographic ab plane and consequently a decrease of the
c-axis orientation were observed. A structural model of these changes is
proposed.

INTRODUCTION

Recent developments in electronic industry would not be possible
without new developments and applications of materials with good insulation
properties. Among other polymers also polyethylene found a wide field of
application in this industry.

Polyethylene belongs to the dielectric materials in which a strong
electric field leads to the production of the constant electric charge
[1 - 2]. Although the structure of polyethylene belongs to the most investigated features of polymers [3 - 4] there is little information concerning
the influence of electric field on its structure. Martin and Stupp [5] observed changes of both the melting enthalpy and the parameters of unit cell
caused by E-polarization of copolymers. In another paper [6] a following
suggestion was put forward: in polar polymers the rubilding of structure
may occur first when the energy of the E-field is higher than the energy

of the thermal motions of macromolecular chains. On the basis of short-
circuit thermally stimulated current experiments with polyethylene [7] the
correlation between electric field and polarization was ascertained.

The thermal charging method consists of the application of a electric
field to a dielectric material at an elevated temperature followed by cooling while the field is still applied. When the macromolecular chains are
adequately mobile the E-field leads to the alignment of polar groups and
consequently to the ordering of the chains. Upon fast cooling all charge
motions may be frozen in.

The aim of the presented work was to study the influence of electropolarization on the crystall structure of low density polyethylene.

MATERIALS AND METHODS

Low density polyethylene (Lupolen 1810 H, BASF) was pressed at the
temperature 423 K during 15 minuts and cooled slowly under presseure. The
constant electric field E was applied to the samples after heating, and
was kept during forming temperature (400 K) as well as during cooling to
the room temperature.

Structure investigations were done using WAXS with the CuK_α radiation.
It was possible to record the diffraction intensity as a function of the
diffraction angle Θ , and also as a function of the azimuthal angle φ , at
constant Θ angle, respectively.

RESULTS

With the help of the wide angle X-ray scattering (WAXS) it was found
out that the solidification of LDPE in electropolarization conditions
effects in higher values of Θ angle for both (110) and (200) maxima. Consequently the relative values of the interplanar distances decrease
(table 1). The decrease of the relative value $\Delta ab/ab$ for the plane ab of
the unit cell, versus electric field strength, is presented in table 1.
This change means more dense packing of the macromolecular chains in the
unit cell of polyethylene. The parameter a varies in the range from
7.462 Å to 7.540 Å, the parameter b in the range from 4.927 Å to 4.995 Å.

The macromolecular orientation was characterized by the Hermans
orientation function $f=(3 \cos\varphi -1)/2$ for the mean value of $\cos\varphi$.
Figure 1 presents the run of the function $I(\varphi) \cos\varphi$ versus φ angle for

TABLE 1

The changes of the parameters of the structure of low density polyethylene
solidificated in electric field

Electric field strength	Relative changes of the interplanar distances		Relative changes of the plane ab of the unit cell
	$\Delta d/d_{(110)}$	$\Delta d/d_{(200)}$	$\Delta ab/ab$
kV/cm	$\times 10^{-3}$	$\times 10^{-3}$	$\times 10^{-3}$
20.0	-3.42	-3.61	-6.70
29.8	-2.28	-1.53	-4.16
41.6	-6.34	-5.73	-12.28
45.3	-4.59	-3.12	-4.48

starting material, and for the samples after the same thermal treatment
but without and with E-polarization, for the planes (110) (Fig. 1a), the
plane (200) (Fig. 1b) and the plane (020) (Fig. 1c). The runs of the function $I(\varphi)\cos\varphi = f(\varphi)$ for both planes (h00) and (0k0) implie a- and b- axis
orientation of the macromolecules forced by the electric field. The respective Hermans orientation coefficients are presented in table 2. The
orientation function f_c was evaluated on the basis of the cosinus law.

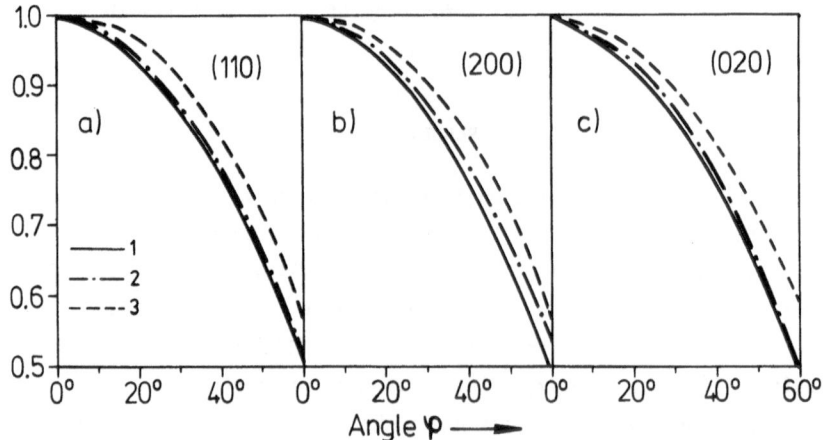

Figure 1. The function $I(\varphi)\cos\varphi$ versus φ angle for: 1 - starting
material, 2 - sample after thermal treatment, 3 - after thermal treatment
and electropolarization; a) plane (110), b) plane (200), c) plane (020)

TABLE 2

Coefficient of the Herman's orientation function for LDPE samples

Orientation function	starting material	without electric field	with electric field
	$\times 10^{-3}$	$\times 10^{-3}$	$\times 10^{-3}$
f_{110} $\quad (f_a)$	-3.9	2.2	13.9
f_{200}	-6.9	9.03	19.75
f_b	-2.9	-6.1	27.6
f_c	9.8	-3.0	-48.35

The experiments showed that a certain orientation of the c-axis, which exists in the starting material, disappears as a result of heat treatment and E-polarization. Especially E-polarization is a factor causing a high decrease of f_c and an increase of the orientation in the ab plane (perpendicular to the main c-axis of the zig-zag chain).

DISCUSSION

The so-called Frederiks transitions are related to the rotation of the molecular chains of liquid crystals in electric or magnetic fields. These rotations are enabled by the anisotropy of the distribution of the van der Waals forces in the unit cell. According to the same mechanism, it may be assumed that in the molten state – when the chains have a higher degree of freedom– the chains of LDPE rotate around the c-axis. Because of it, a decrease of the parameters in the ab plane of the unit cell is possible. This kind of rotation is more difficult during cooling, so probably these effects are not decisive for the changes of the parameters a and b.

Another interpretation, based on the electromechanical effects (5), similar to the one for liquid crystalline copolymers, is proposed. The E-field may lead to the distention of more flexible segments and, consequently, to the orientation. Analogous interpretation may be done for the ordering in short branching LDPE. Because the H and C atoms have different electric charges, the external E-field may cause dipole formation followed by their rotation. This leads to the average preferred ordering of the structure. On the other hand short branching represents a kind of structure defects, which under polarizing conditions are shifted to the amorphous domains (Fig. 2). In this way, distended macromolecular chains form a more perfect structure with average smaller unit cell.

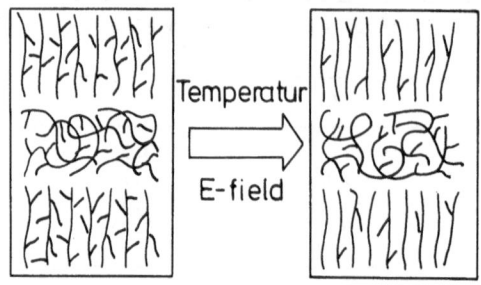

Figure 2. The electrical alignment under polarizing conditions. The short chain branching are shifted to the amorphous domains and the unit cell are of smaller magnitudes

CONCLUSIONS

The action of E-field on polyethylene (LDPE) in molten state and during its solidification results in the decreasing of the average value of the parameters a and b of the unit cell. A preferred orientation of the plane ab, caused by E-polarization, is accompanied by a decrease of the c-axis orientation.

ACKNOWLEDGMENT

The work was done during the stay in the Philipps University of Marburg (F.R.Germany), supported by the Alexander von Humboldt Scholarship, what is greatly acknowledged.

REFERENCES

1. Block H., Adv. Polymer Sci., 1979, 33, 93.

2. van Turnhout J., Thermally stimulated discharge of polymer electrets, Elsevier, Amsterdam, 1975, pp. 1-5.

3. Stein R.S., Norris F.H., J.Polymer Sci., 1956, XXI, 381.

4. Fischer L., Ruland W., Colloid Polymer Sci., 1983, 261, 717.

5. Stupp S.J., Martin P.G., Polymer, 1985, 26, 682.

6. Tynenska B., Gałęski A., Kryszewski M., Polymer Bulletin, 1981, 4, 171.

7. Fischer M., J.Electrostat., 1977/78, 4, 149.

RAMAN MICROPROBE SPECTRA OF A
SPIN-ORIENTED AND DRAWN PET FILAMENT

Fran Adar
Instruments SA
Metuchen, New Jersey   08840
USA

and

Herman D. Noether
Textile Research Institute
Princeton, New Jersey   08542
USA

ABSTRACT

Microprobe Raman spectra of an amorphous spin-oriented filament (1500 m/min TUS) and its room-temperature drawn derivative (D.R.=2.5:1) are presented. Their conformational features in the glycol and carbonyl band regions are compared with DTA and x-ray diffraction data. It is apparent that the three methods of measurement record different aspects of the morphology of these filaments.

INTRODUCTION

The Raman spectra of a series of spin-oriented and drawn PET filaments, obtained with a Microprobe Raman spectrometer, were analyzed in a previous publication [1]. The range between 600 and 1800 cm^{-1} was studied with special emphasis on the glycol bands at 900 to 1200 cm^{-1} and the carbonyl band at about 1730 cm^{-1}.

The bands at 999 and 1096 cm^{-1} in crystalline samples seem to correlate with trans conformation or crystallinity. The change of the band at 1175 to two bands at 1175 and 1182 cm^{-1} confirms the Stokr [2] assignment that 1182 cm^{-1} is due to crystallinity. However weak bands at 999 and 1096 cm^{-1} are already visible in fibers that are considered amorphous by x-ray diffraction pattern. Thus at least part of these bands are due to trans conformation rather than crystallinity.

DATA AND DISCUSSION

The amorphous sample spun at 1500 m/min TUS can be drawn at room temperature to a draw ratio of 2.5:1. Its Raman spectrum gives medium strong bands at 999 and 1096 cm^{-1} and an additional peak at 1080 cm^{-1} [1]. Its x-ray diffraction, while indicating high orientation, shows good order in the fiber direction but no crystalline regularity in the lateral arrangement. This suggests the 999 and 1096 cm^{-1} bands to be due to glycol motions in their trans conformation.

According to Melveger [3] the half width of the carbonyl band at about 1730 cm^{-1} correlates with crystallinity, the half width decreasing with increasing crystalline order. Here both the 1500 m/min extruded sample and its room-temperature drawn derivative have identically broad halfwidths [Table 1, Fig. 1a, 1b] and thus seem equally amorphous.

The Microprobe Raman spectrometer allows accumulation of series of runs and can indicate structural changes as a function of exposure time to the laser beam. This occurred in a series of sequential runs for the room temperature drawn filament in the 950 to 1200 cm^{-1} range [Fig. 2] The relatively broad peak at 999 cm^{-1} after heating by the laser turned more intense and sharper, the peak at 1080 cm^{-1} disappeared, the one at 1096 cm^{-1} became stronger. Thus the 999 and 1096 cm^{-1} bands are due to motions of the glycol group in trans conformation in both amorphous and crystalline environments.

Differential scanning calorimetry (DSC) reveals structural differences. The "as spun" yarn shows typical crystallization and melting peaks. The drawn sample has a very small broad crystallization area and a very large melting peak located at a higher temperature compared to the undrawn material. This suggests some "crystalline-like" order after drawing which during the heating cycle leads to crystallites of much better size and order [Fig. 3a, 3b].

CONCLUSIONS

Raman spectra and x-ray diffraction provide structural information on different dimensions of the morphology of the PET filaments. X-ray diffraction primarily sees degrees of order in the regularity of interchain arrangements.

Raman and infrared bands, however, relate to intrachain vibrational motions, based on the conformations within each chain. Thus crystallinity bands in vibrational spectra need independent proof that their shapes are modified from normal conformational band widths or their frequency is shifted by strong interchain interaction effects. The DSC data provide an insight into the crystallization and melting kinetics as function of the original morphology.

REFERENCES

1. Adar, F. and Noether, H., Raman microprobe spectra of spin-oriented and drawn filaments of poly(ethylene terephthalate). Polymer, 1985, 26, 1935-1943.

2. Stokr, J., Schneider, B., Doskosilova, D., Lovy, J. and Sedlacek, P., Conformational structure of poly(ethylene terephthalate), Infra-Red, Raman and N.M.R. spectra, Polymer, 1982, 23, 741-721.

3. Melveger, A.J., Laser-Raman study of crystallinity changes in poly(ethylene terephthalate), J. Polym. Sci., A-2 1972, 10, 317-322.

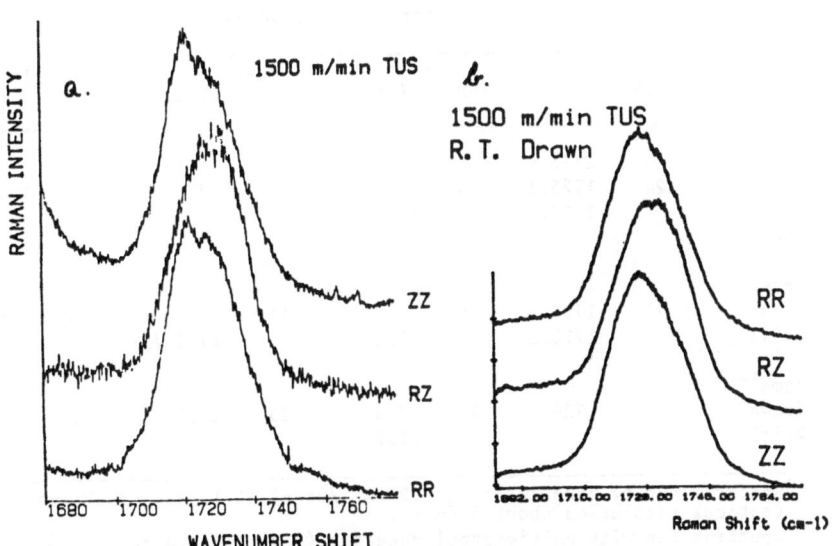

Figure 1. Polarized Raman spectra in the carbonyl region. All spectra are displayed with the bands fully expanded in the intensity direction. (a) Spin-oriented at 1500 m/min, "as is". (b) same as (a) after room temperature drawn 2.5:1.

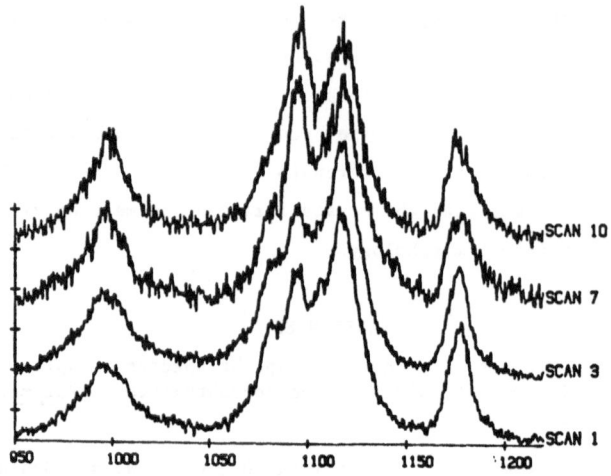

Figure 2. Sequential Raman runs for the spun and drawn filament in the 950 to 1200 cm^{-1} region. After run 3, sample was inadvertently heated by the laser beam.

TABLE 1
Data for Carbonyl Bands

Sample	ZZ FWHM	ZZ FREQ.	ZR or RZ FWHM	ZR or RZ FREQ.	RR FWHM	RR FREQ.	Rel. Int. ZR/ZZ	Rel. Int. RR/ZZ
1500* m/min "as spun"	24	1721.4 1725.1 1730.1	24	1726.0 1731.1	25	1721.1 1727.4	.4	1.02
same* drawn 2.5:1	27	1722.8 1728.3	24	1728.3	24	1731.7	.3	.55
same** drawn 2.5:1	25	1726	23	1728 1732	23	1726	.2	.35

*Spectrum run by step scanning (.2 cm^{-1}/step) 10/15/84. 300μ slits (optical resolution about 3 cm^{-1}).
**Spectrum run with multichannel detector (7 runs at 20 seconds each) 2/10/87. Opt. resolution about 6 cm^{-1}
All runs with a single filament and identical microscope optics.
FWHM = full width at half maximum intensity
FREQ. = Raman frequency in cm^{-1}

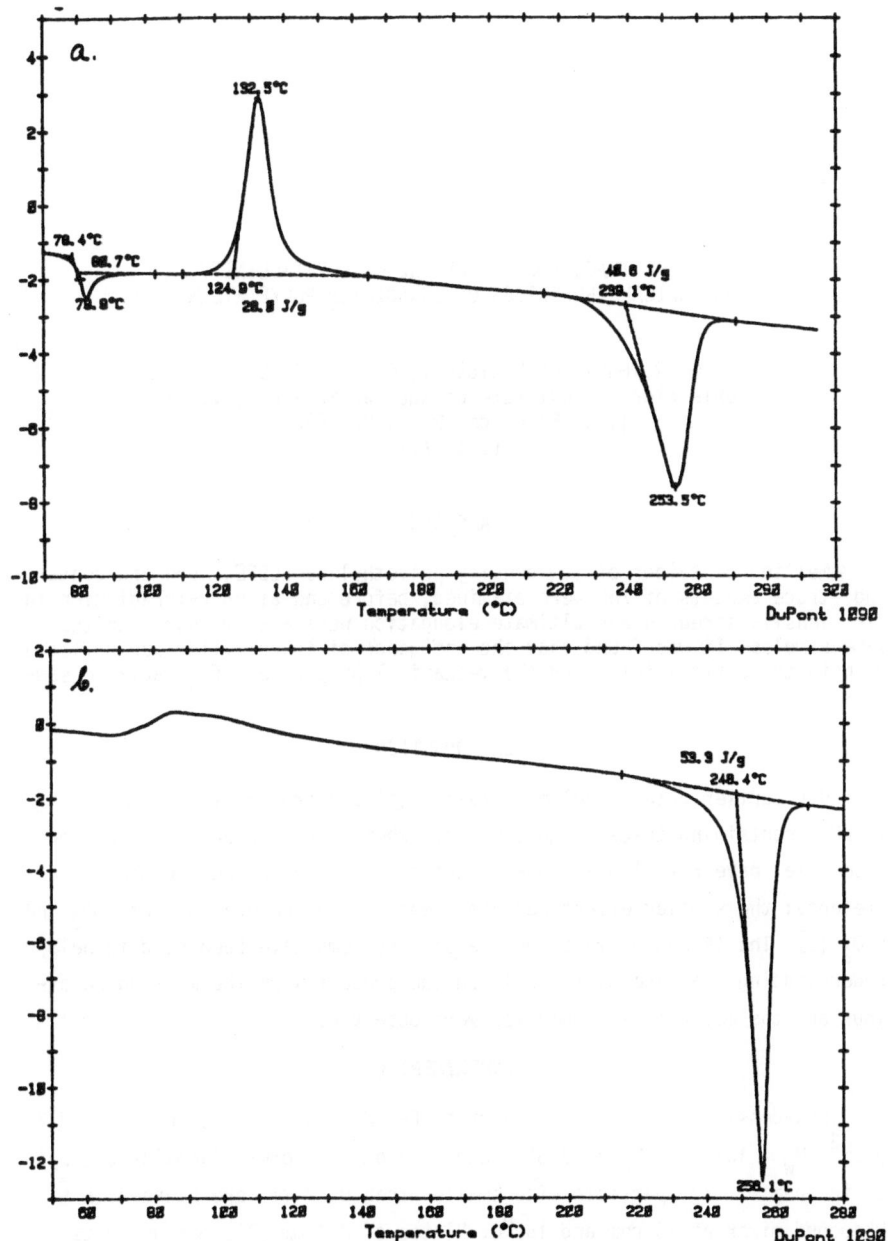

Figure 3. DSC curves for the "as spun" and drawn yarn samples;
heating rate: 20°C/min, Dupont #1090. (a) "as spun",
(b) after drawing.

EFFECT OF TRACE AMOUNTS OF ZnO ON CRYSTALLINITY AND MECHANICAL PROPERTIES OF IRRADIATED POLYETHYLENE

G. R. Ghassemi Mahidasht, O. Gal, D. Babić
"Boris Kidrič" Institute of Nuclear Sciences, Vinča,
11001 Belgrade, P. O. Box 522,
Yugoslavia

ABSTRACT

Melting behaviour of low density polyethylene (LDPE) pure and containing trace amounts of ZnO were examined, before and after γ-irradiation in air. Tensile strength and ultimate elongation were also measured on the same samples. It was found that the influence of trace filler was reflected on both the crystallinity and the mechanical properties of irradiated samples.

INTRODUCTION

Our earlier data on gel measurement [1] and free radical formation [2] in LDPE containing trace amounts of ZnO, when γ-irradiated in air at low dose rate, have revealed an enhancement of crosslinking due to the filler presence; the similar effect has been reported in the presence of TiO_2 and SiO_2 [3]. The IR measurements on the same systems have been used to help understanding that phenomenon [4]. In the present work the melting behaviour and the mechanical properties were observed.

EXPERIMENTAL

Low-density polyethylene, Lotrene CD-0302, CdF Chimie, density 0.922 g/cm^3, M_w = 103,000, M_n = 20,800, and zinc oxide, Zorka (Yugoslavia), specific area 2 m^2/g, particle size smaller than 0.12 mm were blended in a Rheocord mixer at 15 rpm and 150°C. Plates of 0.4 mm (DSC measurements) or 1 mm (for the stress-strain tests) were moulded under press at 140°C and cooled in air to the room temperature.

Differential scanning calorimetry (DSC) measurements were performed in a Perkin-Elmer DSC-2 apparatus by heating under nitrogen at 10 degrees//min up to 420 K (run I) and after isothermal crystallization at 373 K for

10 min (run II). From DSC curves the corresponding melting parameters were derived.

Tensile strength and ultimate elongation were measured by a Zwick Test Machine at room temperature using the extension rate of 90 mm/min.

Irradiations were performed on a Co-60 source at 0.5 kGy/h dose rate, in air at room temperature.

RESULTS AND DISCUSSION

Data on the melting properties of the pure and filled LDPE samples were presented in Table 1.

TABLE 1

Peak temperature (T_m), heat of fusion (ΔH) and crystallinity (x) before and after isothermal crystallization of LDPE, pure and containing ZnO

Weight % of ZnO in LDPE	Dose kGy	run I			run II		
		T_m K	ΔH J/g	X %	T_m K	ΔH J/g	X %
0.0	0.0	382.2	100.3	36.9	383	108.0	39.7
0.12	"	381.7	101.8	37.5	384	106.8	39.3
2.4	"	381.0	102.8	37.8	383	106.1	39.1
0.0	300	382	111.3	40.9	382	107.1	39.4
0.12	"	382	108.3	39.9	382	113.4	41.7
2.4	"	382	110.7	40.7	382	100.2	36.9
0.0	700	381	123.6	45.5	383	107.0	39.4
0.12	"	382	121.2	44.6	382	114.2	42.0
2.4	"	382	124.9	46.0	382	104.5	38.5

(Each experimental point is the mean of four measurements)

For unirradiated samples practically no difference was noticed in melting temperature as well as in heat of fusion between pure and filled LDPE, before (run I) and after isothermal crystallization (run II). However, both the melting temperature and the heat of fusion are higher in the latter case, most probably due to the better conditions for crystallization. After irradiation an increase of crystallinity in the samples before isothermal crystallization by increasing the absorbed dose showed up. This can be explained as an effect of oxidative degradation occuring in the course of rather long irradiation in air (very low dose rate of 0.5 kGy/h); during it the rearrangement of the degradated polymer molecules could take place.

After isothermal crystallization, run II, heats of fussion are of lower values than in run I, as it can be expected because of crosslinking. The exceptions are the values of ΔH^{II} for the samples containing 0.12 wt% ZnO. One of the explanations for this phenomenon might be the additional nucleation due to the presence of ZnO as it has been reported for TiO_2 [3,5].

Tensile strength is affected by the presence of trace amount of ZnO; it increases as the gel content does, but up to a higher ZnO content (> 0.12 wt%), Fig. 1,b. This further increasing of tensile strength can be explained by its dependence not simply on crosslinking and crystallinity but on the other factors as interface forces between polymer and filler. The influence of trace filler is also found on the ultimate elongation, more visible at the lower doses, Fig. 1,c.

The present data confirm that even a small amount of ZnO induces significant changes in the irradiated LDPE.

REFERENCES

1. Ghassemi Mahidasht, G.R., Gal, O., Sharifi, N., Sang, M., Vakilinejad, F. and Chashmi, R., Effect of trace fillers on radiation crosslinking of polyethylene. Radiat. Phys. Chem., 1985, 25, 349-357.

2. Ghassemi Mahidasht, G.R., Gal, O. and Stojić, M., Enhanced formation of free radicals in irradiated LDPE containing ZnO, Radiat. Phys. Chem., 1987, 29, 45-49.

3. Gordienko, V.P., Kartel, N.T., Suprunenko, K.A., Doroshenko, V.N. and Katakchi, A.M., Isledovanie radiacionnova sshivaniya polietilena v geterogennih sistemah, Visokomol. Soedin., 1975, A17, 1737-1940.

4. Ghassemi Mahidasht, G.R. and Gal, O., Influence of trace amount of ZnO on the radiation changes in LDPE, in preparation for publication.

5. Grigorev, V.I., Gordienko, V.P. and Tynnyi, A.N., Isledovanie struktury i mekhanicheskih svoistv napolnennova i polietilena, Fiz-khim. Mekhan. Mater., 1973, 9(6), 55-61.

449

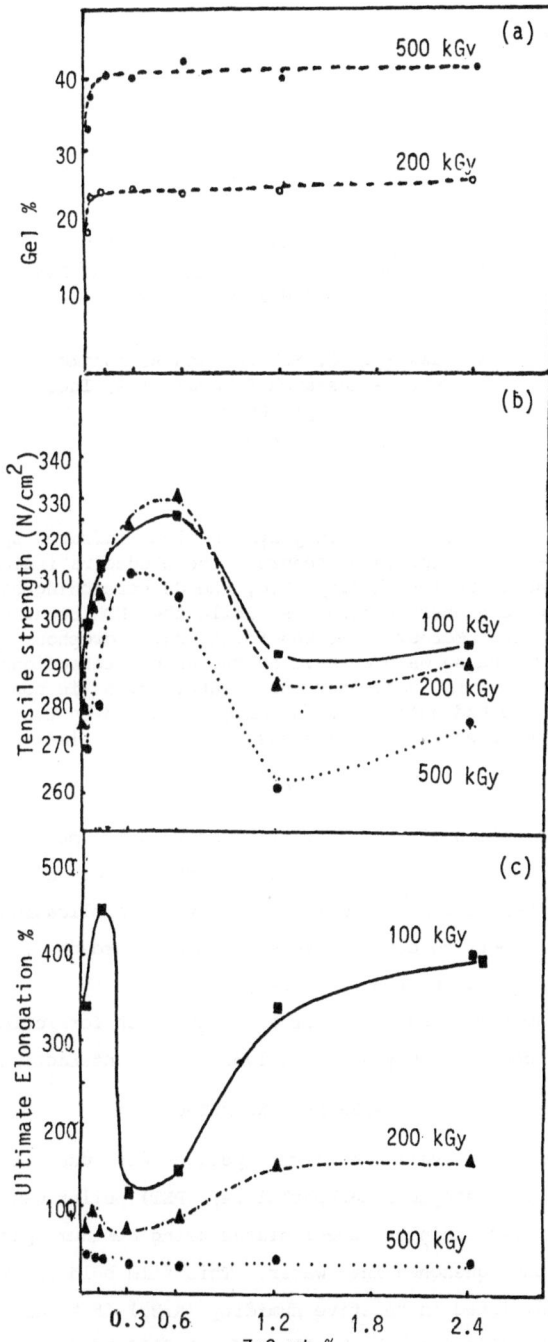

Fig 1. Effect of trace amount of ZnO on gel fraction (a),
tensile strength (b) and ultimate elongation (c),
gamma irradiated in air, room temperature.

A FOURIER-TRANSFORM INFRARED STUDY
ON THE POLYMER-WATER MOLECULAR INTERACTION
FOR HYDROPHOBIC POLYMERS

H. Kusanagi, S. Yukawa and A. Ishimoto
Unitika Research Laboratories, Inc.
Uji, Kyoto
JAPAN

ABSTRACT

For water molecules sorbed in polymers clear infrared spectra could be obtained by FT-IR subtraction procedure. These spectra indicate that as absorbent polymers are more hydrophobic, the OH stretching bands of water molecules shift to higher frequencies, while the HOH bending frequency moves to lower one. Moreover, it was found that hydrophobic polymers of ETFE, PVDF, PET, and so on even have appreciable interactions with the water molecules, and the range for their interactions is considerably large. These obserbed data are reasonably explained by the calculation of normal vibration of water molecules based on a simplified model.

INTRODUCTION

The investigation of the interaction between polymers and water molecules in molecular order is fundamental importance for understanding the water sorption behavior of polymers for practical applications. Although there are many papers concerned with structural approaches, only few reports with infrared spectroscopy [1-2].

In this paper we wish to report a new approach for Fouier-transform infrared spectroscopic study on the polymer-water interaction.

MATERIALS AND METHOD

The commercially available polymer pellets were used in this study. For an example poly(ethylene terephthalate) (PET) pellet was melt and pressed between two stainless steel plates using alminum spacer with 100 μm thickness and then quenched into water. This film held on FT-IR sample holder was conditioned in relative humidity (R.H.) 65 % for 48 hours. After moving the film into the Nicollet Model 7199 FT-IR Spectrometer, the first spectrum was immediately measured. Then drying the film completely in flowing N_2 gas, and thereafter the second measurement was done.

The difference spectrum obtained by computer subtraction of the second spectrum from the first one represents the three absorption bands ν_a(OH), ν_s(OH), and $\delta(\angle HOH)$, corresponding to the normal vibration modes of water molecule. Thus a spectrum on the water molecule sorbed in PET was obtained.

The calculation of normal vibration frequencies for water molecules sorbed in polymers was carried out by using Willson's GF matrix method [3]. For the water molecule, $m_h = 1.008$, $m_O = 16$, $r = 0.957$ Å, and $\phi = 104.5°$ are fixed. Three intramolecular valence force constants of the bond stretching K_r(OH), valence angle bending $H_\phi(\angle HOH)$, and repulsive term F_{rr} are adopted. Intermolecular interaction between water and polymer molecules was estimated within these intramolecular force constants.

RESULT AND DISCUSSION

Excellent FT-IR spectra of sorbed water on several polymers are shown in Figure 1. As polymers are more hydrophobic, stretching frequencies ν_a and ν_s shift to higher frequencies, while valence angle bending δ moves to lower one. Another characteristic feature is the fact that hydrophobic polymers imply more sharp bands than hydrophilic polymers. These are the direct reflection of the difference in the interaction strength between

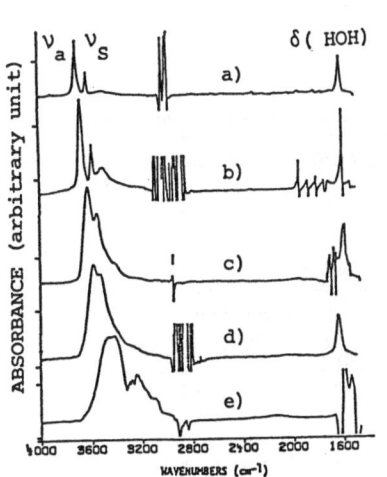

Figure 1. FT-IR spectra of water sorbed in polymers, a) poly(vinylidene fluoride), b) polystylene, c) poly-(ethylene terephthalate), d) poly(ethylene oxide), and e) nylon 6.

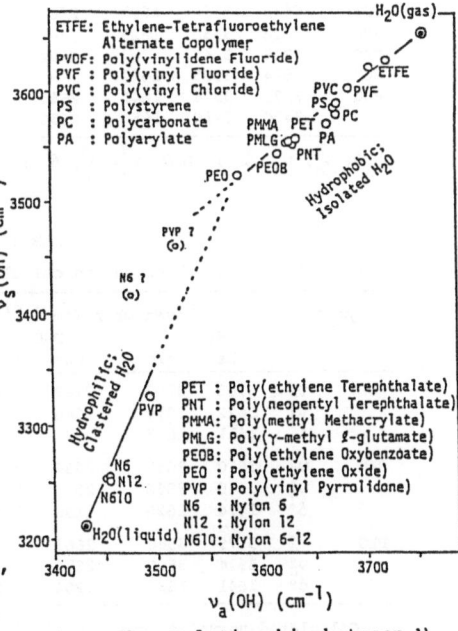

Figure 2. Relationship between ν_a and ν_s for sorbed water.

polymers and water molecules.

Figure 2 indicates the relationship between ν_a and ν_s. Sorbed water in hydrophobic polymers gives the data points of higher wavenumber near that of H_2O gas, and their points lay on a straight line (a), may be corresponding to the isolated H_2O state. On the other hand, hydrophilic polymers form the other line (b), linking to the point of liquid water (clastered H_2O). It was suggested that hydrophobic polymers have also appreciable interaction with water molecules, and the range for interactions $\Delta\nu_a = \nu_a(ETFE) - \nu_a(PET)$ is considerably large 86 cm^{-1} (= 1.0 kcal/mol).

Obserbed frequencies of H_2O and D_2O for various polymers together the

TABLE 1

Observed frequencies of water molecules sorbed in various polymers and the ratio r = H_2O/D_2O

Polymer	ν_a^h cm^{-1}	ν_s^h cm^{-1}	δ^h cm^{-1}	ν_a^d cm^{-1}	ν_a^d cm^{-1}	δ^d cm^{-1}	$r(\nu_a)$	$r(\nu_s)$	$r(\delta)$
$H_2O^{a)}$	3756	3655	1595	2789	2666	1179	1.347	1.371	1.353
ETFE	3721	3630	1608	-	-	-	-	-	-
PVDF	3705	3622	1616	2749	2646	-	1.348	1.369	-
PVF	3684	3602	1620	2734	2633	-	1.347	1.368	-
PVDC	3672	3547	1609	-	-	-	-	-	-
PVC	3673	3589	1606	2728	2624	-	1.346	1.368	-
PS	3670	3585	-	2723	2625	-	1.348	1.366	-
PAN	3630	3540	1624	2694	2590	1198	1.347	1.367	1.356
PET	3635	3556	1621	2701	2602	1199	1.346	1.367	1.352
PMMA	3628	3555	1633	-	-	-	-	-	-
PEOB	3718	3543	-	2192	2591	-	1.344	1.367	-
POM	3614	3535	1627	-	-	-	-	-	-
PEO	3579	3524	1641	2663	2574	1203	1.344	1.369	1.364
$H_2O^{b)}$	3430	3210	1650	2630	2512	1220	1.304	1.278	1.352

H_2O: ν_a^h, ν_s^h, δ^h ; D_2O: ν_a^d, ν_s^d, δ^d ; a) H_2O(gas) ; b) H_2O(liquid)[4].

TABLE 2

Comparison between observed and calculated frequencies for sorbed water

Polymer		Wavenumber $(cm^{-1})^{a)}$				Force constants		
		H_2O obs	calc	D_2O obs	calc	K_r mdyn/Å	H_ϕ mdyn/Å	F_{rr} mdyn/Å
PVDF	ν_a^o a	3705	3702	2749	2745	8.2	0.84	-0.10
	ν_s^o	3622	3614	2646	2645			
	δ^o	1611	1617	-	1191			
PAN	ν_a^o	3630	3635	2694	2695	7.9	0.85	-0.10
	ν_s^o	3540	3546	2590	2595			
	δ^o	1624	1626	1198	1199			
PEO	ν_a^o	3579	3588	2663	2660	7.7	0.86	-0.10
	ν_s^o	3524	3500	2574	2562			
	δ^o	1641	1637	1204	1206			

a) Calculated normal frequencies (ν_a^o, ν_s^o, δ^o) were evaluated using the correction constants for anharmonicity: $\nu^o/\nu^e = 0.955$, $\delta^o/\delta^e = 0.964$ for H_2O and $\nu^o/\nu^e = 0.967$, $\delta^o/\delta^e = 0.974$ for D_2O [4].

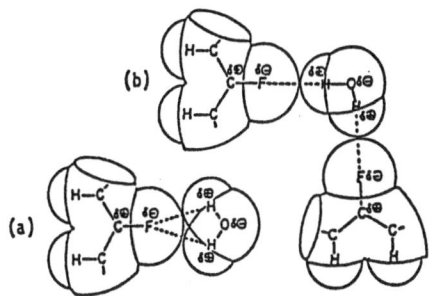

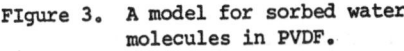

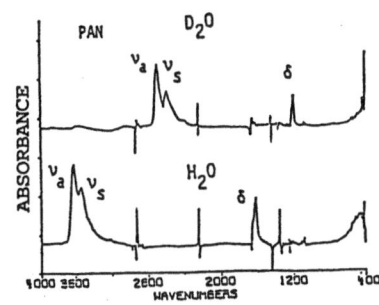

FIgure 3. A model for sorbed water   Figure 4. FT-IR spectra of H_2O and
molecules in PVDF. D_2O sorbed in PAN.

frequency ratio $r = \nu(H_2O)/\nu(D_2O)$ are listed in Table 1. The ratio r for
stretching vibration modes is close to that of H_2O gas. Table 2 shows the
comparison between the obserbed and calculated frequencies. For H_2O and D_2O
of each polymer, six obserbed frequencies are in a good agreement with the
calculated normal frequencies by using only three force constants determined
by trial and error procedure. Although the hydrophobicity of PVDF, PAN, and
PEO decreases in that order, it was found that in the same order K_r(OH)
decreases and $H_\phi(\angle HOH)$ increases. For the interpretation of this relation,
a structure model for the interaction between PVDF molecule and water molecule may be illustrated in Figure 3: water molecules coordinate to defined
atoms of polymers. It is understandable from this figure that the attractive force of the F...H weakens the K_r, while a pair of the F...H increases
the H_ϕ. It seems also reasonable from a fact that the rotational libration
mode may be obserbed at 400 cm^{-1} on FT-IR spectrum of PAN (Figure 4).

These are very interesting new facts found by FT-IR spectroscopy, and
further investigation is in progress.

REFERENCES

1. Jellinek, H.H.G., Water Structure at the Water-Polymer Interface, Plenum
   Press, New York, 1972.

2. Rowland, S.P., Water in Polymers, ACS Symp. Series 127, Am. Chem. Soc.,
   Washington D.C., 1980.

3. Willson, E.B., Molecular Vibrations, McGraw-Hill, New York, 1955.

4. Herzberg, G., Molecular Spectra and Molecular Structure, II. Infrared
   and Raman Spectra of Polyatomic Molecules, D. Van Nostrand Company, Inc.
   1956, pp. 280-2.

NOVEL MODIFIED HEMODIALYTIC FILM.
THE STUDY ON THE PERMEABILITY AND STRUCTURAL CHARACTERISTIC
OF POLY[(ETHYLENE-VINYL ALCOHOL)/ACRYLAMIDE] GRAFTING FILM

Zhufang Liu, Jing Shen, Kangde Yao
Department of Material Science and Engineering,
Tianjin University,
PRC

and

Hanqin Gu
Tianjin Institute of Urologic Surgery
PRC

ABSTRACT

In this paper the characteristic of the chemical structure and morphology of poly[(ethylene-vinyl alcohol)/acrylamide] (EVAL-g-AM) film was studied by the Scanning Electron Microscope(SEM), the Transmission Electron Microscope (TEM) and the Differential Scanning Calorimetry (DSC), etc. The permeability of urea through the film was examined. The results showed that the chemical structure and morphology of the film varied with grafting percentage(G%), so that its permeability was affected and superior to that of original poly(ethylene-vinyl alcohol) (EVAL) film.

INTRODUCTION

As a hemodialytic film, polymer material should be of higher permeability, better blood compatibility and outstanding dimentional stability, in addition to stronger mechanical strength. These properties depended on the structure and polarity of a film[1], therefore, its chemical structure and morphology should be considered in the course of design or selection of a high permeable hemodialytic film[2]. And so we have developed acrylamide-grafted EVAL (EVAL-g-AM) film in order to improve its permeability[3]. In this paper, the effect of structural characteristic on permeability of EVAL-g-AM film was studied.

MATERIALS AND METHODS

2-1 Materials

EVAL film (ethylene/vinyl alcohol = 55/45 molar, 0.06mm thick) was from Kori Company, Japan. EVAL-g-AM film was synthesized by the authors[3].

2-2 Observation of Film Surface

Film specimen was splashed with platinum in vacuum, the coated specimen was observed under HITACHI-450 SEM.

2-3 Observation of Film Cross-section

A film specimen was stained with a saturated palladium chloride aqueous solution. Morphology of cross-section of the stained film was examined under PHILIPS EM-400ST TEM. The dark color region indicated acrylamide-grafting domains.

2-4 Crystallinity

The melting and crystalline peaks of a film specimen were obtained by Perkin-Elmer DSC. According to the peak area the crystallinity was evaluated.

2-5 Measurement of Permeability Coefficient of urea through Film

A glass diffusion device, which consisted of two cells that were separated by EVAL-g-AM film, one with distilled water and the other with 0.1M urea aqueous solution, was immersed in a water bath at 40°C for 8 hrs. Then difference in urea concentration between both cells was determined by nitrogen element analysis. The permeability coefficient (p) of urea through the film was calculated by eq(1)[4]

$$P = Ql / \triangle c \qquad (1)$$

where Q was transfer rate of urea in the film, l was the thickness of the film, c was a difference in concentration of both cells for some intervals.

RESULTS AND DISCUSSIONS

3-1 Effect of Surface Characteristic of Film on Permeability

AS for a hemodialytic film, its surface characteristic could affect

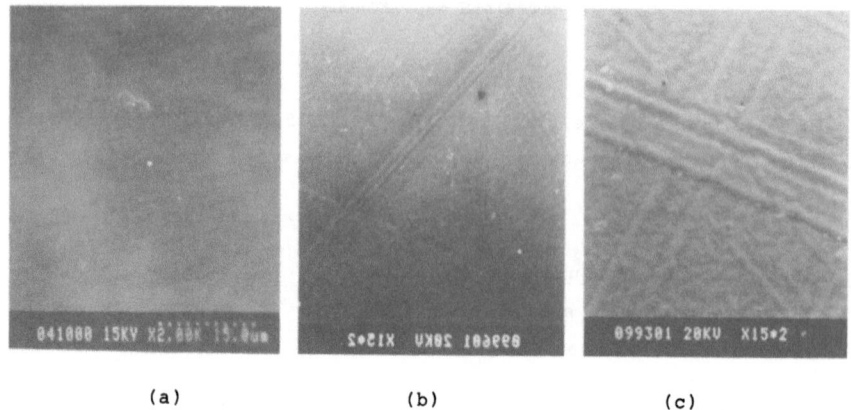

(a)   (b)   (c)

Figure 1. SEM view of EVAL-AM (b:G%=37;c:G%=111.3) and EVAL (a)

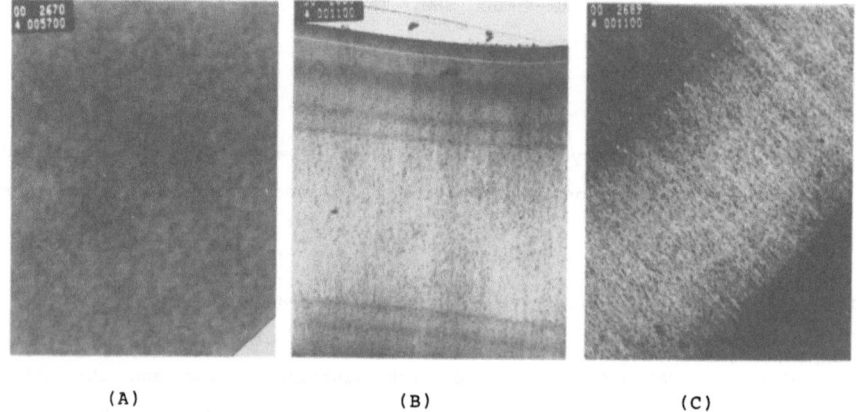

(A)   (B)   (C)

Figure 2. TEM view of EVAL-g-AM (B:G%=75; C:G%=111.3) and EVAL (A).

not only its blood compatibility, but also permeability of a solute
through the film. It appeared that the transfer of urea through EVAL-g-AM
film would greatly relate to the surface characteristic of the film.
The view of SEM (Fig 1) and TEM(Fig 2) and the spectrum of ATR-FI-IR
show that the film surface is composed of soluble grafted-polyacrylamide
chains and that the higher G%, the thicher grafting layer that implys
more grafted chains. It is further confirmed by the contact angle data
of the film. As the film comes in contact with an aqueous solution, the
surface should contain a great amount of water forming the 'diffusion
layer' proposed by Y. Ikada[5].   Within   the   layer   free   ends   of

grafted-polyacrylamide chains solubilize in the solution while the others are covalently bound to the film surface. So it could promote urea to partition and solubilize on the surface, and would have urea maintain higher concentration, resulted in urea easy to diffuse and permeat through the film. Similarly, on the other surface urea could de-partition and de-solubilize, and then urea was apt to diffuse into low concentration aqueous phase. It seems that the permeability of EVAL-g-AM film could be improved by polyacrylamide-grafted surface layer, and superior to that of EVAL film, as shown in Table 1.

TABLE 1

Permeability coefficient of urea through EVAL-g-AM and EVAL film

Film	EVAL-g-AM			EVAL
G%	111.3	75.3	37.5	0
P (cm^2/sec)	3.82×10^{-6}	6×10^{-7}	1×10^{-7}	5×10^{-9}

3-2 Influence of Crystalline Characteristic on Permeability

The results of DSC and X-ray Diffraction analysis indicated that EVAL-g-AM is a crystalline polymer and its crystallinity is reduced with an increase in grafting percentage. It appears that grafted polyacrylamide chains could lead to more amorphous region, diffusiable area increasing . It should be noted that polyacrylamide grafted onto EVAL-g-AM with much higher grafting percentage could result in the film swelling so significantly that the "screen effect" of crystallinites should be reduced, so improving the permeability of the film.

3-3 Impact of Hydrophilicity / Hydrophobicity on Permeability

Pore structure had not found either on the surface of wet or dry EVAL-g-AM film and EVAL film under the Phase-Contrast Microscope or SEM, nor in the cross-section under TEM. It indicated that the films were of compact morphology. So transfer of urea through the films should depend on free volume offered by the film substrate and be currently affected by polarity factors. In comparision with EVAL film, hydrophilicity of EVAL-g-AM film was greater and the latter contained a great amount of water within network providing greater free volume, which could improve urea to transfer through the film. Hence, permeability of EVAL-g-AM film was superior to that of EVAL film.

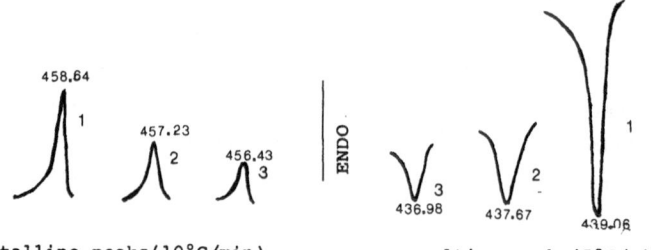

crystalline peaks(10°C/min)   melting peaks(5°C/min)

Figure 3. DSC spetrum of EVAL-g-AM (2:G%=37;3:G%=75) and EVAL(1)

TABLE 2

Water absorption and contact angle of EVAL-g-AM film

G%	10	25	52	72	98
water absorption(%)*	7.4	24.1	31.4	42.4	58.8
contact angle**	82	73	54	46	31

*To be determined from the difference in weight, at 28°c,24hrs
**To be measured by a ERM-422 Gonimeter.

CONCLUSION

1 The permeability of EVAL-g-AM film was superior to that of EVAL film and affected by chemical structure and morphology.
2 The surface of EVAL-g-AM film have a "diffuse layer" composed of grafted polyacrylamide chains. The greater grafting percentage, the thicker grafting layer, the smaller crystallinity and the higher hydrophilicity.

REFERENCE

1. OHYA Haruhiko, Kagaku Supplement (JP), 1981,92,p21.
2. N. A. Peppas, D. L. Meadows, J.Membr. Sci, 1983,16, 361-377.
3. The authors, J. Macromol. Sci., -chem, in press.
4. Cokio H. Progr. Org. Coating, 1982,10 1.
5. Ikada Y. Adv. Polym. Sci. , 1984,57, 103-140,
   Springer-Verlag Berlin Heidelberg.

Part 7
RHEOLOGY/PROCESSING

A New Reptation Model for Polymeric Liquids

Robert J.J.Jongschaap and Bernard J.Geurts
Department of Applied Physics
Twente University
P.O.Box 217
Enschede 7500 AE
The Netherlands

Abstract

A model is presented in which the motion of a polymer in a concentrated solution or melt is represented by the reptating motion of a flexible elastic rope in a tube under the influence of Brownian, elastic and frictional forces. In this model the results of Doi and Edwards [1] and of Curtiss and Bird [2] are obtained as special cases. In fast processes the elasticity of the rope may give rise to an important contribution to the stress tensor.

1.Introduction

The concept of reptation, first introduced by deGennes [1], has proved to be very useful in modeling the rheological behavior of concentrated polymer solutions and melts. The two main approaches are the theories of Doi and Edwards [2], based on a tube model, and of Curtiss and Bird [3] in which the reptation motion was treated by introducing a tensorial Stokes' law. Recently [4] a new model was introduced from which the results of both theories can be derived. In this model, which was called the reptating rope model, the polymeric chain was represented by a smooth space curve of contour length L. This "rope" was confined in a tube of the same shape and about the same cross-section. The rope was assumed to be inextensible and the tube was thought to move affinely with the macroscopic motion.

Using the reptating rope model it was shown [4] that the Brownian forces pulling at the ends of the rope give rise to stresses as predicted by the Doi and Edwards theory [2]. The frictional forces due to the relative motion of the rope with respect to the tube give rise to the extra term first derived by Curtiss and Bird [3].

In the present paper some aspects of the reptating rope model are reconsidered. In particular we are concerned with the nature of the Brownian - and the (averaging of the) frictional contributions to the stress tensor. Furthermore we consider the rope to be elastic instead of inextensible.

2.The constitutive equation for the reptating rope model

The derivation of the constitutive equation involves several steps. First we relate the stress tensor **T** to the tension σ along the rope. Next the tension σ is derived by considering Brownian and frictional force densities. Finally the general form for **T** is obtained by inserting σ in the above mentioned relation.

Let the rope have a correlation length $a \ll L$. Any conformation of the rope can be represented by $N \equiv L/a$ ($\gg 1$) statistically independent segments of length a. The polymer contribution to the stress tensor **T** can then be written as [4]

$$T = nN\int_0^L <\sigma ee>ds \qquad (1)$$

where n is the number density of polymers and **e** is the unit vector tangent to the rope at s at time t. The brackets denote an average per segment using the one segment distribution function $\Psi(e,t,s)$ This function is given by [2,3]

$$\Psi(e,t,s)=\int_{-\infty}^t \chi(t-t',s)<\delta(e-e(e',t',t))>'dt' \qquad (2)$$

in which $e(e',t',t)= F_{t'}(t)\cdot e'/|F_{t'}(t)\cdot e'|$ where $F_{t'}(t)\equiv \partial x(t)/\partial x(t')$ is the macroscopic deformation gradient. The operation $<..>'$ is an average over the unit sphere with respect to the isotropic distribution. The function $\chi(t-t',s)$ is given by [2]

$$\chi(t-t',s)= \sum_{k=0}^{\infty} \frac{4(2k+1)}{\pi\tau_d}\sin\left[\frac{(2k+1)\pi s}{L}\right]\exp\left[-\frac{(t-t')(2k+1)^2}{\tau_d}\right] \qquad (3)$$

where $\tau_d \equiv L^2/D\pi^2$ is the disengagement time, introduced by Doi and Edwards. Here D is a diffusion constant characterizing the reptation motion of the chain along its contour. A segment, confined in a tube element that was created at a time t' will be called a t'-activated segment. The function $\chi(t-t',s)$ can then be interpreted as the probability density along the rope at time t of a t'-activated segment. We will also use the probability density

$$\Phi(t-t',s)=\int_{-\infty}^{t'} \chi(t-t'',s)dt'' \qquad (4)$$

of a segment, activated at any instant before the time t'.

The function $\chi(t-t',s)$ is determined by a diffusion equation, so the average motion of a t'-activated segment can be treated as being governed by a Brownian force

$$F_\chi^B = -kT\frac{\partial \ln\chi(t-t',s)}{\partial s}$$

(5)

The corresponding Brownian force density acting on a t'-activated segment per unit length is then given by $f_\chi^B(t-t',s) = (\chi/L)F_\chi^B(t-t',s)$ and the total Brownian force density acting on any segment at the present time t

$$f^B(s) = \int_{-\infty}^{t} f_\chi^B(t-t',s)\,dt' = -\frac{kT}{L}\frac{\partial \Phi(0,s)}{\partial s}$$

(6)

The function Φ in this expression if given by (4). In this case we have $\Phi(0,s)=1$ for $0\leq s\leq L$ and $\Phi(0,s)=0$ elsewhere. So the Brownian force density $f^B(s)$ consists of two δ-peaks at the ends of the rope.

The frictional force density is given by Stokes' equation

$$f^H = \zeta(v-u)$$

(7)

where u and v are the velocities of the rope and the tube respectively at some point along the rope and ζ is a friction coefficient. Under the assumptions of affine motion of the tube segments one has

$$\frac{\partial v}{\partial s} = \mathbf{e}\cdot\mathbf{D}\cdot\mathbf{e}$$

(8)

in which $\mathbf{D}$ is the macroscopic rate of strain tensor. Let ε be the relative extension of the rope then

$$\frac{\partial u}{\partial s} = \frac{\partial \varepsilon}{\partial t}$$

(9)

Hence the frictional force density f^H is determined by the equation

$$\frac{\partial f^H}{\partial s} = \zeta(\mathbf{e}\cdot\mathbf{D}\cdot\mathbf{e} - \frac{\partial \varepsilon}{\partial t})$$

(10)

The tension σ is determined by the total density of forces $f=f^B+f^H$ acting on the rope, through the equation of motion

$$\frac{\partial \sigma}{\partial s}+f=0 \tag{11}$$

with boundary conditions $\sigma(t,0)= \sigma(t,L)=0$.

On using (6), (10) and (11) we obtain for the tension

$$\sigma(s,t)=\frac{kT}{L}+\zeta \int_0^s \int_0^{s''}\left[e(s',t)\cdot D(t)\cdot e(s',t)-\frac{\partial \varepsilon(s',t)}{\partial t}\right]ds'ds'' \tag{12}$$

The term $\partial \varepsilon/\partial t$ in this expression can be obtained from the equation of motion as a functional of the history of the stretching of tube segments. The averaging of this result, when substituted in (1), requires special techniques and will be discussed elsewhere. In this paper we restrict ourselves to the case that the term $\partial \varepsilon/\partial t$ is neglegible. This will be the case in flows in which appreciable changes in the strain rate only occur on a time scale which is large in comparison with the characteristic times of the so called first relaxation process, described by Doi and Edwards [2].

In determining T from (1) the second term in (12) gives rise to averages of the following type

$$<\int_0^s \int_0^{s''}e(s',t)e(s',t)ds'ds''e(s,t)e(s,t)>$$

$$=\int d^2e \int_{-\infty}^t \chi(t-t')\int \delta(e-e(e',t',t))\frac{d^2e'}{4\pi}$$

$$\left[\int_0^s \int_0^{s''}e(s',t)e(s',t)ds'ds''\right]e(s,t)e(s,t)dt' \tag{13}$$

If we assume that the variables $e(s',t)$ and $e(s,t)$ can be considered as a stochastic variable e taken at particular positions of the rope at time t then the averaging with respect to e can be performed and after that also the integrations with respect to s' and s''. (The extended analysis without this assumption will be published elsewhere). In this way, from (1) and (12) we obtain the following constitutive equation

$$T=n Nk T\int_{-\infty}^t \left[\mu(t-t')<ee>'+\varphi(t-t')D(t):<eeee>\right]dt' \tag{14}$$

where **e** denotes the function $e(e',t',t)$, $<..>'$ is the average defined after eq.2,

$$\mu(t-t')=\frac{1}{L}\int_0^L \chi(t-t',s)ds \tag{15}$$

and

$$\varphi(t-t')=\frac{\zeta}{2kT}\int_0^L s(L-s)\chi(t-t',s)ds \tag{16}$$

3.Discussion

The result (14) is identical with a constitutive equation obtained by Curtiss and Bird (3) if their link tension coefficient is taken to be $\frac{1}{2}$. Their derivation was based however on a phase-space kinetic theory of a freely jointed bead-rod chain with anisotropic friction, instead of a tube model. On the other hand the first term in (14) is (apart from a factor 3) the constitutive equation first derived by Doi and Edwards [2]. The essential difference between their treatment and ours is that in our case by considering Brownian forces, stress calculations based upon Gaussian chain statistics have been avoided and a simple "rope model" could be employed. Finally, in contrast to Doi and Edwards, in our treatment the friction between the rope and the tube was taken into account.

As stated before, the effect of the elasticity of the rope may become signigicant in rapidly changing flows. This is a subject for further study and will be published elsewhere.

References

[1] de Gennes, P.G., J.Chem.Phys.,55,572 (1971)
[2] Doi, M. and S.F.Edwards, J.Chem.Soc. Faraday Trans., 74 (1978), 1789, 1802, 1818; 75 (1979) 38
[3] Curtiss, C.F. and R.F.Bird, J.Chem.Phys., 74 (1981), 2016,2026
[4] Jongschaap, R.J.J., 2nd Conference of European Rheologists, Prague, (1986), Submitted for publication in Rheologica Acta

MELT STRENGTH BEHAVIOUR OF POLYETHYLENES AND POLYETHYLENE BLENDS AND ITS RELATION TO BUBBLE STABILITY IN FILM BLOWING

A. Ghijsels,   J.J.S.M. Ente
Chemical Research Centre, Shell Louvain-la-Neuve
Shell Research S.A.
c/o Koninklijke/Shell-Laboratorium, Amsterdam
(Shell Research B.V.)
The Netherlands
and
J. Raadsen
Koninklijke/Shell-Laboratorium, Amsterdam
(Shell Research B.V.)
The Netherlands

ABSTRACT

Melt strength data are presented on the three major classes of commercial polyethylenes (LDPE, LLDPE and HDPE) and some of their binary blend systems. Melt strength was assessed from uniaxial tensile experiments with a Rheotens apparatus. The melt strengths of both LLDPEs and HDPEs are much lower than those of LDPEs at equal melt index or zero-shear viscosity level. For LLDPE/LDPE blend pairs synergistic effects are observed for the melt strength and other rheological properties. Molecular factors controlling the melt strength of pure polyethylenes are discussed as well as possible mechanisms for the observed synergism in LLDPE/LDPE blends. A high melt strength is identified with a good bubble stability in the film-blowing process.

INTRODUCTION

Tubular film production is with 70 % the most important application area for conventional low density polyethylenes (LDPE) and is likewise an important outlet for the new family of linear low density polyethylenes (LLDPE). The use of polyethylene blends, in particular LLDPE/LDPE, in blown-film processing is also rapidly advancing mainly because of two reasons, viz. it allows the production of novel materials with improved processing and/or end-use characteristics, and recycling of scrap material. An important processability factor in the film-blowing process of polyethylenes is the "melt strength" or the resistance to draw-down of the tubular melt. In the case of a low melt strength only a small axial take-up force is required to stretch the melt, whereby the low-tension bubble becomes sensitive to all sorts of air flows (air from cooling ring, draughts) and also to gravity forces. This then leads to undesired bubble instabilities (e.g. runaway, tumbling) and also to sagging phenomena.

The present study focusses on the melt strength behaviour of the various classes of polyethylenes and some of their binary blend systems and the structural factors controlling this behaviour.

EXPERIMENTAL

The materials used in this study were commercially available grades of LLDPE, high-density polyethylene (HDPE) and LDPE. The LDPE samples were all polymerized by tubular reactor technology. The LLDPE samples were produced with different α-olefins as comonomer.

In total four different blend systems were studied, designated as LLDPE/LDPE-1, LLDPE/LDPE-2, LLDPE/HDPE and LDPE-1/LDPE-2. Table 1 lists the blend components and some of their characteristic properties.

TABLE 1

Characteristic properties of blend components

Sample	Density (g/cm^3)	Melt index[a] $(g/10 \text{ min})$	Zero-shear viscosity $(Pa.s)$	Melt strength (cN)
LLDPE[b]	0.922	1.45	8.6×10^3	10.0
HDPE	0.949	0.10	1.7×10^5	34.5
LDPE-1	0.921	0.24	7.7×10^4	35.5
LDPE-2	0.922	2.30	6.6×10^3	17.5

a   at 190 °C with 2.16 kg
b   comonomer: butene-1

The blends were prepared by melt mixing the polymer parents in a single screw laboratory extruder (manufacturer: Göttfert) at 190 °C. The pure polymers were subjected to the same treatment. Using the Differential Scanning Calorimetry (DSC) technique, the blends were found to be homogeneous at least on a milligram scale.

The melt strength was measured with an experimental set-up, consisting of a laboratory extruder provided with a capillary die and a Rheotens apparatus (Göttfert) as take-up device. With this set-up the polyethylenes are stretched in uniaxial extension under nearly isothermal conditions. The steady value of the tensile force attained during the test is defined as the melt strength. The extrusion temperature was 190 °C.

The Rheometrics System-4 was employed for measuring the zero-shear viscosities at 190 °C. An Instron capillary rheometer was used for determining the rheological characteristics (viscosity, entrance pressure losses and extrudate swell) at high shear rates and a temperature of 190 °C.

RESULTS

Fig. 1 summarizes the melt strength data of the pure polyethylenes as a function of the melt index.

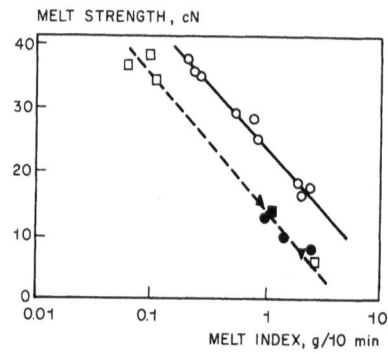

Figure 1 Melt strength of pure polyethylenes at 190 °C
(o) LDPE, (□) HDPE,
(●,■,▲,▼) LLDPE, different
comonomers

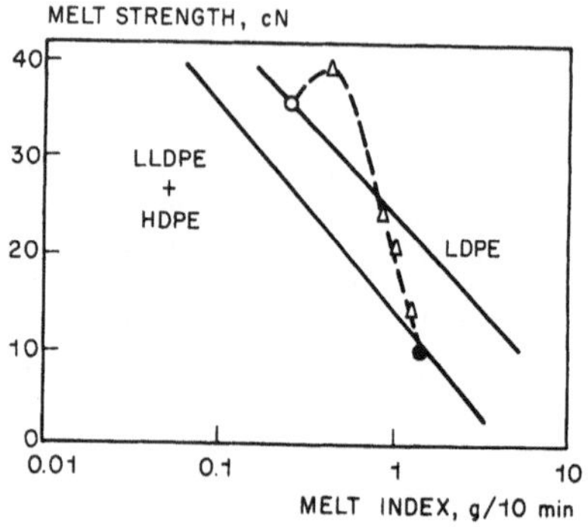

Figure 2 Melt strength of LLDPE/
LDPE-1 blends at 190 °C
(Δ) blends, (●) LLDPE,
(o) LDPE-1

It is evident that for a single class of polyethylenes (e.g. LDPE) the melt strength shows a strong linear decrease with increasing melt index. The melt strength of LLDPEs is considerably lower than that of conventional LDPEs of the same melt index level and is comparable to that of HDPEs. The major cause of this difference is the presence of long chain branching in LDPEs, leading to strain hardening [1] in elongational flow experiments. The type of short chain branching, whether in terms of chain length (C_2, C_4 or C_6) or in terms of structure (iso-C_4 versus n-C_4), has no significant effect on the melt strength of LLDPEs.

Similar conclusions can be drawn from the relation between the melt strength and the zero-shear viscosity, η_o. Straight lines were obtained on logarithmic scales. Of particular interest is the slope of these lines, which is equal to 1/3.4. Since for linear polyethylenes and probably also for branched polyethylenes (with comparable branching level), η_o is proportional to $M_w^{3.4}$, it follows that the melt strength of polyethylenes is directly proportional to the weight average molecular weight, M_w.

Fig. 2 shows the melt strength data of the LLDPE/LDPE-1 blends as a function of the melt index. The melt strength-melt index relationships established for the pure polyethylenes are also given in Fig. 2. Fig. 3 shows the melt strength data of the LLDPE/LDPE-1 and LLDPE/LDPE-2 blends as a function of blend composition. In Fig. 2 it is seen that the addition of small amounts - 10, 20 and 30 %w - of low-melt index LDPE-1 to LLDPE results in blends with a high melt strength. The gap in melt strength between LLDPE and LDPE can thus be easily bridged. At higher LDPE contents (70 %w) we even enter an area which is not accessible with pure polyethylenes made by tubular reactor technology.

MELT STRENGTH, cN

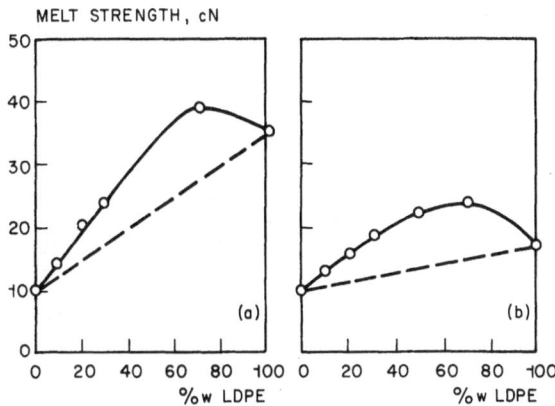

Figure 3 Melt strength versus blend composition for LLDPE/LDPE blends at 190 °C
(a) LLDPE/LDPE-1, (b) LLDPE/LDPE-2

As is evident from Fig. 3a, the melt strength shows strong positive deviations from a linear mixing rule and exhibits a maximum around a blend ratio of 30/70 (w/w) LLDPE/LDPE. Similar results were obtained for the blend pair LLDPE/LDPE-2. As shown in Fig. 3b, for this blend pair a very pronounced maximum is observed at about the same composition. Quite similar plots (to those in Fig. 3) were obtained for the entrance pressure drop at constant shear rate versus blend composition, thus demonstrating that the observed synergism is an intrinsic material characteristic.

In contrast to LLDPE/LDPE blends no melt strength synergism was observed with LLDPE/HDPE and LDPE/LDPE blends. For these blends the melt strength response was found to be nearly additive.

The LLDPE/LDPE blends not only exhibit synergistic effects in the melt strength (and the related entrance pressure drop) but also in some other rheological properties. Fig. 4 shows as an example the zero-shear viscosity of the blends plotted as a function of composition. For the blend pair LLDPE/LDPE-2 a clear maximum is observed at about 60 %w LDPE.

Similar results for the melt strength of polyethylenes and their blends were recently reported by Acierno and coworkers [2,3].

DISCUSSION

As to the melt strength of pure polyethylenes, it follows from this work that the molecular weight and the long-chain branching are the two

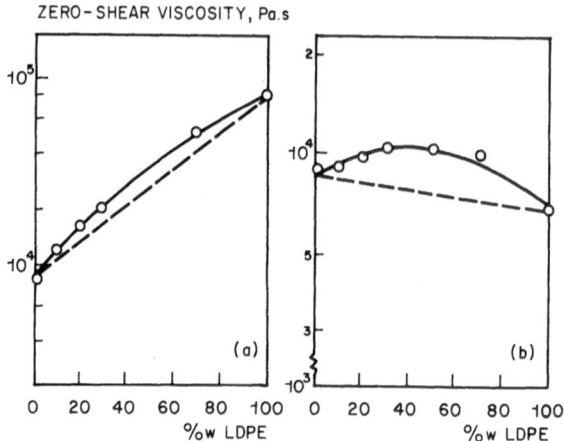

Figure 4 Effect of blend composition on zero-shear viscosity of LLDPE/LDPE
blends at 190 °C
(a) LLDPE/LDPE-1, (b) LLDPE/LDPE-2

most important molecular factors governing the melt strength. The effect
of molecular weight distribution is considered to be small. Widening of
the distribution e.g. by blending low- with high-molecular weight LDPE does
not significantly affect the melt strength.

In explaining the synergistic effects observed in LLDPE/LDPE blends
one can draw an analogy to the work of Bersted et al. [4] on the effects
of long chain branching on the low- shear melt viscosity of polyethylenes.
Depending on the branching level the viscosity can be either substantially
elevated or depressed relative to the viscosity of linear polymers of
identical molecular weight. At low branching levels the viscosity enhancement, caused by an increased amount of intermolecular entanglements,
appears to predominate, while at high branching levels molecular size reduction, which reduces melt viscosity, becomes more important. By considering the LLDPE/LDPE blends as whole polymers of varying branching level
the Bersted model is applicable to these systems as well. In the case of
the LLDPE/LDPE-2 blend system, where the components are of comparable
molecular weight, the addition of highly branched LDPE to LLDPE indeed
first leads to viscosity enhancement at low LDPE (branching) concentrations and after passing a maximum ultimately leads to the lower viscosity
for pure LDPE. In the LLDPE/LDPE-1 blend system the above effects are
overruled by the fact that the molecular weight of LDPE-1 is much higher
than that of LLDPE, thus leading to much higher viscosity levels than for
LLDPE. As expected, the melt strength, being an extensional viscosity
related property, qualitatively shows the same response as the zero-shear
viscosity, except that for LLDPE/LDPE-1 blends a maximum (synergism) in
the melt strength is still just observed.

It will be clear that on the basis of the Bersted model no synergistic effects in zero-shear viscosity or melt strength are expected for
blends of linear polyethylenes such as LLDPE/HDPE. For LDPE/LDPE blends,
composed of two highly branched species, no synergistic effects are expected either. For melt strength synergism a mixture of linear and branched
species of similar molecular weights is required.

The melt strength results as measured in simple laboratory experiments are in line with the observed bubble stability in actual film blowing [1,5]. With LDPE resins generally a good stability is observed, with LLDPEs and low molecular weight HDPEs usually show a poor stability. In practice, when the film blowing of LLDPE is limited by bubble instability, the addition of small amounts of LDPE (20-30 %w) to LLDPE also rapidly increases the bubble stability.

REFERENCES

1. Furumiya, A., Akana, Y., Ushida, Y., Masuda, T. and Nakajima, A., Pure Appl. Chem., 1985, 57, 823.

2. La Mantia, F.P. and Acierno, D., Polym. Eng. Sci., 1985, 25, 279.

3. Acierno, D., Curto, D., La Mantia, F.P. and Valenza, A., Polym. Eng. Sci., 1986, 26, 28.

4. Bersted, B.H., Slee, J.D. and Richter, C.A., J. Appl. Polym. Sci., 1981, 26, 1001.

5. Speed, C.S., Plast. Eng., 1982, 38, 39.

RHEOLOGICAL PROPERTIES OF
THERMOPLASTIC ELASTOMERS BY HYDROGEN BONDING

Reimund Stadler, Liane de Lucca Freitas, Josef Burgert

- Hermann Staudinger Haus -
Institut für Makromolekulare Chemie
Stefan Meier Strasse 31 - D-7800 Freiburg

ABSTRACT:
By polymer analogous reaction polar urazole groups are introduced into poly(butadiene). These functional groups form hydrogen bond complexes and thus give rise to the formation of a
thermoreversible network. The rheological as well as the
stress-strain behaviour of these networks is charakterized.
Although no equilibrium network modulus is observed, the time
independent strain function is similar to rubber networks.

INTRODUCTION

Thermoplastic elastomers are either based on multiphase
block copolymer systems or on one phase systems in which the
thermoreversible junctions are introduced by means of secondary
valence interactions. In ionomers the junctions are formed by
clusters of ion pairs. The degree of association may vary with
the type of counter ion, the matrix polarity and the chain topology (ionic sites linked at the chain ends or statistically
along the chain).

Simpler association behaviour can be obtained if hydrogen
bonds are used as network junctions. In addition, the hydrogen
bond complexes can be investigated independently by IR-
spectroscopy. Thus the mechanical properties (linear viscoelastic and stress-strain behaviour) should be correlated to
the molecular association behaviour. Special attention will be
focused on the question of whether such reversible networks
show elastic equilibrium behaviour similar to that of covalently crosslinked systems.

SYSTEM

In order to investigate the relations between the formation and the properties of thermoreversible networks the following model system has been prepared: by a polymer analogous reaction, 4'-phenyl-1,2,4-triazolidine-3,5-dione (urazole) groups were attached to a polymer backbone [1,2]. As the polymer, any type of diene polymer can be used. The rheological data were obtained on polybutadienes of narrow molecular weight distribution. For the stress-strain experiments a technical grade cis-1,4-poly(butadiene) (CB-10) supplied by Chemische Werke Hüls has been used. By means of IR-spectroscopy the fraction of complexed groups can be determined[3]. A typical example for the temperature dependent >C=O stretching vibrations is shown in Figure 1. As is indicated by the occurence of an isosbectic point, only two structures - the free carbonyl group and the complex - are present in equilibrium.

$$U + U \longrightarrow U_2$$

The concentration of complexes will thus be given by the degree of modification (total concentration) and the equilibrium constant K. The properties of the modified polybutadienes also depend on the rate of separation of functional groups, which may vary under shear. The rate of decomplexation and the number of groups per chain determine the longest relaxation time in such a transient network.

Fig.1: Temperature dependent IR-spectra (>C=O stretching vibration) poly(butadiene) 2% modified ;

free group: 1720 cm^{-1}
complexed group: 1702 cm^{-1}

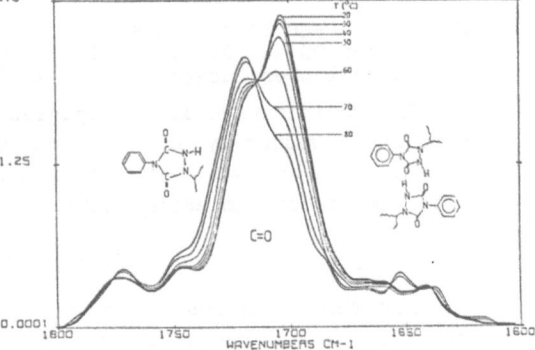

RESULTS and DISCUSSION

BEHAVIOUR IN THE LINEAR VISCOELASTIC REGION

Detailed studies were performed on the linear viscoelastic behaviour of modified polybutadienes in the rubbery and terminal region[4,5]. The dependence of G', G", η' and J' as a function of the primary molecular weight and the concentration of urazole groups was studied. The number of urazole groups per chain has been found to be the quantity that determines the changes in the rheological behaviour compared to the unmodified polybutadienes. As a typical example, the storage modulus master curves $G'(\omega)$ are shown in Figure 2 for a polybutadiene of molecular weight 48,500 and samples modified with 1 and 2% phenylurazole. By analysis of the relaxation time spectra[5] it has been shown, that the hydrogen bond junctions do not result in formation of an equilibrium network modulus as is observed for covalently crosslinked elastomeric networks.

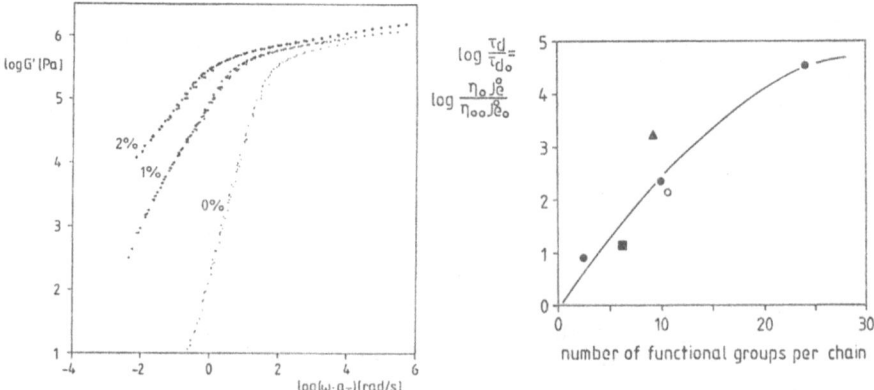

Fig.2: Storage modulus master-curves for PB 48,500 modified with 0,1,2% phenylurazol

Fig.3: Relative increase in the terminal relaxation time τ_d for various primary chain length

The modification is accompanied by a large increase in the zero shear viscosity η_o and recoverable shear compliance J_e^o. From the product of these two quantities the terminal relaxation time τ_d can be estimated[6].

$$\tau_d = \eta_o * J_e^o$$

For those samples in which the steady state values were reached, the increase in the terminal relaxation time with respect to the unmodified polymer is plotted as a function of the number of urazole groups per chain. For different primary chain lengths, the same dependence is observed. This increase in the terminal relaxation times with increasing degree of modification can be attributed to a hindered reptational motion in such polymer melts with attached interacting functional groups[7).

In Figure 4 the room temperature stress-strain curves are shown for samples with varying amounts of urazole groups. The stress- strain curves were obtained at a crosshead speed of 20 mm/min. With increasing degree of modification the modulus increases, while the permanent set decreases. The 2% and 5% modified samples behave more like covalently crosslinked rubbers, while flow behaviour dominates in the samples with lower modification. For larger strains considerable flow is also observed for the higher modified samples. In Figure 5 the stress strain behaviour of the 1% modified poly(butadiene) is shown for various crosshead speeds. According to the transient character of the thermoreversible junctions a very strong strain rate dependence of the stress strain behaviour is observed. The fraction of stored energy obtained as the ratio of the area below the hysteresis curve and the area of the stretching curve, do not change with stretching rate. In addition the permanent set does not vary strongly with crosshead speed for this degree of modification.

Fig.4: Stress-strain plots for polybutadienes with various amounts of phenylurazol (PTD)

Fig.5: Stress strain plots for various strain rates (1% modified CB-10)

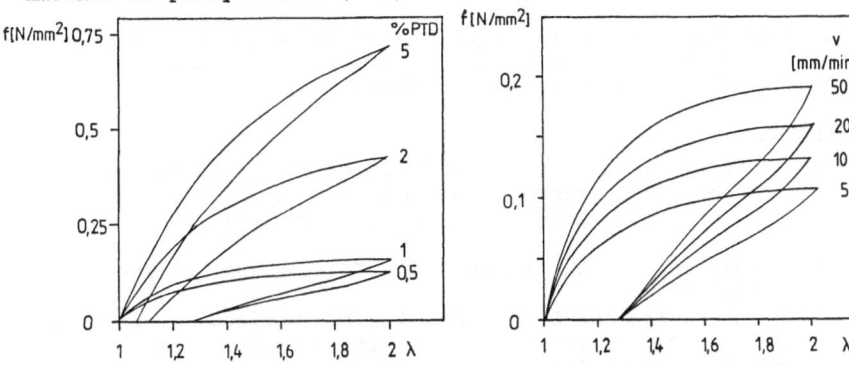

To describe the deformation behaviour of such thermoplastic elastomers, viscoelastic relaxation processes must be taken into account. The simplest approach is to assume independence of the time and deformation dependent contributions to the stress[8,9):

$$\sigma(\lambda,t) = G(t) * S(\lambda)$$

where $\sigma(\lambda,t)$ is the stress at a given strain λ and time t, G(t) is the time dependent shear modulus and $S(\lambda)$ is a deformation function. If such a factorization is applicable, a plot of $\log(\sigma(\lambda,t)_{\lambda=const}$ versus $\log(t)$ should give parallel curves for different values of λ. As is shown in Figure 6, this is observed experimentally. Consequently the strain dependent function $\sigma(\lambda,t)$ at constant time t can be obtained from Figure 6. This is shown in Figure 7 for two samples modified with 1% and 2% phenylurazole at t=25s. Comparison of these curves with theoretical functions to describe the strain dependence are presently in progress.

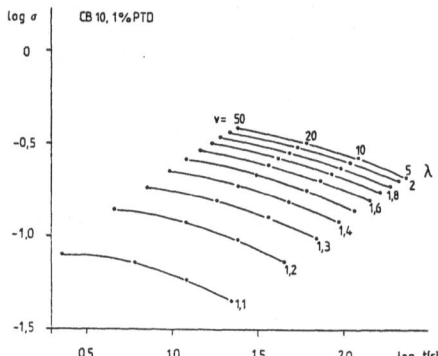 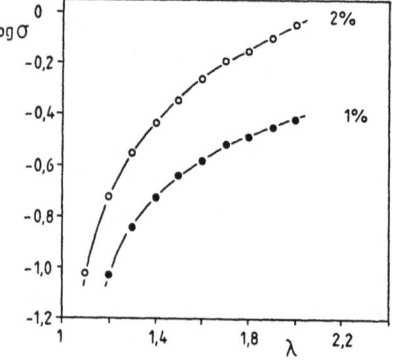

Fig.6: $\log(\sigma)/\log(t)$ for various strains λ obtained from different strain rates

Fig.7: $\log(\sigma)/\lambda$ at 25 s for 1% and 2% modification

ACKNOWLEDGEMENT: Financial support through fellowships from DAAD (L.F.) and from Landesgraduiertenförderung Baden Württemberg is gratefully acknowledged.

REFERENCES
1) Leong K.W., Butler G.B., J.Macromol.Sci. **A-14** (1980) 287
2) Stadler R., Burgert J., Makromol.Chem.**187** (1986) 1681
3) Stadler R., de Lucca Freitas L., Polym.Bull. **15** (1986) 173
4) Stadler R., de Lucca Freits L., Coll.&Polym.Sci.**264** (1986) 778
5) de Lucca Freitas L., Stadler R., Macromolecules, in press
6) Ferry J.D., "Viscoelastic properties of polymers" 3rd edition, Wiley Interscience, New York 1980
7) Stadler R., in preparation
8) Scholtens B.J.R., Leblans P.J.R. Booij H.C., J.Rheol.**30** (1986) 301
9) Scholtens B.J.R., Leblans P.J.R., J.Rheol.**30** (1986) 313

A CLASSIFICATION OF CRYSTALLIZATION PHENOMENA
FOR CONDITIONS OF POLYMER PROCESSING

H. Janeschitz-Kriegl, Inst.Chemistry, Linz Univ.(ICL), A-4040 Linz,
P. Zipper, E. Wrentschur, Inst.Phys.Chem., Graz Univ. (IPCG), J. Koppelmann,
E. Fleischmann, G. Leitner, Inst.Chem.& Phys.Technol., Leoben Mining Univ.
(ICPTL), H. Muschik, M. Radax, Austrian Lab. for Plastics Technol. and Inst.
Appl.Phys., Vienna Univ.of Techn. (IAPV), W.F. Geymayer, Res.Inst.Electron
Microscopy, Graz Univ. of Techn. (RIEMG)*).

ABSTRACT

Special attention is paid to the nature of a highly oriented surface
layer, as formed in injection moulded parts of polypropylene. A classification of this phenomenon is tried within the frame of (previously published
and shortly reviewed) perceptions as gained from quenches of quiescent isotropic melts.

INTRODUCTION

Inhomogeneous crystallization is a consequence of the interaction between heat transfer and crystallization kinetics and sometimes of shear. In
this paper the highly oriented surface layer, as caused by shearing , is given special attention. It manifests itself in several ways, optically, mechanically and thermally. These various types of characterization can help us a
little further in our understanding.

CHARACTERIZATION OF THE ORIENTED SURFACE LAYER

A disk was moulded from an industrial polypropylene Daplen KS 10

*) Members of the working party S 33, "Kunststoff-Formteile", sponsored by
the Austrian "Fonds zur Förderung der Wissenschaftlichen Forschung".

(Petrochemie Danubia, M_w = 290 000, M_w/M_n = 5,7) at Ludwig Engel KG (Injection Moulding Machines, Schwertberg, Austria) in cooperation with Mr. W. Friesenbichler and Prof. W. Knappe (Mining University, Leoben, Austria, Institute of Plastics Processing) under usual processing conditions. In a radial direction two small test bars were cut from the disk at a certain distance from the centre, their thickness corresponding with that of the disk. One of these bars was stretched on a mechanical testing machine until crazes became visible (ICPTL).

The mechanically maltreated and the virginal bar were investigated with the aid of X-ray small-angle scattering (IPCG). For the purpose use is made of a Kratky camera (1) with a line focus. In Fig.1 the scattering intensity profiles are shown for both bars at a fixed scattering angle of 1 mrad (Cu-K_α-radiation). The mechanically untreated bar does not produce big differences in scattering intensity. The mechanically maltreated bar, however, clearly portrays the oriented layers near the two disk surfaces.

In pursuing this phenomenon very thin test bars of a thickness of only 0,05 mm were cut from the disk in the radial direction and at the previous location. In Fig. 2 the stretch at fracture of these very thin test bars is plotted together with the birefringence profile normal to the disk surfaces (ICPTL). It is seen that the oriented layer as characterized by a relatively large birefringence, shows a remarkably lower stretch at fracture than the core of the disk.

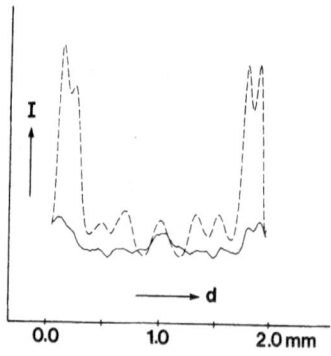

Fig. 1: Small angle X-ray scattering profile at an angle of
1 mrad over the cross-section of a radial cut:
full line ... virginal sample, dashed line ... sample
after loading according Zipper et al.

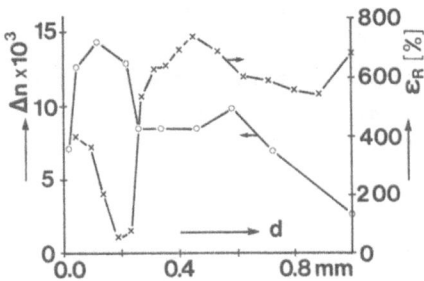

Fig. 2: Birefringence profile and stretch at fracture over the cross-section of a radial cut according to Koppelmann et al.

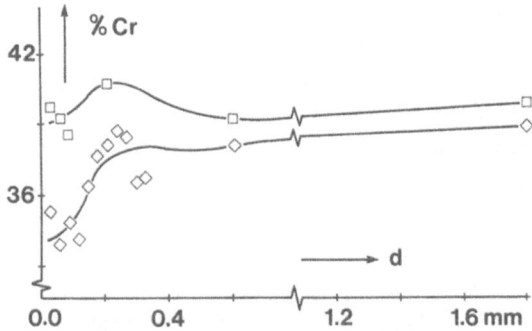

Fig. 3: Degree of crystallinity as a function of the distance from the sample surface; ◊ ... after 3 h, ◘ ... after 7 days of ageing, according to Muschik et al.

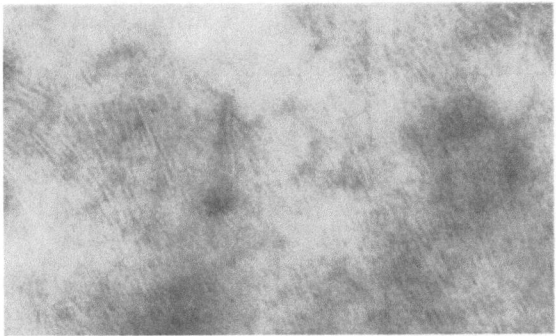

Fig. 4: Transmission electron micrograph of section from surface layer, according to Geymayer et al., enlargement of 68.000.

The presence of a surface layer of different physical properties is also documented by DSC-measurements (IAPV). With these measurements the degree of crystallinity as well as the ageing was investigated. For the purpose slices of a thickness of 30 μm were successively cut from the sample with a microtom. From these slices tablets were punched and heated in a Perkin-Elmer DSC II. In Fig. 3 the degrees of crystallinity thus obtained are shown as functions of the distance from the surface for two samples, one investigated after 3 hours of ageing, the other after one week of ageing. It appears that the oriented layer near the surface shows a lower crystallinity than the core after 3 hours of ageing. Later the boundary between the surface layer and the core seems to have the highest crystallinity.

For the characterization of the oriented layer also a fourth method was tried. A sample was prepared for transmission electron microscopy (RIEMG). This sample was treated first with chlorosulphonic acid according to Kanig (2), in order to chemically modify the amorphous phase*) and then cut by cryoultramicrotomy and stained with uranyl acetate. In Fig. 4 the fine ribs (bright areas = crystalline lamellae) have a thickness of the order of 10 nm.

A REVIEW OF CLASSIFICATION FOR QUIESCENT MELTS

As has been shown in earlier investigations (3) (4), there are two characteristic ranges of the (mould) wall temperature T_w:

$$a) \quad T_p \leqq T_w \leqq T_m \quad ; \quad T_g \leqq T_w \leqq T_p \qquad \qquad ...(1)$$

where T_m is the thermodynamic melting point, T_g is the glass transition temperature and $T_p^{\bullet}$ is the temperature halfway between T_g and T_m, where the linear growth speed of spherulites shows a maximum.

Range a: As there are only a few nuclei in the melt, the wall acts as a predominant nucleation area. A supercooled crystallization front departs from this area. Pertinent experimental work is presented in refs. (3) (4) (5). For polypropylene the birefringence of the layer behind the crystallization front is very low in contrast to the oriented surface layer described in the previous section. Theory is developed in refs. (6)

*) Foot Note: The assistance of Dr. H.G. Braun of BASF is gladly acknowledged.

and (7). In the experiments on polypropylene at temperatures below 100°C
the propagation of the supercooled crystallization front was seriously ham-
pered by nucleation in the (supercooled) bulk of the fluid.

Range b: In the supercooled melt diffuse nucleation occurs. A remarkable theory for this diffuse crystallization was developed by Malkin
et.al. (8) in connection with experiments on Nylon 6. Independently, Berger
and Schneider (9) of our working party developed a theory in which they proposed the use of a "zone" model (in contrast to the "front" model of refs.
(6) and (7)). If a half infinite space is considered, one observes a relatively broad zone, in which crystallization takes place, instead of a crystallization front. If the thickness of the moulded part is smaller than that
of the zone, one has diffuse crystallization as just defined. The classical
theory by Neumann (10), and Stefan (11), is a limiting case of this zone model. This has recently been shown also by Astarita and Kenny (12) in an ele-
gant study (Stefan/Deborah → ∞).

CLASSIFICATION OF THE SURFACE LAYER

With polypropylene the oriented surface layer is formed in the wall
temperature range b, where no crystallization front is formed in a quiescent melt. For an experiment in which the melt is sheared (in a slit die)
previous to the quench (on a chill role) (3) (4), the thickness of the oriented surface layer is independent of the conditions of heat transfer
(chill role temperature, final bath temperature). This is in contrast to
the pronounced dependence of the growth speed of the quiescent layer on
temperature. The conclusion is that flow is precursory of the oriented surface layer, which is merely fixated by the cooling. For the investigated
Daplen KS 10 a fluid element which solidifies in the manner characteristic
for the surface layer, first has to undergo a tremendous total shear of the
order of 500. (Eder et al. (13)). It goes without saying that the mentioned
condition is highly dependent on the molar mass distribution of the polymer.
By the nature of the experiment, no kinetics of the crystallization process
are available for the oriented surface layer. It seems, however, that the
rate of crystallization is much faster than in an unoriented melt.

DISCUSSION

From the enormous total shears which are necessary for the formation
of the surface layer one may deduce that the melt which undergoes this
type of solidification is free of entanglements. Such an assumption is
supported by the low stretch at fracture of the surface layer (Fig. 2).
However, the rather low degree of crystallinity, as observed in this
layer after short ageing times, is not directly in favour of this assumption (Fig. 3). It may be that a very high rate of crystallization is responsible for this initially low degree of crystallinity. The picture obtained with transmission electron microscopy shows details of the order
of the magnitude of dimensions of single macromolecular coils. As similar details are found in samples of polyethylene obtained from dilute
gels by evaporation of the solvent (14), our electron micrographs are
at least not in contradiction to the assumption of no entanglements.
After a more detailed study of the X-ray small angle scattering we will
certainly be able to know more about the size and the shape of the structures (or the holes between the structures).

REFERENCES

1) O. Kratky, Z.Elektrochem. 62 , 66(1958), O. Kratky and Z. Skala,
   ibid. 62 , 73(1958).
2) G. Kanig, Prog.Coll.Polymer Sci., 57 , 176(1975).
3) H. Janeschitz-Kriegl, G. Eder, G. Krobath, S. Liedauer, J. non-Newtonian Fluid Mech. 1987, in press.
4) H. Janeschitz-Kriegl, R. Wimberger-Friedl, G. Krobath, S. Liedauer,
   Kautschuk & Gummi, Kunststoffe 1987, in press.
5) G. Krobath, S. Liedauer, H. Janeschitz-Kriegl, Polym.Bull. 14 ,
   1(1985).
6) H. Janeschitz-Kriegl and G. Eder, Plastics & Rubber Proc. & Appl. 4 ,
   145(1984).
7) G. Eder and H. Janeschitz-Kriegl, Polym.Bull. 11 , 93(1984).
8) A.Yu.Malkin, V.P. Beghishev, I.A. Keapin, S.A. Bolgov, Z.S.Andrianova,
   Polym.Engng.& Sci. 24 , 1386, 1396(1984).
9) J. Berger and W. Schneider, Plastics & Rubber Proc.& Appl. 6 ,127(1986).
10) See H.S. Carslaw and J.C.Jaeger, "Conduction of Heat in Solids",
    2nd ed., Oxford 1959, p.283.
11) J. Stefan, Ann.Phys. und Chem.(Wiedemann), N.F. 42 , 269(1891).
12) G. Astarita, J.M. Kenny, Chem.Eng.Commun. 1987, in press.
13) G. Eder et al., to be published.
14) C.W.M. Bastiaansen, P.Froehlich, A.J. Pijpers and P.J. Lemstra,
    "Integration of Fundamental Polymer Science and Technology, L.A.
    Kleintjes, P.J. Lemstra eds., Elsevier 1986, p. 508.

PROCESSING AND PROPERTIES OF UHMW-PE

W. Payer
Ruhrchemie Aktiengesellschaft
Kunststoff-Forschung
D-4200 Oberhausen 11

ABSTRACT

Ultrahigh molecular-weight polyethylene with an average mol mass of $4\text{-}6\times10^6$ g x mol^{-1} is generally processed by compression moulding or ram extrusion to sheets and profiles from which finished parts are machined. The high zero shear viscosities in the range of $2\text{-}10\times10^8$ Pa x sec at 190 °C do not permit applications of the standard thermoplastic processing techniques such as extrusion and blow moulding. The injection-moulding process, which is particularly economic for the manufacture of complicated parts in large series, was also hitherto only applicable to a limited extent.

The different processing techniques and the attainable properties of the finished parts are discussed. The implementation of the injection-moulding process with UHMW-PE is presented.

INTRODUCTION

Linear polyethylenes with an average mol mass of more than 2×10^6 g x mol^{-1} (M_w) are generally called ultrahigh molecular weight polyethylene(UHMW-PE). In contrast to most polyethylene types these products are marketed in the form of a fine powder with an average grain size of about 100 µm. Some typical data of such commercial products are compiled in table 1. They show UHMW-PE to be a partly crystalline material with a degree of crystallinity of more than 80 % and with a relatively narrow mol mass distribution.

TABLE 1

Characterisation of UHMW-PE

		Hostalen GUR 412	Hostalen GUR 415
Molecular Weight, M_W	(g · mol^{-1})	$4 \cdot 10^6$	$6 \cdot 10^6$
Molecular Weight Distribution M_W/M_n		5 – 6	5 – 6
DSC T peak	(°C)	147	147
ΔH	(J · g^{-1})	246	235
Cristallinity	(%)	84	80
Branching	(CH_3/1000 C)	0,9	1,5
Medium Average Particle Size	(µm)	100	100
Bulk density	(g · cm^{-3})	0,4 – 0,5	0,4 – 0,5
Density	(g · cm^{-3})	0,956	0,954

The physical data determined on finished UHMW-PE products are shown in table 2.

TABLE 2

Physical Properties of
UHMW-PE Hostalen GUR

Density, moulded	$g \cdot cm^{-3}$	0,94
Yield stress at 23 ºC	$N \cdot mm^{-2}$	22
at 120 ºC	$N \cdot mm^{-2}$	3,6
Ultimate tensile strength at 23 ºC	$N \cdot mm^{-2}$	44
at 120 ºC	$N \cdot mm^{-2}$	25
Notched impact strength (with V-notch)	$mJ \cdot mm^{-2}$	>120

Two things are particularly noticeable from this table. Firstly, the high strength even at a temperature of 120 °C and secondly the notched bar impact strength which is far higher than that of other thermoplasts.

Fig. 1 shows the dependency of the notched bar impact strength on the temperature. UHMW-PE is an extremely impact-resistant material over a very wide temperature range which permits its use both at -200 °C and 80-100 °C.

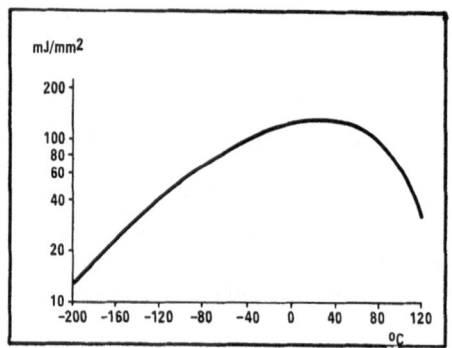

Figure 1. Notched Impact Strength of Hostalen GUR
as Function of Temperature

Another characteristics which distinguishes UHMW-PE from the rest of the thermoplasts is the abrasion resistance, which, for example, makes it suitable as an implantation material for hip endoprostheses.

Material	Abrasion Rate
UHMW-PE	
a) Molecular Weight 4 x 10^6	100
b) Molecular Weight 6 x 10^6	80
Polyacetal	700
PVC	920
Nylon 6.6	160
PTFE	530
PP (H 2065)	440
Beech Wood	2700
Steel (St 37)	160

Figure 2. Comparison of wear
behaviour
(sand-slurry method)

The results obtained with the sand-slurry method (1) show clear superiority of UHMW-PE even over carbon steel. A dependency on the average mol mass can be observed even within the UHMW-PE group itself. As the mol mass increases so does the abrasion strength.

Another characteristic of UHMW-PE is also the enormous viscosity of the melt which is around 2 to 8×10^8 Pa x sec at 190 °C (2).

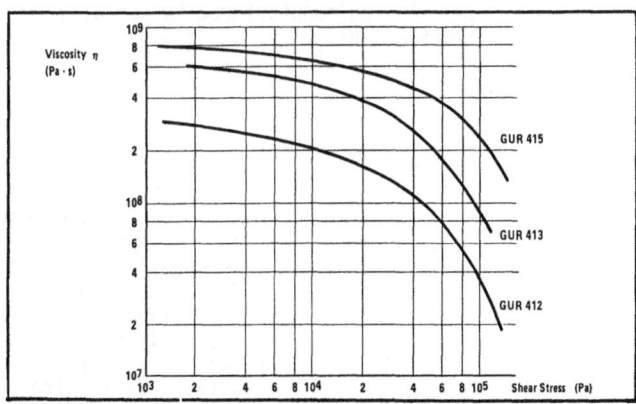

Figure 3. Viscosity of a UHMW-PE Melt at 190 °C

This enormous viscosity is responsible for the extremely difficult processing of the material. Thus, it is not possible to use standard thermoplast processing methods such as screw extrusion or hollow-body blow moulding with UHMW-PE without severely impairing the properties. Instead sintering techniques are employed such as

TABLE 3
Processing of UHMW-PE

Compression moulding	Sheets, blocks
Ram-Extrusion	Profiles
Sintering	Porous parts
Injection moulding	Finished parts

sintering under pressure at temperatures above the crystallinity melting point, ram extrusion which is really a form of continuons sintering, sintering without pressure which produces porous parts and finally injection moulding which can be used under certain technical and material conditions.

In the following the sintering technique under pressure and injection moulding are to be discussed in more detail.

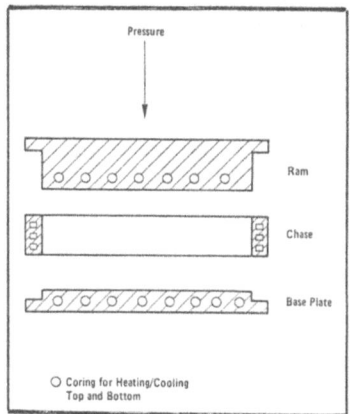

Fig. 4 is a sketch of the principle and shows the apparatus necessary. It consists of a frame into which the powder is filled as well as a base plate and a ram. The UHMW-PE powder is first compressed in a cold state and then heated under pressure to above the crystallite melting point. The heat is supplied through the base plate and ram but heating and cooling channels can also be accommodated in the frame. With this method sheets 1 to 6 m long, 1 to 2 m wide and 1 to 150 mm thick are manufactured. Blocks up to 400 mm thick can also be manufactured for special applications.

Figure 4. Compression Moulding of
UHMW-PE

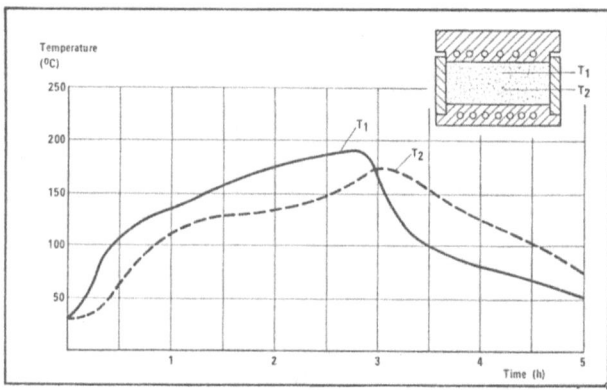

Fig. 5 shows the temperature behaviour in the UHMW-PE layer during the sintering process. The arrest point near the crystallite melting point is very marked. It is also clear what a long space of time is required to manu- facture an 80 mm- thick plate as in this example. Typical conditions for compression sintering processing

Figure 5. Compression Moulding of UHMW-PE
Temperatures as function of time

are a temperature of 200 to 250 °C and pressures of between 4 and 10 MPa. The required sintering times naturally depend on the heating temperature and the plate thickness. However, the quality of the product is also marked by the processing conditions. Fig. 6 shows the dependency of the elongation at tear on the compression temperatures and the compression pressure. An increase in temperature leads to better fusion of the individual polymer grains and thus a higher elongation at tear whilst an increase in pressure has the opposite effect.

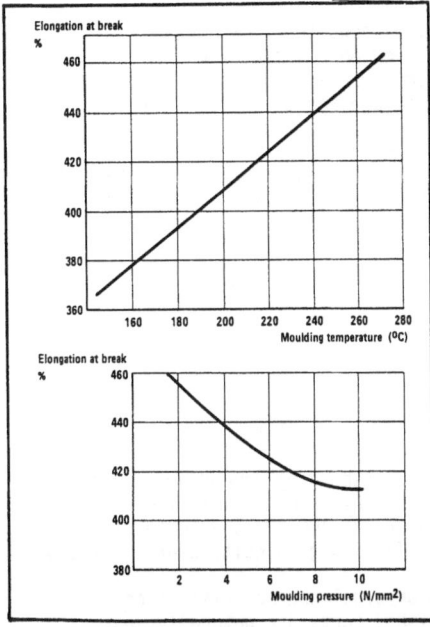

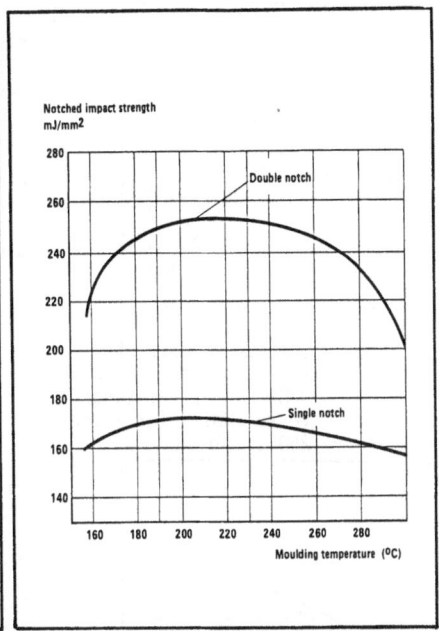

Figure. 6   Compression Moulding
            of UHMW-PE

Figure 7.   Compression moulding of
            UHMW-PE (Moulding pressure
            10 N/mm^2) Notched impact
            strength as function of
            moulding temperature

As far as the notched bar impact strength is concerned, it is at its
maximum at compression temperature of around 200 to 220 °C.

The abrasion strength is also at its highest at compression temperatures
of 200 °C.

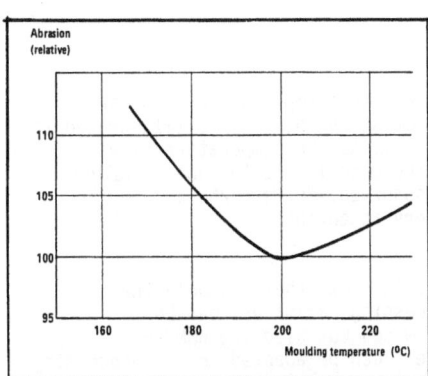

About 60 % of all UHMW-PE is processed
to semi-finished products using the
compression sintering technique.
Finished products are then machined
from them. This method is relatively
expensive and really not suitable for
a thermoplast. In particular it is
almost impossible to manufacture
small complicated parts
economically. Therefore there have
been many attempts to make finished
parts from UHMW-PE using the injection
moulding method.

This has now been achieved. Beforehand,
however, both the injection-moulding
machine and the ultrahigh molecular
weight polyethylene had to be
modified.

Figure 8. Compression moulding of
          UHMW-PE (Moulding
          pressure 10 N/mm^2)
Abrasion(sand-slurry)as function of moulding temperature.

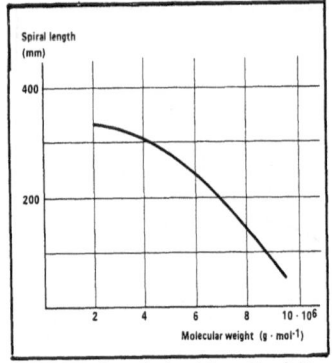

Fig. 9 shows the flow distance
measured on a 4 mm-thick spiral
mould of square shapes as a function
of the average mol mass at an
injection pressure of 120 MPa and
a temperature of 250 °C. Both the
high pressure and the temperature are
necessary to achieve good filling
of the mould even with complicated
moulds.

Figure 9.   Injection Moulding of
            UHMW-PE Spiral length
            versus Molecular Weight
            250°C, 120 MPa

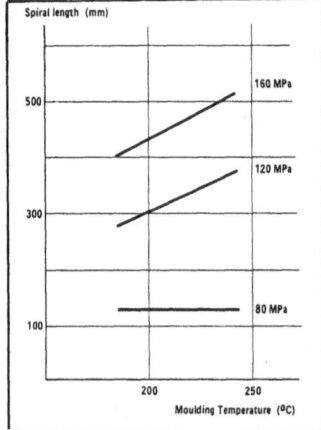

The corresponding dependencies are
shown in Fig. 10. The minimum
pressure of 100 MPa and temperatures
above 200 °C give quite useful results.

The physical characteristics of the
finished parts are naturally just
as dependent on the processing
conditions as with compression
sintering techniques.

Figure 10.   Injection Moulding of
            UHMW-PE Hostalen GUR 812
            Spiral length versus
            Temperature

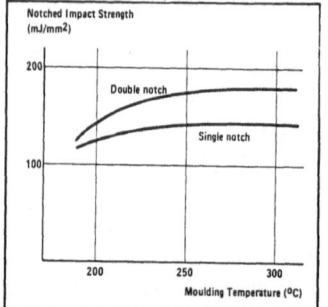

Taking the example of notched bar
impact strength, this is illustrated in
Fig. 11. As the temperature rises,
an increase in the toughness values
is observed. Between 250 and 300 °C
a very acceptable level is reached.

Fig. 12 shows that not only the
processing parameters can be
optimised but also the UHMW-PE
itself can be adapted to the processing
technique.

Figure 11.   Injection Moulding of
            UHMW-PE Hostalen GUR 812
            Notched Impact Strength
            versus Temperature

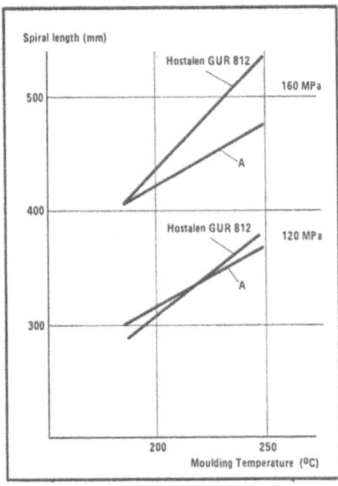

In comparison to a conventionell UHMW-PE (product A) Hostalen GUR 812, an optimised type specially developed for injection moulding, exhibits a clearly improved flow characteristic with a sharper rise in the flow path length as a function of the temperature. Moreover, a very good surface quality is achieved with this material.

Figure 12. Injection Moulding of UHMW-PE
Spiral length versus Temperature

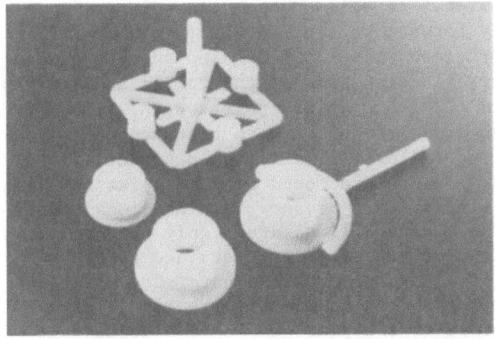

Figure 13. Bearing Bushes in Film Developers

Fig. 13 shows a typical application, bearing bushes made of Hostalen GUR 812 for use in film developers. In this application the excellent resistance to chemicals, the excellent abrasion strength and the very good gliding behaviour are used to the full.

REFERENCES

1. Berzen, J. , Chemie-Technik 4, 1974, 129-134

2. Fleißner, M., Hoechst AG, unpublished results 1986

ROLLTRUSION: DOUBLE-ORIENTATION OF POLYMERS USING A SINGLE STEP PROCESS

J.H. Magill and D.C. Sun

School of Engineering, University of Pittsburgh,
Pittsburgh, PA 15261 USA

ABSTRACT

The properties of commercial polymers (plastics) have been considerably improved by rolltrusion. In this process the polymer is deformed stepwise or continuously between rollers* where it drawn simultaneously. A variety of workpiece shapes may be used and the sample size is only limited by the practical dimensions of the roller and wind-up assembly used. Quality materials of high modulus and strength, high transparency, good creep resistance and toughness have been produced in a single step operation without using fillers. The finished product is non-spherulitic. Real draw ratios of X60 are found for some plastics that are comprised of anisotropic crystallites of measurable dimensions from small angle X-ray scattering. These crystallites are highly oriented in a connecting "amorphous" matrix whose orientation is determined by the processing conditions. Wide angle X-ray measurements have shown that the crystallites are frequently disposed with a low index crystallographic plane coinciding with the roll plane of the plastic. Some molecular weight dependence of this roll plane has been noted especially for polyethylene.

The morphology - property relationships are found to depend upon processing conditions, molecular weight and polymer chemistry. The technique is applicable to all crystalline or potentially crystallizable homo-polymers, copolymers and liquid-crystal-forming polymers. The technique provides a useful means of upgrading many common commercial plastics to an acceptable level for engineering applications, without using additives.

The wear behavior has been determined with respect to processing conditions and specimen anisotropy.

---

*The rollers may be heated or cooled, rotating or fixed and the workpiece may also be preheated if required.

INTRODUCTION

For many years now,[1-6] we have pioneered and developed a novel
polymer processing technique which is now named Rolltrusion[+]. It is a
single step procedure for producing doubly oriented materials. The
functional value of this process has been well demonstrated as an
effective method for producing transparent, high modulus and high strength
plastics[*] without using fillers or other additives or impairment of
specimen molecular characteristics. Rolltrusion has been used and also
described in the literature in the case of nylon 6,[1] nylon 66,[2,3]
polyethylene,[3,4,5] polypropylene,[4,5] polyethyleneterephthalate,[6] to
produce quality high strength materials, some of which are useful for
engineering applications. Other work[7] is in progress on PVF_2, PEEK,
ethylene/propylene copolymers etc., including blends and also liquid
crystal forming polymers.

The double orientation procedure is carried out under tension in a
single operation (using a static or moving roller assembly). A variety of
experimental processing conditions that are commensurate with the
properties of several materials have been employed. The actual
rolltrusion operation can be conducted batchwise or continuously depending
upon the dimensions of the workpiece being processed. The advantages of
the method are outlined in Table I and also contrasted with other better

---

[+]Note that this rolltrusion procedure contrasts with the well-known two-
step double orientation technique exploited by Bunn and Garner[8] in their
classical crystallographic structure studies on nylon 66 and nylon 610 of
many decades ago. It should not be confused with biaxial orientation.

[*]Processing is coupled to many primary variables and is naturally a
function of molecular weight, but quality polymer(s) having properties in
line with well-publicized ultraoriented polymers have been produced from
relatively lower molecular weight samples.

known techniques that have been and are currently employed to produce high strength polymers over the past decade.

A commercial polymer* may be converted into a high quality doubly-oriented transparent material in a single operation. The morphology or texture of the processed specimen is different in the three axial directions X, Y and Z (see X-ray diffraction diagrams in Figures 1 and 2). Correspondingly, the molecular orientation is also considerably enhanced. The improvement in molecular chain alignment and interchain cohesion is responsible for the enhanced mechanical properties, tensile modulus and strength.

For a particular polymer physical properties are coupled with processability in a complex manner,[4,5] eg.

Property = function   thermal treatments

texture

crystallinity

stress and strain history

draw ratio (DR)

Consequently, it is necessary to establish whatever relationships exist between processing – structure – morphology if a basic understanding is to be established.

The goal of the research described here has been to improve properties of commercial plastics‡ in a way that expands their usage for

---

*Since we first published and developed our technique many years ago a related roller drawing technique has been described for processing polyethylene.[9]

‡Of diverse molecular weights

TABLE I.

Solid State Deformation Techniques and Their Relative Merits(4)

	Tensile Drawing (10, 11)	Ram (11a) or Hydrostatic Extrusion (12)	Die Drawing (13)	Rolling (14)	Rolltrusion (1-7) and Roller Drawing (9)
Process mode	batch	batch	batch/ continuous	batch or continuous	batch or continuous
Product form	thin films	films, fibers, etc.	rods, sheets, tubes, fibers, etc.	rods, sheets, tapes, etc.	rods, sheets, tapes, etc.
Product dimension (a) length	limited	limited	unlimited	unlimited	unlimited
(b) width	narrow	limited by die shape	limited by die shape	limited by roller dimensions	limited by roller length
(c) thickness	thin	limited by die shape	limited by die shape	thin or thick adjustable	thin or thick (adjustable)
Applied load	uniaxial tension	multi-axial stresses with shear stress	multi-axial stresses with shear stress	multi-axial stresses with shear stress	multi-axial stresses with shear stress
Product Morphology	uniaxial orientation	uniaxial orientation	uniaxial orientation	biaxial orientation	double orientation

494

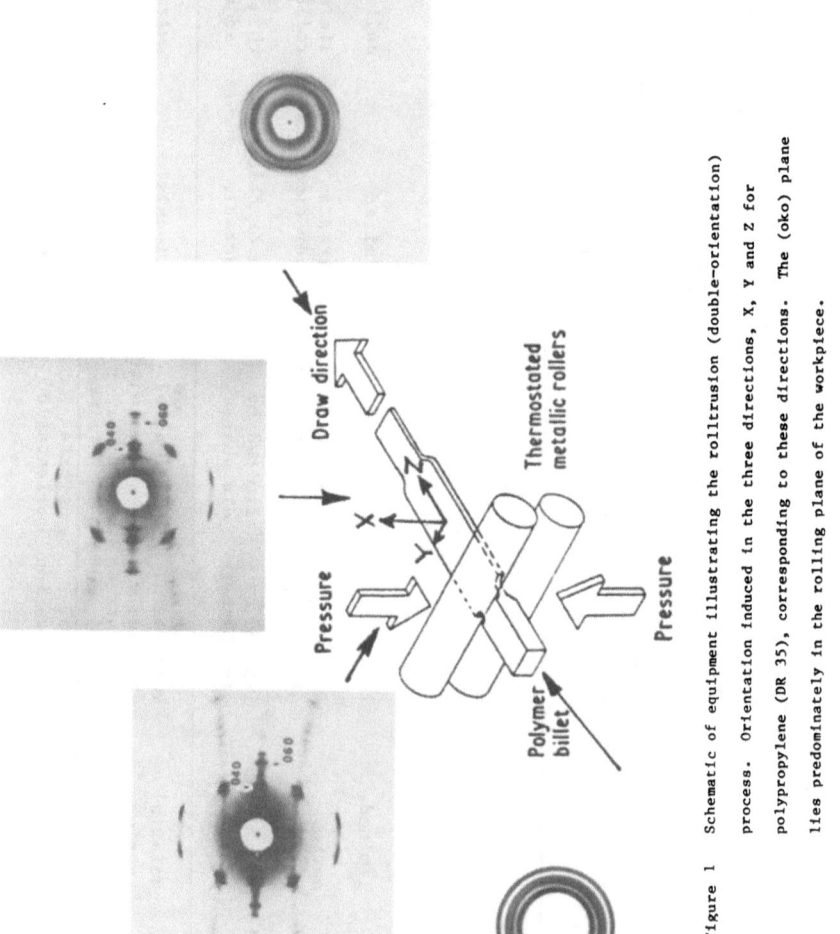

Figure 1   Schematic of equipment illustrating the rolltrusion (double-orientation)
process. Orientation induced in the three directions, X, Y and Z for
polypropylene (DR 35), corresponding to these directions. The (oko) plane
lies predominately in the rolling plane of the workpiece.

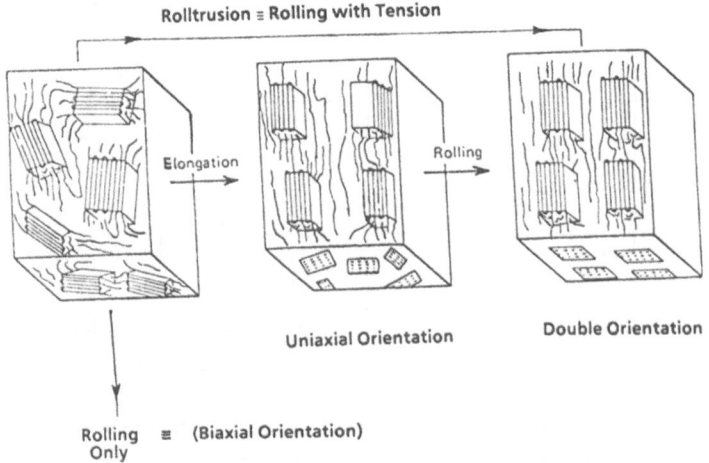

Figure 2   Morphologies illustrated by various tensile and roller deformation
techniques emphasizing particularly the single step rolltrusion process.

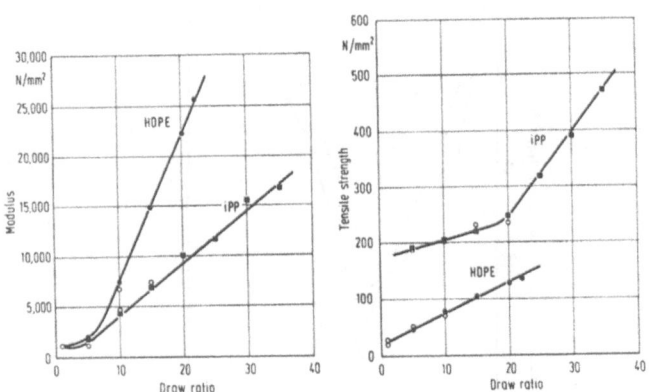

Figure 3   The effect of tensile strength and tensile modulus for high density
polyethylene ($\bar{M}_w$ = 91,900, $\bar{M}_n$ = 13,900) and polypropylene ($\bar{M}_w$ = 413,000 and
$\bar{M}_n$ = 65,600) respectively.

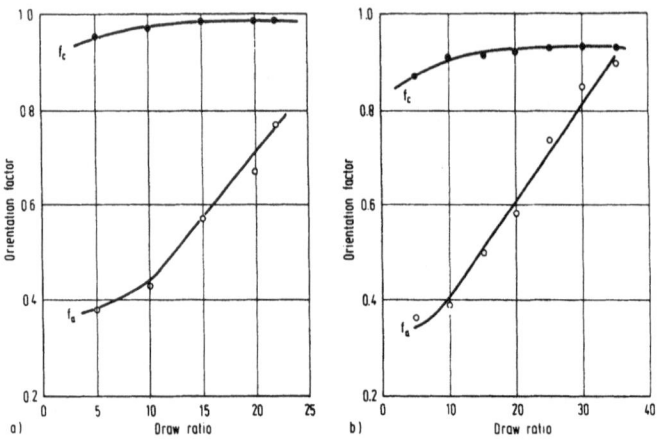

Figure 4   Amorphous and crystallite orientation factors f_a and f_c respectively

illustrated as a function of draw ratio for polyethylene and polypropylene.

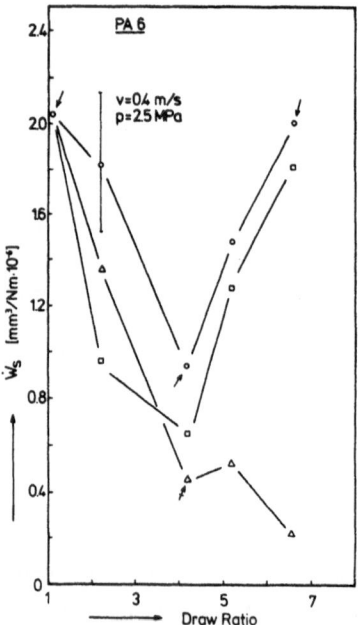

Figure 5   Wear rate $\dot{W}$ in doubly-oriented nylon 6 as a function of directions X, Y and

Z for different draw ratios (15).

many engineering applications. Currently, rolltrusion is being used to gain a better understanding and insight into microscopic (molecular) and macroscopic deformation in polymeric systems. A double-orientation or triaxially oriented system lends itself to more direct morphological and mechanical property investigations than does uniaxial or biaxially deformed polymer of comparable orientation in many instances.

At this point in our work, we have developed two prototype rolltrusion devices and a third much more versatile computer interfaced apparatus now being constructed, will soon be in operation for rapid materials processing.

RESULTS AND DISCUSSION

The relationship between processing conditions and properties is well illustrated in the tensile strength and tensile modulus dependence upon deformation ratio shown in Figure 3(a)-(d) for polyethylene and polypropylenes of only moderate molecular weights. The stipulaton of ultrahigh molecular weight characteristics is not a necessary condition to acquire good mechanical properties. Corresponding changes in crystallite and amorphous orientation are illustrated in Figure 4(a) and 4(b) for the same polymers. Changes in other properties are already documented[4,5,16].

It is also noteworthy that some important engineering properties such as wear rate are quite different in the three mutually perpendicular directions for rolltruded samples[15,16]. This is important evidence that is illustrated for nylon 6 in Figure 5. Other polymers such as polyethylene and polypropylene have also been investigated[16] in this manner for the first time in relation to textural changes corresponding to different stages of deformation.

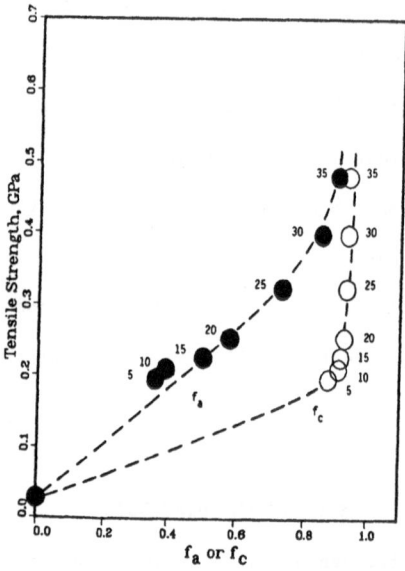

Figure 6   Tensile strength versus orientation function for rolltruded polypropylene.

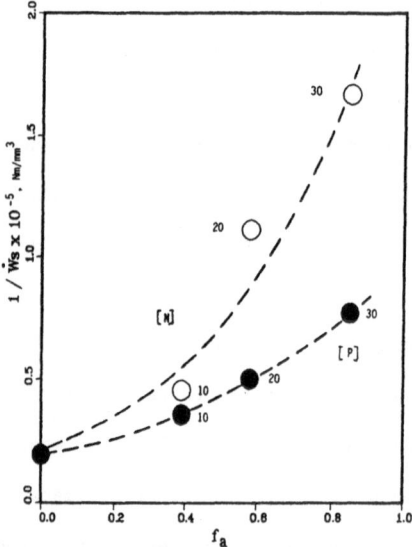

Figure 7   Inverse wear rate, $\dot{W}_s^{-1}$ as a function of orientation factor for
polypropylene (16).

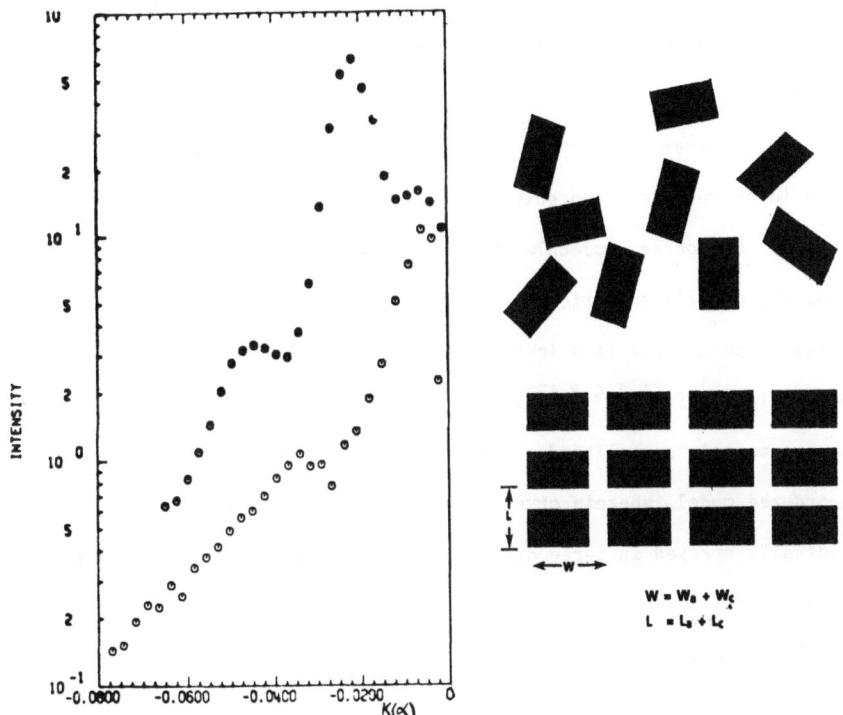

Figure 8   (a)   SAXS intensity vs scattering angle K (a) along the draw direction and

              (b) in the transverse direction in rolltruded polypropylene.

        (b)   Schematic of a rolltruded polymer, indicating the relative disposition

              of crystallites with respect to non-crystallite material (blank

              regions) interdispersed between them. The illustration does not

              attempt to show the tie molecules (taught and relaxed to varying

              degrees), chain folds and other intercrystallite connections. The

              upper region with staggered blocks depicts the crystallites for

              instance in a spherulitic or other morphology before rolltrusion; the

              ordered assembly of blocks (lower part of diagram) is in keeping with

              the rolltruded polymer in which considerable order prevails in

              different directions.

It is interesting and important to see that tensile strength correlates better with f_a than f_c (Figure 6) and that this type of correlation is bourne out against reciprocal wear rate too (see Figure 7).

The spherulitic morphology of different polymers disappears at some draw ratio. For example, nylon 6 becomes non-spherulitic about a DR=5 whereas in polyethylene and polypropylene the DR value lies closer to 10 than to 5. Of course, molecular weight plays a significant role too. Anyhow, the overall scenario of rolltrusion indicates that a low index crystallographic plane lies in the rolling plane with two other directions perpendicular to it[1,2,4,6,17]. The latest small angle X-ray measurements made on rolltruded materials[18] supports this relatively well-ordered model, wherein crystallites and amorphous regions are periodically arrayed in three dimensions; the largest periodic spacing lying along the molecular chain direction of the oriented crystallites. Figure 8 illustrates this type of rolltruded, high strength transparent polymer for instance in polypropylene or polyethylene where many orders of the SAXS long period are found.

In an effort to optimize the rolltrusion procedure using process parameters related to the processed polymers themselves, finite element method (FEM) has been employed. Process simulation using the high speed CRAY computer facilities in Pittsburgh have demonstrated the feasibility of this procedure[19]. Process optimization is the goal of this project aimed at upgrading the physical properties of commercial plastics generally.

ACKNOWLEDGEMENTS

The authors thank the National Science Foundation (Polymers Materials Program) for partial support of this work.

501

REFERENCE

1. Magill, J.H., unpublished; Moore, L.J., Master's thesis and Z. Frund Jr., Master's Thesis, Univ. of Pittsburgh (1984).

2. Magill, J.H., Kojima, M., Pollack, S.S. and Haller, M.N., Amer. Chem. Soc. Coatings and Plastics Preprints, 34, #2, 201 (1974).

3. Magill, J.H., Amer. Phy. Soc. (DHPP) March Meeting, Abs. JH9, p. 409, Washington, DC (1978).

4. Shankernarayanan, M.J., Sun, D.C., Kojima, M. and Magill, J.H., J. Int. Polymer Processing, 1, 65 (1987).

5. Magill, J.H., Sun, D.C. and Shankernarayanan, M.J., J. Applied Poly. Sci (in press 1987).

6. Ehnot, J., Thesis, Univ. of Pittsburgh (1986).

7. Magill, J.H. et al., unpublished results.

8. Bunn, C.W. and Garner, G.V., Proc. Roy. Soc. London, A 189, 39 (1947).

9. Kiato, A., Nakayama, K. and Kanetsuna, H., J. Applied Poly. Sci., 30, 1241 (1985).

10. Capaccio, G. and Ward, I.M., Polymer, 15, 233 (1974).

11. Clark, E.S., Scott, L.S., Poly. Eng. Sci., 10, 682 (1974).

11.a Southern, J.H., Porter, R.S., J. Appl. Sci, 14, 2305 (1970).

12. Gibson, A.J., Ward, I.M., Cole, B.N. and Pearsons, B., J. Mat. Sci., 9, 1193 (1974).

13. Richardson, A., Hope, P.S., Ward, I.M., J. Poly. Sci., Poly. Phys. Ed., 21, 2525 (1983).

14. Bigg, D.M., Smith, E.G., Epstein M.M. and Fiorentino, R.J., Polym. Proc. Soc. Proceedings Ab. 617, p. 35, Second Annual Meeting, Montreal Canada, Apr. 1-4 (1986).

15. Voss, H., Magill, J.H. and Friedrich, K., J. Applied Polym. Sci., 33, 1745 (1987).

16. Voss, H., Magill J.H. and Friedrich, K., J. Applied Polym. Sci., 33, XXXX (1987).

17. Matsubara, I. and Magill, J.H., J. Polym. Sci., (Poly Phys Ed.) 11, 1173 (1973).

18. Magill, J.H., Shankernarayanan, M.J. and Lin, J.S., Amer. Phys. Soc (DHPP) Meeting Abs. OV6 p. 837, March 16-20, New York (1987).

19. Sun, D.C. and Magill, J.H. (work still in progress).

RHEOLOGY OF CHEMICALLY REACTING POLYMERS

A. Silberberg
Department of Polymer Research
Weizmann Institute of Science
Rehovot 76100, Israel

ABSTRACT

Relaxation of stress involves molecular re-arrangements around, or over, energy barriers imposed by secondary bonds with neighbors, or by the covalent bonds of the system. Covalent bonds are not in general subject to breakage. The presence of a chemical reaction, however, has the effect of lowering the barriers which protect some of these bonds and an alternative relaxation path is created. The reaction also permits the re-distribution of stresses over the mechanisms of the system and, in principle, permits the utilization of the energy of the chemical reaction to perform useful mechanical tasks. Input of mechanical energy, in turn, can affect the rate and specificity of a chemical reaction.

INTRODUCTION

When stresses are caused to appear in a material and deformation and flow is induced, the thermodynamic state of the material is altered. The work that has been performed upon the system will have raised its potential energy above the level corresponding to thermodynamic equilibrium. Unless the work input is maintained, however, stresses will relax, as the system tends to return to equilibrium. A fluid in particular cannot indefinitely support a system of non-isotropic stresses. At thermodynamic equilibrium only a hydrostatic pressure can exist in a fluid.

In materials which are not fluids there will have to be permanent links, which tie the molecules to each other, such that externally applied non-isotropic (that is off-diagonal) stresses can be stored in permanent, macroscopic structures linked to the surface of the body. At least part of the system will have to be a covalently linked super-molecular network which permanently adheres to and connects the confining walls through which the induced stress pattern is administered.

It is clear of course that covalent bonds are only conditionally permanent. No link between atoms is stable over an infinite time, but covalent bonds are generally taken to have life times which by far exceed the time scale of experimental, or even technical interest. If flow and deformation occurs it is generally taken to involve the making and breaking of the cohesive interactions which hold molecules in a condensed state by secondary bonds. In addition there is, in general, rotational freedom about covalent bonds. Long polymer chains, for example, exist in a large number of energetically roughly equivalent conformations. Spontaneous transitions between these conformations permit the slippage and passage past each other of these long chain-like structures.

Although the individual cohesive interaction is weak there are systems where good alignment of neighboring molecules produces a group of contacts which function collectively and thus tend to produce links of extended life time. This is particularly obvious in the case of crystals, but sufficiently long stretches of paired macromolecular strands, for example, can produce structures which collectively act like covalent bonds.

The organization of atoms in condensed systems thus results in bonds between them, possessed of a variety of life times, some falling well beyond the experimental time scale and others interacting much more commensurately in time with the deformational events mechanically imposed on the system. Fluid like behavior will always occur, if the "permanently" linked structures of the system are smaller than the confines of the body.

In general, therefore, an applied deformation will raise the free energy of the body by straining all compliant mechanisms. Relaxation occurs by slippage of molecules, or parts of molecules, past each other. The activation energy for driving these processes is thermal. Each exploitable degree of freedom can count upon an energy of order kT to be available on the average. Much higher amounts of energy can, however, be concentrated into a mechanism from time to time. The larger the energy the longer the time between such events. The larger the energy required to alter the state of a mechanism the longer lived this state will be. Long-lived structures are thus protected from relaxation by high energy barriers.

In this sense no distinction is made between the nature of the barrier that has to be overcome and covalent, "permanent" bonds also fall into this scheme. In the case of covalent bonds it is simply assumed that the confining barriers are so high that the bonds are essentially permanent. This is not true, however, when a chemical reaction can occur in the system. In such a case the energy barriers are lowered and the covalent bonds in question become labile. Their life time is determined by the rate at which they will react at the temperature of the system. It follows that stresses and constraints that depend upon the covalent link being maintained will be released at a rate commensurate with the rate of chemical reaction.

CHANGES IN RHEOLOGICAL RESPONSE INDUCED BY CHEMICAL REACTION

When the amplitude of deformation (out of an original state of thermodynamic equilibrium) is kept small enough, the response of almost any system is linearly viscoelastic, i.e. each mechanism for mechanical relaxation is independently addresseable. The rate $\dot{\gamma}$ at which new deformations are produced and the rate at which stresses, induced in the past, will relax will operate in each mechanism separately in linearly additive fashion. The stress σ_j in a mechanism of relaxation time τ_j, for example, will change in time according to the following differential equation

$$d\hat{\sigma}_j/dt = -\hat{\sigma}_j/\tau_j + \dot{\gamma}G_j \tag{1}$$

Relaxation occurs thermally at a rate proportional to the level of stress $\hat{\sigma}_j$ in the mechanism with a proportionality factor $(1/\tau_j)$ characteristic of the mechanism. New stresses are induced at a deformation rate $\dot{\gamma}$ with a proportionality constant G_j. G_j is a modulus of instant response. The mechanism j is in general associated with some molecular species m of the system. If there are H_{jm} mechanisms j per molecule of m and there are n_m molecules of m per unit volume of the system the stress σ_j due to j in unit volume will be given by

$$\sigma_j = \hat{\sigma}_j \, H_{jm} \, n_m \tag{2}$$

If we now sum over all the mechanisms contributed by m and all the species m in the system we arrive at an expression for the total stress σ. We have

$$\sigma = \sum_m \sum_j \sigma_j \ . \tag{3}$$

If, therefore, the system is subject to a chemical reaction the quantities n_m are also functions of time and we have

$$d\sigma/dt = \sum_m \sum_j (d\sigma_j/dt)$$

$$= \sum_m \sum_j [H_{jm}n_m \, (d\hat{\sigma}_j/dt) + \hat{\sigma}_j H_{jm}(dn_m/dt)]. \tag{4}$$

The dependence of (dn_m/dt) on the concentration of the other species present can obviously be very complex, in general, but by way of illustration we shall assume that n_m degrades monomolecularly and forms a species p, say, which contributes no mechanical relaxation times, i.e. all the quantities H_{jp} are zero, or at any rate so small that terms with H_{jp} can be neglected in (4). We thus have

$$H_{jm}(dn_m/dt) = -k_m H_{jm} n_m, \tag{5}$$

where k_m is the reaction rate constant. Equation (4) can, therefore, be rewritten, using (1), (2) and (3), as

$$d\sigma/dt = -\sum_m \sum_j (1/\tau_j + k_m)\sigma_j + \dot{\gamma}\sum_m \sum_j G_j H_{jm} n_m \tag{6}$$

The mechanical relaxation time τ_j appears alongside the chemical rate constant k_m. We can, thus, define a new relaxation time τ'_j given by

$$1/\tau'_j = 1/\tau_j + k_m. \tag{7}$$

If, therefore, the rate of chemical degradation of m is fast enough (large k_m) the effective mode of relaxation would be predominantly chemical.

Hennenberg and Silberberg [1,2] have looked at somewhat more complex, but still linear, reaction schemes in place of (5). Their analysis involves, in part, the transfer of stress from one mechanism to another in the reaction.

The stress so transferred was assumed to be part, or all, of the stress carried by the molecules which have reacted away. This limitation is not necessary. The species created can have been formed with mechanical energy directly transferred to them from the reaction. They could possess either kinetic, or potential, energy whose source was the energy released by the reaction (mechanochemistry). This happens when we burn a fuel in an internal combustion engine or when we tense our muscles. The reaction produces molecular species p which in this case are created such that some of their mechanisms j carry stresses whose relaxation can be exploited to perform some useful task.

EFFECT OF MECHANICAL DEFORMATION ON CHEMICAL REACTION.

So far we have viewed the chemical reaction purely as a device which can compete with secondary bond relaxation by reducing the life time of certain covalent links. We have also considered the possibility of forming new links and thus trapping chemical energy in newly created mechanism. The stresses and strains introduced mechanically can, however, also influence the state of the system chemically.

In principle mechanical action can lower, or raise, energy barriers and change life times most easily by altering the orientational organisation of macromolecules. Deformations which lead to strand alignment can favor the formation of co-operatively stregthened bonds and thus increase τ_j. Deformations of this kind can also increase the steric factor of a chemical rate constant and thus increase the reaction rate. Mechanical stress, particularly if it can be concentrated at specific molecular sites, can also lower the effective energy barriers protecting a covalent bond. For example the degradation of long macromolecules in dilute solution by shear produces breakage at their midpoint due to an accumulation there of the elongational force. Taken singly, however, direct mechanical effects are too small to influence covalent bond stability seriously.

CONCLUSION

The rheological response of a material will be modified in the presence of a chemical reaction. Energy can be made to pass, as a result of the reaction, from mechanism to mechanism and the energy of the chemical reaction can in part be utilised to achieve mechanical effects and to perform useful tasks. Mechanical energy and the deformation it produces can in turn influence the rate and the course of a chemical reaction [3].

REFERENCES

1. Hennenberg, M. and Silberberg, A., J. Rheology, 1985, 29, 379-414.

2. Hennenberg, M. and Silberberg, A., J. Rheology, 1985, 29, 415-430.

3. Silberberg, A. and Hennenberg, M., Nature, 1984, 312, 746-748.

RHEOLOGY OF REACTING POLYMER SYSTEMS NEAR THE GEL POINT

Horst Henning Winter
Max-Planck-Institut für Polymerforschung
Jakob-Welder-Weg 11
D - 6500 Mainz, West Germany

Permanent address: Department of Chemical Engineering,
University of Massachusetts, Amherst, MA 01003, USA

Chemical gelation of polymers has been studied extensively, for applied and for basic scientific reasons. Applications are governed by the change of properties as a function of reaction extent. Basic scientific interest comes from the fact that gelation is a critical phenomenon and that very simple properties are expected at the critical point (gel point.)

POLYMERS AT THE GEL POINT (CRITICAL GELS)

Experiments on crosslinking polydimethylsiloxane[1-3] and on polyurethane[4,5] show that the mechanical behavior at the gel point is given by a power law relaxation modulus

$$G(t) = S\, t^{-n} \ , \ p = p_c.$$

The strength S and the relaxation exponent n characterize the critical gel. The relaxation exponent may adopt values in the range, $0 < n < 1$. Experimental values were between 0,5 and 0,66. It is important to note that the power law is not valid at very short times at which the sample response would be glassy (interference of vitrification with gelation). The relaxation modulus can be used to formulate an equation for the stress, the so called "gel equation"[2]

$$\underline{\tau}(t) = S/n \int_{-\infty}^{t} (t-t')^{-(1+n)} \underline{c}^{-1} \, dt', \quad p = p_c,$$

where $\underline{c}^{-1}$ (t, t') is the Finger strain tensor.

Correspondingly, the complex modulus of small applitude oscillatory shear follows a power law

$$G^*(\omega) = a \, S \quad (i\omega)^n, \qquad p = p_c$$

with $a = \pi / [\Gamma(n) \sin(n\pi)]$

$\Gamma(n)$ = Gamma Function

The power law behavior is an expression of mechanical self-similarity (fractal behavior)[6,7]. The range of selfsimilarity is defined between an upper and a lower frequency limit. The lower frequency limit (reciprocal of characteristic relaxation time) corresponds to an upper scaling length, the correlation length, which is of the order of the size of the largest molecular cluster (of pre-gel) or of the largest remaining percolation cluster (of post-gel). The upper frequency limit (glass frequency) corresponds to a lower scaling length, which is given by the dimension of the molecular network units responsible for glassy behavior. The correlation length and, hence, the characteristic relaxation time increase in the approach of the gel point, diverge to infinity at the gel point, and then decrease again with increasing extent of crosslinking. The critical gel has no characteristic length or time scale. All observations are restricted to polymers at temperature much above the glass transition temperature and at frequencies much below the glass frequency.

VICINITY OF CRITICAL STATE

The rheological behavior in the critical region around the
gel point is given by power laws with critical exponents s
and z:[8]

$$\text{pre-gel } (p<p_c): \quad \eta_0 \sim (p_c - p)^{-s}$$

$$\text{post-gel } (p>p_c): \quad g_\infty \sim (p - p_c)^z$$

Before the gel point, $p<p_c$, the polymer is a viscoelastic
liquid with a zero shear viscosity and beyond the gel point,
$p>p_c$, the polymer is a viscoelastic solid with an equilibrium
modulus g_∞. The characteristic relaxation time then also follows a power law.[9]

$$\text{pre-gel } (p<p_c): \quad \lambda \sim (p_c - p)^{-s/(1-n)}$$

$$\text{post-gel } (p>p_c): \quad \lambda \sim (p - p_c)^{-z/n}$$

At the gel point, $p = p_c$, the characteristic relaxation time
diverges, $\lambda \longrightarrow \infty$. Values for the critical exponents are
not available with sufficient accuracy.

RANGE OF VALIDITY

The experiments were all done with endlinking polymers. It
is likely that randomly crosslinking polymers behave similar,
however, experiments on such systems are still needed. Additionally, all our data were taken at temperatures much above

the glass transition temperature and at frequencies much below the glass frequency, i. e. interference of gelation and vitrification was avoided. This limits the application of our results to polymers far away from vitrification. The effect of physical entanglements was minimized by choosing prepolymers with molecular weight below the entanglement limit.

REFERENCES

1. Chambon, F., and H. H. Winter, Polymer Bull., 13, 499 (1985).
2. Winter, H. H., and F. Chambon, J. Rheol., 30, 367 (1986).
3. Chambon, F., and H. H. Winter, J. Rheol., 30 (1987) to appear.
4. Chambon, F., Z. S. Petrovic, W. MacKnight, and H. H. Winter, Macromolecules, 19, 2146 (1986).
5. Winter, H. H., Chambon, and P. Morganelli, to be submitted.
6. Muthukamar, M., and H. H. Winter, Macromolecules, 19, 1284 (1986).
7. Vilgis, T., and H. H. Winter, J. Phys. Chem., 1987, submitted.
8. Stauffer, D., A. Coniglio, and M. Adam, Adv. Pol. Sci., 44, 74 (1982).
9. Winter, H. H., Progress Coll. Polym. Sci., 1987, to appear.

ACKNOWLEDGEMENT

The support of the National Science Foundation, grant number MSM - 860 1595, and of the Max-Planck-Institut für Polymerforschung, Mainz, is gratefully acknowledged.

THE REACTIVE PROCESSING OF COMPOSITE MATERIALS:
STRUCTURE-PROPERTY RELATIONS IN RRIM COPOLYURETHANES

J.L. STANFORD

Wolfson Polymer Research Unit,
Department of Polymer Science and Technology,
The University of Manchester Institute of Science and Technology,
Manchester, M60 1QD, England.

ABSTRACT

Segmented copolyurethanes have been formed by RIM using commercial polyisocyanates reacting with different polyol blends. Various proportions of
high and low molar mass polyether triols in admixture with ethylene glycol
were used as compatible or incompatible blends, and their effects on the
degree of phase separation and subsequent copolyurethane physical properties
were determined.

Filled copolyurethanes containing up to 10% w/w of glass and wollastonite fibres were also formed using RIM equipment fitted with a special
dosing unit for processing discontinuous fibre/polyol slurries. The effects
of fibre type, loading, size and surface treatment on slurry rheological behaviour at low and high shear rates have been evaluated in relation to the
RIM process. Composites containing up to 50% w/w of continuous fibres have
been formed using glass and carbon mats, pre-placed in the RIM mould.

Both filled and unfilled RIM-materials were characterised using tensile,
dynamic mechanical-thermal and impact properties. Materials ranged from
flexible elastomers to stiff composites with Young's moduli of 0.1 and 15.6
GPa, respectively.

INTRODUCTION

Reaction injection moulding (RIM) of polyurethane-based materials is a
rapid, one-stage forming process in which two liquid reactant streams, one
a polyol blend and the other a polyisocyanate, are accurately metered, imp-
ingement-mixed and forced at low pressure into a mould cavity. Mixing the
catalysed reactants initiates the highly-exothermic polymerisation which
continues in the mould, leading to gelation and solidification [1]. Postcuring completes the polymerisation process and establishes the final
phase-separated structure and morphology [2]. The physical properties [2,
3] of these RIM-segmented copolyurethanes are determined by the relative
proportions of and degree of compatibility [3] between the hard and soft
segment phases.

Improved properties and reduced temperature-dependence can be obtained
by using inorganic fillers to form copolyurethane composites [3]. However,
incorporating the fillers into the reactant streams to form slurries,
increases viscosity and complicates the RIM process particularly in terms
of transporting and mixing of the reactants. An understanding of slurry
rheological behaviour, giving the interrelationships between fibre loading,

packing fraction, aspect ratio and viscosity [4], is essential if the fibre content and property enhancement in RIM-composites are to be maximised whilst retaining acceptable processability.

This paper considers firstly the effects on copolyurethane properties of the relative proportions, compatibilities and molar masses of polyether polyols used to form the soft segment phase. Secondly, the rheological behaviour of slurries containing development wollastonite fibres is presented in comparison with earlier data [4] on polyol/glass fibre slurries. Finally, the mechanical properties of RIM-copolyurethane composites containing wollastonite fibres are compared with those of structural composites containing pre-placed glass and carbon fibre mats.

EXPERIMENTAL

The polyols used, T32/75 and LHT240, are polyethers and the polyisocyanates, VM10 and RMA 400, are liquid variants of 4,4'-diphenylmethane diisocyanate (MDI): characterisation data of all reactants are given in Table 1.

TABLE 1. Characterisation data for polyol and polyisocyanate reactants.

Reactant	En(g mol^{-1})	Mn(g mol^{-1})	fn(Mn/En)
Polyisocyanate VM10 (ex. ICI)	166.0	454.8	2.7
Polyisocyanate RMA400 (ex. Upjohn)	158.0	331.8	2.1
Polyol T32/75 (ex.ICI)	1930	5260	2.7
Polyol LHT240 (ex. Union Carbide)	228.4	685.5	3.0
Chain extender, ethylene glycol EG	31.0	62.0	2.0
Catalysts TED/DBTDL (ex. Air Products)	0.028/0.004 w/w mixture in DPG		

En,equivalent weight (acetylation). Mn,molar mass (GPC). fn,functionality.

For polyol compatibility studies, various polyol blends containing different weight proportions of T32/75:LHT240:EG, given by the sequence of code numbers describing the copolyurethanes PU821 to PU401 in Table 2, were reacted with polyisocyanate VM10. Reactant structures, RIM equipment, processing and moulding conditions have been described in detail elsewhere [3].

Rheological studies were made on slurries comprising wollastonite and polyol blend 401. Wollastonite (ex. Blue Circle) a development grade of naturally-occurring, acicular calcium metasilicate fibres, had mean fibre length and diameter 33 and 6μm, and an aspect ratio of 6. Some of the wollastonite was surface-treated with either stearic acid, an aminosilane or a titanate. Viscosity data were compared with those previously obtained on slurries comprising similar polyols and glass fibres [3,4].

RESULTS AND DISCUSSION

Polyol Compatibility Studies.

The incompatible polyol blend T32/75 and EG, used to form PU401 with 59% w/w hard segment (HS) content and 5,260 g mol^{-1} soft segment (SS) molar mass, becomes a compatible blend on addition of LHT240. For materials PU821 to PU221, formed using increasing proportions of LHT240, HS content increases from 34 to 53%, and SS molar mass decreases from 2,300 to 1,270 g mol^{-1}. The effects of these structural variations on dynamic mechanical behaviour are clearly seen in figure 1. PU401 is well phase-separated with well defined SS and HS glass transition temperatures, Tg^S(-60°C) and Tg^H (125°C), respectively. In direct contrast, PU421 and PU221, containing the highest proportions of LHT240, are virtually phase mixed, each with a single merged Tg^S/Tg^H peak at about 80°C. Intermediate phase separation is observed for PU821 and PU621 which show broad but split Tg^S/Tg^H peaks at 0/40°C. The respective shear modulus/temperature plots show transitions and inflexions corresponding to the peaks and shoulders in the relaxation spectra.

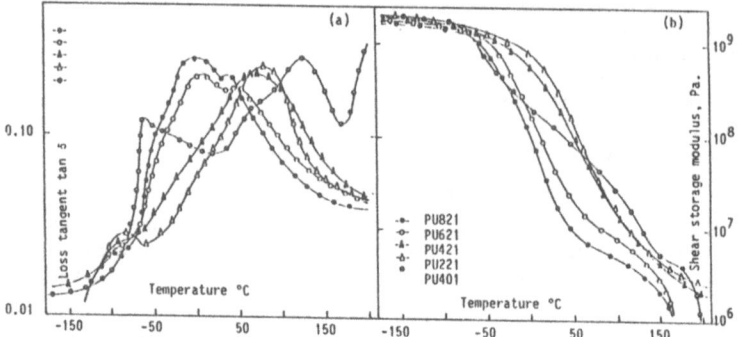

Figure 1. Dynamic mechanical spectra (torsion pendulum, 1Hz) of RIM-copolyurethanes showing the effects of polyol compatibility on phase separation.

Low values for the modulus ratio at -30 and 65°C indicate good phase separation. For PU401, this ratio was 4.3, whilst for PU221 to PU821, the ratio increased from 8.8 to 50.0.

Considering Table 2, the best combination of properties and minimal temperature dependence is exhibited by PU401. By comparison, PU221 and PU 421 being essentially glassy at 23°C, are yielding plastics due to the dominating effects of higher crosslink densities in the SS phase. In contrast, PU821 and PU621 are in their Tg^S/Tg^H region at 23°C and exhibit soft elastomeric behaviour. Although materials PU521 and PU621 have tensile properties similar to PU401, they show a much greater temperature dependence. Overall, the results indicate that polyol incompatibility is an important factor in promoting good phase separation in segmented copolyurethanes.

Polyol Slurry Rheology Studies.

The rheological behaviour of reactant slurries was studied over a range of shear rates using specially-constructed apparatuses [4]. Additional studies on the various slurries showed that in the case of cylindrical glass fibres, the packing fraction (ϕ_o) is uniquely related to the weight-average aspect ratio: this relation, however, does not hold for acicular wollastonite fibres. Rheological measurements on slurries at low shear rates

TABLE 2. Tensile properties (23°C) of RIM-copolyurethanes formed from either compatible (c) or incompatible (i) polyol blends.

Property \ Material	PU821 (c)	PU621 (c)	PU521 (c)	PU421 (c)	PU221 (c)	PU401 (i)
Modulus, E(MPa)	109	221	464	818	1280	325
Yield Stress, σ_y(MPa)	–	–	–	23	33	–
Yield Strain, ε_y	–	–	–	10	13	–
Strength, σ_u^*(MPa)	16	22	24	35	38	26
Elongation, ε_u^* (%)	155	152	130	100	75	147
Toughness (MJ m^{-3})	17	23	23	27	24	28

($<10^4$ s^{-1}) using modified cone and plate geometries, show that the relative

viscosity (η_r) is determined by the volume fraction (ϕ) of fibres in relation to the packing fraction for a given value of ℓ/d. These data are
shown in figure 2 and are of significant technological importance to filled
-RIM processing. The type and level of fibre surface treatment are shown
to have significant effects in reducing slurry viscosity. In the case of
polyisocyanate slurries, η_r increases much more rapidly compared to corresponding polyol slurries containing the same fibres. Analysis of the data
in figure 2 shows that the relative fluidity ($1/\eta_r$) of a glass fibre slurry
decreases linearly (with unit slope) as relative volume fraction (ϕ/ϕ_0) increases. Measurements on wollastonite polyol slurries at high shear rates

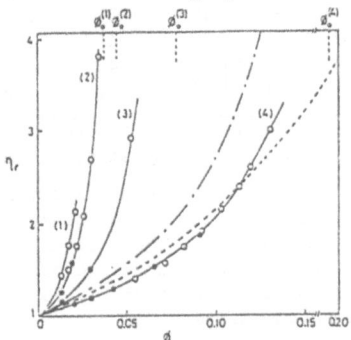

Figure 2. Relative viscosity versus volume traction of fibres with diff-
           erent ℓ/d and ϕ_0. Polyol-wollastonite -----; polyisocyanate-
           wollastonite -.-.-.Curves (1) to (4) are glass fibres (ℓ/d 162
           to 38) in polyol.

(up to 10^6 s^{-1}) have been carried out using a capillary viscometer attachment on the RIM machine. The results show that slurry viscosity decreases
as shear rate increases. The viscosity at 25°C of unfilled polyol blend
401 at zero shear rate ($\dot{\gamma}$) is 11.1 Poise, which is reduced to 5.5 Poise at
$\dot{\gamma} = 10^3$ s^{-1}. The higher initial viscosities of all slurries are also reduced to 5.5 Poise, but only at higher values of $\dot{\gamma}$(10^4 to 10^5 s^{-1}) which
depend on surface treatment and weight fraction % of fibres. $\dot{\gamma}$ increases
in the order stearate (18%) < titanate (18%) < uncoated < aminosilane (34%).

Structure-Property Studies on RIM-copolyurethane composites.

Composites, based on PU401 (formed using RMA 400) as matrix, were characterised in terms of their tensile stress-strain, flexural and impact
properties in relation to fibre type and loading, aspect ratio and surface
treatment. Table 3 summarises the averaged tensile and fracture properties
of composites, expressed as relative values with respect to unfilled PU401
matrix. For the wollastonite-PU401 composites, improved tensile properties
are only obtained when the fibres are surface treated with an aminosilane
coupling agent. However, reduced ultimate elongation causes an overall
loss of material toughness and consequently inferior impact properties.
Instrumented impact data, analysed using fracture mechanics to give relative fracture energies (G_C^c/G_C^m), show the hammer-milled glass-PU401 composites to be superior to those containing wollastonite, due to the higher
mean aspect ratio of the glass fibres.

The most significant enhancement of RIM composite properties is achieved by using continuous-fibre reinforcement rather than discontinuous
fibres such as HMG and wollastonite. Despite the higher fibre contents
shown in Table 3 (a distinct advantage of using pre-placed reinforcement in
the mould rather than using reactant slurries), the fibre mat, reinforced

RIM-composites exhibit very high values of stiffness and strength, although these are achieved only at the expense of drastically reduced elongations.

TABLE 3. Relative tensile and fracture properties of RIM-PU40 composites.

Reinforcement \ Property	W_f	E_c/E_m	σ_c^*/σ_m^*	$\varepsilon_c^*/\varepsilon_m^*$	G_c^c/G_c^m
Wollastonite, untreated	10	1.24	0.81	0.71	0.51
Wollastonite, silane-treated	10	1.93	1.16	0.75	0.59
Hammer milled glass fibres	10	1.79	1.12	0.98	1.14
Carbon fibre, 0/90° 2-mats	19	31	4.33	0.02	–
Carbon fibre, 0/90° 4-mats	38	43	6.96	0.03	–
Glass fibre, 0° longitudinal mats	49	48	16	0.02	>4
Glass fibre, 90° transverse mats	49	3.24	1.22	0.02	0.34

G_c = critical strain energy release rate (fracture energy, impact).
W_f = weight fraction of fibres. c = composite; m = matrix.

Nevertheless the impact properties of these composites are superior to the unfilled matrix. However, in the case of the glass mat materials in which all the fibres are uniaxially aligned, there is a high degree of anisotropy shown by comparing longitudingal and transverse properties. Anisotropy may be reduced by forming composites with cross-ply laminates as in the case of the 0°/90°, carbon fibre-PU composites.

ACKNOWLEDGEMENTS

The author wishes to thank Dr. S.R. Bentley and Mr. A.N. Wilkinson for the experimental data on wollastonite and pre-placed mat composites. Supply of development wollastonites and part-financial support from Blue Circle Industries are kindly acknowledged.

REFERENCES

1. Stanford, J.L., Still, R.H. and Stepto, R.F.T., In Reaction Injection Moulding and Fast Polymerisation Reactions, ed. J.E. Kresta, Plenum Publishing Corporation, New York, 1982, pp. 31-54.

2. Camargo, R.E., Macosko, C.W., Tirrell, M.V. and Wellinghoff, S.T., Polymer, 1985, 26, 1145-1154.

3. Barksby, N., Dunn, D., Kaye, A., Stanford, J.L. and Stepto, R.F.T., In Reaction Injection Moulding, ed. J.E. Kresta, ACS Symposium Series 270, American Chemical Society, Washington D.C., 1985, pp. 83-96.

4. Cross, M.M., Kaye, A., Stanford, J.L. and Stepto, R.F.T., In Reaction Injection Moulding, ed. J.E. Kresta, ACS Symposium Series 270, American Chemical Society, Washington D.C., 1985, pp. 97-110.

STRUCTURE-PROPERTY RELATIONSHIPS
IN DIAMINE-EXTENDED, RIM POLY(URETHANE-UREA)S

A.J. RYAN, J.L. STANFORD AND R.H. STILL

Wolfson Polymer Research Unit,
Department of Polymer Science and Technology,
The University of Manchester Institute of Science and Technology,
Manchester M60 1QD, England.

ABSTRACT

A series of poly(urethane-urea) materials, based on a liquid polyiso-cyanate and a polyether triol in admixture with different hindered aromatic diamines, were prepared by reaction injection moulding (RIM). All reactants were fully characterised prior to use in formulations. Materials were formed on in-house development RIM equipment using similar processing conditions.

RIM-materials properties were determined from dynamic mechanical thermal analysis (DMTA), tensile stress-strain data and differential scanning calorimetry (DSC). The materials obtained ranged from translucent, flexible elastomers to opaque, brittle plastics depending on the chemical nature of the diamine used and the hard segment content. The results indicated that phase separation occurred in all materials.

INTRODUCTION

Polyurethane-based RIM systems, incorporating aromatic diamine chain extenders, have enhanced properties [1,2] over those based on an aliphatic diol chain extenders [3,4]. RIM polyurethanes have been extensively investigated [3,4,5] but little systematic work has been reported on structure-property relationships for poly(urethane-urea) (PUU) materials. This paper presents systematic data on RIM PUU's based on a commercial, liquid poly-isocyanate reacted with a polyether polyol in admixture with (a) diethyl toluene diamine (DETDA) a commercially-available hindered aromatic diamine chain extender [6], or (b) methylene bis diisopropyl aniline (M-DIPA) or (c) methylene bis methylisopropyl aniline (M-MIPA). The latter two are novel-hindered aromatic, diamine chain extenders.

EXPERIMENTAL

Materials Formation

PUU materials were formed using in-house RIM equipment with the initial reactant temperatures, ratio of functional groups ($r = 1.03$), catalyst content and post-curing conditions held constant for each formulation. The

development RIM equipment is based on a two-stream Viking Engineering HP90 machine. One stream (A) consisted solely of a liquid polyisocyanate Isonate RMA 400 (ex. UpJohn/Dow). RMA 400 is based on 4,4'-diphenyl methane diisocyanate (MDI) I and is a straw-coloured, low viscosity liquid (1.4P at 25°C) with an equivalent weight of 158g mol^{-1} of isocyanate groups (end-group analysis). The other stream (b) comprised three components:-
(i) Polyol: Daltocel T32-75 (ex ICI) II, with an equivalent weight of 1930g mol^{-1}(end-group analysis), $\overline{Mn}$ (VPO) = 5220 g mol^{-1} and a number average functionality $\overline{fn}$ = 2.7.
(ii) Chain Extender: either DETDA, III (an 80:20 mixture of the 2,6 and 2,4 isomers.) or M-DIPA, IV or M-MIPA, V. All chain extenders were characterised by combustion analysis and N.M.R. spectroscopy prior to use.
(iii) Catalyst: pure dibutyltin dilaurate was used at a constant concentration of 1.05 x 10^{-3} mol/mol isocyanate functional groups.

The two streams, A and B, were mixed in a Krauss Maffei MK-12-4KF mixing head, which is fitted with a 12mm diameter, self-cleaning control piston. Reactant ratios and throughputs were determined from stoichiometry and the RIM machine pumps were set accordingly. The mixing head was adjusted to give an impingement pressure of 200 bar to give efficient mixing. During PUU processing, the mixing head is fixed to a rectangular steel mould (700 x 400 x 3mm) clamped in a hydraulic press. Heating and cooling was effected by thermostatted water, and the initial mould temperature was 60°C.

For a typical formulation, gel times were of the order of 5 seconds with mould filling occuring in less than 2 seconds. RIM plaques were removed from the mould within 30 seconds, though some formulations gave materials which were brittle and cracked in the mould. On demoulding plaques were postcured at 100°C for approximately 18 hours.

Materials are designated using a code in which the letters refer to the chain extender, D=DETDA, MD=M-DIPA and MM=M-MIPA; the number refers to the percentage hard segment content defined as the mass of chain extender plus the stoichiometric equivalent of polyisocyanate divided by the total mass of the system.

Materials Characterisation

Differential Scanning Calorimetry (DSC) was performed on a DuPont 990 Thermal Analyser equipped with a DSC cell. Samples (10-15mg) and an inert reference material, glass beads (10mg), were encapsulated in aluminium pans and cooled to -120°C in the cell which was then heated at 20°C min^{-1} to 300°C.

Dynamic mechanical measurements were made at 1Hz in the temperature range -100°C to 250°C at a heating rate of 5°C min^{-1} using a Polymer Laboratories Dynamic Mechanical Thermal Analyser (DMTA). A double-cantilever bending geometry was used for beam samples (3 x 10 x 45mm).

Tensile stress-strain data at 23°C (ASTM D638M-81) were obtained on an Instron 1122 universal testing machine using specimens of gauge length 75mm at an extension rate of 10mm min^{-1} (2.22 x 10^{-3}s^{-1} initial strain rate).

RESULTS AND DISCUSSION

Thermal properties are presented in Table I. DSC curves yielded the soft segment glass transition temperatures (Tg^S), but hard segment glass transitions (Tg^H) were not observed even up to 250°C, when degradation commenced. Tg^S for all the materials is insensitive to hard segment content, but varies with the diamine structure in the series D<MM<MD.

TABLE I   THERMAL AND MECHANICAL PROPERTIES OF RIM PUU

PUU	Tg^S/°C		Tg^H/°C	$\frac{E'(-30°C)}{E'(+65°C)}$	Tensile Properties		
	DSC	DMTA	DMTA		E/MPa	σ_u/MPa	ε_u/%
D35	-58	-51	185	3.7	258	12	320
D42	-57	-47	183	4.3	322	23	287
D46	-57	-49	190	2.9	423	27	228
D51	-58	-46	194	3.5	497	26	187
D56	-57	-45	195	3.2	689	28	37
D61	-55	-48	196	2.8	735	28	17
MM46	-55	-40	151	9.5	162	5	25
MM51	-56	-41	182	5.2	-	-	-
MM56	-57	-40	186	4.2	592	13	5
MD52	-50	-35	176	4.9	483	18	163
MD56	-51	-33	174	5.1	594	16	70
MD62	-52	-30	175	3.9	811	22	40

E = Young's modulus, σ_u = tensile strength, ε_u = ultimate elongation, values are averages of at least five determinations.

The dynamic mechanical spectra (Fig. 1) in terms of flexural storage modulus (E') and mechanical damping (tanδ) versus temperature (°C) show distinct differences in materials properties. RIM materials based on DETDA have a smaller modulus-temperature dependence than materials based on either M-DIPA or M-MIPA as shown by the lower ratios for E'(-30°C)/E'(+65°C) in Table 1. Lower ratios are attributed to differences in phase mixing, as well as to the obvious differences in urea group concentration. The damping curves show an interesting feature in the 50-100°C region coinciding with an inflexion in the modulus, which may be attributed to a mixed phase transition. This occurs prior to Tg^H, which is seen as a more significant drop in modulus, followed by a small rubbery plateau after which, loss of all physical integrity occurs due to degradation. The Tg^H values vary systematically with hard segment content for DETDA and M-MIPA materials while the M-DIPA series show a very broad Tg^H with maxima in tanδ occuring at approximately 175°C.

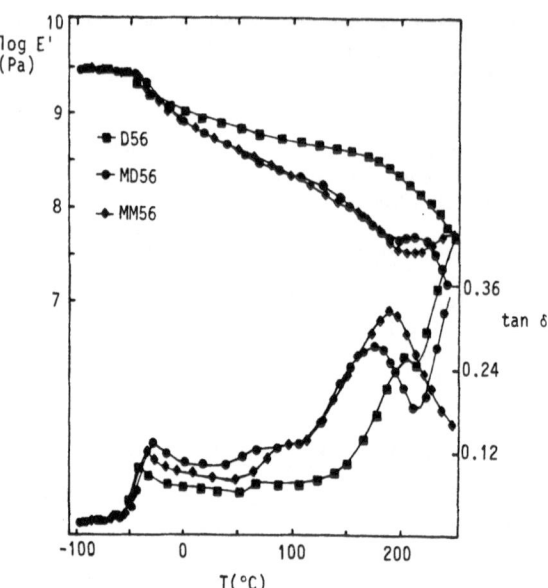

Figure 1. Representative Dynamic Mechanical Spectra showing the effects of hard segment structure and content on the modulus and damping behaviour

The tensile properties of the RIM PUU's are also presented in Table 1 and were derived from stress strain curves similar to those shown in Fig. 2. In the DETDA series two types of mechanical behaviour are observed. Lower hard segment materials (D35-D51) are elastomeric, showing high toughness and ultimate elongation. The PUU's, D56 and D61, show properties

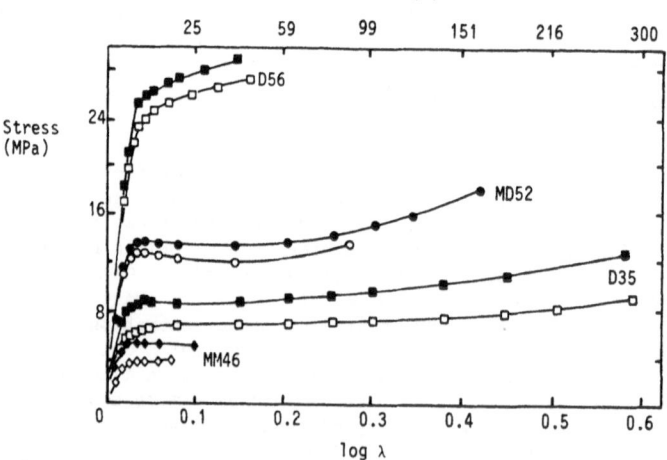

Figure 2. Mean Tensile stress-strain curves showing the effects of hard segment content (D56,D35), hard segment structure (D, MD,MM) and postcuring. Closed symbols refer to postcured samples, open symbols to non-postcured samples.

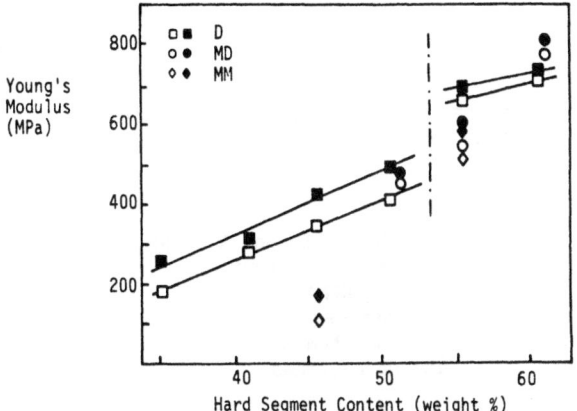

Figure 3. Variation of Young's modulus with hard segment content showing the effects of hard segment structure and postcuring. Closed symbols refer to postcured materials, open symbols refer to non-post-cured materials.

typical of a toughened, rigid plastic with a much higher modulus and a lower (<40%) ultimate elongation. M-DIPA materials show a gradual change from a highly extensible elastomer (MD52) to a tough ductile plastic (MD62). The M-MIPA material, MM56, is a very brittle polymer with a Young's modulus almost identical to MD56. MM46 is a soft material with a very low tensile strength. Fig. 2 also shows the effect of postcuring; increases in Young's modulus and ultimate tensile strength are apparent whilst the effects on ultimate elongation are less pronounced. In PUU's of low HS content (<55%), the polyether-urethane is the continuous phase. Above 55%, however, these PUU's tend to phase invert and the polyurea hard segments become the continuous phase. This behaviour is illustrated in fig. 3 which shows a clear transition in the modulus-HS plots (see dashed line) for DETDA systems. Increases in modulus following postcuring are also evident for all materials.

ACKNOWLEDGEMENT

The authors acknowledge the kind gift of the aromatic diamine chain extenders DETDA, M-MIPA and M-DIPA from Lonza Ltd., Basle, Switzerland.

REFERENCES

1. L.J. Lee, Rubber Chem.Technol. 53, 541, (1980)

2. D. Nissen and R.A. Markovs, J.Elast.Plast. 15, 96, (1983)

3. R.B. Turner, H.L. Spell and J.A. Vanderhider, Vol. 18. Polymer Science and Technology; Ed. J.E. Kresta, p.63 (1982)

4. R.E. Camargo, C.W. Macosko, M. Tirrell and S.T. Wellinghoff, Polymer, 26, 1145, (1985)

5. N. Barksby, D. Dunn, A. Kaye, J.L. Stanford and R.F.T. Stepto, ACS Symposium Series (172) Ed. J.E. Kresta, p.83 (1985)

6. U.K. Patent 1, 534, 258, (1978)

INFLUENCES OF PROCESSING ON THE ELECTRICAL DEGRADATION OF QUARTZ-FILLED EPOXY RESIN

J.J. Smit, W.S.M. Geurts and R. Ross
Department of Physics
N.V. KEMA
Utrechtseweg 310
6812 AR ARNHEM
The Netherlands

ABSTRACT

The electrical endurance of quartz-filled anhydride-cured epoxy resin has been investigated in view of high voltage applications. Process-property correlations have been studied by comparing the influence of processing modifications upon electrical endurance, as well as on dielectrical and mechanical properties. Significant influences on a.c. endurance are curing agent concentration, mixing time, gelation temperature and wet storage conditions, whereas cooling rate, postcure and filler concentration have marginal effects.

INTRODUCTION

In The Netherlands long distance electricity transport operates at high voltages up to 380 kV. For the insulation a long service life of 20 years is often required. Because of unpredictable cases of long term breakdown more knowledge about electrical ageing phenomena is necessary. The way in which ageing is influenced by fluctuations in the production process of insulators is of practical interest.

This report focusses on the electrical endurance of a conventional resin-system, which is widely used for indoor high voltage insulators, for example in metal enclosed gas-insulated switchgear. The investigation concerns the intrinsic material, whereby casted samples instead of real insulators are studied.

MATERIALS AND METHODS

The investigated resin system is composed of bisphenol-A diglycidylaether (100 pbw) cured with phtalic anhydride (30 pbw) and as a filler ground quartz-flour (200 pbw) with a median particle size of 14 μm. Using commercial grades of components, batches of 0.01 m^3 have been produced with several modifications, which are listed in table 1.

The standard way of processing has as main features: preconditioning of components, mixing and degassing of the quartz/resin melt (2h/150°C/1 mbar), reaction stage (10 min/135°C/20 mbar), vacuum casting (140°C/20 mbar), gelation (3,5h/146°C), demoulding, postcure (9h/145°C) followed by cooling at a rate of 5 K/h.

For the electrical testing, series of identical samples have been produced, with a recessed test layer (see insert in fig. 1) and vacuum deposited thin gold electrodes.

Table 1

Listing of investigated modifications

case nr.	modified parameter	standard value	modified value
1	none	–	–
2	cooling rate (K/h)	5	25
3	cure schedule (h-°C)	12.5-145	48-125
4	postcuring temperature (°C)	145	120
5	filler (p.h.r.*)	200	0
6	curing agent (p.h.r.)	30	40
7	mixing, degassing time (h)	2	0.5
8	gelation temperature (°C)	146	150
9	storage conditions	dry	wet**
10	filler (p.h.r.)	200	240

* parts per 100 parts by weight of resin   ** 2 years at 100% R.H.

To measure the 50 Hz-electrical lifetime curve (electrical field-strength at breakdown versus time to breakdown) over a wide range of time two experimental acceleration methods have been used. Firstly progressive a.c. stress levels with rates of rise between 10^{-6} and 10^{-1} $MVm^{-1}s^{-1}$ have been applied. Secondly in some experiments the frequency of the high voltage has been increased to 500 Hz. More details can be found in [1].

Furthermore dynamic mechanical thermal analysis (DMTA) and dielectrical thermal analysis (DETA) have been carried out over a temperature range from -150 to +160°C using Polymer Labs apparatus. DMTA runs at 1 Hz and 2 K/min with sample sizes of $35 \times 10 \times 1$ mm^3. DETA is operated at 4 K/min, however 5 frequencies at the time have been measured (1, 3, 10, 30, 100 kHz). The samples ($1 \times \emptyset$ 40 mm) have vacuum deposited gold electrodes with a guard ring configuration.

Both techniques show the α and β relaxation phenomena, which are attributable to the mobility of respectively the polymeric main chain segments and certain side groups [2,3].

RESULTS

The accelerated endurance tests, described above, have been applied to 140 samples of case 1. Using the empirical power law dependence [1] figure 1 shows the statistical result. From the 500 Hz data the equivalent 50 Hz time-to-breakdown has been plotted, roughly assuming an acceleration factor of roughly 10. The life-time curve shows a remarkable decline after 10^5s. The sections before and after the decline have been fitted to empirical life-time equations [1]. In the middle of the curve data points for 50 and 500 Hz join together nicely within the 68% confidence limits. For testing times above 120 days only 500 Hz data are available. After 3 years (50 Hz equivalent) the breakdown stress goes down to 8 MV/m, thus approaching the service stress (3-5 MV/m). From breakdown statistics it has been shown that electrical ageing is present in the second region [1].

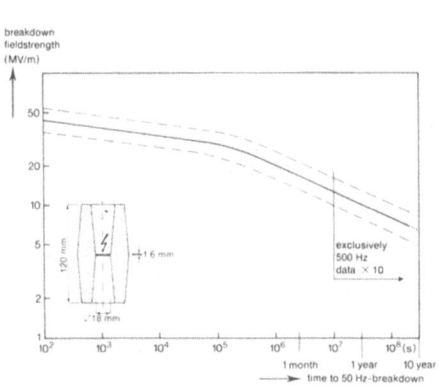

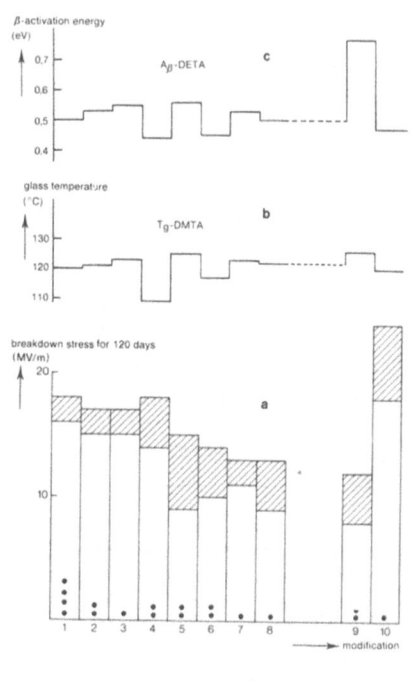

Fig. 1   Electrical life-time curve
         with 68% bandwidth (dashed)

Fig. 2   Influences of modifications
         on T_g, A_β ad electrical
         breakdown fieldstrength with
         68% bandwidth (dashed) and
         number of samples: 24 per
         circle

To study the long term differences on electrical ageing it is necessary to measure breakdown beyond the decline in the life-time curve. For that purpose progressive 50 Hz tests have been carried out for different cases (table 1). From these results the breakdown fieldstrengths for 120 days stressing time have been derived, in a way similar to figure 1. The values are shown in figure 2a in a sequence, arranged in order of descending breakdown levels, except for case 10.

To study possible correlations, some results from DMTA and DETA measurements are shown in figure 2b, c. Since for many applications polymers are selected on basis of their glass temperature, the α-transition temperature from DMTA is included (plotted as T_g). Furthermore the β-activation energy A_β has been obtained from the frequency dependence of DETA-scans. Figure 3 shows a typical example of one thermal run for a sample of case 2. The high temperature maxima in $\tan\delta_{el}$ concern the α-transition, while at low temperatures the β-transition occurs. From the Arrhenius plot, the activation energy of $A_\beta = 0.53$ eV can be calculated. The Arrhenius plot does not apply to the co-operative phenomena at the α-transition, however the frequency interval is too small to observe deviations from linearity (the W.L.F. theory describes this).

Taking into account estimated errors of 2K and 0.04 eV for T_g and A_β respectively, cases 4, 5 and 9 have significant changes in T_g and only case 9 changes A_β significantly. For the present the other effects remain marginal trends.

Considering the breakdown results in more detail, it can be observed from figure 2 that only cases 5 to 9 differ significantly from case 1. The fieldstrength values decrease and the largest effect has a 2 year storage at 100% R.H. (case 9). For this case the DMTA and DETA data are determined after the conditioning. The T_g-value increases 6K and A_β shifts 0.3 eV upwards. In addition it is noted that T_β, measured by DETA at 100 kHz, has been increased to -3°C. In a consecutive DETA-scan T_β decreases to -24°C, which comes closer to the T_β value for case 1 of -42°C at the same frequency. On the other hand the mechanical β transition determined by DMTA (1 Hz, first scan) is not affected: T_β = -66°C for case 9 and -65°C for case 1.

The influence of composition is apparent, comparing the unfilled and over-filled resin system.

DISCUSSION

Processing effects on breakdown do not correlate to trends in the material properties shown: cases 7 and 8 have only small and opposite effects on T_g and A_β, while the reverse happens for case 4. The incomplete cure at 120°C of case 4 results in a lower degree of cross-linking. The higher main and sidechain mobilities decrease T_g and A_β (109°C and 0.44 eV respectively). There is no real effect on breakdown strength, however an insulator may fail mechanically. Process technology for case 7 is such that no optimal contact between quartz particles and resin is obtained. For case 8 demoulding after 3.5h at 150°C introduces mechanical stresses.

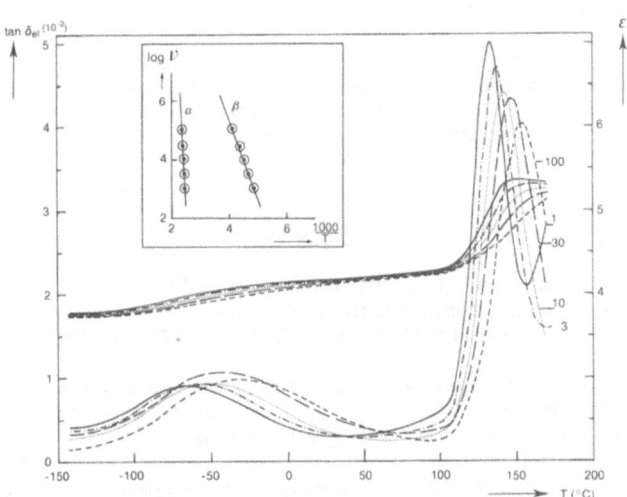

Fig. 3 DETA. Permittivity ϵ' and loss factor $\tan\delta_{el}$ at five frequencies, indicated in kHz. Insert shows Arrhenius plot.

The unfilled system (case 5) reaches higher cross-link density, more so as the exothermal reaction is not moderated by the filler. However, localized internal temperature differences and shrinkage stresses will increase, which may explain the larger scatter in breakdown data.

From reaction kinetics it is expected that for case 6 branching increases. This plasticizing effect decreases T_g and the dipolar activation level. Possibly electrical voidgrowth is easier in this structure.

Moisture that has been avoided in the other experiments, shows large effects for case 9. The location of the dielectric β-relaxation in the Arrhenius plot is characteristic for absorbed water [3]. Since there is no plasticizing effect on the mechanical relaxation temperature T_β, the relatively high dielectric T_β values are probably caused by a lowering of the freezing temperature of water. The solidified water-polymer system hinders the motions of polar groups, resulting in a high activation energy of $A_\beta = 0.77$ eV. The second DETA scan shows the return of the relaxational behaviour of the epoxy-matrix.

The presence of water-filled voids or channels, probably at the quartz/resin interface, is electrically important. The lower breakdown level confirms this.

Since breakdown itself is a localized phenomenon, models based on local heterogenities are optional. Sub-T_g annealing effects [2], microvoids near the quartz-resin interface [1] and impurities, e.g. water, may be involved.

CONCLUSIONS

Without adhering to a particular model, the correlations between the electrical and material properties form a complex and unresolved network. Across it the breakdown results give a selection of effective influences from processing.

As far as the number of samples allows to conclude, the long term breakdown strengths are easier decreased than increased by the modifications. In general no drastic effects on correlated properties have been observed, except for moisture contamination.

REFERENCES

1   Smit, J.J. and Ross, R., "Manufacturing influences on the H.V.-A.C. endurance of epoxy resin insulation", Symp. "New and improved materials for electrotechnology", Vienna, 5-7 May 1987, CIGRE, Paris

2   Smit, J.J. and Geurts, W.S.M., "Endurance tests on a quartz-filled epoxy resin system", 4th Int. Conf. on Diel. Mat. Meas. and Appl., Lancaster, September 1984, IEE 239 (1984) 68

3   Hedvig, P., Dielectric spectroscopy of polymers, Hilger, Bristol (1977), 293 & 332.

DEFORMATION AND BREAK-UP OF DROPLETS IN ELONGATIONAL FLOW

C. van der Reijden-Stolk, A.S. van Heel,
J. Schut and J. van Dam,
Laboratory of Polymer Technology,
Delft University of Technology,
Julianalaan 136, 2628 BL Delft,
The Netherlands

ABSTRACT

Polymer blends and other two-phase systems are formed in blenders, in which shear and elongational flows induce deformation and break-up of the dispersed phase. For elongational flow, stationary as well as time-dependent, theory predicts deformation and break-up quite well. The shape of the deformed particles, however, does not agree with theoretical calculations.

INTRODUCTION

Polymer blends are produced by mixing incompatible polymers in the liquid state. The gap between industrial blending operations and their theoretical description is stil very wide. One reason is the complex rheological behaviour of polymer melts. Another problem is that in real blenders shear as well as elongational flow fields are present with varying and often time-dependent rates.
In this contribution experiments on the deformation and break-up behaviour in elongational flow of several model systems are reported. The results are compared with existing and newly developed theories.

ELONGATIONAL FLOW FIELDS

An elongational flow field can easily be created by means of a contraction. Contractions are often encountered in real blenders. One of the most simple geometries is a cone (Figure 1a). However, the interpretation

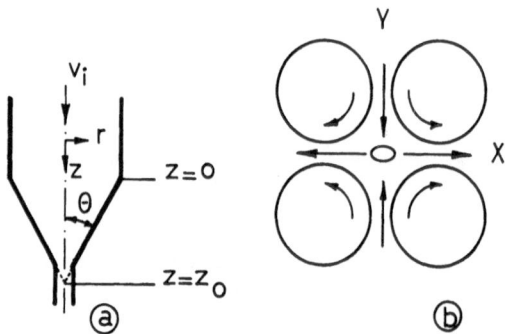

Figure 1. (a): Cylinder-cone set-up; (b): Four-roller apparatus.

of the results is complicated by the fact that the elongation rate, $\dot{\varepsilon}$, is
not constant. As a result the deformation of the dispersed particles is
time-dependent, it lags behind the change in elongation rate. At the other
hand this makes the cone very well suited for studying time-dependent
behaviour. Purely elongational, extensional flow is only present at the
center line. Here the rate is given by:

$$\dot{\varepsilon} = \frac{\partial v_z}{\partial z} = \frac{2\,v_i}{z_o(1 - z/z_o)^3} \tag{1}$$

where v_i is the velocity in the cylindrical section.
This elongation rate can also be written as a function of time t, elapsed
since entering the cone at z = 0:

$$\dot{\varepsilon} = \frac{2}{3\,(t_o - t)} \tag{2}$$

in which t_o is the time needed to travel through the cone from z = 0 to
z = z_o.
Because the cone is not infinitely high a correction should be applied for
the discontinuity between the cylindrical and conical part [1].
A stationary elongational flow field is not so easy to realize
experimentally. The classical solution is the four-roller apparatus
(Figure 1b). In the stagnation point and its immediate surroundings a

hyperbolic flow field exists of which the components are:

$$v_x = \dot{\varepsilon}\, x \quad , \quad v_y = -\dot{\varepsilon}\, y \quad , \quad v_z = 0 \tag{3}$$

DEFORMATION AND BREAK-UP THEORIES

In the course of a blending process the length scale of the dispersed phase decreases. Due to the increasing surface between the phases the influence of interfacial tension becomes more important. Finally an equilibrium situation is reached in which the droplets are small enough to withstand the disruptive hydrodynamic forces. The complexity of the equations which govern deformation and break-up is such that no complete analytical solutions are available. All analytical approaches are based on perturbation methods. Starting point is a spherical drop for which the internal and external flow fields are calculated. From these flow fields the normal force difference at the surface is obtained and equated to the force resulting from interfacial tension. In this way one obtains the shape of the deformed particle in first order approximation. The procedure described above can be repeated for the now non-spherical drop, it then leads to a second order approximation of the droplet shape.

As early as 1932 Taylor [2] derived a first order theory for systems consisting of Newtonian droplets in a Newtonian continuous phase, subjected to shear and hyperbolic flows. His result for the latter type of flow reads:

$$D = We \ \frac{19\ p + 16}{8\ (p + 1)} \tag{4}$$

where D is a measure for the deformation defined by $D = (L - B) / (L + B)$ (see Figure 5). The deformation D is proportional to the Weber (or Capillary) number $We = \eta_c \dot{\varepsilon} R / \sigma$; here η_c is the viscosity of the continuous phase, R the radius of the initial, spherical droplet and σ the interfacial tension. The viscosity ratio $p = \eta_d / \eta_c$ (η_d = viscosity of the dispersed phase) plays a minor rôle, the factor in D containing p varies between 2 and 2.375.

A first order approximation for the deformation of purely elastic particles in a Newtonian matrix can be derived from a theory recently published by Brunn [3] . For hyperbolic flow the result is:

$$D = \frac{15}{2}\, We_E \tag{5}$$

where $We_E = \eta_c \dot{\epsilon}/E$ is the elastic Weber number; E is the modulus of elasticity.

It can be shown that the deformation as a result of a stepwise change in the elongation rate behaves exponentially:

$$D = D_\infty \left(1 - e^{-t/\tau}\right) \qquad (6)$$

where D_∞ is the equilibrium deformation reached after infinite time. Theories for the time constant τ exist for Newtonian/Newtonian systems [4] and elastic/Newtonian systems [3] . The retardation times are given by:

$$\tau = \frac{19\ \eta_d\ R}{20\ \sigma} \quad \text{and} \quad \tau = \frac{9\ \eta_c}{2\ E} \qquad (7)$$

respectively. Application to cone flow is possible by applying the superposition principle.

Break-up theories are still scarce. The only analytical approach is Taylor's theory [5] which states that break-up occurs when the normal force difference exceeds the force due to interfacial tension. This leads to a deformation at break D = 0.5 if one neglects deviation from the spherical shape of the dispersed drop.

EXPERIMENTAL RESULTS

Model systems were studied with simple rheological properties: Newtonian, elastic and shear thinning. The advantage of this approach is that the effect of different parameters can be determined separately. Parameters varied include flux, viscosity ratio, cone angle, interfacial tension, drop size and elasticity. All experimental evidence indicates that the results can be expressed in terms of the quantities introduced in the previous section. As an example, Figure 2 proves that the radius is not a parameter in itself , its influence is fully accounted for by the Weber number. Figure 2 also demonstrates that the second order approximation predicts the experimentally measured deformation to higher values of We than the first order approximation. Break-up occurs at D = 0.5, in accordance with Taylor's theory.

Figure 3 depicts results of stationary experiments on elastic spheres of widely different moduli of elasticity.

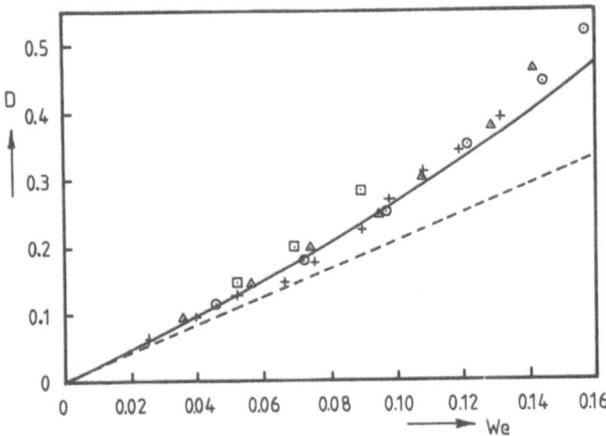

Figure 2. Deformation of Newtonian droplets (corn syrup) in a Newtonian
continuous phase (silicon oil), subjected to hyperbolic flow.
Radii varying between 0.2 (+) and 0.6 mm (⊡).
---: 1st order approximation, ——: 2nd order approximation.

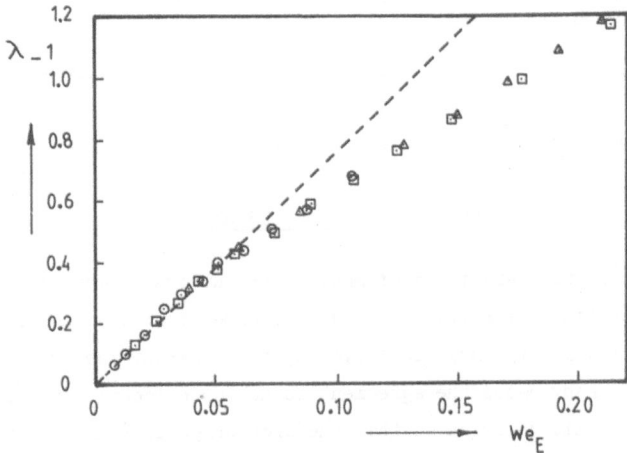

Figure 3. Deformation of elastic particles (gelatin/water) in hyperbolic
flow. Continuous phase : silicon oil. ▲: $E = 140$ N/m^2,
⊡ : 690 N/m^2, ⊙ : 1400 N/m^2.

In this case not D, but λ-1 is plotted as a measure of deformation, in which λ = L/2R is the stretch. The (first order) theory predicts the experimental results up to values of λ-1 of 0.5. The deviation at higher deformation is probably largely due to strain hardening. In Figure 4 an example of a time-dependent deformation experiment is shown.

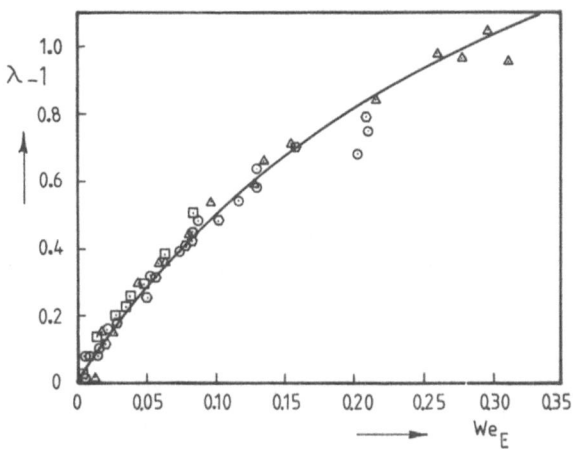

Figure 4. Deformation of elastic droplets (gelatin/water, E = 1500 N/m^2) in cone flow. Continuous phase: silicon oil. Fluxes were varied between $4\cdot10^{-7}$ and $19.5\cdot10^{-7}$ m^3/s.

Noticeable is the fact that for these time-dependent experiments the agreement between experimental and theoretical deformations holds to much higher deformations than for the stationary experiments.

DISCUSSION AND CONCLUSIONS

From our results and those of many other authors it appears that the theories describing deformation predict the experimental values far beyond their range of validity, sometimes even up to break-up. The question is, however, if in many cases the agreement is not coincidental or artificial. A weak point of all theories is that the drop shape is predicted very poorly. In practice one always finds almost perfect ellipsoids. An example of a calculated shape is depicted in Figure 5. This peculiar shape also explains why in some cases D is used as a measure of deformation and sometimes λ-1, the choice often depends on which one agrees better with the experimental values.

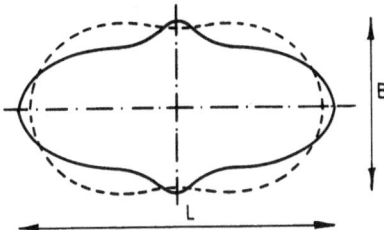

Figure 5. Theoretical shapes of Newtonian droplets in hyperbolic flow,
p = 1, We = 0.15. ---: 1st order approximation; ___: 2nd order
approximation.

In our opinion the perturbation method leads to unrealistic predictions for
values of D greater than 0.2. It is even questionable if it is a converging iteration method. For deformations above 0.2 a new theory is needed in
which the fact that the actual shapes are almost perfect ellipsoids is
incorporated.

REFERENCES

1. Van der Reijden-Stolk, C. and Sára, A., Deformation of Newtonian drops
   submerged in another Newtonian fluid flowing through a converging cone.
   Pol. Eng. Sci., 1986, 26, 1229-39.

2. Taylor, G.I., The viscosity of a fluid containing small drops of
   another fluid, Proc. Roy. Soc. (London), 1932, A138, 41-8.

3. Brunn, P.O., The deformation of a viscous particle surrounded by an
   elastic shell in a general time-dependent linear flow field,
   J. Fluid Mech., 1983, 126, 533-44.

4. Cox, R.G., The deformation of a drop in a general time-dependent fluid
   flow, J. Fluid Mech., 1969, 37, 601-33.

5. Taylor, G.I., The formation of emulsions in deformable fields of flow,
   Proc. Roy. Soc. (London), 1934, A146, 501-25.

Part 8

FIBRES/COMPOSITES

A NEW DEFORMATION MODEL FOR PETP FIBRES

M.G. Northolt and A. Roos
Akzo Research Laboratories
P.O. Box 60, 6800 AB ARNHEM
The Netherlands

ABSTRACT

The experimental relation between the birefringence and the dynamic compliance of PETP fibres is explained by two complementary versions of the series aggregate model. For the low-oriented fibres the classical series model provides a satisfactory interpretation. A model developed earlier for aramid fibres explains the experimental results for the well-oriented fibres. A further development of this model incorporating visco-elastic deformation shows that the creep and stress relaxation of these fibres are caused mainly by a progressive contraction of the chain orientation distribution.

INTRODUCTION

Attempts have been made to explain the extension of fibres made from semi-crystalline polymers, using various models. In the Takayanagi model [1] the two-phase nature of the fibre is emphasized, while the orientation distribution of the chains is not taken into consideration. On the other hand, a single phase is assumed in the series aggregate model developed by Crawford, Kolsky and Ward [2] [3]. In this model the fibre is considered to be built up of a series of elements. Each element is a transversally isotropic and elastic body of cubical shape having five elastic constants e_1, e_3, g, υ_{12}, υ_{13}, and whose values are those of the fully oriented fibre (for a definition of the symbols, see the end of this paper). According to this model the compliance of the fibre is given by

$$S = \frac{1}{e_1} \langle \sin^4\varphi \rangle + \frac{1}{e_3} \langle \cos^4\varphi \rangle + (\frac{1}{g} - \frac{2\upsilon_{13}}{e_3})\langle \sin^2\varphi\cos^2\varphi \rangle \qquad (1)$$

where the expressions in φ are the second and fourth moments of the chain orientation distribution. The interpretation of the relation between the birefringence Δn and S observed for PETP and polypropylene fibres was not successful, however [3].

A modified series model was developed by Northolt and Van der Hout describing the elastic extension of well-oriented aramid and cellulose fibres [4] [5]. In this model the elements have an oblong shape with the longest dimension parallel to the chain axis. Due to a different definition of the strain contribution of each element the compliance of the fibre is now given by a modified expression

$$S = \frac{1}{e_3} + \frac{\langle \sin^2\varphi \rangle}{2g} \tag{2}$$

This study will demonstrate that the two versions of the series model, as outlined above, can be used in a complementary way for the interpretation of the relation between Δn and S observed for a wide range of PETP yarns as shown in Figure 1. The curve unfolds itself into two branches: a linear relation for the well-oriented fibres and another linear relation with a smaller slope for the low-oriented fibres.

According to expression (2) the dynamic compliance S of a well-oriented fibre is proportional to the second moment of the orientation distribution of the chains. Hence, by monitoring S during creep and stress relaxation the change of the orientation distribution during these visco-elastic phenomena can be studied. It will be shown that the modified series model can be extended to incorporate visco-elasticity, yielding simple relations for the creep and the stress relaxation in these fibres.

THEORY

For the orientation distribution range

$$0.25 < \langle \sin^2\varphi \rangle \leq \frac{2}{3}$$

the fourth moment of gaussian and similar kinds of distributions may be approximated by

537

$$\langle \sin^4\varphi \rangle \sim \langle \sin^2\varphi \rangle - 0.15 \tag{3}$$

The Hermans relation for the birefringence is

$$\frac{\Delta n}{\Delta n_1} = 1 - \frac{3}{2} \langle \sin^2\varphi \rangle \tag{4}$$

where Δn_1 is the birefringence value of the cubical elements.
For $e_3 \gg e_1$ and g it follows from (1), (3) and (4) that

$$\frac{\Delta n}{\Delta n_1} \approx 0.77 + 0.23 \frac{e_1}{g} - \frac{3}{2} e_1 S \tag{5}$$

which applies only to the low-oriented fibres.

From the Hermans relation for $\Delta n/\Delta n_2$, where Δn_2 is the birefringence of the oblong shaped elements in the modified series model, and relation (2) it follows for well-oriented fibres that

$$\frac{\Delta n}{\Delta n_2} = \left(1 + \frac{3g}{e_3} \right) - 3gS \tag{6}$$

The assumption is now made that the creep strain of a well-oriented fibre is solely caused by shearing of the elements in the series aggregate model, resulting in a rotation of the chains towards the fibre axis. Then at a time $t > t_0$ the creep strain is given by

$$\varepsilon(t) - \varepsilon(t_0) = \frac{1}{2} \left(\langle \sin^2\varphi(t_0) \rangle - \langle \sin^2\varphi(t) \rangle \right) \tag{7}$$

Because at any time during creep the dynamic compliance is given by (2), it follows that

$$\frac{\varepsilon(t) - \varepsilon(t_0)}{S(t_0) - S(t)} = g \tag{8}$$

In a similar manner we have derived a relation for the stress relaxation

$$\frac{\sigma(t_0) - \sigma(t)}{S(t_0) - S(t)} = ge_3 \tag{9}$$

Note that no assumptions have been made with respect to the functional dependence on the time. The derivation holds for any kind of time-dependent shear deformation of the elements during creep and stress relaxation.

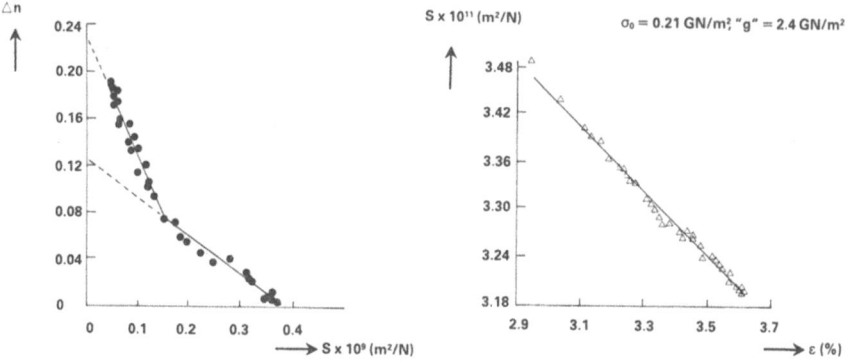

Fig. 1. Initial birefringence vs. initial dynamic compliance for PETP yarns.

Fig. 2. Dynamic compliance vs. strain of a well-oriented PETP yarn during creep caused by a load of 0.21 GNm^{-2}.

RESULTS AND DISCUSSION

Here only some exemplary experimental results are presented. The curve in Figure 1 is explained by the relations (5) and (6) of the two complementary versions of the series model as discussed above. Assuming that at high frequencies the elastic constants are the same for both kinds of elements the values derived for g and e_1 are 1.6 and 1.9 GNm^{-2}, respectively. X-ray diffraction patterns indicate that only above a birefringence value of about 0.06 the presence of oriented crystallites in the PETP yarns is observed, which appears to correspond with the point of inflection of the curve.

Figure 2 shows an example of the creep results. The straight line is predicted by relation (8); only a slightly higher value of 2.4 GNm^{-2} is observed for g. As will be shown in a future report this increase can be attributed to the fact that the derivation of (8) and (9) given here does not consider the effect of the finite width of the chains. An analysis of the shear deformation shows that there has to be some slip between adjacent chains. This effect is easily visualized by a row of books which is slowly falling over.

We believe that this investigation shows that visco-elasticity in well-oriented PETP and other (semi)crystalline fibres can be interpreted by means of a time-dependent shear deformation of oblong shaped elements. This approach explains the progressive contraction of the orientation distribution during creep and stress relaxation.

REFERENCES

1. M. Takayanagi, K. Imada and T. Kajiyama, J. Polym. Sci. C15 (1966) 263.

2. S.M. Crawford and H. Kolsky, Proc. Phys. Soc. B64 (1951) 119.

3. D.W. Hadley, P.R. Pinnock and I.M. Ward, J. Mat. Sci. 4 (1969) 152.

4. M.G. Northolt and R. van der Hout, Polymer 26 (1985) 310.

5. M.G. Northolt and H. de Vries, Angew. Makrom. Chem. 133 (1985) 183.

LIST OF SYMBOLS

e_1	modulus normal to the chain axis
e_3	modulus parallel to the chain axis
g	shear modulus in a plane parallel to the chain axis
υ_{12}, υ_{13}	Poisson ratio's
φ	orientation angle of the chain axis with the fibre axis
S	compliance along the fibre axis
Δn	birefringence
ε	strain
σ	stress
t	time

THE STRUCTURAL MECHANICS OF NYLON AND POLYESTER FIBRES : APPROACHES TO THEORETICAL UNDERSTANDING

J.W.S. Hearle, R. Prakash, M.A. Wilding
Department of Textiles
U.M.I.S.T.
Manchester M6O 1QD
England

and

H.A. Davis
Textile Fibers Department
E.I. Du Pont de Nemours
Wilmington, DE, 19898
U.S.A.

ABSTRACT

Despite their commercial importance and scientific origins, nylon and polyester fibres are poorly understood in their structural mechanics. Views of structure are discussed, and the parameters necessary to characterise a basic structural model, consisting of crystals and connectors are given. The mechanical behaviour of components are indicated, and the limitations of a two-phase composite approach are referred to. An outline is given for a computational treatment of a connector network, involving minimisation of chain energies and volume energy. This is yielding promising predictions of properties above the glass transition temperature.

UNSOLVED PROBLEMS

It is over 50 years since nylon fibres were first made, to be followed a decade later by polyester. Since then, these materials have moved through a succession of stages of scientific advance, technical innovation, and industrial activity: from expensive "miracle" fibres, through a wide range of highly profitable products made by a few companies, to a common commodity with all the problems of a mature industry dispersed throughout the world. But although these advances came from science laboratories, and were carried on in an engineering environment, it is remarkable how little is known about the structural mechanics of the fibres.

During the same 50 years, quite detailed models of the structure of the natural fibres, cotton and wool, have been established with reliable quantitative specification; and their stress-strain curves have been predicted from mathematical analysis of the mechanics of the models [1-3]. However, with these natural fibres, there is only limited opportunity to change the fine structure and properties. In

the melt-spun synthetic fibres, the situation is reversed :
great diversity of structure and properties can be achieved by
varying the thermo-mechanical sequence from melt to commercial
fibre; but we have no clear view of the form of fine
structure, and there have been little more than crude
qualitative accounts of the mechanics of deformation. Another
contrast is the fact that, more recently, the structures of
some of the highly oriented rigid chain polymer fibres have
been well worked out, and the mechanics clearly analysed (4).

Two reasons for the lack of progress in understanding the
mechanics of nylon and polyester fibres are: (1) the
difficulty of characterising stuctures which lie midway
between order and disorder; (2) the lack of a strong applied
mechanics presence in polymer research. Consequently the
industrial control of product has been through empirical
variation of the melt-spinning and subsequent processes,
without the benefit of rational scientific understanding, such
as is found, for example, in relation to the process of
polymerisation. There are two motivations for attempting to
change this. The first is that the ignorance is an offence and
a challenge to a scientist. The second is the potential value
of a more knowledgeable empiricism, or ultimately true
engineering design with calculation of what occurs. The
structure/property relations are central to the engineering
approach, though this would also have to include process
modelling to predict what structure formed and performance
modelling to determine the effect of properties.

STRUCTURE

The current uncertainty about fine structure is
graphically illustrated by comparing the two representations
of the structure of PET fibres wound up at 5000 m/min shown
by Heuvel and Huisman and by Shimizu, Okui and Kikutani in
the same book [5]. However, the most commonly accepted view
of the fine structure of nylon and polyester fibres consists
of a pseudo-fibrillar assembly of linked crystalline
micelles, although fibres which have not been annealed may
have a more uniform structure, which Hearle [6] has called a
dynamic crystalline gel.

Before the mechanics can be analysed, we need to define
a list of parameters which are necessary and sufficient to
describe the structure. Surprisingly, this has not been done
before. It is necessary to be simpler and more precise than
the usual subjective impressions, and a basic model is shown
in Figure 1. The following set of independent variables
describes the geometrical features of the simplest
reasonable model, which has features of uniform size, square
perpendicular to the axis of orientation:

```
polymer molecule - mass of repeat unit
                 - aspect ratio of repeat unit
                 - degree of polymerisation
density - crystal
        - amorphous
crystallite - mass   and Crystallinity
            - aspect ratio
fraction of folds
superlattice - aspect ratio
             -stagger (0.5 or 0)
```

Relatively simple equations enable other dependent variables
to be derived. Real fibres will show more complicated
shapes, distributions of values of the parameters and
imperfect orientation, but the model offers a suitable
starting point for analysis.

Figure 1. Schematic representation of basic structural
model, with typical connectors shown in two
dimensions instead of three.

Specification of the listed parameters gives the
complete location of polymer in the crystalline regions,
together with necessary information on the distribution of
positions from which chains emerge into amorphous regions,
the number of free ends emerging, and the mass of polymer in
the amorphous regions. However some important features of
the amorphous network remain to be settled, namely: the
pairing of positions among re-entrant chains and different
categories of connectors between crystals; and the
distribution of chain lengths among free ends, loose folds,
and the connectors.

MECHANICS

Estimates of the anisotropic elastic constants of the crystal lattice can be made. With moduli of the order of 100 GPa, the crystals can often be approximated as undeformable. There are three states for the amorphous material (with two transition zones). At very low temperatures, when freedom of chain rotation is inhibited, the material will be a glass with the modulus likely to be around 10 GPa. At room temperature, sequences of about 6 $-CH_2-$ groups in the polyamides or $- CO.O.CH_2CH_2.O.CO -$ in PET will be flexible, but effective cross-links between $-CO.NH_2-$ groups or benzene rings make the material a fairly highly cross-linked rubber, with a modulus around 0.1 GPa. Above the glass transition, the inter-molecular bonds will be mobile and the material will be freely rubbery, with a modulus around 0.01 GPa.

One approach to the mechanics is to treat the system as a two-phase composite in which each phase is a continuum with defined elastic constants. We have shown [7] that available approximate treatments can predict the right properties, but contain severe errors in the mechanics and are undiscriminating or false in regard to structural parameters. The alternatives are then either to use a more exact computational finite element approach to the mechanics of the composite system, or to relate more closely to reality by treating the system as a connector network. Certainly for the upper temperature state the latter approach is better. We have developed a computational method, involving energy minimisation, but only a brief outline can be given here.

Unless the undesirable assumption of constant volume is introduced, two energy forms must be considered. The first is the energy of chain extension. Starting from an assumed reference state, the stuctural model would give the lengths of connector chains L_i and the direct distance ℓ_i between the ends emerging from crystals, as illustrated in Figure 1. Rubber elasticity theory then gives the chain energy $U_c(L_i,\ell_i)$ for a specified size of equivalent random link and temperature. The second is the volume energy $U_v = \frac{1}{2}K[(V/V_0)-1]^2$, where K is the bulk modulus, V is the volume, and V_0 is the volume when $dU_v/dV = 0$, namely the volume which the material would take up if there were no forces acting within the chains. The actual stress-free dimensions, which will differ slightly from the starting reference state will be given by equations of the form $\partial U/\partial x_j = 0$, where $U = \Sigma U_c + U_v$ and x_j is an external dimension of the system.

In deformation, under forces, F_j, it is assumed that there is affine displacement of crystal locations, from which the changes in the set of values ℓ_i and the volume can be calculated. The force-deformation relations are then derived from equations of the form $F_j = \partial U/\partial x_j$.

CONCLUSION

A network approach seems to be the best way to handle
the structural mechanics of nylon and polyester fibres, and
our computational method is yielding promising results in
predicting properties above the glass transition.

However there are other effects to be considered beyond
what is outlined in this paper. These include alternative
forms of basic structural model; the treatment of the system
at room temperature with intermolecular cross-links; and the
treatment of orientation. Last, but not least, there is the
need to improve experimental techniques to determine values
of the relevant parameters.

ACKNOWLEDGEMENTS

The financial support of Du Pont in a jointly funded
research project with UMIST is gratefully acknowledged,
together with valuable discussions with others working on
the project at UMIST and scientists from Du Pont.

REFERENCES

1. Hearle, J.W.S. and Sparrow, J.T., Mechanics of the
    extension of cotton fibres, II. Theoretical modelling.
    J. Applied Polymer Sci., 1979, **24**, 1857-1874.

2. Chapman, B.M., A mechanical model of wool and other
    keratin fibres. Textile Res. J., 1969, **39**, 1102-1109.

3. Hearle, J.W.S., and Susutoglu, M., Interpretation of
    the mechanical properties of wool fibres. Proc. 7th Int.
    Wool Textile Res.Conf. Tokyo, 1985, **1**, 214-223.

4. Northolt, M.G. and Hout, R.v.d., Elastic extension of an
    oriented crystalline fibre. Polymer, 1985, **26**, 310-316.

5. Heuvel, H.M. and Huisman, R., Figure 12, page 310;
    Shimizu, J., Okui, N. and Kikutani, T., Figure 12, page
    444. In High-Speed Fiber Spinning, ed. A. Ziabicki and
    H, Kawai, John Wiley, New York, 1985.

6. Hearle, J.W.S., On structure and thermomechanical
    responses of fibres, and the concept of a dynamic
    crystalline gel as a separate thermodynamic state.
    J. Appl. Polymer Sci.: Appl.Polymer Symp., 1977,
    31, 137-161.

7. Hearle, J.W.S. Prakash, R and Wilding, M.A. Prediction
    of mechanical properties of nylon and polyester fibres as
    composites. Polymer, 1987, **28**, 441-448.

COMPRESSIVE FAILURE IN HIGH MODULUS POLYMERIC FIBRES

S. van der Zwaag and G. Kampschoer

Akzo Corporate Research Laboratories

PO Box 60, 6800 AB Arnhem, the Netherlands

ABSTRACT

The compression strength of aramid fibres is considerably higher than that of aromatic polyester or UHMW polyethylene fibres. In aramid fibres the compression strength is a function of the elastic modulus but does not depend on the tensile strength.

INTRODUCTION

Axial compression of well-oriented, high modulus polymeric fibres results in the formation of so-called kinkbands [1-3]. These kinkbands are very narrow regions of (sub) micron thickness where extensive crystallite or molecular rotation takes place. This is the normal compressive failure mode in anisotropic materials. In this work the compressive strength, i.e. the stress at which kinkbands start to form, of an UHMW polyethylene fibre, an aromatic polyester fibre and an poly(p-phenylene terephthalamide) or aramid fibre, all having the same tensile properties are compared. The compressive strength of aramid fibres is studied in more detail as a function of the tensile strength and the elastic modulus.

EXPERIMENTAL

In this work the 'elastica' test was used to determine the compressive strength. It is an extremely simple test where a filament is wound into a single loop, which is then gradually contracted. During contraction the lengths of the two axes, c and a, are measured using an optical microscope. For an uniformly deforming fibre the c/a ratio has a constant value of 1.34 [4]. At the onset of compressive failure the c/a ratio increases: the kinkbands at the top of the loop act as plastic hinges. Using simple elastic bending beam theory the following equation for the compressive strength can be derived :

$$\sigma_{compression} = 2.68 * E * r / c_{crit} \quad\quad (1)$$

where E is the elastic modulus, r the fibre radius and c_{crit} the length of the c-axis at which the c/a ratio starts to deviate from the elastic value.

RESULTS

Comparison of Aramid, Aromatic Polyester and Polyethylene Fibres.

A typical result of the elastica test on a medium modulus aramid fibre is shown in figure 1. In this figure the ratio of the major and minor axes is plotted versus the length of the major axis. The individual data points are indicated by an * or o depending on whether kinkbands were observed or not. The figure shows that the c/a ratio remains constant up to a c-value of 2.1 mm and then increases rapidly. The increase in the c/a ratio occurs just before the first observation of kinkbands. The first kinkbands appear at the inner side of the top of the loop. A further contraction of the loop leads to the formation of new kinkbands away from the top and existing kinkbands grow towards the centre of the filament and become more visible. The kinkbands are rather evenly distributed along the top of loop. No signs of tensile failure were observed.

The results for the aromatic polyester fibre (figure 2) are qualitatively the same as for the aramid fibre. Hoewever, the loop dimensions at which the first kinkbands were observed were considerably larger, indicating a much lower compressive strength. Elastica tests with the polyethylene fibre were less successful due to the very low compressive strength of polyethylene, which resulted in the formation of kinkbands at very large loop sizes, where no accurate measurements could be made.

Using equation 1 the compressive strengths were calculated from the critical loop dimensions. The results for aramid, aromatic polyester and polyethylene were 0.7, 0.2 and 0.05 GNm^{-2} respectively. It should be stressed that these large differences in failure stress were not found in the tensile testing of the fibres. All three types of fibre had a modulus of 90 GNm^{-2} and a tensile strength of about 3 GNm^{-2}. Evidently there is no simple relation between compressive and tensile strength.

Compressive strength versus tensile strength in aramid fibres

This aspect was investigated further for a set of aramid fibres having the same modulus but different tensile strengths, ranging from 2 to 4 GNm^{-2}. All fibres had a regular and uniform microstructure. The results are plotted in figure 3, which indicates that for this wide range of tensile stresses the compressive strength is surprisingly constant at 0.73 $\pm$ 0.05 GNm^{-2}.

Compressive strength versus elastic modulus for aramid fibres

The elastica test was also applied to a range of aramid fibres of about equal tensile strength but with different elastic moduli. The results are plotted in figure 4. The compressive strength increases with increasing modulus. For a low modulus fibre the compressive strength is about 0.5 GNm^{-2} and for a high modulus fibre about 0.9 GNm^{-2}. On the other hand, the critical strain for kinkband formation decreases continuously from about 1 % for a low modulus fibre to about 0.5 % for a high modulus fibre.

DISCUSSION

The results on aramid, polyester and PE fibres have shown that the compressive strength values can differ widely between different polymeric fibres, notwithstanding comparable tensile properties. This is due to the combined effects of differences in bending stiffness of the chains and differences in lateral cohesive binding forces. In the case of aramid fibres we have an intrinsicly stiff molecule and strong lateral cohesion due to a regular network of hydrogen bonds. In polyethylene fibres we have a flexible chain and only relatively weak Van der Waals forces between the chains. In the same way the higher compressive strength of inorganic fibres such as glass, carbon and Al_2O_3 fibres can be attributed to the strong covalent bonds in the lateral directions. However, in contrast to polymeric fibres, compressive failure in these brittle fibres leads to complete failure with no residual tensile strength left. Earlier work [3] has shown that in the case of aramid fibres the loss in tensile strength due to the presence of kinkbands is only of the order of 10%.

The lack of correlation between tensile and compressive strength in aramid fibres suggests that these failure processes depend on different microstructural factors.

The dependence of the compressive strength on the elastic modulus can be explained on the basis of a critical shear stress. The shear stress τ in a desoriented crystallite due to a stress σ is given by

$$\tau = \sigma * \sin \Phi * \cos \Phi \tag{2}$$

The modulus E is related to the average desorientation angle of the crystallites through the following relation

$$1/E = 1/e_3 + \langle \sin^2 \Phi \rangle /2 \; g \tag{3}$$

where e_3 is the crystallite modulus and g the shear modulus. Combination of (2) and (3) leads to the following expression for the compressive strength

$$\sigma_{comp} = \tau_{crit} * \{ 2 \; g \; (1/E - 1/e_3) \}^{-1/2} \tag{4}$$

Application of this equation to the experimental data yields a value of τ_{crit} of 0.1 GNm^{-2}. This equation is only applicable for E-moduli less than 130 GNm^{-2}. The compressive strength of very high modulus fibres can be derived in the following model, which is also based on the critical shear stress concept. Consider two crystallites under an applied load F as indicated below

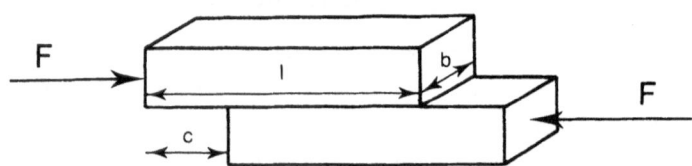

The equilibrium conditions yield : $\sigma = \tau * (1-c) / b$. Since $\langle c \rangle = 0.5 \; l$ and $l/b \approx 20$ a compression strength of 1 GNm^{-2} is calculated for very high modulus aramid fibres.

REFERENCES

1. Greenwood, J.H. and Rose, P.G., Compressive behaviour of aramid fibres and composites. J. Mater. Sci., 1974, 9, 1809-1814

2. Dobb, M.G., Johnson, D.J. and Saville, B.P., Compressional behaviour of Kevlar fibres. Polymer, 1981, 22, 960-965

3. DeTeresa, S.J., Porter, R.S. and Farris, R.J., A model for the buckling of extended chain polymers, J. Mater. Sci., 1985, 20, 1645-59

4. Sinclair, D., A bending method for measurement of the tensile strength of glass fibres, J. Appl. Phys., 1950, 21, 380-85

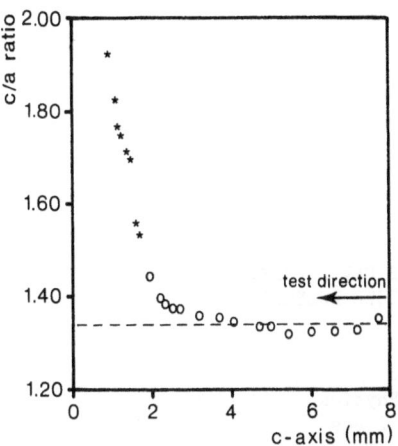

Figure 1: 'Elastica test' results
for an aramid fibre.
Dashed line indicates
elastic limit.
o : no kinkbands observed
* : kinkbands observed

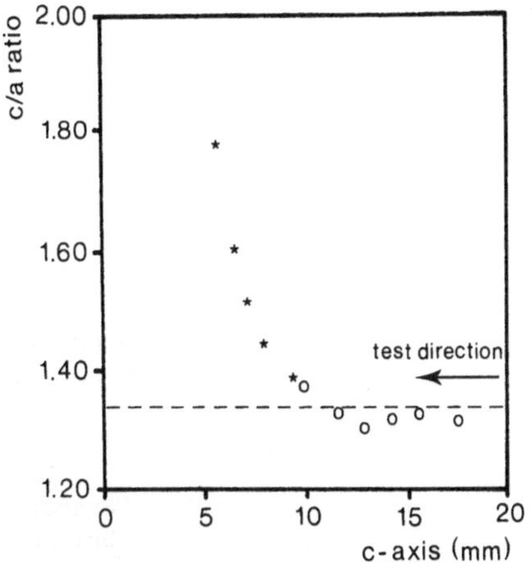

Figure 2: 'Elastica test' results
for an aromatic polyester
fibre. Dashed line indi-
cates elastic limit.
o : no kinkbands observed
* : kinkbands observed

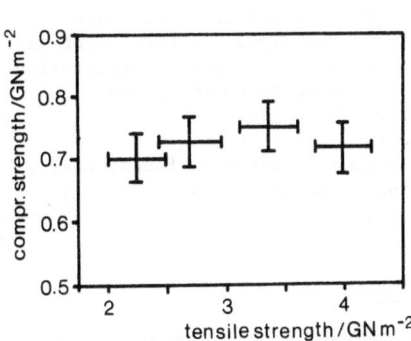

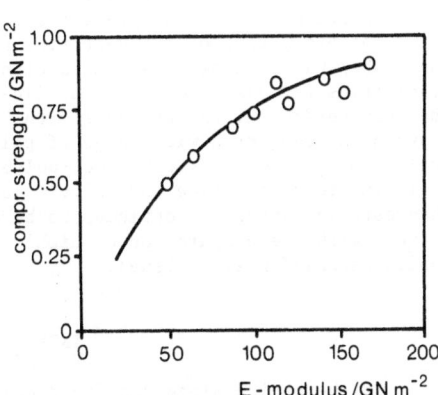

Figure 3: Compressive strength of
aramid fibres versus the
monofilament tensile
strength.

Figure 4: Compressive strength of
aramid fibres versus the
elastic modulus.

RECENT PROGRESS IN THE DEVELOPMENT
OF HIGH MODULUS FLEXIBLE POLYMERS

Ian M Ward
Department of Physics
University of Leeds
Leeds LS2 9JT
UK

ABSTRACT

Recent research at Leeds University on high modulus polymers is
reviewed. With the development of practical processes for melt spun and
drawn ultra-high modulus polyethylene fibres, attention has been directed
to improvements in creep behaviour, especially by electron beam
cross-linking treatment, and in tensile strength by changes in molecular
weight. The high ductility of the fibres suggests potential applications
in fibre reinforced composites. Guidelines have been established for the
effectiveness of surface treatments designed to improve fibre/resin
adhesion, and a wide range of composites produced, in some instances
incorporating both polyethylene fibres and either glass or carbon fibres.
The production of solid-section high modulus materials (rod, sheet
and tube) has seen major advances with the development of large scale
facilities for die-drawing. The die-drawing technique has been shown to
be very versatile, and products with both uniaxial and biaxial orientation
have been made for a wide range of polymers including polyethylene,
polypropylene, polyoxymethylene, polyethylene terephthalate,
polyvinylidene fluoride and polyvinylchloride. The technique of
hydrostatic extrusion continues to be of some interest at a fundamental
level, with the preparation of highly oriented materials from
chain-extended polyethylenes.

INTRODUCTION

In the period since the first Rolduc Polymer Meeting two years ago,
research at Leeds University on high modulus polymers has developed in two
major directions. First, with the licensing of the melt spinning/drawing
process for ultra high modulus polyethylene fibres, co-ordinated activity
on the fibre process has been linked with the requirement to establish
firm guidelines for improvements in creep behaviour and tensile strength.
The application to fibre reinforced composites has also emerged as a firm
possibility, so that the preparation and properties of such composites is
being actively studied. The development of surface treatments to improve
fibre/resin adhesion is a key area of research.

Secondly, die-drawing has become established as a viable process for the manufacture of large section materials. In this case the technology is not limited to polyethylene, but commercially important degrees of property enhancement can be achieved in a very wide range of polymers. This technology is now also being developed under licence from the British Technology Group (presently by BP Chemicals PLC and Metal Box PLC in UK).

TENSILE DRAWING

During the last few years research on the tensile drawing of polyolefins to produce high modulus fibres has been consolidated and extended in terms of new routes to spectacular materials and the development of practical commercial processes.

In terms of fundamental understanding, it is now well-established that the key concept is the stretching of a molecular network, and to obtain the high draw ratios required for high modulus and strength a comparatively low concentration of physical entanglements is required. This was recognised in the melt-spinning/drawing route by choice of molecular weight and initial thermal treatment [1] and in the gel spinning/drawing route by production of a gel from a polymer solution [2]. Both these routes now form the basis for commercial developments, the former by Celanese and SNIA FIBRE, and the latter by DSM/Toyobo and Allied Chemical.

Recent developments at a research level include spectacular properties in terms of modulus and strength obtained by Porter and co-workers [3] using co-extrusion of polymer crystal mats, and the drawing of in situ polymerized films (virgin ultra high molecular weight polyethylene) by Smith and co-workers [4]. Both these discoveries emphasise that alternate routes can exist to obtain the optimum initial structure for high draw. It has also been found possible to extend the gel spinning/drawing route to produce polypropylene films of outstanding stiffness and strength, somewhat similar research being undertaken independently by Peguy and Manley [5], and by Matsuo and co-workers [6].

Finally, there is the interesting development by Mitsui Petrochemicals of a melt spinning/drawing process based on blending polyethylene of weight average molecular weight 700,000 with low molecular weight paraffin of molecular weight 500 in the proportion 70:30. This process aims at retaining the advantages of melt spinning, and at the same time making a product of high strength as well as high modulus.

To date, the tensile drawing route for flexible polymers has only shown spectacular results for polyethylene, polypropylene and polyoxymethylene, although high draw can lead to enhanced properties for many polymers. For example, in polyvinylidene fluoride an increased piezoelectric response can be obtained. As will be discussed later, these more modest improvements are often more valuable for solid section products such as sheet or tube, whereas in fibres very great enhancement is required, because the comparison is often to be made with glass, carbon and Kevlar. It is therefore of some interest to end this section with a league table (Table 1) showing the mechanical properties of a range of oriented fibres as well as glass and carbon fibre.

TABLE 1

Properties of reinforcing fibres (room temperature)

Property Fibre	Tensile Modulus (GPa)	Tensile Strength (GPa)	Elong- ation at Break %	Density ρ g/cm³	Specific Modulus GPa/ρ	Specific Strength GPa/ρ
Carbon	250	3.6	1.5	1.80	139	2.0
Glass	75	3.0	2.5	2.54	30	1.2
Kevlar 49	125	3.0	3.0	1.45	85	2.1
Polyethylene (melt spun)	40-100*	1-1.5**	4-18**	0.96	42-104	1-16
Polyethylene (gel spun)	100	2.5**	5	0.96	104	2.7

* Dependent on draw ratio   ** Dependent on strain rate

Creep Behaviour

The creep behaviour of oriented polyethylene has been studied in
detail, because permanent flow or even extensive recoverable creep can be
a severe limitation on the use of high modulus polyethylene in permanent
and intermittent loading situations.   It was shown by Wilding and Ward
[7] that a most useful way of examining the creep data for oriented
polyethylene is to plot the creep rate as a function of total creep strain
[8].   Wilding and Ward [9] showed that these constant creep rates (termed
plateau creep rates $\dot{\epsilon}_p$) were affected by stress and temperature in the
manner expected for a single thermally activated process.   Moreover the
activation energy of this process was about 30 kcals/mole which is very
similar to that found for the α-relaxation process in polyethylene, and
the activation volume was in the range of 100Å which could correspond to
the movement of a Reneker-type defect in the crystalline regions.   It was
further shown that there is a direct correspondence between the creep and
yield behaviour of oriented polyethylene.   In practical terms, the high
temperature drawing process, which is so effective in producing highly
oriented material, relates to crystal slip processes which can then be a
source of weakness in the final product properties.   Just as the drawing
behaviour, however, relates not only to the crystal slip processes, but
also to the stretching of a molecular network, the creep behaviour can
also be regarded as relating to two thermally activated processes acting
in parallel.   Furthermore at low stresses, where the creep behaviour will
be of most relevance, the network process will have the predominant
effect.

The modelling of the creep behaviour of polyethylene in terms of two
thermally activated processes has recently been extended to polypropylene
with equivalent success [10].   Again the smaller activation volume
process can be identified with the α-relaxation process, and there are
comparable effects due to draw ratio and molecular weight.   The major
interest, however still remains with polyethylene, where the situation for
a series of melt spun and drawn fibres can be best illustrated by
examining plots of plateau creep rates $\dot{\epsilon}_p$ versus applied stress.   As shown

in Figure 1, a set of approximately parallel curves results.   The lowest
molecular weight homopolymer shows the worst behaviour, in terms of high

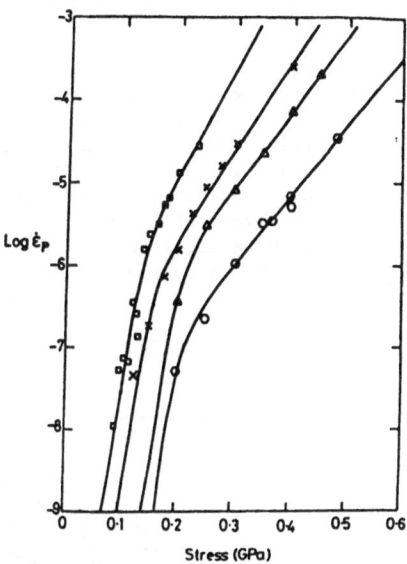

Figure 1.   Effect of molecular weight and cross-linking on
             creep behaviour. ☐ Low molecular weight
             x Intermediate molecular weight
             ○ Higher molecular weight
             Δ Cross-linked before drawing.
             Reproduced from J.Polym.Sci., Polym.Phys.Edn.,
             22, 561, (1984) by permission of the publishers,
                   John Wiley & Sons, Inc. (C)

creep rates at high stresses, with the creep rate only becoming negligible
at a very low stress level.   Increasing the molecular weight, introducing
a small degree of branching, and cross-linking the initial polymer before
drawing, all serve to improve the creep behaviour by providing a more
effective molecular network.

    For melt spun and drawn fibres, all these methods of improving the
creep behaviour have been shown to reduce the final draw ratios which can
be achieved (because the more effective molecular networks are of reduced
extensibility) and hence limit the property enhancement in terms of
initial modulus and strength.   Where solid section material is being
produced, for example by the die-drawing process discussed later, a
compromise between ultimate properties in terms of short time stiffness
and strength and long time response may be very acceptable.   One
illustration of a successful compromise is the development of Tensar grids
for soil reinforcement using a polyethylene copolymer.

    Where high levels of initial modulus and strength are mandatory, an
alternative approach is to introduce cross-linking after drawing.   It has
been shown that effective cross-linking of ultra-high modulus polyethylene
fibres can be achieved by electron-beam irradiation [11].   In general,

electron beam irradiation produces both cross-linking and chain scission, and the changes in chemical composition are very well described for monodispersed systems by the Charlesby-Pinner equations. Because chain-scission will lower the tensile strength of the fibres, it is essential that it is reduced as much as possible. In the first instance this can be achieved by irradiation in the absence of oxygen; more effectively the cross-linking can be accelerated by the use of chemical reagents. Figure 2 shows some comparative results for the electron beam irradiation of fibres in acetylene and in vacuum, together with results for unirradiated fibre. It can be seen that for vacuum irradiation, there is an improvement in the creep behaviour and low levels of stress but higher creep rates at high levels of stress. The introduction of acetylene not only reduces the dose level required to obtain significant cross-linking, it also produces an improvement in the creep behaviour at all levels of stress.

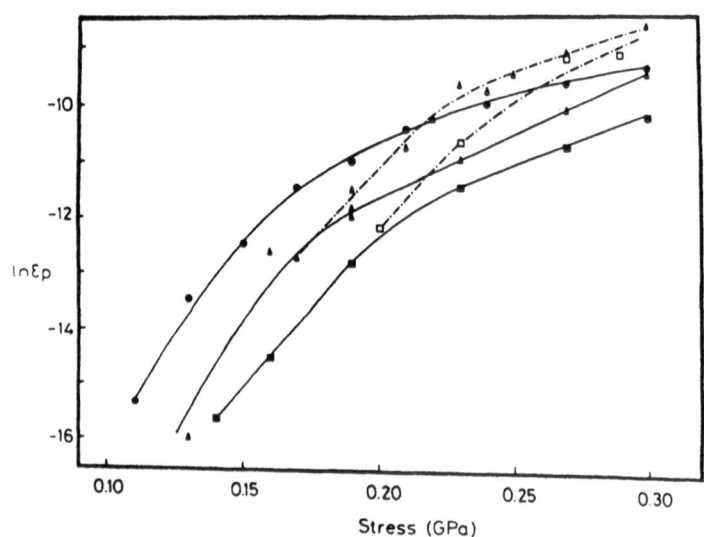

Figure 2.   ℓn ε̇ₚ versus stress for melt spun fibre, unirradiated (●), and irradiated with 9Mrad in vacuum (Δ), 17 Mrad in vacuum (□), 0.27 Mrad in acetylene (▲) and 0.70 Mrad in acetylene (■). Reproduced from J.Polym.Sci., Polym.Phys.Edn., (in press) by permission of the publishers John Wiley & Sons, Inc. (C)

Simple intuitive reasoning, borne out by the Charlesby-Pinner analysis, suggests that electron-beam irradiation will become less effective in terms of the balance between cross-linking and chain scission as the polymer molecular weight is increased. This is illustrated by the results obtained for gel spun fibres, where even acetylene irradiation in produces an increase in the creep rate compared with the unirradiated control, and in both cases the critical stress for permanent load bearing applications is ~ 0.25 GPa, which is comparable to the best that can be achieved by cross-linking the melt-spun and drawn fibres. It is interesting that cross-linking of the melt spun and drawn fibre [9], under similar conditions, produces an increase in strength at all strain rates, from those corresponding to long term creep, to those corresponding to high speed impact.

Tensile Strength

The tensile strength of a fibre is of considerable practical importance, and for ultra-high modulus polyethylene fibres values are obtained which have raised considerable interest. There are two aspects of the behaviour of the polyethylene fibres which are worthy of emphasis. First, the properties are extremely time dependent. Whereas the conventional tensile test is usually made in the strain rate range ~ $10^{-1}sec^{-1}$, the practical interest in the properties is often either at ballistic rates or, for permanent or cyclic load bearing applications at long times i.e. very low strain rates. It is therefore very important to compare different materials e.g. polyethylene and Kevlar over a wide range of loading conditions, so that the comparison of Table 1 could be very misleading. Secondly, for many practical applications it is the combination of tensile strength and elongation at break which must be considered. The elongation at break is important because it determines the conversion efficiency when a rope or multifilament yarn is produced, and also because the energy to rupture depends on both tensile strength and elongation at break.

Table 1 shows that ultra-high modulus polyethylene fibres have a comparatively high elongation to break. This suggests that their conversion efficiency will be higher than for other fibres in the table. It also suggests that they will show good damage tolerance and high energy to break. For ballistic applications the comparison between the different fibres is likely to be more favourable to high modulus polyethylene fibres than is implied by Table 1. Even low molecular weight melt spun/drawn fibres show tensile strengths of about 1.5 GPa.

Accepting that comparable draw ratios and moduli can be obtained for a wide range of polyethylenes, there has been considerable interest in the influence of molecular weight and molecular weight distribution on the tensile strength. Complimentary studies of this facet of behaviour have recently been undertaken by Smith and co-workers [12] and by Ward and co-workers [13]. Smith et al, showed that the theoretical stochastic model for failure of oriented fibres predicted that the tenacity of blends of different molecular weights would follow a simple weight-average summation of the component properties. This result was confirmed experimentally by measuring the tensile strengths of oriented polyethylene filaments prepared by blending ultra-high molecular weight polymer with low molecular weight polymer. Hallam, Pollard and Ward have shown that there is good agreement between experimental and theoretical values of

556

tensile strengths for a wide range of melt spun and drawn polymers, using a weight average summation and knowledge of the molecular weight distribution for commercially available polymers.

ULTRA-HIGH MODULUS POLYETHYLENE FIBRE COMPOSITES

Recent research at Leeds University on polyethylene fibre composites has developed in two related areas; towards a greater understanding of surface treatments designed to improve fibre/resin adhesion, and in carrying out studies of composites incorporating polyethylene fibres in different resins, and of hybrid composites incorporating polyethylene fibres with carbon or glass fibres.

Surface Treatment and Adhesion

Following earlier work by Ladizesky and Ward [14], Nardin and Ward [15] have recently undertaken a detailed study of the influence of surface treatment on the adhesion of polyethylene fibres to epoxy resin. The pull-out adhesion strength was determined for untreated, chromic acid treatment and plasma etched monofilaments with different draw ratios and thermal annealing treatments. The majority of the plasma treatments were carried out using a Plasmaprep 300 (Nanotech Ltd) with oxygen as the carrier gas. In some instances additional chemical treatments were applied to the plasma treated fibres prior to the pull-out test. These chemical treatments included immersion in glycol or sulphuric acid in attempts to reduce adhesion and borane reduction or silane treatment in attempts to produce increased adhesion. The polyethylene surface energy was also determined by measurement of contact angle using either glycerol or liquid epoxy resin.

The results of these studies showed that the adhesion depends on three factors. First, there is the wettability (or physico-chemical interaction) which is affected by the fibre draw ratio as well as by the nature of the surface treatment. It is particularly interesting that the contact angle falls markedly with increasing draw ratio. Secondly, after plasma etching in the presence of oxygen gas, a honeycomb structure is produced which provides mechanical keying between the fibres and the resin. Finally, in the case of borane/silane treatments chemical bonds are formed between the fibre surface and the resin. The results obtained suggested that very simplistically these three factors can be regarded as additive, and under optimum conditions produce in each case a contribution of about 2 MPa to the total maximum pull-out adhesion strength of about 6 MPa.

Preparation and Properties of Composites

Fibre composites have been prepared by three methods [16]

(1) The leaky mould technique where a bundle of fibres is placed in a rectangular mould, fully wetted with the liquid resin, and then compressed with a smooth fitting top to the mould.
(2) The wet lay-up technique where layers of square fabric are laid down between layers of liquid resin in a rectangular mould.
(3) Lamination of pre-impregnated sheets of fibre.

In all cases the fibre volume fraction was about 55%, and composites were produced with a range of epoxy and polyester resins. Composites containing either untreated polyethylene fibres, or fibres subjected to either plasma etching in oxygen gas or chromic acid treatment, were examined.

These extensive studies have been described in a number of publications [16-19], and only the principal results will be summarised here.

First, the composites where polyethylene fibres were the only reinforcing phase showed very satisfactory values of stiffness and strength, and high energy absorption in Charpy impact tests. The high ductility of the fibre made the production of the composites (either with fibres or fabric) relatively straightforward, and also reflected itself in the high damage tolerance of the composites, which retained their integrity after impact or flexural tests.

Secondly, the interlaminar shear strength of the composites could be significantly increased by surface treatment of the fibres. Plasma etching in oxygen gas was most effective and raised the interlaminar shear strengths of the fibre polyethylene composites from 14 to 28 MPa. Hybrid composites incorporating polyethylene fibres with either glass or carbon fibres showed correspondingly higher values of interlaminar shear strength reaching 40 MPa for 50/50 mixtures. The chromic acid treatment was found to be about half as effective as plasma etching in oxygen, as anticipated from the fundamental study of adhesion discussed above. Surface treatment reduced resin cracking in flexural and impact tests, but did not reduce the impact energy absorption very greatly because the latter is primarily associated with plastic deformation of the fibres.

Thirdly, the results for polyester and epoxy resins with a range of chemical composition and different breaking elongation showed no major differences in mechanical behaviour. There were however significant changes in the impact energy and compressive strength of different formulations. For example use of a highly ductile epoxy resin reduced the level of resin damage in flexural and impact tests, but also the energy absorption, as would be expected.

Finally, hybrid composites have been produced with a range of sandwich structures incorporating polyethylene fibres with either glass or carbon fibres. Such hybrid composites can give a good balance of mechanical properties in terms of tensile modulus and strength, compressive strength and impact energy, including the capability of withstanding high levels of deformation without disintegrating. This last feature of the hybrid composites is illustrated in Figure 3.

HYDROSTATIC EXTRUSION AND RAM EXTRUSION

The hydrostatic extrusion process has been developed to a production scale by Bethlehem Steel, manufacturing "BeXor" sheet. This is a thick biaxially oriented polypropylene sheet made by hydrostatic extrusion of a biaxially stretched tube which subsequently slit and flattened. The product shows outstanding impact behaviour. In the UK, hydrostatic extrusion has been superceded in commercial developments by die-drawing (to be discussed below) but has still been the subject of some fundamental investigations. In particular, there has been interest in the hydrostatic extrusion of chain-extended polyethylene, where it has been shown that high modulus products can be produced by the imposition of comparatively modest draw ratios [20]. It is conceivable that such

558

materials might be of interest for special applications, where the
relatively high cost of the hydrostatic extrusion process is of less
importance.

Split billet ram extrusion [3] also continues to produce very highly
oriented materials of considerable fundamental interest, but there are no
reports of commercial exploitation.

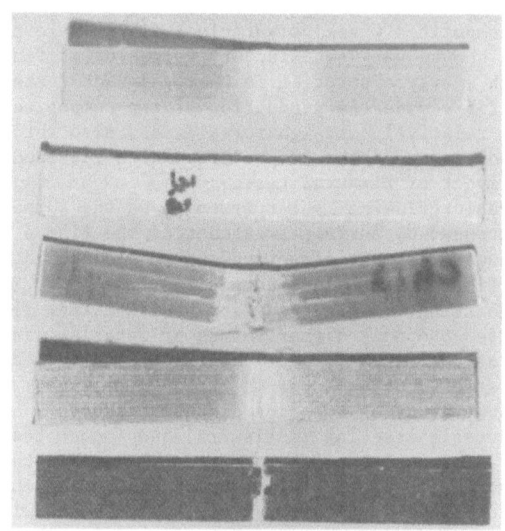

Figure 3. Unidirectional composites after flat Charpy test:
a, UHMPE, 7 impacts; b, UHMPE/E-type glass, 7 impacts;
c, E-type glass, 1 impact; d, UHMPE/EXAS carbon,
7 impacts; e, EXAS carbon, 1 impact. Reproduced from
Composites Science & Technology, 26, 199 (1986) by
Permission of the publishers Elsevier Applied Science
Publishers Ltd (C)

DIE-DRAWING

Die-drawing [21,22] combines the best features of free tensile
drawing and solid-state extension, so that oriented polymers can be
produced in large sections at economically viable production rates.

A schematic diagram of the die-drawing process developed at Leeds
University is shown in Figure 4 for a circular cross-section product.
The heated polymer billet is drawn through a heated conical die by the
haul-off force exerted on the oriented product.   There are three
deformation zones

Zone 1.   Conical die flow where the polymer is in contact with the
heated die
Zone 2.   Free tensile drawing within the heated die
Zone 3.   Further tensile deformation as the polymer cools down to

ambient temperatures.

The total deformation is defined by the actual deformation ratio or draw ratio R_A, in terms of the initial and final product diameters, d_o and d_f respectively, as $R_A = (d_o/d_f)^2$. The initial billet size can be characterised by R_N, the nominal deformation ratio, given by

$$R_N = (d_o/d_1)^2$$

where d_1 is the die exit diameter.

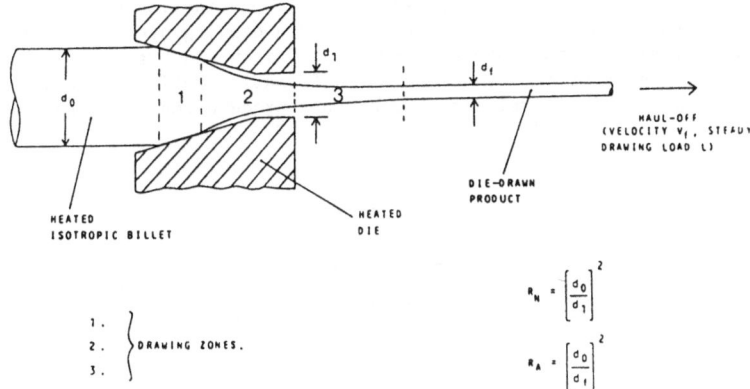

Figure 4.   Schematic diagram of the die-drawing process; 1,2, 3 are the drawing zones. Reproduced from Plastics & Rubber Processing & Applications, 6, 347 (1986) by permission of the publishers, Elsevier Applied Science Publishers Ltd. (C)

The first large scale die-drawing facility constructed at Leeds University [23] is capable of producing drawn products up to 11 m lengths at a maximum draw speed of 2 m/min and a draw load of 2 tonnes. The billet and die-holder assembly is approximately 2 m long and comprises a tubular container and a die-block which have separate heating controllers so that billets up to ~ 10 cm in diameter can be maintained at temperatures up to 200°C with a temperature uniformity of ±2°C. Instrumentation permits the continuous measurement of temperature, drawing speed and drawing force.

To produce oriented sheet, a rectangular slot die of exit cross-section 63.5 mm x 3.18 mm with a 15° semi-angle taper on the broad face, has often been used.

For production of oriented tubes [24], the three tooling configurations illustrated in Figures 5(a), (b) and (c) were employed, each giving very different deformation patterns. Figure 5(a) shows the simplest, which is akin to the die-sinking process normally applied to metals, and involves the drawing of a tubular billet through a die of much smaller bore than the outside diameter of the billet. A mandrel is not essential in this case. Figure 5(b) shows the configuration where the initial tubular billet is drawn over a mandrel which has the same diameter

as the bore of this billet i.e. drawing to constant bore.   Figure 5(c)
shows the third configuration where the mandrel has a larger diameter than
the bore of the initial tubular billet.

In recent research at Leeds University the production of all these
different sections, rod, sheet and tube, including the three
configurations for tube are being actively pursued.   Some preliminary
reports have already been given, and a wide range of possible applications
are being explored for many polymers.   It is therefore only possible to
give a brief summary of some of the key results.

Die-drawing of rods was undertaken for many polymers, including
polyethylene, polypropylene, polyoxymethylene, polyvinylidene fluoride and
polyetheretherketone.   In one respect, the die-drawing process was found
to be very flexible, and there was usually a fairly wide range of
conditions in terms of choice of nominal deformation ratio R_N, draw
temperature and haul-off velocity which would produce the same product.

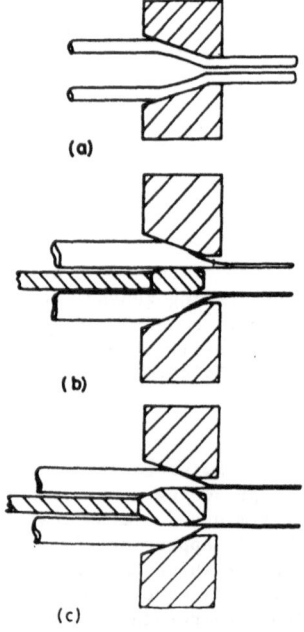

Figure 5.   Tooling configurations for die-drawing tubes
Reproduced from Plastics & Rubber Processing &
Applications, (in press) by permission of the
Publishers Elsevier Applied Science Publishers. (C)

In all instances the Young's modulus of the oriented product relates only
to the actual deformation ratio or draw ratio R_A.   From a practical
viewpoint, however, the rate of production may be a very important factor,
and it is therefore necessary to obtain a cost-effective compromise
between effective draw ratio R_A and haul-off velocity v_f. Richardson,
Parsons and Ward have reported results for the die-drawing of linear

polyethylene (LPE).   The production rate can be increased by increasing the operating temperature, which is to be expected, because drawing involves thermally activated processes, but also in polyethylene by increasing molecular weight and introducing a small amount of chain branching, and by changing the morphology of the initial billet.   These findings are consistent with previous knowledge of tensile drawing.   It has, of course, to be borne in mind that the ultimate available draw ratio and hence the final product properties may also be reduced by such process changes, so that the final decision on a practical production process will usually involve a compromise.

The unifying theme throughout such exploratory research remains the unique relationship between modulus and draw ratio R_A.   Figure 6 shows results for polypropylene [25], where data for both small scale and large scale die-drawing are combined.   Although in this instance the maximum modulus of 20 GPa was obtained by small scale drawing to R_A = 23.2, on the large scale it was possible to produce excellent products at speeds up to 2 m min^{-1} (the maximum capability of the available equipment) at a temperature of 155°C and R_A = 19.3.

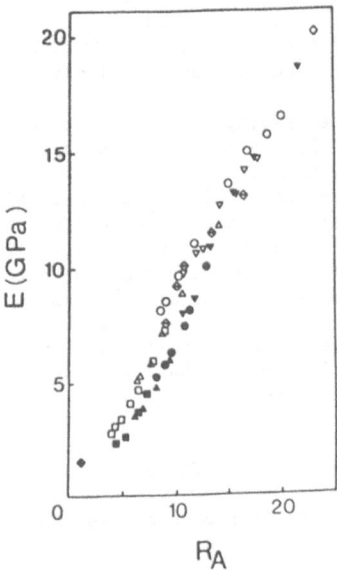

Figure 6.   Variation of product three point bend modulus E with draw ratio R_A for various draw temperatures.
(◆) isotropic; R_N = 3 (■ T_N = 155°C, ☐ T_N = 110°C); R_N = 5 (Δ 155°C, ▲ 110°C); R_N = 7 (● 155°C, ○ 110°C, ◈ 90°C); R_N = 9; (▼ 155°C, ▽ 110°C, ◇ 140°C). Reproduced from J.of Applied Polymer Science (in press) by the publishers John Wiley & Sons.Inc. (C)

The die drawn rods of polyethylene and polypropylene have been supplied to the Civil Engineering Department at Leeds University for extensive studies of the use of high-modulus polymers to reinforce concrete. In some instances, square cross-section rods were produced using a square-section converging die. The drawn products retained the square cross section shape of the starting billet with no distortion, and permanent twist could be incorporating without significant changes in stiffness and strength, to give a 'barley sugar' geometry for improved physical bonding between the reinforcing rod and the concrete. Full accounts of this research are given elsewhere.

Earlier research on the hydrostatic extrusion and die-drawing of polyoxymethylene and glass-filled polyoxymethylene had given indications of the improvements in toughness as well as stiffness which can be obtained by orientation. An illustration of these improvements is given in Table 2, which shows results for impact tests on flat, unnotched specimens of die-drawn sheets of polyoxymethylene (POM) and Hizex 7000F LPE sheet, and, for comparison, an epoxy resin fibre reinforced composite, reinforced with ultra high modulus polyethylene fibres (melt spun and drawn), arranged in a parallel array matching the draw direction of the die-drawn sheets. The table shows results for repeated tests where the materials all showed sufficient ductility to be bent back into shape for successive tests. All these materials show high initial energy absorption, and the POM sheet can absorb similar energy in the repeated impacts.

In the case of tube drawing, results have been obtained for the three configurations shown in Figure 5, and it is only possible to select a few examples for presentation here.

Results for the production of tube drawn to a constant bore of about 22 mm, with a 4.5 mm wall thickness have been obtained for a series of LPE copolymers of increasing branch content. It was found that for each polymer the draw ratio increases with draw speed to give products with draw ratios R_A in the range 6-10 at draw speeds in the range of 50 cm/min. These tubes have been characterised with regard to stiffness, equilibrium uptake and diffusion coefficient for liquids, and in long term internal pressure tests. In all cases a substantial degree of property enhancement has been recorded.

TABLE 2

Sheet energy absorption over six consecutive impacts, Charpy test

Sample	Absorbed energy per unit area (kJm^{-2})					
	1	2	3	4	5	6
POM, R_A = 7.5	152	152	152	152	152	152
LPE, R_A = 14.7	159	99	84	76	71	68
Composite	170	72	51	44	39	37

As an example of a tube drawn to reduced bore, a billet of LPE
homopolymers (BP Chemicals 006-60 grade) was slow cooled from the melt and
machined into a tube of 21 mm outside diameter and 6 mm bore.   This was
drawn over a mandrel of 2 mm outside diameter and through a die of 8 mm
diameter at 110°C.   The product was a tube of 6 mm outside diameter and
1.5 mm bore, with an axial draw ratio of 16 and a measured axial modulus
of 28 GPa.

As an illustration of drawing tubes to greatly expanded bore, the
drawing of clear amorphous polyethylene terephthalate (PET) tube will be
cited.   One example is the production of clear transparent tube of 55 mm
outside diameter and a wall thickness of 0.45 mm from an initial tube of
25 mm outside diameter and 18 mm bore.   The final dimensions imply an
axial draw ratio of about 3 and a hoop draw ratio of about 2.5.   The
structure and properties of such a tube are similar to those of a
biaxially drawn PET film i.e. improved stiffness and strength in both the
axial and hoop directions, and satisfactory gas barrier properties.
X-ray diffraction measurements show the expected biaxial orientation where
there is preferred orientation of the plane of the terephthalate residues
in the direction tangential to the radius of the tube.

Tube drawing to expanded bore has also been undertaken for
polyethylene copolymer.   Comparative results for Phillips 47100
copolymer, where tubes of constant bore were compared with tubes of
expanded bore, showed the anticipated improvements in hoopwise behaviour.
Moreover the expanded bore tube showed a greater resistance to axial
splitting, and could be squashed completely flat in a vice, before the
outer layers of the tube wall began to fail.   Again, X-ray diffraction
measurements showed a significant degree of biaxial orientation.

CONCLUSIONS

The development of practical processes for the production of
ultra-high modulus polyethylene fibres has moved the interest in this area
to seeking satisfactory solutions to subsidiary problems, notably
improving creep behaviour and devising satisfactory surface treatments to
ensure adequate fibre/resin adhesion in composites.

Die-drawing has emerged as a serious contender for large scale
production of highly oriented polymers in the form of rod, tube and sheet.
The die-drawing technology offers a broad range of options, both in terms
of property enhancement (stiffness, strength, impact resistance,
permeability to gases and liquids, chemical resistance, thermal stability,
etc) and in terms of the very wide range of polymers to which it is
applicable.

REFERENCES

1. Capaccio, G. and Ward, I.M., Preparation of ultra-high modulus linear
    polyethylenes; effect of molecular weight and molecular weight
    distribution on drawing behaviour and mechanical properties.
    Polymer., 1974, 15, 233-238.

2. Smith, P. and Lemstra, P.J., Ultra-high-strength polyethylene
    filaments by solution spinning/drawing. J. of Mater.Sci., 1980, 15,
    505-514

3. Griswold, P.D., Zachariades, A.E. and Porter, R.S. Solid state coextrusion: A new technique for ultradrawing thermoplastics illustrated with high density polyethylene. Polym.Eng.Sci., 1978, 18, 861.

4. Smith, P., Chanzy, H.D. and Rotzinger, B.P., Drawing of virgin ultrahigh molecular weight polyethylene: an alternative route to high strength fibres. Polymer Comm., 1985, 26, 258-260.

5. Peguy, A. and St.John Manley, R., Ultra-drawing of high molecular weight polypropylene. Polymer Comm., 1984, 25, 39-42.

6. Matsuo, M., Sawatari, C. and Nakano, T., Ultradrawing of isotactic polypropylene films produced by gelation/crystallization from solutions. Polymer Journal., 10, 759-774.

7. Wilding, M.A. and Ward, I.M., Tensile creep and recovery in ultra-high modulus linear polyethylenes. Polymer., 1978, 19, 969-976.

8. Sherby, O.D. and Dorn, J.E., Anelastic creep of polymethyl methacrylate. J.Mech.Phys.Solids, (1958), 6, 145-162.

9. Wilding, M.A. and Ward, I.M., Creep and recovery of ultra high modulus polyethylene. Polymer., 1981, 22, 870-876.

10. Duxbury, J. and Ward, I.M., The creep behaviour of ultra high modulus polypropylene. J.Mater.Sci., (in press)

11. Woods, D.W., Busfield, W.K. and Ward, I.M., Improved mechanical behaviour in ultra-high modulus polyethylenes by controlled cross-linking. Plast. Rubb. Proc. Appl., 1985, 5, 157-165.

12. Termonia Y., Greene, W.R., Jr. and Smith P., Tensile strength of oriented filaments of binary mixtures of high and low molecular weight polyethylenes. Polym.Comm., 1986, 27, 295-298.

13. Hallam, M.A., Cansfield, D.L.M., Ward, I.M. and Pollard, G. A study of the effect of molecular weight on the tensile strength of ultra-high modulus polyethylenes. J.Mater.Sci., 1986, 21, 4199-4205.

14. Ladizesky, N.H. and Ward, I.M. A study of the adhesion of drawn polyethylene fibre/polymeric resin systems. J.Mater.Sci., 1983, 18, 533-544.

15. Nardin, M. and Ward, I.M. (to be published).

16. Ladizesky, N.H. and Ward, I.M., Ultra-high-modulus polyethylene fibre composites: I. The preparation and properties of conventional epoxy resin composites. Composites Science and Technology., 1986, 26, 129-164.

17. Ward, I.M. and Ladizesky, N.H., Ultra high modulus polyethylene composites. Pure & Appl. Chem., 57, 1641-1649.

18. Ladizesky, N.H., Sitepu, M. and Ward, I.M., Ultra-high-modulus polyethylene fibre composites: II. Effect of resin composition on properties. Composites Science and Technology., 1986, 26, 169-183.

19. Ladizesky, N.H. and Ward, I.M., Ultra-high-modulus polyethylene composites: III. An exploratory study of hybrid composites. Composites Science and Technology., 1986, 26, 199-223.

20. Sahari, J.B., Parsons, B. and Ward, I.M. The hydrostatic extrusion of linear polyethylene at high temperatures and high pressures. J.Mater.Sci., 1985, 20, 346-354.

21. Coates, P.D. and Ward, I.M. Drawing of polymers through a conical die. Polymer, 1979, 20, 1553-1560.

22. Gibson, A.G. and Ward, I.M. The manufacture of ultra-high modulus polyethylenes by drawing through a conical die. J.Mater.Sci., 1980, 15, 979-986.

23. Richardson, A., Parsons, B. and Ward, I.M. Production and properties of high stiffness polymer rod, sheet and thick monofilament oriented by large-scale die drawing. Plast. & Rubb. Proc. & Appl., 1986, 6, 347-361.

24. Selwood, A., Ward, I.M. and Parsons, B. The production of oriented polymer tube by the die drawing process. Plast.& Rubb. Proc. & Appl., (in press).

25. Taraiya, A.K., Richardson, A. and Ward, I.M. Production and properties of highly oriented polypropylene by die drawing. J.Appl.Polym.Sci., (in press).

THERMOTROPIC LIQUID CRYSTAL POLYMERS WITH FLEXIBLE CHAIN ELEMENTS.
SYNTHESIS, SPINNING AND MECHANICAL PROPERTIES

Doetze J Sikkema
Enka bv
Research Institute Arnhem
P O Box 60
6800 AB Arnhem, the Netherlands

ABSTRACT

A thermotropic LC aliphatic polyether inspired by cellulose, featuring
flexible comonomer isomorphous with the rigid monomer was prepared and spun,
as well as an LC polyester incorporating heteromorphous flexible comonomer.
Upon spinning, both materials exhibit high molecular orientation; the
polyether with isomorphous flexible units shows extremely rapid crystallization, whereas the polyester forms an oriented glass - a solidified
nematic melt.
Both types of yarns showed disappointing mechanical properties. We rationalize these observations by chain folding, resulting in a pseudo-lamellar
structure of the liquid crystalline melt which cannot form in all-aromatic
polyamide solutions or all-aromatic polyester melts. In the case of the more
ordered polyether system, experimental evidence for lamellar organization
was secured.

INTRODUCTION

Over the past ten years, much effort has been directed at the production of high-modulus, high tenacity yarns by melt spinning. Almost always
such work was concerned with thermotropic liquid crystal polyesters, the
most successful example being the material developed by workers at Celanese
[1]. A major disadvantage of all-aromatic polyesters rendered fusible by
copolymerizing with ill-fitting comonomers is the annealing necessary to
improve the yarn crystallinity, necessary to develop its tenacity
potential.[1,2]. An alternative way to arrive at thermoplastic behaviour of
polymers consisting mainly of stiff chains is the incorporation of flexible
elements.[3] Striving for isomorphicity of the flexible with the stiff
elements, with high crystallization rates as the result, seemed a possibility to escape the costly yarn annealing while retaining the advantages
of melt processability.

Model studies suggest that tetrahydrofuran (THF) fragments may well he isomorphous with 7-oxabicyclo[2.2.1]heptane (OBCH) derived fragments in a polymer - poly-THF crystallizes in a zigzag conformation [4].

As it turned out, the OBCH-THF system produced extremely disappointing yarns. As a comparison with these polyether results, we studied a potentially cheap thermotropic polyester featuring flexible elements non-isomorphous with the rigid units.

EXPERIMENTAL

OBCH monomer was synthesized by heterogeneous gas phase ring closure of 1,4-cyclohexanediol prepared by hydrogenating hydroquinone, as described earlier [5]. High molecular weight polymer was obtained by careful cationic polymerization of OBCH-THF mixtures, with further cyclic ether additives at low levels if desired, as reported recently [6].

4-Hydroxybenzoic acid-hydroquinone-azelaic acid copolymers were prepared by acidolysis: the monomers are charged, a small excess of acetic anhydride relative to phenolic groups is added; the mixture is heated to 280 ^{o}C in about 90 min; then vacuum is applied and 60 min of high vacuum (< 1 mbar) completes the reaction. We found it convenient to isolate prepolymer after a mild vacuum cycle, at a relatively low melt viscosity, and postcondense (1 h, 250 ^{o}C, 1 mbar) to avoid problems in product collection from the autoclave.

Spinning was performed using a miniature spinning machine allowing melting the product under vacuum and extrusion by piston action; a small extruder (30 mm) spinning machine with an evacuated hopper was also used. A poor vacuum - or a nitrogen blanket - resulted in gas bubbles in the yarns: the extreme shear thinning in nematic melts makes degassing in an extruder very difficult.

RESULTS AND DISCUSSION

Polymerization: OBCH copolymers of RV 2.5-3.5 (1 %, m-cresol) could routinely be prepared, at 90-95 % conversion.[6] GPC work suggests molecular weights in the range $4-8x10^{4}$ (polystyrene standards) at these viscosities.

The polyester showed inherent viscosities of about 4 (0.1 %, pentafluorophenol). The K-N melting point is 240 ^{o}C; the isotropic phase is inaccessible. The at-rest viscosities of the polyethers were higher than those of the polyesters; the rheologies were characterized qualitatively only; both types of polymers showed strong shear thinning.

Early work with OBCH-THF copolymers employing the maximum level of OBCH to retain thermoplasticity, viz. 80-20 random copolymers, showed immediate crystallization upon exit from the spinning plate: a few millimeters into the air the yarn would become highly opaque. This crystallization proved to be isotropic, independent of shear rate in the spinning channel. We theorized that the isomorphous and fairly flexible 80-20 OBCH-THF system, which indeed showed but small nematic melt undercooling before crystallization occurred in hot stage microscopy, was too much of a good thing in terms of crystallization rate: orientation in elongational flow seemed experimentally inaccessible by the fast crystallization, sufficiently energetic to destroy any orientation achieved by way of spinning channel shear. Thus, we prepared similar copolymers adding small amounts of ethylene oxide (EO) as crystallization retarder. The terpolymers incorporating 1 - 5 % of EO showed crystal-nematic transitions 10-20 K below the related copolymers (Table 1) and allowed substantial undercooling in hot stage microscopy (as much as 35 K with EO vs. ~ 7 K without EO).

Terpolymers, including material with as high as 84 % of rigid monomer, could be spun into yarns showing a high molecular orientation and significant crystallinity by X-ray diffraction, although in all cases, drawdown was limited to values <2.4, independent of extrusion rate, spinning temperature, or the use of a hot air zone below the spinneret at temperatures up to 400 ^{o}C. The yarn properties were disappointing (Table 2).

Similar results were obtained with the polyester, spun at 300 ^{o}C. Drawdown was limited to 8.8x. The as-spun yarns show pronounced orientation but no proper crystallinity - an oriented nematic structure seems to be frozen in. WAXS measurements gave orientation angles of 17.5 - 20.5^{o} in various yarns. Very poor yarn properties were recorded (Table 3).

All of the yarns collected resisted cold or hot drawing.

This experience at first seemed very puzzling: in the literature, liquid crystal behaviour is suggested to aid high-performance fiber preparation [7] - see, however, work showing no benefit at all from nematic behaviour per se in solutions of poly(phenylene terephthalamide) [8].

Viewed on the 10 - 100 nm scale, the liquid crystal arrays of macromolecules that consist of relatively rigid sequences interspersed with flexible units are likely to be organized into structures analogous to the lamellae encountered in the classical polymer crystallite: in regions where the interchain alignment begins to degrade, the flexible units permit chain folding. Although poly(THF) crystallizes with an extended zigzag conformation, high temperatures favour gauche conformers by an entropy term of RT ln 2. Such chain folding decreases the concentration of tie molecules between the primary domains of aligned and oriented material.

A lamellar structure of the polyether yarns could be ascertained by recording a sharp SAXS pattern corresponding to a period of about 20 nm in various yarns. No SAXS pattern could be distinguished in the polyester yarns - we think this is due to the much lower crystallinity compared with the polyethers.

REFERENCES

1. Calundann, G.W., US Patent 4 161 470, 1979; Guttierez, G.A., Blackwell, J. and Chivers, R.A., Polymer 1985, 26,348

2. Schaefgen, J.R., US Patent 4 118 372, 1978

3. Jackson, W.J., Jr. and Kuhfuss, H.F., J. Polym. Sci., Polym. Chem. Edn. 1976, 14, 2043; Ober, C.K., Jin, J.-I. and Lenz, R.W., Makromol. Chem., Rapid Commun. 1983, 4, 49

4. Cesari, M., Perego, G. and Mazzei, A., Makromol. Chem. 1965, 83, 196

5. Sikkema, D.J., Hoogland, P., Bik, J. and 't Lam, P., Polymer 1986, 27, 1441

6. Sikkema, D.J. and Hoogland, P., ibid., 1443

7. Preston, J. in "Liquid Crystalline Order in Polymers", (Ed. Blumstein, A.), Academic Press, New York, 1978, p. 141; Morgan, P.W., Chemtech 1979, 316

8. Weyland, H.G., Polym. Bull. 1980, 3, 331

Table 1 Comparison of phase transition temperatures of some co- and terpolymers

Copolymer composition OBCH/THF/EO	K→N (°C)	N→I (°C)
80/20	270–290	>330, d
80/17/3	260–270	>330, d
79/17/4	259–273	>330, d
84/8/8	270–290	>330, d
75/20/5	231–256	>330, d
75/25	250–258	>330, d
70/30	220–230	>320, d
60/40	165–178	220–250
46/54	70–90	153–187
40/60	<25	104–120
30/70	<25	38–55

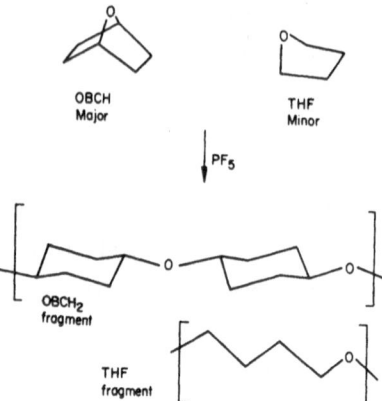

Table 2 Yarn tensile properties of OBCH/THF(EO) copolymers, monofilament spun on a piston extrusion machine

Polm. comp. OBCH/THF/EO	Spinning temp., °C	Drawdown	Tenacity cN/tex	Elongation %	Max. modulus cN/tex
70/30	270	2.4	16.4	1.4	1360
60/40	210	1,6	16.7	6.1	510
80/17/3	300	1.4	2.8	0.9	327
75/20/5	288	1.6	2.5	0.9	243

Table 3 Yarn tensile properties of LC polyester, ten filaments, spun on an extruder spinning machine. 4-hydroxybenzoic acid-hydroquinone-azelaic acid 2:1:1 molar, spinning temperature 300°C

Drawdown factor	Count dtex	Tenacity cN/tex	Elongation %	Max. modulus cN/tex
7.5	92	5.1	0.8	655
7.1	89	5.8	1.1	658
4.9	123	5.0	1.1	466

MICROMECHANICS OF DEFORMATION OF HIGH MODULUS POLYMER FIBRES AND COMPOSITES

R. J. Day, I. M. Robinson, M. Zakikhani and R. J. Young
Department of Polymer Science and Technology
UMIST
Manchester
M60 1QD, UK

ABSTRACT

A Raman microscope has been used to investigate the deformation of high-modulus polymer fibres such as poly(p-phenylene terephthalamide) (Kevlar 49), substituted polydiacetyelene single crystal fibres and poly(p-phenylene benzobisthiazole) (PBT) fibres. Well-defined intense Raman spectra can be obtained from individual fibres and the spectra for the different fibres show characteristic bands due principally to the stretching modes of back-bone bonds. It has been found that for each type of fibre most of the bands are strain sensitive moving by up to -20 cm^{-1}/% strain. The reasons for this strain sensitivity are discussed in relation to the deformation of bonds in the individual molecules. The way in which the strain-induced band shift can be used to follow the micromechanics of deformation in composites reinforced with high modulus fibres is described.

INTRODUCTION

Over recent years there have been important developments in the production of polymer fibres with high degrees of strength and stiffness [1-6]. Efforts have been made to produce ultraorientation in flexible-backboned polymers with a high crystal modulus, such as polyethylene. This has been achieved by tensile drawing [1], extrusion [2] and gel-spinning [3]. High modulus polymer whisker single crystal fibres can be produced from certain substituted polydiacetylenes by solid-state polymerisation of single crystal monomers [4]. High modulus fibres of aromatic polyamides such as poly(p-phenylene terephthalamide) (Kevlar) can be produced by the spinning of liquid crystalline polymer solutions [5]. The most impressive mechanical properties reported so far have been for fibres of heterocylic rigid-rod polymers such as poly(p-phenylene benzobisthiazole) (PBT), again by spinning from liquid crystalline solutions. Such fibres have been reported to have moduli of the order of 110 GPa in the as-spun condition but this can be increased to 300 GPa by heat treatment [6].

In our laboratory we have embarked upon a large programme of research concerned with the micromechanics of the deformation of high modulus fibres and composites using Raman spectroscopy. It has been found that strong, well-defined Raman spectra are generally obtained from high-modulus, highly-oriented polymer fibres. Since when such polymers are deformed the bonds in the molecular backbone are strained and some of the frequencies of Raman-active bands decrease by an amount $\Delta\nu$ dependent upon the band under consideration. For polymer single crystals stressed parallel to the molecular axis, the shifts can be considerable. Values of $\Delta\nu$ in the order of 20cm^{-1}/% strain have been reported [7] for the triple bond stretching band in substituted polydiacetylenes. Shifts to a lower frequency of approximately 5cm^{-1}/% strain are found for the 1610cm^{-1} band in Kevlar fibres [8]. However, the stress-induced shifts have been found to be somewhat smaller in lower-modulus, less-oriented aramid fibres since in this case the polymer chains experience less strain because of the difference in morphology [9].

EXPERIMENTAL

The PBT fibres used in this present study were supplied by the Materials laboratory of the Wright-Patterson Air Force Base, Ohio, USA and were based on the molecule

Raman spectra were obtained from individual PBT fibres using a Raman microscope system. This is based upon a SPEX 1403 double monochromator connected to a modified Nikon optical microscope. Spectra were obtained at a resolution of the order of 5cm^{-1} using the 632.8nm line of a 10mW He/Ne laser. A x40 objective lens with a numerical aperture of 0.65 was used and this gave a 2μm spot when focussed (although the objective lens was generally defocussed to reduce the possibility of damage through excessive heating). The laser beam was always polarised parallel to the fibre axis for all measurements.

Spectra were obtained from fibres during deformation in a small straining rig which fitted directly on to the microscope stage. Individual fibres were fixed onto the straining rig using a cyanoacrylate adhesive, giving a gauge length of about 10mm which was measured accurately using the light microscope. Raman spectra were obtained during deformation by scanning strong individual peaks between loading steps of 0.1% strain.

DEFORMATION OF INDIVIDUAL FIBRES

The stress/strain curve for an individual heat-treated PBT fibres is approximately linear up to fracture. Mean values of Young's modulus, fracture stress and elongation to failure obtained from five sets of specimens are 250 GPa, 1 GPa and 1.6% respectively.

Raman spectra could be readily obtained from different positions on the PBT fibres using relatively low powers of laser radiation. Typical spectra from deformed and undeformed, heat-treated fibres are shown in Figure 1. The spectra consisted of a few well-defined intense peaks on a strong fluorescent background. Their assignment is discussed elsewhere [10]. It can be seen from Figure 1 that the spectra change with deformation. Certain peaks shift to lower frequency and peak intensities also change. Detailed changes in the spectra are described below.

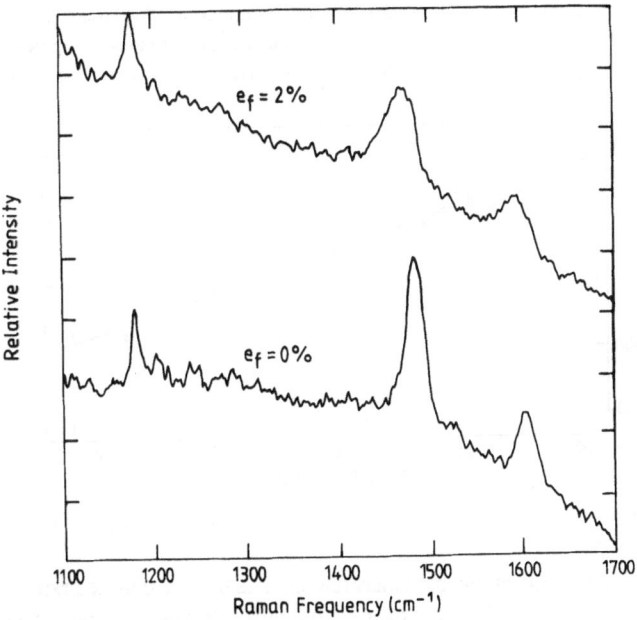

Figure 1.   Raman spectra in the range 1100-1700cm^{-1} for individual fibres of PBT obtained at fibre strains (e_f) of 0% and 2%

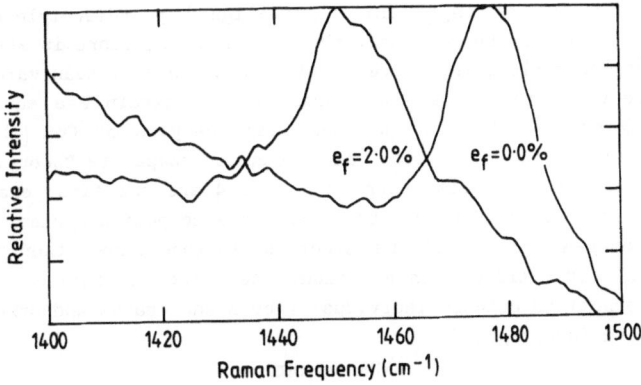

Figure 2.   Raman spectra in the region of the 1480cm^{-1} band for a PBT fibre obtained at fibre strains (e_f) of 0% and 2% showing the peak shift.

Figure 2 shows the spectra for the PBT fibre in the 1400–1500cm^{-1} region in the undeformed state and following deformation of the fibres to 2% strain. It can be seen that there is a significant change in the 1480cm^{-1} band with the peak position shifting by almost -30cm^{-1} for the high modulus heat-treated fibre. It can also be seen that there is also some broadening of the peak during deformation.

The effect of deformation upon the 1480cm^{-1} peak can be seen in more detail in Figure 3. It can be seen that there is a linear shift in Raman frequency up to failure. Since the Raman technique follows the molecular strain directly, this demonstrates that it is an extremely powerful tool for observing molecular processes involved in the deformation of polymer fibres.

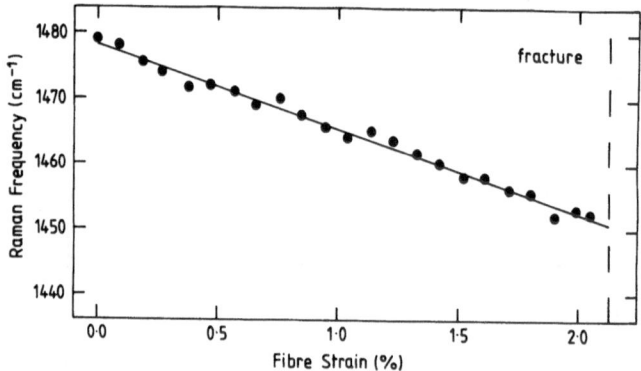

Figure 3.   Variation of the position of the 1480cm^{-1} peak with fibre strain for a PBT fibre.

MICROMECHANICS OF DEFORMATION OF FIBRES IN COMPOSITES

As well as being able to follow the deformation of the molecules in polymer fibres, the Raman technique can be used to measure the point-to-point variation of strain in individual fibres in composites [11]. An example of this is shown in Figure 5 for an epoxy resin tensile specimen containing a single uniaxially-aligned polydiacetylene single crystal fibre. It can be seen that the strain in the fibre is zero at the ends and increases along the fibre until it reaches a plateau value at the fibre centre where the fibre and overall matrix strain are equal. The behaviour is consistent with the shear lag analysis of Cox [12] which forms the basis of the theoretical background of composite theory. As far as we are aware, the results shown in Figure 4 are the first examples of the direct measurement of fibre strain within a composite system. Work is currently underway to extend the study to include many other types of fibre/matrix combinations. In addition, the effect of surface treatment and the interactions between individual fibres and cracks and other fibres are also being investigated.

CONCLUSIONS

It has been found that Raman microscopy is a powerful technique for the investigation of the deformation of PBT fibres at the molecular level. As well as gaining an important insight into response of the fibre molecules to strain it is clear that the Raman technique can be applied generally to the study of the deformation of high modulus polymer fibres and composites. The investigation is currently being extended to other types of high modulus fibres and to the study of the micromechanics of their deformation in composites.

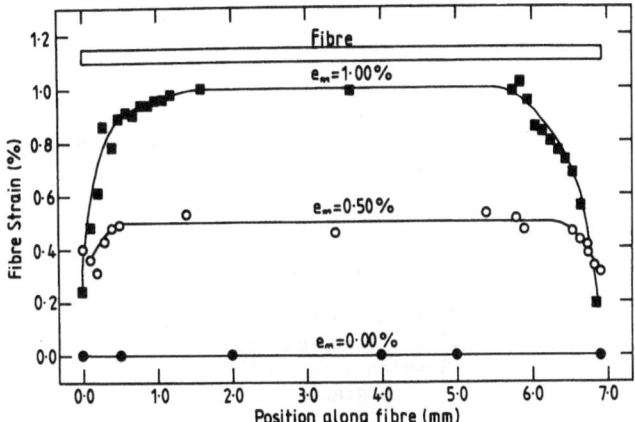

Figure 4.   Epoxy resin tensile sample containing a polydiacetylene single crystal fibre. Variation of fibre strain with position along the fibre for three different levels of applied matrix strain e_m.

REFERENCES

1. Capaccio, G and Ward, I. M., J. Mater. Sci., 1975, 15, 219.

2. Zachariades, A. E., Mead, W. T. and Porter, R. S. in 'Ultra-High Modulus Polymers', Ed. Ciferri, A. and Ward I. M., Applied Science, London, 1979, p.77.

3. Lemstra, P. J. in 'Developments in Oriented Polymers-2' Ed. Ward, I. M., Applied Science, London, 1987.

4. Galiotis, C. and Young, R. J., Polymer, 1983, 24, 1023.

5. Schaefgen, J. R. in 'Strength and Stiffness of Polymers', Ed. A. E. Zachariades and R. S. Porter, Marcel Dekker, New York, 1983, p.327.

6. Allen, S. R., Filippov, A. G., Farris, R. J. and Thomas, E. L., in 'Strength and Stiffness of Polymers' Ed. A. E. Zachariades and R. S. Porter, Marcel Dekker, New York, 1983, p.357.

7. Galiotis, C., Young, R. J. and Batchelder, D. N., J. Polym. Sci. Polym. Phys. Ed., 1983, 21, 2483.

8. Galiotis, C., Robinson, I. M., Young, R. J., Smith, B. J. E. and Batchelder, D. N., Polymer Comm., 1985, 26, 354.

9. van der Zwaag, S., Northolt, M. G., Young, R. J., Galiotis, C., Robinson, I. M. and Batchelder, D. N., Polymer Comm., submitted for publication.

10. Day, R. J., Robinson, I. M., Zakikhani, M., Young, R. J., Polymer, submitted for publication.

11. Day, R. J., Galiotis, C, Robinson, I. M. and Batchelder, D. N., J. Mater. Sci., in press.

12. Cox, H. L., Brit. J. Appl. Phys., 1952, 3, 72.

DEVELOPMENTS AND APPLICATIONS OF CONTINUOUS FIBRE REINFORCED THERMOPLASTICS

David Leach

ICI Advanced Materials
Wilton, Middlesbrough
Cleveland, TS6 8JA
England

ABSTRACT

The current status of continuous fibre reinforced, thermoplastic matrix composites is discussed. These materials contain high proportions (>50%) of reinforcing fibres in either a unidirectional or woven form, and are being used in highly demanding applications such as the aerospace industry. Aspects discussed include the chemical structure and properties of the polymer matrices, and properties, processing and fabrication of continuous fibre reinforced thermoplastic composites.

INTRODUCTION

Continuous fibre reinforced composite materials are now used in a wide range of applications including aeropsace, automotive, industrial and sports goods. The materials normally contain a thermosetting polymer matrix (usually an epoxy or polyester resin) reinforced with carbon, glass or aramid fibres. These materials have higher strength and stiffness per unit weight and often have lower finished component costs than the metals they replace.

It has been recognised for some time that thermoplastic matrix composites would also have potential in these applications. There are a number of resons for using a thermoplastic composite and these are summarised below:

Handling and Storage - Thermosetting matrix composites must be kept under special conditions, such as refrigerated storage, and can only be kept for a certain length of time.

577

Properties - Linear chain thermoplastic polymers typically have
higher toughness than cross-linked thermosetting resins.

Processing and Fabrication - Thermoplastic polymers require no
chemical changes during processing, and it is only necessary to
heat the polymer to the melt, shape the component and then cool
the component. There is therefore potential for more rapid
fabrication and a greater confidence in reproducibly
manufacturing high quality components with a thermoplastic
matrix composite.

During the 1970's there was research and development on
thermoplastic matrix composites (1) and the potential for component
cost reduction was demonstrated (2). However the development and
commercialisation of continuous fibre reinforced thermoplastics did
not take place because there were a number of limitations to the
materials then available:

Polymer Properties - The thermoplastic polymers then available
had lower stiffness and strength than the thermosetting
polymers, and this was reflected in lower composite properties.

Impregnation - In the melt thermoplastic have much higher
viscosities than the monomers or pre-polymers used in
thermosetting systems and hence it is more difficult to
impregnate thermoplastics to give well wetted, high fibre volume
composites.

More recently a number of new polymers have been developed which
combine high stiffness and strength with high toughness, good solvent
resistance and higher temperature performance. Proprietary
impregnation processes have also been developed which can give very
good wetting out of the fibres by the thermoplastic polymers. The
rest of this paper will review current status of continuous fibre
reinforced thermoplastic matrix composites and discuss some of the
areas which need further investigation.

THERMOPLASTIC POLYMERS AS COMPOSITE MATRICES

The chemical structures of some of the thermoplastic polymers
which are being considered as composite matrices are shown in
Figure 1. The polymers are arranged in order of ascending glass-
rubber transition temperature (Tg). A common feature of the polymers
is the aromatic nature of the polymer backbone which provides high
temperature stability and a rigid segment. The properties of the
polymer also depend on the linking groups, with for example sulphone,
and ketone units giving stiff linkages. Several of the polymers
include imide units which also give a rigid linkage in the polymer
chain. Another important consideration is the ability of the polymer
to crystallise as this typically results in superior solvent
resistance and retention of a proportion of properties above Tg. Of
the polymers in Figure 1 only poly(phenylene-sulphide) (PPS) and
poly(ether-ether-ketone) (PEEK) are able to crystallise during
normal processing.

578

Figure 1   Chemical Structures and Glass-Rubber Transition
            Temperatures of various Thermoplastic Polymers.

Polymer	Glass Transition Temperature °C	Structure
Poly(Phenylene Sulphide)	85	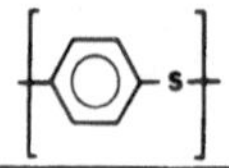
Poly(Ether Ether Ketone)	143	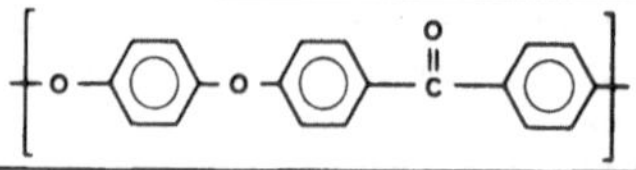
Poly(Sulphone)	190	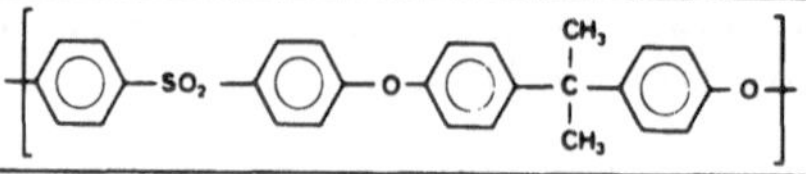
Poly(Ether Imide)	216	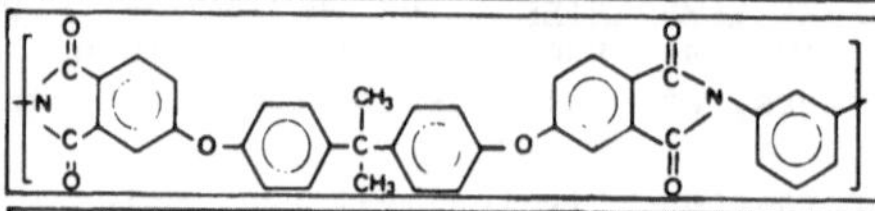
Poly(Phenyl Sulphone)	220	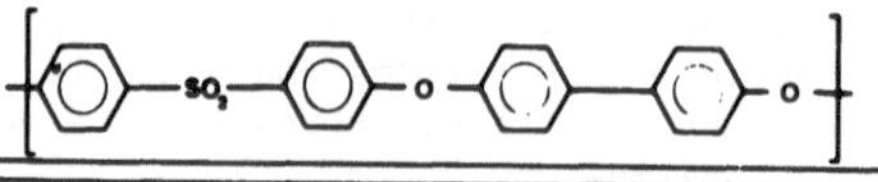
Poly(Ether Sulphone)	230	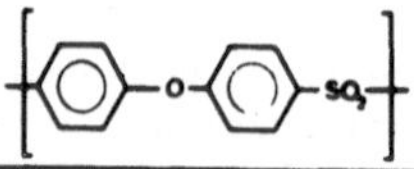
Poly(Amide Imide)	249-288	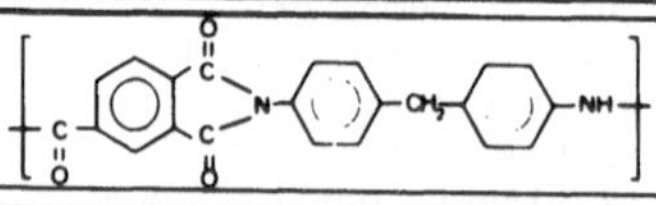
Poly(Imide)	256	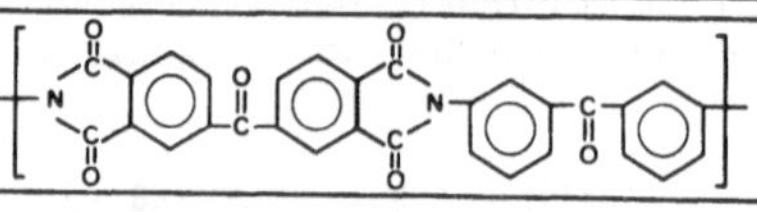

The tradenames of some of these thermoplastic polymers and
manufacturers are listed in Table 1 and a summary of the properties of
some of the polymers is given in Table 2. The target values are based
on experience with the effect of polymer properties on composite
properties (3). The data on the thermoplastics are taken from the
manufacturers data sheets whilst the comparative data on a typical
current epoxy resin, Hercules 3501-6, is taken from Ref 4. One of
the major differences between the thermoplastic polymers and typical
epoxy resins is the higher toughness of the thermoplastics as shown by
the fracture toughness and failure strain.

TABLE 1

Tradenames and Manufacturers of Thermoplastic Polymers

Manufacturer	Polymer	Tradename
Amoco	Poly(phenyl sulphone)	Radel
	Poly(amide imide)	Torlon
	Poly(sulphone)	Udel
BASF	Poly(ether ketone)	Ultrapek
DuPont	Poly(imide)	K-Polymers (K-I etc)
General Electric	Poly(ether imide)	Ultem
Hoechst	Poly(ether ketone)	Hostatec
ICI	Poly(ether ether ketone)	Victrex PEEK
	Poly(ether ketone)	Victrex PEK
	Poly(ether sulphone)	Victrex PES
Mitsui Toatsu	Poly(imide)	LARC-TPI
Phillips	Poly(phenylene sulphide)	Ryton

COMBINING MATRIX AND REINFORCING FIBRE

The process of impregnating the relatively high viscosity polymer into the fine diameter reinforcing fibres has been one of the major challenges in producing high quality thermoplastic matrix composites. It is necessary to impregnate the polymer between all the fibres to give a uniform distribution, and this must be achieved at fibre contents of greater than 50% by volume. This stage also creates the fibre-matrix interface which determines many of the composite properties, and hence the impregnation stage is vital in determining the properties of the composite. Some of the possible impregnation techniques are summarised in Table 3.

There are now a number of commercially available thermoplastic matrix composites which are usually manufactured in the form of a tape with unidirectional or woven fibre reinforcement. A polished section through a unidirectional PEEK/carbon fibre tape (Aromatic Polymer Composite) is shown in Figure 2 illustrating the high degree of wetting that can be achieved in thermoplastic composites.

TABLE 2   Properties of Various Thermoplastic Polymers.

Property	Target	Semi-Crystalline Polymers		Amorphous Polymers							Typical Current Epoxy
		'Victrex' PEEK 450G	'Ryton'	'Victrex' PES 4100G	'Udel' P1700	'Ultem' 1000	'Radel'	K-I	K-II	'Torlon'	Hercules 3501-6
Specific Gravity		1.27-1.32[1]	1.36	1.37	1.27	1.27	1.29	1.27		1.28-1.40[1]	
Modulus GPa	2.8	3.7	3.4	2.6	2.6	3.0	2.1	2.4	2.9	3.8-4.8[1]	3.8
Tensile Yield Strength MPa	90	92	79	84	70	105	72	104		152-193[1]	46[3]
Failure Strain %		50	2-20[1]	40-80	50-100	60	35	6.4		10-15[1]	1
Fracture Toughness kJ/m^2	1.9-3.0	4.8		3.0	2.5			6.3			0.08
Izod Notched Impact Strength J/m		83	16	84	69	60			14.0	80	
Glass Transition Temperature °C		143	85	230	190	216		210	277	249-288[1]	190
Heat Distortion Temperature @ 1.82 MPa °C		140	137	203	174	200	204			274	

(1) Properties dependent on crystallinity
(2) Properties dependent on post-cure
(3) Brittle failure occurs prior to yield

Data on Thermoplastics from manufacturers
Data on Epoxy from Augl(1979)

Figure 2   Polished section through Aromatic Polymer Composite
            PEEK/Carbon Fibre Impregnated Tape.

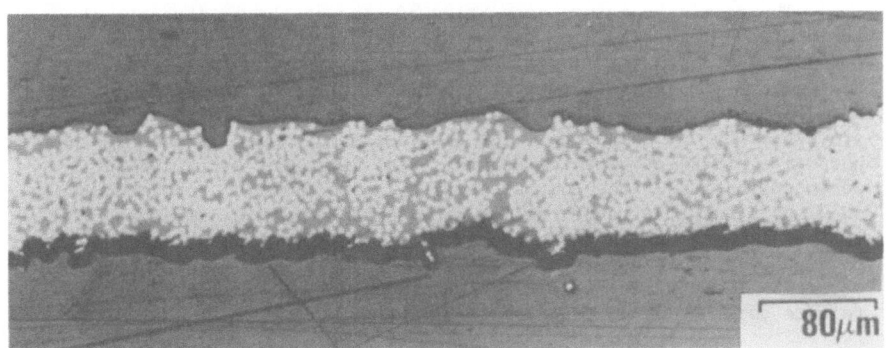

TABLE 3

Methods of Combining Fibre and Matrix

Technique	Method	Comments
Film Stacking	Layers of polymer film interleaved with reinforcing fibres.	Requires high pressures and long processing times.
Solution	Polymer dissolved in solvent bath and reinforcing fibres pulled through.	Polymers which dissolve typically have poor solvent resistance.
Co-Woven/ Co-mingled fibres	Polymer fibres woven or mingled with reinforcing fibres	Difficult to obtain intimate mix. Long processing times.
Hot Melt	Reinforcing fibres pulled through bath of molten polymer.	Polymer viscosities generally too high to obtain good impregnation.
Slurry	Impregnate from fine powder in slurry.	Difficult to make fine powders. Long processing times.

PROPERTIES OF CONTINUOUS FIBRE REINFORCED THERMOPLASTICS

The room temperature properties of some of the unidirectional carbon fibre reinforced composites are summarised in Table 4, the data being taken from Refs 5-10. The 0° tensile properties should be dominated by the fibre properties and the fibre content but the matrix and the fibre-matrix interface will have a significant effect on the properties of the composite. The compressive strength of some of the composites is relatively low and this will represent a limitation in some applications. The composites with low compressive strength also show low flexural strength and it is assumed that a compressive failure is occuring in these cases. The short-beam shear test is frequently used to assess composites, but with thermoplastic matrix materials its usefulness is questionable as an interlaminar shear failure does not usually occur.

The interlaminar fracture toughness (G_{1c}) of the composite is important in comparing the delamination resistance and therefore resistance to defects damage of the composite. The G_{1c} values of the thermoplastic composites are all relatively high in comparision with values of $0.2kJ/m^2$ for a 'standard' epoxy matrix composite and $0.5kJ/m^2$ for a toughened epoxy matrix composite. A fracture surface from the interlaminar toughness test is shown in Figure 3 for APC-2/AS4, PEEK/carbon fibre composite. This illustrates the high matrix ductility and strong fibre-matrix interfacial adhesion that are responsible for the high fracture toughness.

In the case of the semi-crystalline polymers the composite properties will depend on the level of crystallinity and the morphology. Initial work on carbon fibre/PEEK has shown that spherulites nucleate from the surface of the reinforcing fibres and from fibre junctions (11). An etched surface showing the spherulite texture in this material is shown in Figure 4. To date only preliminary work has been carried out to examine the effect of spherulite texture on mechanical properties (12) and further work is needed in this area.

Two other aspects which are of considerable importance are the effects of temperature and of fluid environments on mechanical properties. The uppper temperature limit for composites is usually defined by the Tg of polymer matrix, and although semi-crystalline polymers do retain properties above Tg, it still represents the limit for practical purposes. Further data on the effect of temperature on mechanical properties for some thermoplastic matrix composites is given in Refs 13, 14. Aerospace structures come into contact with a wide range of fluids including water, jet fuels, hydraulic fluids, de-icing fluids and paint strippers, hence any composites used in these applications must possess a high resistance to a wide range of organic solvents. The semi-crystalline polymers in particular have very good chemical resistance (13, 15) but some of the amorphous polymers also show adequate resistance to a wide range of agressive fluids.

TABLE 4 Room Temperature Properties of Unidirectional Carbon Fibre/Thermoplastic Composites.

COMPOSITE DETAILS		'Victrex' PEEK		'Ryton' PPS			'Torlon'	'K-Polymer'	
Polymer									
Composite		APC-2/AS4	APC-2/D/IM6	AC40-60	AC40-66	—	—	K-I/AS4	K-II/AS4
Fibre		AS4	IM6	AS4	AS4	IM6	AS4	AS4	AS4
Status		Commercial	Development	Commercial	Commercial	Research	Commercial	Research	Research
PHYSICAL PROPERTIES									
Fibre Content (Volume)	%	61	61	—	—	51	—	60	60
Void Content	%	<0.5	<0.5	<1.0	<1.0	1.1	—	<2.0	<2.0
MECHANICAL PROPERTIES									
Tensile Modulus	GPa	134	176	121	135	137	—	—	—
Tensile Strength	MPa	2130	2700	1379	1655	1979	—	—	—
Compressive Strength	MPa	1100	1050	621	655	—	1379	—	—
Flexural Modulus	GPa	121	151	103	124	132	128	117	124
Flexural Strength	MPa	1880	2170	1276	1310	1090	2070	1366	1531
Short-Beam Shear Strength	MPa	105	116	69	69	42	110	110	94
Interlaminar Fracture Toughness	kJ/m²	2.7/2.1	3.3/2.5	0.5-0.9	—	1.0	1.1-1.2	—	—

Data from Manufacturers

Figure 3   Fracture Surface from Interlaminar Toughness Test for
            APC-2/AS4, PEEK/Carbon Fibre Composite.

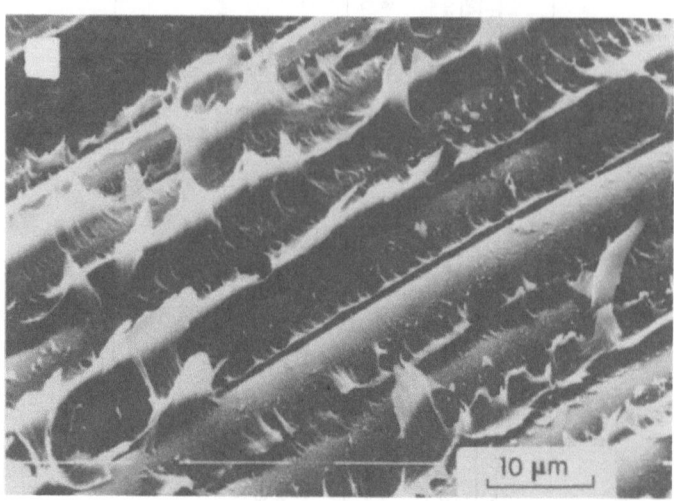

Figure 4   Spherulite texture in PEEK/Carbon Fibre Composite
            (Permanganic etching technique).

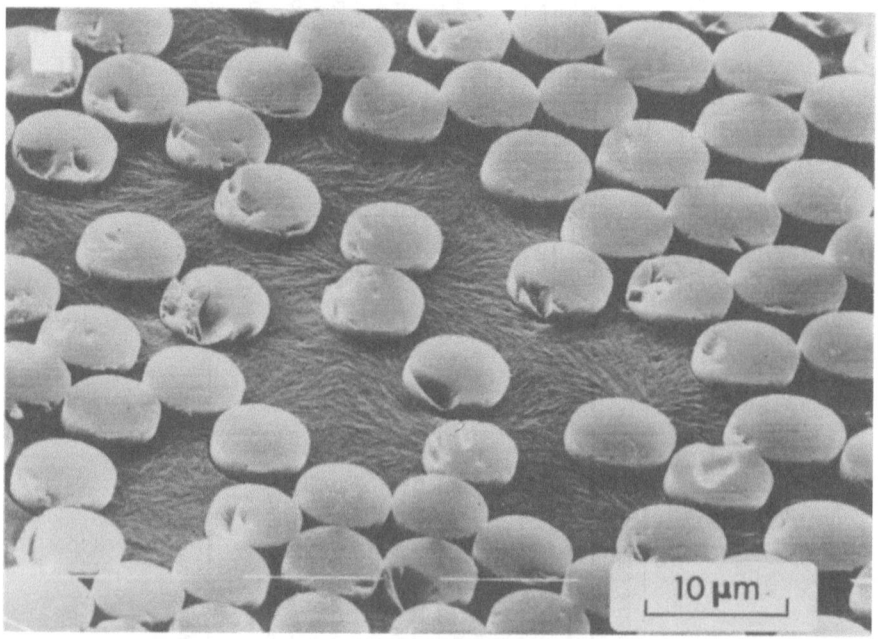

In summary the thermoplastic matrix composites can show equivalence in basic mechanical properties to thermosetting composites. However the thermoplastic composites have considerably higher toughness and hence much lower sensitivity to damage. The influence of matrix morphology on composite properties requires more detailed examination.

PROCESSING AND FABRICATION

In order to make panels or components it is necessary to heat the polymer to its melt state and as the polymers must have high Tg then the melt temperatures also tend to be high. Thermoplastic matrix composites are typically processed in the range 300-400°C, though lower temperature processing may be possible in cases where the pre-impregnated tape also contains a solvent. For the materials that are pre-impregnated then short processing times and low pressures can be used, but for some materials such as film stacked, co-woven, co-mingled or powder impregnated, then longer times and higher pressures are needed. For example the pre-impregnated PEEK/carbon fibre composite APC-2 can be consolidated at 380°C for 5 minutes (5) whereas for a film stacked PEEK/carbon fibre material conditions of 400°C for 2 hours must be used to give good impregnation and consolidation (16). Some of the polymers also require a 'post-cure' which imparts a slight degree of cross-linking and hence increases the stiffness and yield stress of the polymer, with benefits in composite strength properties. For the semi-crystalline matrix composites the cooling rate must be selected to give the appropriate crysallinity (17). Some work has also been carried out to examine the effect of cooling rate on spherulite texture for PEEK/carbon fibre composites (11).

Thermoplastic matrix composites can be fabricated into components using a wide range of techniques, and these have been discussed by Barnes and Cattanach (18). A novel technique is diaphragm forming which builds on the experience of superplastic forming of metals with the layers of the composite being held between two superplastic diaphragms during the shaping process. Complex shapes can be made readily using this technique and components made by this process are shown Figure 5. One aspect of fabrication that requires further attention is the understanding of the rheological processes, though some preliminary work has already been reported (19). The possible flow processes include intra-ply slip, in-plane slip and transverse spreading, the latter two occuring within a single pre-preg layer, and these flow processes are illustrated in Figure 6.

Thermoplastic matrix composites are now being used in a number of aircraft applications, one of these being the manufacture of radomes. The UK version of Hercules transport aircraft is now fitted with a Poly(ether-sulphone)/glass fibre reinforced radome and this is illustrated in Figure 7. The components, made by Specmat, are fabricated using a film stacking process and have proved to be very successful in this application. A wide number of demonstration components are being developed by aerospace companies and it is anticipated that there will be many flying components in the near future.

Figure 5   Components made from APC-2/AS4 using the diaphragm
               forming process.

Leading Edges

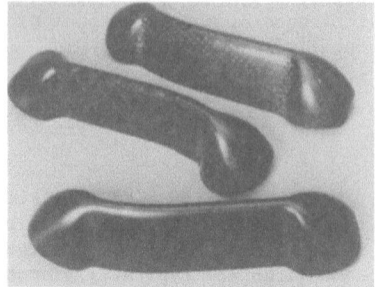

Variations

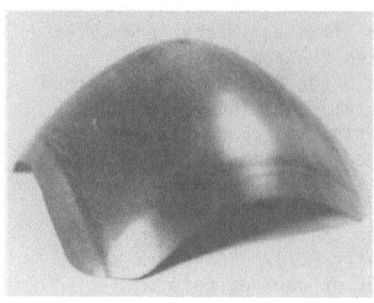

Double Curvature

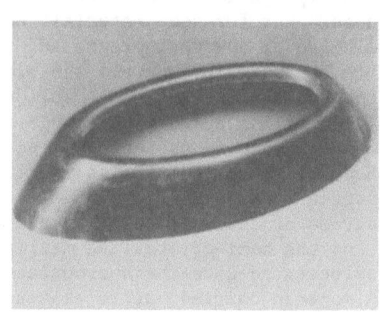

Lip Skin

Figure 6   Flow processes in Continuous Fibre Reinforced
               Composites.

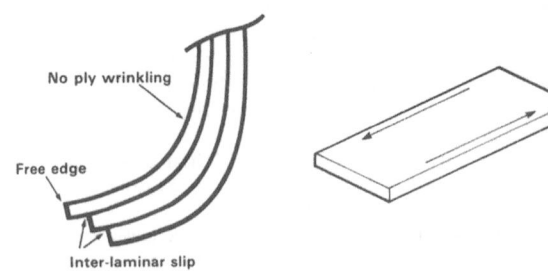

Inter-laminar slip   **In-plane shear of prepreg**   **Transverse components W**
                                                                  give transverse extension

Figure 7   Radome for Hercules transport aircraft manufactured from
              Poly(ether sulphone)/glass fibre by Specmat.

CONCLUSIONS

A range of thermoplastic polymers are now available which
provide the required properties for advanced composite matrices.
Impregnation techniques have been developed which can give very low
voidage despite the relatively high viscosity of thermoplastic
polymers. The composites show equivalence with the existing
thermosetting composites in basic mechanical properties with
considerably improved toughness and hence damage tolerance. Further
work is needed to establish the long term, high temperature and
environmental properties of the materials and to relate mechanical
properties to matrix morphology. A wide range of fabrication
techniques have been demonstrated and considerable reductions in cost
are being realised with demonstration components. There are already
a number of components in service on aircraft and further applications
in aerospace, automotive industrial and sporting goods fields are
iminent.

REFERENCES

1   Phillips, L N, ' The properties of carbon fibre reinforced
       thermoplastics moulded by the film stacking method', RAE
       Technical Report 76140, 1976.

2   Hoggat J T, 'Thermoplastic resin composites', 20th SAMPE
       Symposium, May 1975.

3   Johnston N J, 'Synthesis and toughness of properties of resins
       and composites', ACEE Composite Structures Technology
       Conference, Seattle, August 1984.

4   Augl J M, "Moisture effects with mechanical properties of
       Hercules 3501-6 epoxy resin", Naval Surface Weapons Center
       Report TR 79-41, 1979.

5   Aromatic Polymer Composites Data Sheets, ICI/Fiberite,
        Monchengladbach, W. Germany, 1986.

6   Ryton Advanced Composites Data Sheets, Phillips Petroleum,
        Bartlesville, Oklahoma, USA, 1986.

7   O'Connor J E, Lou A Y, Beever W H, 'Polyphenylene sulfide - A
        thermoplastic polymer matrix for high performance composites',
        Proc ICCM V, p963-970, 1985.

8   Hardesty G H, 'Torlon-C advance composites', SAMPE Los Angeles
        Chapter Meeting, August 22, 1985.

9   Gibbs H H, 'The processing and properties of damage tolerant
        composites based on K-polymer composite materials, Proc ICCM V,
        p971-993, 1985.

10   Chung T S, McMahon P E, 'Thermoplastic polyester amide-carbon
        fiber composites', J. Applied Polymer Science, 31, p965-977,
        1986.

11   Peacock J A, Fife B, Nield E, Crick R A, 'Examination of the
        morphology of Aromatic Polymer Composite (APC-2) using an
        etching technique', 1st Conference on Composite Interfaces,
        Ohio, May 1986.

12   Crick R A, Leach D C, Meakin P J, Moore D R 'Interlaminar
        fracture morphology of carbon fibre/PEEK composites', Accepted
        for publication J.Mat.Sci.

13   O'Connor J E, Lou A Y, Brady D G, 'Polyarylene sulfide
        composites', Proc 1st Conference American Society for
        Composites, p21-35, Oct 1986.

14   Cogswell F N, Leach D C, Nield E, 'High temperature performances
        of thermoplastic Aromatic Polymer Composites', 31st SAMPE
        Symposium, p434-448, April 1986.

15   Horn W J, Shaikh F M, Soeganto A, 'The degradation of advanced
        composite materials exposed to aircraft in-service environment'
        AIAA Conference, p353-361, 1986.

16   Hartness J T, 'Polyetheretherketone matrix composites' 14th
        SAMPE Technical Conference, 1982.

17   Blundell D J, Osborn B N 'Crystalline morphology of the matrix
        of PEEK-carbon fiber Aromatic Polymer Composites II,
        Crystallisaton behaviour', SAMPE Quarterly, 17, 1, p1-6, 1985.

18   Barnes A J, Cattanach J B, 'Advances in thermoplastic composite
        fabrication technology', Proc 'Materials Engineering', London UK
        1985.

19   Cogswell F N, 'The processing science of thermoplastic
        structural composites', Polymer Processing Conference, Akron,
        1986.

Theoretical study of the stress transfer in
fiber-reinforced composites

Yves Termonia
Central Research and Development
Experimental Station
E.I. du Pont de Nemours, Inc.
Wilmington, DE 19898

Abstract

The stress transfer in fiber-reinforced composites is studied theoretically with the help of a finite difference type of approach. The results show that the efficiency of stress transfer between matrix and fiber is a unique function of the fiber aspect ratio and of the mismatch in elastic moduli between the matrix and the fiber. The effect of the adhesion between fiber and matrix is also studied. The predictions of the theory are compared to available experimental data and a good agreement is found.

1. Introduction

There has been recently a growing interest in composites made by incorporating brittle high-modulus fibers in soft epoxy matrices. Since the elastic modulus of the fiber is typically much larger than that of the matrix, the axial elastic displacements of the two components can be markedly different. In order to rationalize the design of these reinforced materials, it is thus of primary importance to have a detailed knowledge of the stress distribution induced by the applied load.

This problem of stress transfer has received considerable attention over the years. Analytical equations for the variation of stress along discontinuous fibers in a cylindrically symmetric model have been derived by Cox [1] and by Dow [2]. Both approaches, however, neglect the adhesion across the end faces of the fibers and they fail to take into account local stress concentration effects near fiber ends. The importance of these assumptions has been demonstrated by finite element approaches [3-5]. Unfortunately, due to the large amount of computer time required, these studies were restricted to very small systems and the importance of finite size effects could not be assessed. In our approach [6], we attempt to tackle these problems of stress transfer using a finite difference type of approach.

2. Model

In our approach, the material is represented by a regular 3-dimensional lattice whose nearest neighbor nodal points are linked by bonds having different elastic constants for the fiber and for the matrix. For a given external strain, these nodes are relaxed towards local mechanical equilibrium with their neighbors by a systematic sequence of operations [7] which steadily reduce the net residual force acting on each node. Figures 1a and 1b give a 2-dimensional representation of the lattice model in a plane parallel (Fig.1a) and perpendicular (Fig.1b), respectively to the direction of loading.

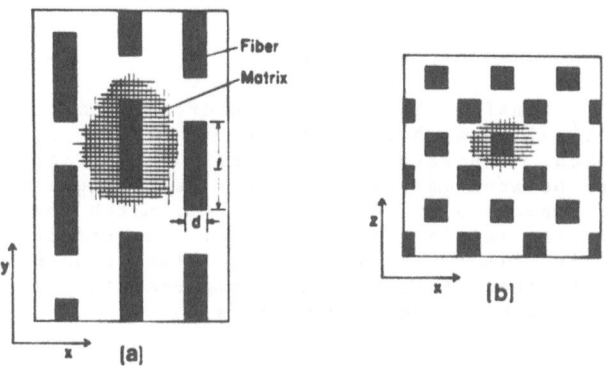

Figure 1: Two-dimensional representation of the model.

The parameters d and l denote the fiber diameter and length, respectively. The Young and shear moduli of the matrix and of the fiber are denoted by (E_m, G_m) and (E_f, G_f), respectively.

3. Results and Discussion

Figure 2 gives a two-dimensional representation of the stress concentration along the loading axis for a single fiber embedded in an infinitely large matrix with $E_f/E_m=40$. The figure shows an important bending of radial lines at regions close to the fiber end. Far away from the fiber end, radial lines recover their original direction perpendicular to the fiber axis, thus indicating that the limit of a very long fiber in an infinite matrix is attained.

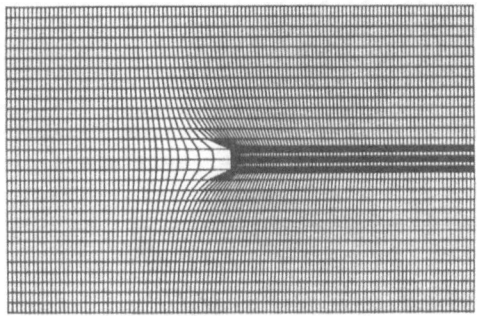

Fig.2: Stress distribution along the loading axis.

Figure 3 shows the effect of Ef/Em on the critical length of the fiber, i.e. the length necessary to build up in the fiber a maximum strain equal to 97% of that for an infinitely long fiber [8].

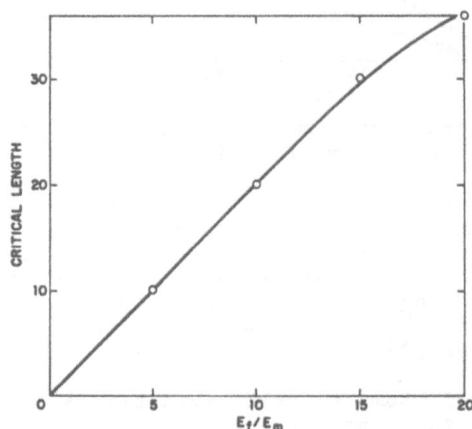

Figure 3: Dependence of the critical length on modulus.

The critical length is in units of the fiber diameter. Inspection of the figure shows that the critical length increases almost linearly with an increase in the fiber Young's modulus. We have also studied the effect of adhesion on the critical length. Changing the adhesion in the model was realized by breaking bonds at the fiber-matrix interface, with probability (1-adhesion factor). The results show that a decrease in adhesion substantially increases the critical

length, that increase being particularly dramatic when the
adhesion factor becomes less than 0.3.

We now turn to application of the model to multi-fiber
composites. Figure 4A shows the dependence of the composite
modulus Ec (in units of Em) on the volume fraction Vf of the
fibers for a ratio Ef/Em=20. The curves are for different
values of the fiber aspect ratio l/d.

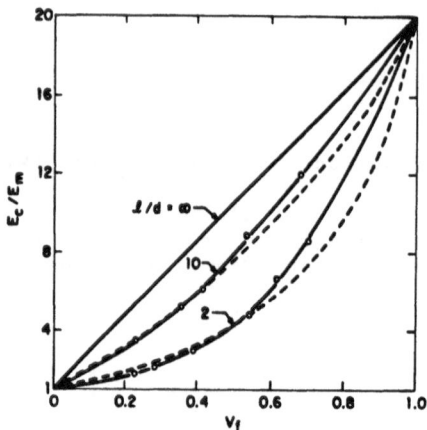

Figure 4A Dependence of composite modulus on aspect
ratio and volume fraction of fibers.

For continuous fibers (l/d=∞), the dependence is linear and
Ec reaches its maximum value given by the so-called Voigt
average. For discontinuous fibers, the curves show a negative
deviation from the ideal continuous case and the modulus
decreases with a decrease in the aspect ratio. Also
represented (dashed line) is the prediction of the widely used
Halpin-Tsai equation [9].

Figures 4B and 5 compare the predictions of our model
(continuous line) to experimental data (symbols) on silica
particles in epoxy (Fig.4B) and on several thermoplastic foams
(Fig.5) [10]. In all cases, the agreement with experiment is
judged excellent and far better than that obtained with the
help of the Halpin-Tsai equation (dashed line).

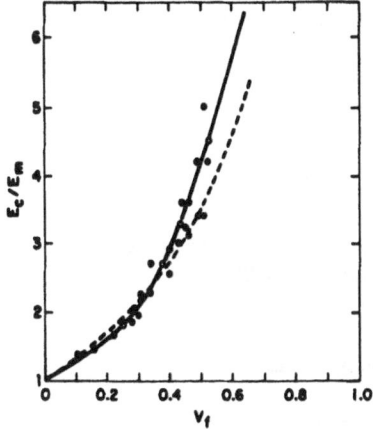

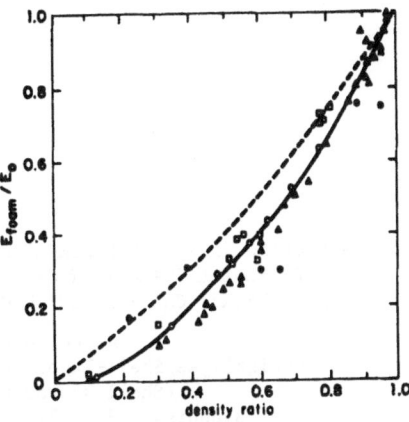

Figure 4B Silica particles   Figure 5   Polymeric foams.
          in epoxy.

To conclude, we have presented a finite-difference type of approach for the study of the elastic properties of fiber-reinforced composites. A major advantage over previous approaches is that the present model is microscopic in nature and allows a study of the effect of the fiber size distribution and of the fiber-matrix adhesion on the mechanical properties. The model also permits an easy evaluation of the stress distribution inside the material and therefore lends itself to a detailed study of the fracture process itself [3,11].

References

[1] H.L.Cox, Brit.J.Appl.Phys., 3, 72, (1952)
[2] N.F.Dow, G.E.C., Rep.R63 SD61 (1963)
[3] A.S.Carrara and F.J.McGarry, J.Comp.Mat., 2, 222 (1968)
[4] R.A.Larder and C.W.Beadle, J.Comp.Mat., 10, 21 (1976)
[5] E.D.Reedy, J.Comp.Mat., 18, 595 (1984)
[6] Y.Termonia, J.Mat.Sci., in press (1987)
[7] Y.Termonia, P.Meakin and P.Smith, Macromolecules, 18, 2246 (1985)
[8] W.H.Sutton and J.Chorne, Fiber Composite Materials, American Society for Metals, Metals Park, Ohio, p.173 (1965)
[9] J.C.Halpin and J.L.Kardos, Pol.Eng.Sci., 16, 344 (1976)
[10] Y.Termonia, J.Mat.Sci., in press (1987)
[11] Y.Termonia, P.Meakin and P.Smith, Macromolecules, 19, 154 (1986);
    Y.Termonia and P.Smith, Polymer, 27, 1845 (1986)

COMPOSITE MODELLING: POLYETHERETHERKETONE/CARBON FIBRE COMPOSITE (APC-2),
EXPERIMENTAL EVALUATION OF CROSS-PLIED LAMINATES
AND THEORETICAL PREDICTION OF THEIR PROPERTIES

A. Cervenka
Koninklijke/Shell-Laboratorium, Amsterdam
(Shell Research B.V.)
Badhuisweg 3, 1031 CM Amsterdam,
The Netherlands

ABSTRACT

Regular symmetrical cross-plied laminates were manufactured by
compression moulding and the angular dependences of their tensile
characteristics – elastic modulus and the stress/strain coordinates of the
failure – were measured in different loading directions. A macromechanical
model that uses the lamina characteristics together with the geometrical
parameters of the laminate has been adapted and used for predicting the
measured properties and conducting a number of parametric studies.

Whilst experiment/theory correlation has revealed an excellent
agreement for the laminate stiffness, the fit for the ultimate properties
is good only under conditions of the linear, elastic composite response.
Otherwise, a qualitative correlation was obtained for the ultimate strength
and no relation has been established between the theoretical and
experimental ultimate strains.

Parametric studies have been conducted in the areas where the model
had proved to be valid and the effects of three parameters – stacking
sequence, number of ply groups and the ply ratio – have been examined.

INTRODUCTION

In the previous communication [1] we reported our studies on the
uniaxial laminates. Experimentally, compression moulding was established as
a manufacturing route to composites of good quality; a testing method that
gives reliable tensile characteristics was developed and the fundamental
building block – the lamina – was characterised in terms of its stiffness
and strength characteristics. Theoretically, micromechanics was used to
derive engineering constants from the properties of the constituents and
macromechanics was employed to relate the off-axis characteristics to those
measured in the principal directions. An agreement better than 10 % was
realised when correlating the theories with the experimental data.

This contribution concentrates on the cross-plied laminates which are
more difficult to analyse theoretically. Four topics are addressed:
1) Adaption of an adequate theoretical apparatus, 2) experimental
determination of the angular dependence of the stiffness and strength
characteristics, 3) experiment/theory correlation, and 4) parametric
studies in the region of model validity.

THEORY

The objective is to formulate a model that relates the two fundamental laminate properties in tension, i.e. stiffness and strength, to the material properties of the lamina and the geometrical characteristics of the laminate test coupon. The term "material properties of the lamina" encompasses the following characteristics: longitudinal and transverse moduli E_1 and E_2, the shear modulus G_{12}, the Poisson ratio ν_{12}, longitudinal and transverse ultimate strengths X and Y and, finally, the shear strength S. The "geometrical characteristics of the laminate" involve information on the lamina construction: the number n of plies used, the orientation α_i of each lamina with respect to a fixed coordinate system, and the direction κ of load applied during testing in relation to the same coordinate system.

The model is formulated on the basis of the "classical lamination theory" utilising the mathematical expressions contained in refs. [2-5]. Although now 50 years old and often used for analysing the performance of e.g. glass/epoxy laminates, the theory does not appear to have been applied to laminates based on a thermoplastic matrix. The following is a brief outline of mathematical operations involved.

The theoretical stiffness is derived in four steps by 1) converting the engineering constants (E_1, E_2, G_{12} and ν_{12}) into elements of the stiffness matrix $|Q|$, 2) converting $|Q|$ into a transformed reduced stiffness matrix $|\overline{Q}|$ via the transformation matrix $|T|$, 3) calculating the extensional matrix $|A|$ using the ply index concept [6], and 4) deriving the laminate stiffness $E(\kappa) = a_{11}$, being the appropriate element of the inverse matrix $|a| = |A|^{-1}$.

The ultimate laminate strength is calculated in five additional steps: 5) deriving the force distribution matrix $|f|$ that relates the stress level of a given lamina to the resulting force N for a symmetrical laminate under simple tension, 6) calculating the force N_i acting on an individual ply i from the Tsai-Hill failure criterion [7], 7) determining the stress/strain coordinates of the "knee-point" that corresponds to the first ply failure, 8) eliminating any of failed lamina f by redefinition $Q_{12}(f) = Q_{22}(f) = Q_{66}(f) = 0$, $Q_{11}(f) = Q_{11}(i)$ and repeating the operations 2-6 with the goal to characterise the post-knee regime in terms of the terminal stiffness and the force balance. Finally, step 9) involves calculation of the ultimate strength $\sigma(\kappa)$ from a simple geometrical consideration by analysing the behaviour of unidirectional laminate.

The ultimate strain $\varepsilon(\kappa)$ is calculated from the coordinates corresponding to the ultimate and first-ply failures.

EXPERIMENTAL

The cross-plied laminates were manufactured using the ICI APC-2 prepreg tape, which was 140 mm wide and 0.125 mm thick, cutting it into 370 mm long strips and stacking them parallel to the edges of the picture frame mould of 1 mm thickness. Stacking of eight plies was sequenced in the order $(+45/-45/+45/-45)_s$. Compression moulding on Bucher presses (Bucher-Guyer, Niederweningen, Switzerland) was used as the manufacturing route with all the experimental variables (time, temperature, pressure) for each manufacturing step (contact, consolidation and cooling) set at the same level as given in ref. [1].

The tensile characteristics were measured at ambient temperature using an Instron 1195 universal testing machine along the lines of the standard

procedure [8] modified with respect to the way of gripping specimens
(sandpaper – cf. [1]). The stress/strain traces were recorded while
straining at a rate of 1 mm/min with the paper movement controlled by 25 mm
extensometers. The test coupons were cut by means of a water–cooled diamond
saw into 12 x 260 mm strips. The loading angle κ was varied in the interval
$0 \leq \kappa \leq 45°$. Defining the angle ψ_i as $\kappa - \alpha_i$, the cutting operation is to
satisfy the orientation of the surface plies (1 and 8) with respect to the
core plies (4 and 5) so that: $\psi_1 - \psi_8 < \psi_4 - \psi_5$.

RESULTS AND DISCUSSION

Experiment/theory correlations

Whilst the theory assumes linear elasticity when the laminate is
loaded in the direction $\kappa - 0$, two distinct linear parts (initial and
terminal) are observed experimentally. On the other hand, no singularity
point has been identified experimentally although its existence can be
predicted on the "first ply failure" basis. Thus we concentrate only on the
tensile characteristics that can readily be identified theoretically (the
elastic modulus, stress and strain corresponding to the failure) and do not
attempt to analyse finer details such as the knee characteristics and the
terminal modulus.

The theoretical analysis was carried out with the lamina
characteristics established in ref. [1]: $E_1 - 135.2$ GPa, $E_2 - 9.2$ GPa,
$G_{12} - 4.9$ GPa, $\nu_{12} - 0.35$, X – 2090 MPa, Y – 73 MPa and S – 78 MPa.

Correlation of the calculated $E(\kappa)$ and the experimentally determined
modulus with the loading direction κ is given in Fig. 1. A very good
quantitative agreement is obtained providing that the "stiffer" ply
constitutes the coupon surface. Unfortunately, this was not always the case
and an error exceeding 10 % can be introduced at $\kappa - 45°$ when the
extensometer aretes are parallel to the fibres in the surface plies.

Concerning the other two tensile characteristics, Fig. 2 shows the
angular dependence of the ultimate strength and Fig. 3 illustrates the
situation for the ultimate strain. Whilst an acceptable agreement has been
obtained for the linear, elastic region ($\kappa - 45°$), an appreciable
theory/experiment deviation is encountered when $\kappa - 0°$. For the ultimate
strength, the theory ranks laminates qualitatively in the order of their
failure stresses; in quantitative terms however, it underestimates the
measured property with the deficit in direction $\kappa - 0°$ amounting to a
factor of two. In the case of the ultimate strain, the theory provides
neither the quantitative nor even the qualitative guidelines: it predicts
the trend opposite to that found experimentally and, consequently,
disagreement in the worst case amounts to a factor of more than ten.
Further experimental effort that will concentrate on the angle–plied
laminates +/– α and examine the strain dependence of the ply angle when
loading in $\kappa - 0°$ is hoped to improve the current, unsatisfactory
understanding in this area.

Parametric studies

We also theoretically analysed the performance of some other
cross–plied laminates constructed in a manner different from that used in
the experimental programme. Being aware of the model limitation, we geared
our modelling to the areas within the model applicability: the laminate

stiffness generally and prediction of the failure characteristics only under conditions of the linear elasticity. Three structural laminate characteristics have been studied: 1) the stacking sequence, 2) the number G of laminate ply groups, and 3) the ply ratio P.

The stacking sequence for symmetrical laminates of ply ratio $P = 1$ has been found not to have any effect on $E(\kappa)$ and $\sigma(\kappa)$ when κ is either 0 or 45°.

Concerning the number of groups, eight plies can be stacked in seven different ways with the coupling matrix $|B| = 0$ for G being an odd number. When G is even, $|B| \neq 0$ as the element $B_{16} = B_{26}$ is finite. One construction, namely that for $G = 6$, is outstanding because zero $|B|$ matrix is obtained inspite of the laminate asymmetry. This is an interesting finding which is in sharp contrast to the classical statement [9-11] of composite science that demands the laminate symmetry to be a prerequisite for decoupling extension from bending.

The effect of the ply ratio for symmetrical laminates was modelled with plies 1 and n orientated in the direction $+45^o$ and varying the number n-2 of the core plies at -45°. All the important cross-plied laminate characteristics are correlated with P in Fig. 4, which clearly illustrates the a $P = 1$ laminate type is the least sensitive construction as regard the possibility to identify the singularity point on the stress/strain dependence.

REFERENCES

1. Cervenka A., PEEK/carbon fibre composites: Experiment/theory correlation for uniaxial laminates, poster No. 14 presented at the Gordon Conference on "Composites", St. Barbara, January 12-16th, 1987; submitted for publication to Polymer Composites.

2. Tsai, S.W. and Hahn, H.T., Introduction to Composite Materials, Technomic Publishing Co., Westport, 1980, pp. 115-166.

3. Jones, R.M., Mechanics of Composite Materials, Scripta Book Co., Washington D.C., 1975, pp. 198-205.

4. Tegtmeier, G., "Berechnung multidirektionaler CFK-Laminate", Studiearbeit S 8305, Institut fuer Kunststoffverarbeitung, Aachen, 1983.

5. Brown, C.P., "Computer aided analysis of laminate composites", B.Sc.thesis Massachusetts Institute of Technology, 1984.

·6. Ref. 2, Fig. 6.2.

7. Ref. 3, Eqn. 2.126.

8. ASTM Standard D 3039-74.

9. Christensen, R.M., Mechanics of Composite Materials, John Wiley & Sons, New York, 1979, p. 166.

10. Tsai S.W. and Massard, T.M., Composite Design, Air Force Aeronautical Laboratories Report TR-84-4183, p. 9-2.

11. Ref. 3, p. 160.

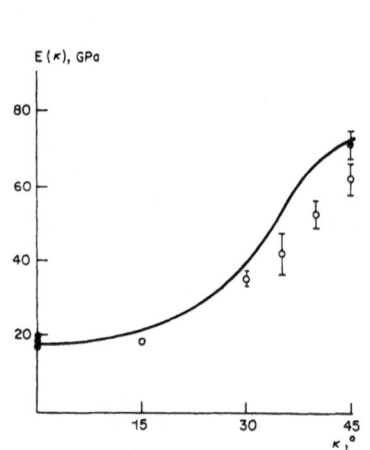

Figure 1. Theoretical (solid curve)
and experimental (o,•)
dependence of the modulus
on the loading direction κ.
(•: $\psi_1 = \psi_8 = 0°$,
o: $\psi_1 = \psi_8 = 90°$)

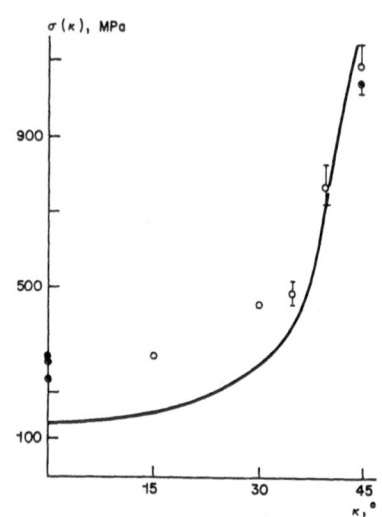

Figure 2. Theoretical (solid curve)
and experimental (o,•)
dependence of the ultimate
strength on the loading
direction κ.
(symbols as in Figure 1)

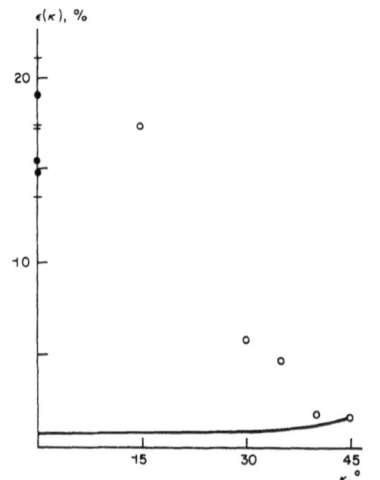

Figure 3. Theoretical (solid curve)
and experimental (o,•)
dependence of the ultimate
strain on the loading
direction κ.
(symbols as in Figure 1)

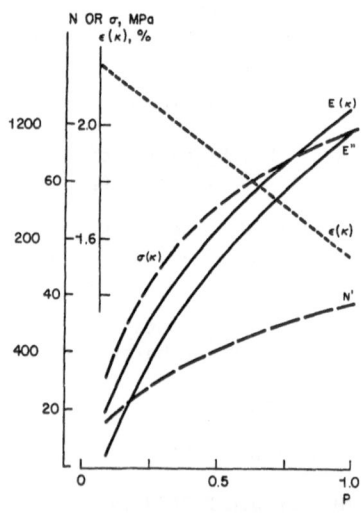

Figure 4. Theoretical analysis of
variation of the stress/
strain characteristics
with the ply ratio
(N' - stress at the first-
ply failure,
E" - post-knee stiffness)

EVALUATION OF COIR AND JUTE FIBRES AS REINFORCEMENT IN ORGANIC MATRIX RESINS

D.S.Varma, Manika Varma, S.R. Ananthakrishnan and I.K.Varma
Department of Textile Technology
and
Centre for Materials Science and Technology
Indian Institute of Technology, Delhi
Hauz Khas, NEW DELHI-110 016, India

ABSTRACT

The present work describes the evaluation of USP resin blended with urethane-acrylate-styrene resin (Crestomer 1080) and jute/coir fibre reinforced composites fabricated using these resin blends. Addition of crestomer 1080 to USP in different proportion (0,25,50,75 and 100%) was done with an aim to develop resins having varying strain-to-failure. Curing characteristics and thermal behaviour of these resin blends was evaluated by DSC and TG. The elongation-at-break of cured resins (USP) having 25 and 50% crestomer was found to be 13.5 and 29.25% respectively and jute and coir fibres were used for reinforcement of these matrix resins. The hybrid composites consisting of glass fibre mat, coir fibre mat (nonwoven)/chopped bristle coir fibre (1cm long) and resin blends were fabricated. Composites having 66% resin (w/w) and 21% w/w coir fibre mat showed ~240% improvement in flexural strength over that of neat resin sheets.

INTRODUCTION

Efforts have been made in the past for utilization of natural fibres such as jute, sisal and palm in fabrication of low cost composites, which may be used in a number of less demanding areas such as low cost housing, grain silos, partition panels and as a substitute of timber [1-3]. In our earlier papers [4-7] we have reported the fabrication of composites based on coir fibres using unsaturated polyester resin (USP) as the matrix polymer. Hybrid composites fabricated with 20% (w/w) chopped bristle coir fibres (1cm long) as core and glass fabric on the outer surface showed ~54% improvement in flexural strength and ~125% improvement in flexural modulus over that of neat USP sheets. One of the problems associated with coir fibres was the higher elongation-at-break (~28%) compared to that of cured USP resin (~2%). Thus there is a significant mismatch in the strain-to-failure of USP and coir fibres. In order to attain the optimum reinforcement of a matrix material, it is necessary that the strain-to-failure of matrix be higher than that of the fibres. It was, therefore, considered of interest to examine the properties of the composites based on coir fibres and a matrix resin having elongation-at-break slightly higher than that of the fibre.

In order to achieve this objective, USP resin was modified by addition of crestomer 1080, which is tough, flexible, unsaturated urethane-acrylate-styrene resin, having an elongation-at-break of 120% [8]. The properties of the blend of USP and crestomer 1080 resin will depend significantly on the ratio of the two constituents. Hence in the present work blends of USP having 0(sample A) 25(sample X), 50(sample Y), 75(sample Z) and 100% (sample B) (w/w) crestomer 1080 were prepared and cured by using free radical initiator. The mechanical properties of these neat sheets were evaluated in order to obtain the appropriate blend composition which will have a strain-to-failure comparable to coir (~28%) or jute fibres. Composites were then fabricated from these resin blends and coir/jute as a reinforcement.

EXPERIMENTAL

Materials : Bristle coir fibres and coir fibre non-woven mat were obtained from Central Coir Research Institute, Allepey, Kerala. Bidirectional woven jute fabric was obtained from Jute Technology Research Institute, Calcutta Unsaturated polyester resin, Acrolite (commercial) grade 310 was obtained from Acropolymers Pvt. Ltd., Faridabad. Crestomer 1080 resin was a gift sample from Scott Bader (U.K). Methyl ethyl ketone peroxide (commercial and cobalt naphthenate (A.G. Fluka) were used as such.

Characterisation of resins : Viscosity of USP, crestomer and their blends was determined at room temperature ($30^{\circ}C$) using Brookfield Synchrolectric Viscometer Model LVT (spindle 2 and 3).

Room temperature gel time of USP, crestomer and their blends was determined in a controlled circulating water bath at $25^{\circ}C$ using 100g resin in a 150ml beaker [9].

Thermal characterisation of cured resin was done by thermogravimetry (TG) and differential scanning calorimetry (DSC) techniques using DuPont 1090 thermal analyser having a 951 TG and 910 DSC module. A heating rate of $10^{\circ}C$/min, sample weight of 10 ± 2 mg was used and experiments were done in a nitrogen atmosphere at a flow rate of $60cm^3$/min.

The thermogravimetric trace was characterised by (1) temperature of onset of degradation (T_i), (ii) final decomposition temperature (T_x). Both these temperatures were obtained by extrapolation of the steep portion of thermogravimetric curves, (iii) temperature of maximum rate of weight loss (T_{max}) was noted from differential thermogravimetric (DTG) curves. In case the decomposition occured in more than one step then T_{max} value for each step was evaluated and has been designated as T_{max-1}, T_{max-2} etc.

An exothermic transition indicative of curing was observed in the DSC trace in temperature range of $40-150^{\circ}C$. This exotherm has been characterised by determining the onset temperature of curing (T_1) (extrapolated), exothermic peak position (T_{exo}) and temperature of end of exotherm (T_2). The heat of polymerisation (ΔH) was evaluated from the area under the exotherm. Activation energy for curing reaction was determined according to Daniels Borchardt kinetic program.

Casting of resin sheets: Neat resin sheet of USP, crestomer and blends were prepared between steel plates using a spacer and "Spreadasil" silicone mould release agent. Resin (50g), methyl ethyl ketone peroxide (2%) as catalyst and cobalt naphthenate (1%) as promoter were used and curing was done for 4h at $60^{\circ}C$.

Composite fabrication : Hybrid composites were fabricated using chopped bristle coir fibres/coir fibre mat together with polyester compatible glass fabric (Fibre glass Pilkington, weighing 360 g/m^2). The details of fabrication and evaluation of composites using ASTM standards have been described elsewhere.

RESULTS AND DISCUSSION

' Viscosity of USP resin changed significantly by the addition of crestomer resin (Table 1). Maximum drop in viscosity occured after mixing USP resin with 25% w/w crestomer resin, beyond which only a marginal drop in viscosity took place. Viscosity of crestomer resin was found to be almost half than that of USP resin.

TABLE 1

Brookfield viscosity of USP, crestomer 1080 and resin blends.

Composition of resin (w/w)		Sample designation	Viscosity (cps)	
USP	Crestomer		Spindle-LV2	Spindle-LV3
100	0	A	1625	2117
75	25	X	1069	1328
50	50	Y	1050	1301
25	75	Z	958	1172
0	100	B	941	1122

The onset of gelation was much faster in USP resin than that in crestomer In blends of USP/crestomer resin (samples X and Y) the onset of gelation occured at about the same time as in USP resin, while in sample Z it was slightly delayed. Temperature of gelation was found to decrease while the time of gelation was found to increase by increasing the concentration of crestomer resin in the blend.

A decrease in ΔH and activation energy of curing reaction was observed on addition of crestomer to USP resin (Table 2). However, Texo and T_2 values increased with an increase in concentration of crestomer in USP resin blends.

TABLE 2

DSC Curing characteristics of various resin samples.

Sample	T_1 (^oC)	T_{exo} (^oC)	T_2 (^oC)	ΔH (J/g)	E KJ/mole
A	44	80	141	192	101
X	46	91	150	154	99
Y	44	101	170	166	79
Z	65	106	155	101	-
B	97	126	181	143	-

A two step decomposition was observed in cured crestomer resin and hence from DTG trace two T_{max} values were noted down (Table 3). An incr- ease in crestomer content in the blend resulted in an increase in T_{max-2}.

TABLE 3

Results of thermogravimetric analysis of various resin sample (cured at 60^oC for 4h)

Sample	T_i (^oC)	T_{max-1} (^oC)	T_{max-2} (^oC)	T_x (^oC)
A	305	-	395	421
X	293	338	392	425
X	322	355	412	439
Z	302	340	413	435
B	306	340	422	439

The mechanical properties of various resin sheets were evaluated using Instron 1121 tensile testing machine. Five specimens were tested for each resin sample and average of these values is given in Table 4. As expected an increase in elongation-at-break was observed on addition of crestomer 1080 to USP resin.

TABLE 4

Mechanical properties of various resin sheets (curved at $60^{\circ}C$ for 4h)

Sample	Tensile strength (MN/m^2)	Elongation-at-break (%)	Flexural strength MN/m^2
A	37.70	2.9	122.9
X	35.35	13.5	74.24
Y	26.33	29.25	36.96
Z	17.98	96	-
B	14.91	122	-

Hybrid composites were fabricated using Glass/Coir fibre mat/Glass reinforced USP/crestomer resin blends (sample X, Y and Z). These composites have been designated as MX, MY and MZ respectively. Chopped coir fibres (1cm long) were also used instead of coir fibre mat (composite sample CY). The results of mechanical property evaluation are given in Table 5.

TABLE 5

Mechanical properties of hybrid composites of glass/coir fibres and resin blends

Property tested	Sample MX	Sample MY	Sample MZ	Sample CY
Coir fibre content (%)	21.20	22.17	22.18	19.21
Glass fibre content (%)	11.41	12.99	12.52	14.87
Resin content (%)	67.39	64.84	65.30	65.92
Flexural strength (MN/m^2)	141.8	126.2	83.02	107.7
Flexural modulus (GN/m^2)	3.33	1.91	1.26	2.45
Maximum strain (%)	3.5	5.29	8.1	4.60
Inter laminar shear strength(MN/m^2)	12.5	12.96	13.25	13.27
Density (g/cc)	1.27	1.26	1.29	1.35

Composites were also fabricated from bidirectional jute fabric and 75:25 USP/crestomer resin blend. Increasing the weight percent of jute fabric from 28 to 43% resulted in an increase in flexural strength. Interlaminar shear strength, however, remained unchanged. Further increase in weight percent of jute fibres deterioted the mechanical properties of the composites. This could perhaps have been due to the fact that jute fibres tend to soak up large amount of resin. As a consequence the wetting of fibres may not be proper on increasing weight fraction beyond an optimum limit.

These results thus indicate that mechanical properties of USP/crestomer blend (50/50) are improved significantly by reinforcing with coir fibre mat/glass fibres. An improvement of ~240% in flexural strength was observed in comparison to neat resin blend. Similar reinforcement of USP resin by coir fibres/glass resulted in an ~54% improvement. Thus a matrix resin having a strain-to-failure greater than that of reinforcing fibres shows better improvement in flexural strength.

ACKNOWLEDGEMENT

The authors wish to thank Scott Bader (U.K) for supplying a gift sample of crestomer 1080 resin.

REFERENCES

1. Parmasivam, T. and Abdul Kalam, A.P.J., Fibre Sci. and Technol., 1974, 7, 85.

2. Chawla, K.K., Aragao, E.A.A., Montario, R.R.C., Fernandes, F.G. and Mores M.M., Advances in Composite Materials, Pergamon Press, Oxford 1980, 1, 444.

3. Prasad, S.V., Pavithran, C. and Rohatgi, P.K., J. Mater.Sci., 1983, 18, 1443.

4. Varma, D.S., Varma, M. and Varma, I.K., J. Reinforced Plastics and Composites, 1985, 4, 419.

5. Varma, D.S., Varma, M. and Varma, I.K., Ind.Eng. Chem. Prod. Res. & Dev., 1986, 25, 282.

6. Varma, D.S., Varma, M. and Varma, I.K., J. Polym. Mater., 1986, 3, 101.

7. Varma, D.S., Varma, M. and Varma, I.K., Text. Res. J., 1984, 54, 827.

8. Crestomer 1080 data sheet, Scott Bader, U.K. technical leaflet number 300, September 1985..

9. Oleasky, S.S and Mohn, J.G., Handbook of reinforced plastics, Reinhold Publishing Co., New York, 1964, P 21.

INDEX OF CONTRIBUTORS